AF588118

Kiank / Fruth
Planning Guide for Power Distribution Plants

Dr.-Ing. Hartmut Kiank, born in 1952, is a principal expert for Power Distribution Solutions in the Siemens Energy Sector. In this professional management position, he is concerned with planning and project management of public and industrial power supply installations. He is a member of the VDE and author of numerous technical articles and reports (CIRED, ICEE). His “etz“ paper “EMC and personal safety in multiply-fed industrial networks“ was included in the “VDE annual report 2007 of electrical engineering“.

Dipl.-Ing. Wolfgang Fruth, born in 1966, is a project planning engineer and head of central consultant support for Totally Integrated Power in the Siemens Industry Sector. He is a co-developer of the network calculation and dimensioning software “SIMARIS design“ and the author of various technical publications.

Planning Guide for Power Distribution Plants

Design, Implementation and Operation of Industrial Networks

by Hartmut Kiank and Wolfgang Fruth

Publicis Publishing

Bibliographic information published by the Deutsche Nationalbibliothek
The Deutsche Nationalbibliothek lists this publication in the Deutsche Nationalbibliografie; detailed bibliographic data are available in the Internet at http://dnb.d-nb.de.

www.publicis-books.de

ISBN 978-3-89578-371-5

Editor: Siemens Aktiengesellschaft, Berlin and Munich
Publisher: Publicis Publishing, Erlangen

Printed in Germany

Foreword

The Power Distribution Market is shifting and undergoing dramatic change. The growing scope of decentralized renewable sources, and interconnection by means of power electronics, just to mention these examples, are adding complexity to network topology. Balancing and safeguarding power, managing the demand, stabilizing voltage and frequency, and – sure enough – optimizing the costs are major drivers that have to be taken into consideration at an early stage in design of an electrical network.

Responding to rapid changes in demand and reconfiguring the network accordingly are bringing the digital world to power distribution. More intelligence in the field and at the component level, in conjunction with digital communication and software, is required in order to meet this challenge. Here again, a well designed network will support the customer in utilizing assets to their full capacity.

Complying with the highest international standards is of course an essential prerequisite.

A thorough understanding of the customer's needs and processes, whether the customer be a utility or an industrial corporation, is essential and instrumental in designing a robust and reliable electrical network and its protection scheme. Optimized integration and deployment of our state-of-the-art products and solutions will enable you to engineer a highly secure and dependable electrical network, crucial in today's world economy.

This book aims to become a reference for those designing and dimensioning electrical networks. It is likewise intended for engineers and technicians working in the energy industry, as well for students who wish to become familiar with this exciting subject matter and for graduates starting their career in this business.

My warm thanks go to Dr. Hartmut Kiank and his co-author Wolfgang Fruth for their meritorious contributions and dedication in the production of this book. They have created an excellent work, balancing theory and practice and placing this complex topic on an understandable and concrete level. All this is the fruit of their experience in the Power Distribution Solutions Business Segment.

One recommendation though: keep this book open on your desk, use it without moderation, dig into it. You will discover a mine of information, unfolding page by page.

Erlangen, July 2011
Jean-Marc Vogel

Preface

Industrial distribution networks must be reliable enough to ensure that the production and process engineering processes they serve can function efficiently, reliably and with the highest possible quality. This is only possible if the planning decisons made for industrial networks meet all the process requirements for power consumption, supply reliability and voltage quality in a technically optimum and efficient way. Because of their complexity and their far-reaching implications for the supply quality and energy efficiency, planning decisions made in the design, dimensioning and operation of networks must be reached in a particularly responsible and judicious way. This is crucial as the true technical risks are often concealed by the complexity of the planning task at hand. If cost-saving potential is also to be exploited, technical risks can only be avoided with competent planning solutions, that is, using the available process expertise and knowledge of the industry technology, technical knowledge about networks and plants, in-depth product knowledge and sound knowledge of the applicable standards and specifications.

With this aim in mind, this guide attempts to convey the solution competence gained in many years of practical work on process-related design, dimensioning and operation of safe and efficient industrial power systems in a simple and understandable way. While Part A discusses the relevant basis of planning, Part B and Part C offer planning recommendations for medium-voltage and low-voltage industrial power systems. These recommendations also provide details of switchgear and protection equipment for networks as well as the interrelationship between the voltage levels (110 kV, MV, LV).

Recommendations for the design and operation of power systems and the selection and parameteriziation of protection equipment are not always stipulated in standards and specifications. In many cases, they have emerged from many years of positive operating experience and practical expertise. Because regulations can only be applied to strategic network planning to a limited degree and planning conditions can vary greatly, some of the recommendations in this guide do offer a certain margin for discretion. It is in the nature of the matter that discrepancies arise within this discretionary margin between the planning recommendations and procedures in specific branches of industry.

This book addresses engineers and technicians working in industrial power engineering, in industrial companies and planning offices. It also helps students and graduates to familiarize themselves with the subject matter.

This planning guide evolved from an idea by the management of the Power Distribution Solutions Business Segment in the Siemens Energy Sector. I would like to thank all involved members of management expressly for their support in the realization of this book project. Many thanks also go to Wolfgang Fruth for his co-authorship of Section C of this book. I am also much indebted to Ursula Dorn who provided competent and committed support with the electronic preparation of the manuscript. And, last but not least, I would like to thank Dr. Gerhard Seitfudem for the fruitful editorial collaboration.

Any critical comments regarding this planning guide are very welcome.

Erlangen, July 2011

Hartmut Kiank

Table of contents

A Fundamentals

B Planning recommendations for medium-voltage systems

A Fundamentals

1 Introduction

1.1 Special aspects of industrial power systems

Power systems are used for the transmission and distribution of electrical energy by means of conduction. Power transmission and distribution is always performed on a number of voltage levels. At the high-voltage, medium-voltage and low-voltage levels, the power systems therefore consist of branches (lines, transformers) and nodes (substations with integrated protection and control equipment). A distinction is made between public and industrial distribution systems because of the differing supply tasks and purposes of the power systems. Industrial distribution systems have features and characteristics that distinguish them from public distribution systems. The distinguishing features of industrial power systems are:

- *High density of loads and switchgear*

  For the distribution of electrical energy in industrial plants, distances between system nodes are relatively short at all voltage levels. For that reason, the ratio of the number of items of switchgear to the total line length is greater in industrial power systems than in public power systems [1.1]. Industrial power systems also exhibit very large loads per unit area. The load per unit area in plants in the metal processing sector, for example, is between 70 and 600 VA/m^2, depending on the structure of the loads in the system. In mechanical workshops, values of between 150 and 300 VA/m^2 can be expected on average. These loads per unit area include the lighting (approx. 20 to 30 VA/m^2) and ventilation (approx. 15 to 20 VA/m^2) [1.2, 1.3].

  Public distribution systems, on the other hand, generally only exhibit a load per unit area of between 2 and 20 VA/m^2.

  Because of the clear difference in density of loads and switchgear, the network structures preferred for public power supplies are normally unsuitable for industrial supplies [1.4].

- *High short-circuit power*

  High short-circuit powers are required to ensure that large motors and groups of motors can be started and restarted. Industrial power systems must therefore exhibit sufficiently low system impedance. However, a low system impedance is associated with a high level of short-circuit currents and correspondingly high dynamic and thermal stress on the equipment. Calculations must always consider the worst-case short-circuit current stress, like in the case of connected asynchronous motors. When a short circuit occurs, asynchronous motors produce additional short-circuit current that is fed back into the network. Because of the comparatively high short-circuit current stress, fast tripping of protection devices is particularly important in industrial systems [1.5]. For such applications, HV HRC fuses and differential protection devices are therefore preferred.

- *High mechanical and electrical stresses on the switchgear*

  In industrial power systems there are applications that make especially high demands of the switchgear [1.6]. For example, switchgear that is used for reactive-power compensation and for operating arc furnaces is subject to greater mechanical stresses. In reactive-power compensation, capacitors or shunt reactors usually have to be connected and disconnected several times a day. In arc furnace operation, the number of operating cycles can even reach 100 per day. Connection and disconnection of the high-current electrodes of furnace transformers can also result in extremely high electrical stresses.

  Furnace transformers are dynamic loads that, on connection, can cause high-frequency transient activity accompanied by dangerous resonance phenomena. On disconnection, on the other hand, high transient overvoltages are possible because of current chopping and multiple re-ignition. Excessively high transient overvoltages usually result in dielectric overloading of the equipment insulation.

  To ensure that all switching duties in industrial systems are reliably performed, special attention must be paid to the choice of switching devices (for example, the necessary number of operating cycles, reliable switching of large short-circuit currents and of small inductive and capacitive currents) and any necessary protective measures against impermissible overvoltages (for example, surge arresters and/or RC and CR protection circuits coordinated for the power system).

- *Pure cable networks with relatively short distances between substations*

  Industrial power systems are pure cable networks with relatively short distances between substations. Because the cable connections between substations are shorter than in public distribution systems, protection concepts using distance protection devices usually have lower priority. On the other hand, selectivity problems can also occur with protection concepts with time-overcurrent protection devices. The cause of such problems may be the distribution of the fault currents due to the switching state of the system or the setting of a short total clearing time for the selective grading of protective devices. Because of possible restrictions in the use of distance and time-overcurrent devices, differential protection devices are preferred as the main protection in industrial cable networks.

- *Stringent requirements for the supply reliability of the low-voltage system*

  The requirements for the supply reliability of industrial low-voltage power systems are much more stringent than for public low-voltage systems. In public low-voltage systems (secondary distribution systems), the focus is on fulfilling the supply mission during normal operation. The $(n-1)$ principle is not applied or is only applied to a limited extent [1.7]. In industrial LV systems, on the other hand, application of the $(n-1)$ principle is an absolute condition for a reliable supply of power to production processes.

- *Serious perturbations in the system caused by dynamic loads*

  In industrial systems, there are many loads that produce reactive power or alter the sinusoidal shape of the current [1.8]. Operation of large asynchronous motors, resistance welding equipment and converter-fed drives can cause serious system perturbations in the form of voltage fluctuations, voltage dips, voltage unbalance and harmonic voltage distortions. In the case of periodic pulse loads, flicker is also produced. All system perturbations must be limited so that the effects on the load causing it, on the other loads and on individual items of equipment can be kept to permissible values. For that reason, adequate design and dimensioning of industrial systems must also include measures to prevent impermissible system perturbations. Such measures include, for example, starting methods for large high-voltage motors, active and passive tuned filter circuits, reactive-power compensation equipment with closed-loop control and dynamic voltage restorer (DVR) systems.

- *Existence of in-plant generation systems*

  If industrial plants include in-plant generation systems, technical constraints for stable interconnected operation of the industrial in-plant generation network with the public network must be defined [1.4]. If instability due to short-circuit-type faults in the external power system or impermissible reversal of power flow is likely, the in-plant generation network must be put into stable island operation. Islanding is performed using a tripping device for network splitting. The tripping criteria are frequency reduction, voltage dips and direction of power flow and current [1.5].

  To ensure stable island operation, an additional automatic load-shedding system is often required. In the event of falling frequency, loads are shed to adapt the power demand required for the main processes to the sole remaining in-plant generation. After the fault in the public network has been eliminated and automatic synchronization has been performed with the in-plant generation network, the two networks are synchronized and interconnected again.

- *Many hours of use of the electrical equipment and installations*

  The optimum utilization of capital-intensive production plants and the necessity for economically viable production are resulting in high numbers of hours of use of the electrical equipment and installations. In some branches of industry, utilization periods of up to 8,000 h/a are reached [1.9]. Due to the many annual utilization hours, especially energy-efficient and low-loss power supplies should be aimed for.

- *Close linking of power transmission, distribution and process control*

  In industry, the two primary functions of an electrical power system, transmission and distribution of electrical energy, are closely associated with the specific production process. For the association of functions close to the process, an integrated flow of information between protection, control and automation systems is required. This requirement is often only met by multifunctional industrial control systems for power distribution and process control.

The special aspects explained above underline the main differences between public and industrial distribution networks. These result in different planning recommendations for the design and dimensioning of industrial power systems.

1.2 Need for complete power system and installation engineering solutions

Most power systems used in industry have been developed over a long period of time. The result of such developments are system configurations that have arisen historically and do not meet all the requirements for

- high cost-efficiency and energy efficiency,
- clear mode of operation,
- sufficient redundancy in case of a fault,
- selective protection tripping and quick fault clearance,
- personal safety according to the rules of the employer's liability insurance association (e.g. accident prevention regulation BGV A3) or the technical regulations for safety at the workplace (e.g. TRBS 2131),
- short-circuit withstand capability of the equipment,

- high electromagnetic compatibility (EMC),
- low environmental impact

equally well and/or in compliance with the standards. It is the task of the system planners to reassess the historically arisen structures and to develop overall solutions for a cost-efficient and reliable power supply.

Every expansion or upgrade of a system offers an opportunity to develop a complete power system and installation solution [1.10]. This can include the following measures:

- reinstallation of cables for system expansions for production reasons,
- connection of additional system distribution substations or transformer load-centre substations for the power supply to new factory halls or production areas,
- replacement of cables that have become unreliable or prone to short circuit,
- replacement of MV switchgear having insufficient or obsolete safety standards,
- restructuring measures at the incoming supply and distribution level (e.g. implementation of a new nominal system voltage).

In industry, too, the pressure to boost efficiency in the reliable operation of distribution systems will force a departure from restrictively handled investments in isolated measures. The necessary efficiency boost and investment security is offered only by sustainable investments based on a complete power system and installation solution. Only with such a solution can a cost-efficient and reliable power supply with lasting customer benefit be ensured. Moreover, increasing electricity costs, lower pay-back times and new legal regulations encourage investment in energy-efficient complete solutions [1.11].

1.3 Task of system planning

In planning the power supply for industrial plants, decisions have to be made about system design, dimensioning and mode of operation. These decisions must be characterized by sufficient quality of supply (= supply reliability + voltage quality) and high efficiency. While the quality of supply is solely determined by the specific requirements of the production process in question, the efficiency largely depends on the available potential for cost reductions. It is up to system planning to resolve the conflict between making use of cost reduction potential and achieving a high quality of the supply [1.12]. The following planning aspects serve to resolve this conflict:

- definition of new and improvement of old system structures,
- selection of switchgear configurations and basic switchgear circuits,
- determining the location for substations and choosing the routes for cables and lines,
- dimensioning the equipment according to current-carrying capacity for load current and fault current,
- method of neutral earthing for operation of galvanically separated MV networks,
- process-dependent use of the $(n-1)$ failure criterion,
- definition of starting methods for large high-voltage motors,
- specification of solutions for putting industrial power systems with in-plant generation and imported power into stable island operation,
- definition of measures for compensating for flicker and dynamic voltage dips,

- definition of measures to limit system perturbations caused by harmonics,
- drawing up reactive-power assessments and derivation of appropriate compensation measures,
- elaboration of selective and reliable system protection and generator protection concepts,
- choice of electrical equipment according to ambient conditions (e.g. climate, pollution degree, fire load, explosion protection).

These planning aspects show how multifaceted and demanding the planning of industrial power systems is. Because of its multifaceted nature and complex effects on the quality of supply and efficiency, planning decisions must be made especially responsibly. Moreover, decisions on system design, dimensioning and method of operation made in the planning phase can only be corrected to a limited extent in the subsequent project planning and processing phases.

Fig. A1.1 shows how system planning and the phases of the renewal process in industrial plants are interlinked. It is evident that system planning and system operation are

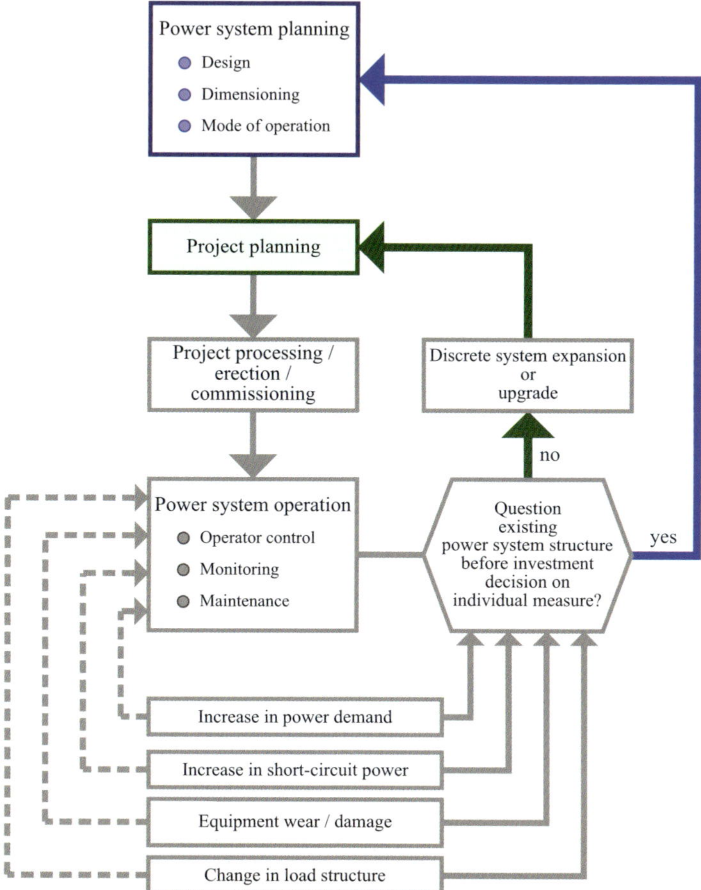

Fig. A1.1 Phases of the renewal process in industrial plants

interactively linked by decisions to be made about the necessary consequential investments. Power system operation after commissioning is characterized by

- operating and monitoring measures and
- maintenance and service measures.

The measures for system operation are subject to external influences. The distinctive influencing factors over the many years of operation of industrial systems are:

- in some cases sudden increase in load due to expansion of production,
- increase in active short-circuit power due to replacement of the transformers with larger rated power or smaller percent impedance voltage in the upstream power system,
- ageing and natural wear of the equipment,
- damage to equipment in case of a fault and
- change in load structure due to the growing proportion of EMC-sensitive loads (e.g. IT equipment and computer systems) and harmonic sources (e.g. replacement of conventional incandescent and fluorescent lamps with energy-saving types, modernization of the drives from variable-speed to static converter technology, preferred use of variable-speed drives with power electronics).

The requirements for reliability of system operation can also be affected by changes in regulations and standards. Standards are acknowledged rules of technology that are constantly adapted to the current state of knowledge. This adaptation of standards to the current state of the art can make new system planning advisable.

New system planning is always recommended when the existing structure of the distribution system has to be reconsidered before the decision on whether to invest in a new system expansion is made (see Fig. A1.1). The task of system planning then includes efficient definitions for design, dimensioning and mode of operation adapted to the modified requirements. The planning definitions for the design and dimensioning in this case must especially consider basing the power system on clear structures and the creation of technically and economically expedient margins for the load and fault current-carrying capacity of the equipment. The industrial power system planning as a whole ensures that today‘s production and process engineering can be managed with efficient use of energy, reliably and with the highest possible quality.

2 Basic workflow for planning

2.1 Top-down principle

The development of network and installation concepts for industrial power supplies requires a systematic and strategic approach. This approach involves taking an overall view across all voltage levels (110 kV, MV, LV) that are important for supplying power to the production process.

The top-down principle is especially suitable for systematic planning because decisions with long-term binding consequences must be made with a very broad view and much experience [2.1, 2.2]. Fig. A2.1 shows the basic planning process for industrial power systems based on the top-down principle.

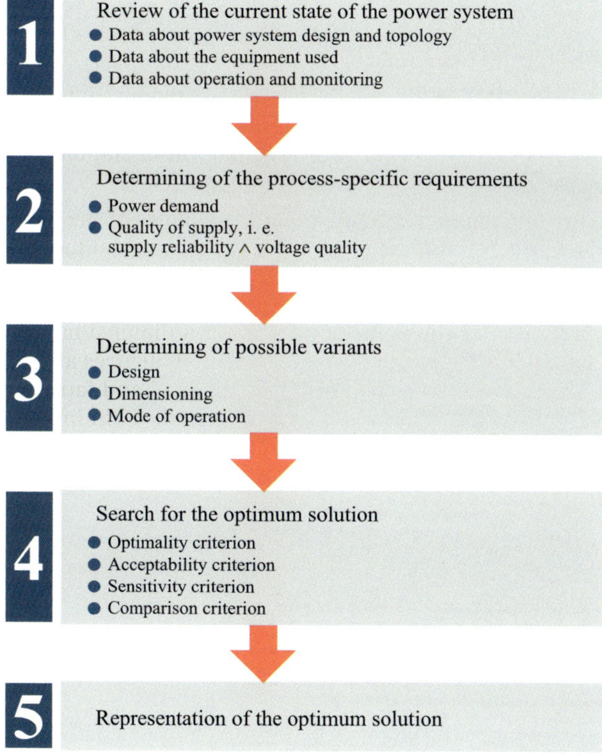

Fig. A2.1
Basic planning process for decisions with long-term binding consequences

2.2 Determining the state of the existing system

The state of the existing system is required to transform historically arisen industrial system structures into clear and simple structures that meet all process-dependent requirements for power demand and quality of supply in a cost-efficient way.

The following data must be recorded to determine the state of the existing system:

a) Data about power system design and topology

These include data that are contained in schematic and topological system plans, development plans and machine installation plans. In particular, these are data about

- the physical location of the incoming supply from the upstream system,
- the locations of substations for power distribution,
- the routing of the MV and LV cables,
- the buildings on the factory site,
- the in-plant production areas with local load centres and
- the general structure of the industrial power system.

b) Data about the equipment used

These data include system and plant data about the incoming supply and the equipment in the system. Table A2.2 shows a general view of the equipment data to be included to determine the basic state for industrial power system planning.

c) Data about operation and monitoring

The system planner requires the following information about operation and monitoring:

- method of neutral earthing of the MV system (isolated neutral (OSPE), resonant neutral earthing (RESPE), low-impedance neutral earthing (NOSPE) or solid earthing),
- type of LV system earthing (electromagnetically compatible TN system with a centrally earthed PEN conductor, TN-S system, TN-C system, TN-C-S system, TT system or IT system),
- relevant circuit states (normal operation (NOP), operation under fault conditions (OPFC)),
- organization of the power system management (e.g. from occurrence to elimination of a fault),
- measured load values of 110-kV/MV transformers, MV/LV transformers and MV cable connections,
- measured voltage values at selected MV and LV nodes (RMS values of the load voltage, values for the magnitude, duration and frequency of voltage dips),
- data about the load structure (presence of large asynchronous motors, converter-fed drives, resistance welding equipment, arc furnaces, large test bays, etc.),
- data from the logbooks of individual items of equipment (commissioning and maintenance times, diagnostic results and faults),
- data about the protection and automation equipment (automatic switchover, UPS, protection relays and their settings, control equipment, alarming equipment, etc.)

Provision of the necessary information for determining the state of the existing system is increasingly supported by information and data processing systems (IT systems) [2.3, 2.4].

Table A2.2 Required equipment data of the basic state (power system data)

Upstream system

Utility company	Nominal operating voltage	System short-circuit power		R / X ratio	Asymmetrical current peak factor	System time constant
		Maximum value	Minimum value			
	U_{nN}	S''_{k-max}	S''_{k-min}		κ	τ
	[kV]	[MVA]	[MVA]	[1]	[1]	[1]

110-kV/MV transformer(s)

Transformer name	Rated voltage		Rated power	Percent impedance voltage		Load losses	No-load losses	No-load current	Vector group	Additional voltage per tap step	Tap setting		Percent impedance voltage at	
	primary	secondary		complex	ohmic						lowest tap	highest tap	lowest tap	highest tap
	U_{rT1}	U_{rT2}	S_{rT}	u_{rZ}	u_{rR}	P_k	P_0	i_0		u_k	s_{min}	s_{max}	u_{uZ}	u_{oZ}
	[kV]	[kV]	[MVA]	[%]	[%]	[kW]	[kW]	[%]		[%]	[1]	[1]	[%]	[%]

MV substations (short-circuit strength and MV/LV transformer data)

Substation or node name	Short-circuit strength parameters					Parameters of the connected transformers								
	Rated voltage	Rated short-circuit breaking current	Rated short-circuit making current	Rated short-time withstand current	Rated short time	Number of identical transformers	Rated voltage		Rated power	Impedance voltage at rated current		Load losses	No-load losses	Vector group
							primary	secondary		complex	ohmic			
	U_m	I_{sc}	I_{ma}	I_{thr}	t_{thr}	n	U_{rT1}	U_{rT2}	S_{rT}	u_{rZ}	u_{rR}	P_k	P_0	
	[kV]	[kA]	[kA]	[kA]	[sec]	[1]	[kV]	[kV]	[MVA]	[%]	[%]	[kW]	[kW]	

MV(LV) substations (load and motor data)

Substation or node name	Load data			Data of the connected motors								
	Maximum power demand	Power factor	Maximum simultaneous motor power	Number of identical motors	Rated voltage	Rated power	Rated current	Efficiency	Power factor	Ratio of starting current to rated current	Synchronous speed	Number of pole pairs
	P_{max}	$\cos\varphi$	$g \cdot a \cdot \Sigma P_{rM}$	n	U_{rM}	P_{rM}	I_{rM}	η_{rM}	$\cos\varphi_{rM}$	I_{start}/I_{rM}	n_{syn}	p_M
	[MW]	[1]	[MW]	[1]	[kV]	[kW]	[A]	[%]	[1]	[1]	[min^{-1}]	[1]

MV(LV) cables

Name of the cable connection	Type	Rated voltage	Current-carrying capacity	Resistance per unit length in the positive-sequence system at		Reactance per unit length in the positive-sequence system	Resistance per unit length in the zero-sequence system	Reactance per unit length in the zero-sequence system	Specific earth-fault current	Rated short-time current		Rated short time
				20 °C	90 °C					Conductor	Screen	
		U_m	I_r	R'_{1-20}	R'_{1-90}	X'_1	R'_0	X'_0	I'_{CE}	I_{thr1}	I_{thr2}	t_{thr}
		[kV]	[A]	[Ω/km]	[Ω/km]	[Ω/km]	[Ω/km]	[Ω/km]	[A/km]	[kA]	[kA]	[sec]

Short-circuit current limiting reactors

Name of the short-circuit current limiting reactor	Rated voltage	Rated current	Rated voltage drop	Reactance	Rated short-time current	Rated short time
	U_{rD}	I_{rD}	u_{rD}	X_D	I_{thr}	t_{thr}
	[kV]	[A]	[%]	[Ω]	[kA]	[sec]

Generators

Name of the generator	Rated voltage	Rated power	Nominal power factor	Subtransient reactance	R/X ratio at the connection point
	U_{rG}	P_{rG}	$\cos\varphi_{rG}$	x''_d	
	[kV]	[MW]	[1]	[%]	[1]

2.3 Determining the requirements

Industrial distribution systems must be planned in such a way that they meet all process-related requirements for

- power demand and
- quality of supply

cost-efficiently and the production process to be supplied with power can run energy-efficiently, reliably and with the highest possible quality. How these process-related requirements for the planning of industrial distribution systems are determined is explained below.

2.3.1 Power demand

The magnitude of the power demand must be determined for each location. The power demand refers to the process-related maximum power of individual loads and groups of loads and the total power demand of the industrial plant. The annual maximum demand for an industrial plant is calculated as follows:

$$P_{\text{max}} = b \cdot \sum_i P_{\text{pr-i}} = g \cdot \sum_i P_{\text{max-i}} \tag{2.1}$$

b demand factor
g coincidence factor
$P_{\text{pr-i}}$ power rating of a load or a group of loads i
$P_{\text{max-i}}$ maximum active power consumption of a load or group of loads i

Guidance values for the demand factor b and the coincidence factor g are listed in Table A2.3. The power demand for the production processes that are performed in a relatively limited space (e.g. factory halls) can also be determined using per-unit-area factors [2.6]. Assuming that the loads are approximately evenly distributed over the production area, the power demand is calculated as follows:

$$P_{\text{max}} = A \cdot P' \tag{2.2}$$

A production area in m^2
P' load per unit area in W/m^2

Table A2.4 lists guidance values for loads per unit area. Depending on the type of production and level of automation, higher or lower loads per unit area must be applied.

Equation (2.2) can be used to calculate the power demand of a modern data centre. The power demand calculation is based on a load per unit area of $P' = 1{,}500\ W/m^2$ [2.10].

To calculate the long-term power demand, the system planner must also consider how the production process may develop in the future. One indicator of this is the present potential for future expansion of production (e.g. spare space that could be used to increase production or productivity in the factory halls or on unbuilt areas of the factory site). To take a possible load increase in the industrial plant into account, [2.11] proposes a power rating for the incoming supply that is 30 % to 50 % larger than that calculated according to Eq. (2.1).

Table A2.3 Demand factor b and coincidence factor g for calculation of the power demand of plants in various industries (guidance values)

Factors for determining the power demand			
Industry	Demand factor b		Coincidence factor g
	acc. to [2.5]	acc. to [2.6, 2.7]	acc. to [2.8]
Machine manufacturing	0.20 ··· 0.25	0.23	0.95 ... 0.99
Automotive industry	0.25	--	0.95 ... 0.99
Paper and cellulose industry	0.50 ··· 0.70	0.34 ··· 0.45	0.95
Textile industry (spinning, weaving)	0.60 ··· 0.75	0.32 ··· 0.62	1.00
Rubber industry	0.60 ··· 0.70	0.45 ··· 0.51	0.92
Chemical industry incl. oil industry	0.50 ··· 0.70	0.60 ··· 0.70	0.95
Cement factories	0.80 ··· 0.90	0.50 ··· 0.84	0.97
Food and beverage industry	0.70 ··· 0.90	--	1.00
Underground hard coal mining	1.0	0.36 ··· 0.64	--
Open-cast brown coal mining	0.7	0.70 ··· 0.80	--
Metallurgy	0.50 ··· 0.90	0.33	1.00
Woodworking industry	--	0.15 ··· 0.30	0.98
Mechanical workshops	--	0.15 ··· 0.30	0.99
Rolling mills	0.50 ··· 0.80	--	--
Foundries	--	0.40 ··· 0.50	0.94
Breweries	--	0.40 ··· 0.50	--
Footwear factories	--	0.40 ··· 0.52	0.99

Table A2.4 Guidance values for loads per unit area P'

Production workshops / loads	Load per unit area P' in W/m^2	
	acc. to [2.6]	acc. to [2.9]
Mechanical workshops	200 ··· 400	50 ··· 250
Toolmaking	50 ··· 100	70 ··· 100
Punch shops	150 ··· 300	80 ··· 120
Press shops	150 ··· 300	300 ··· 450
Welding equipment	300 ··· 600	150 ··· 250
Painting and curing equipment	300 ··· 1,000	200 ··· 400
Electroplating	600 ··· 800	--
Synthetic resin extrusion shop	100 ··· 200	--

The process-related calculation of the power demand also has an impact on the selection and dimensioning of the equipment and the reactive-power compensation. The details of these planning aspects (choice of transformer power rating according to the load carrying capacity or voltage stability criterion, determining of the necessary capacitive power by the multiple coefficient method) will be explained in Chapters 11 and 12.

2.3.2 Quality of supply

The necessary quality of supply is derived from the requirements that the production process has in terms of supply reliability and voltage quality. For that reason, the quality of supply necessarily includes both supply reliability and voltage quality.

The quality of supply is evaluated by the following coordinating conjunction:

$$QS = SR \wedge VQ \tag{2.3}$$

QS quality of supply
SR supply reliability
VQ voltage quality

Only if this AND conjunction is fulfilled (*QS* complies with the process if *SR* and *VQ* comply with the process), can production processes be performed reliably and with the highest possible quality.

2.3.2.1 Supply reliability

The supply reliability (*SR*) is an essential component of the quality of supply (*QS*). Its determinants are the frequency and duration of supply interruptions. A low frequency of supply interruptions is largely achieved by high quality assurance standards in production and assembly of the electrical equipment. Moreover, correct selection and dimensioning of all equipment indirectly contributes to reduction of future supply interruptions.

The technically plannable duration of supply interruptions and the maximum interruption duration throughout which a production process can be continued without damage or costs due to losses are especially important for practical system planning. This interruption duration depends on failure events that can occur with a plausible minimum probability at the various levels of the power system (110 kV, MV, LV). The influence of a failure event on continuation of the production process without damage or outage costs can be analysed in two steps:

- *Step 1*

  Enumeration of failure events, i.e. definition of failure events above a plausible minimum probability [2.12]. In clearly structured systems, failure events with a plausible minimum probability are mainly primary single failures. Double faults and difficult combinations of failures can usually be excluded from consideration. Failures of MV busbar and LV high-current busbar systems are considered highly unlikely.

- *Step 2*

  Examination of the effects of the failure event on the power supply of the production process.

To make a sound *SR* planning decision, clarity must be obtained about the permissible interruption duration of the production process and the duration of supply interruptions that can be planned in technical and economic terms. Table A2.5 shows interruption durations that can be used to plan supply reliability, divided into classes.

Table A2.5 Classification of the interruption duration

SR class	Measure for ending or controlling a disturbance / fault-induced interruption	Interruption duration T_u
1	Repair or replacement of the damaged equipment followed by manual reclosure	$h < T_u \leq d$
2	Local manual load transfer	$min < T_u \leq h$
3	Load transfer by means of remote control	$sec < T_u \leq min$
4	Simple automatic load transfer (e. g. controlled by residual voltage)	$300\,msec < T_u \leq sec$
5	Automatic rapid load transfer	$30\,msec \leq T_u < 300\,msec$
6	Thyristor-based, static high-speed transfer	$T_u \geq \mu sec$
7	Isolation of the fault location by protection equipment	$T_u = 0$ $msec \leq t_{\Delta u'} < sec$
8	Isolation of the fault location by protection equipment and dynamic compensation for voltage dips	$T_u = 0$ $t_{\Delta u'} = 0$ $150\,msec \leq t_{DVR} \leq 600\,msec$
9	Uninterrupted power supply with static or dynamic energy store	$T_u = 0$ $t_{\Delta u'} = 0$ $h \leq t_{UPS/DUPS} < d$

d = day(s), h = hour(s), min = minute(s), sec = second(s), msec = millisecond(s), µsec = microsecond(s)

Class 1	Power supply of the process without transfer and instantaneous reserve
Class 2 to 6	Backed-up power supply of the process by switchover reserve ("cold standby" redundancy for a single fault)
Class 7	Backed-up power supply of the process by the instantaneous reserve ("hot standby" redundancy from the incoming supply to the distribution level)
Class 8	like Class 7 but with additional DVR system for the protection of critical loads from voltage dips $\Delta u'$
Class 9	Doubly backed-up power supply of the process by the instantaneous reserve that is independent of the power system ("hot" standby for complete power system failure and highest power quality using static online UPS or dynamic DUPS systems)
$t_{\Delta u'}$	Duration of voltage dip $\Delta u'$ in case of short circuit
t_{DVR}	Compensation time of a DVR
$t_{UPS/DUPS}$	Backup time of a static online UPS or a DUPS
SR	Supply reliability
DVR	Dynamic Voltage Restorer
DUPS	Diesel UPS
UPS	Uninterruptible Power System

The interruption durations stated in Table A2.5 apply to a single fault and planning decisions made according to the $(n-1)$ criterion. With $(n-1)$-based planning decisions, the not improbable failure of an item of equipment must not result in an impermissible supply interruption. According to this planning principle, the supply reliability is taken into account as a technical constraint. However, optimum planning is achieved by financial evaluation of the supply reliability [2.12]. The financial evaluation of the supply reliability involves explicit consideration of the present value of investment, operating and outage costs and their aggregate minimization. A model for determining the failure costs due to supply interruptions incurred by industrial businesses is presented in [2.13]. Figs. A2.6 and A2.7 show the results of the model calculation.

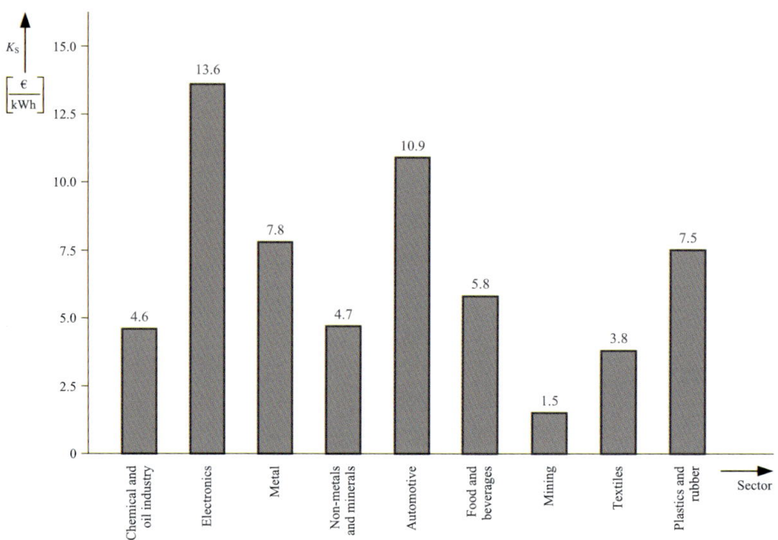

Fig. A2.6 Outage costs K_S due to supply interruptions for various German industries according to [2.13]

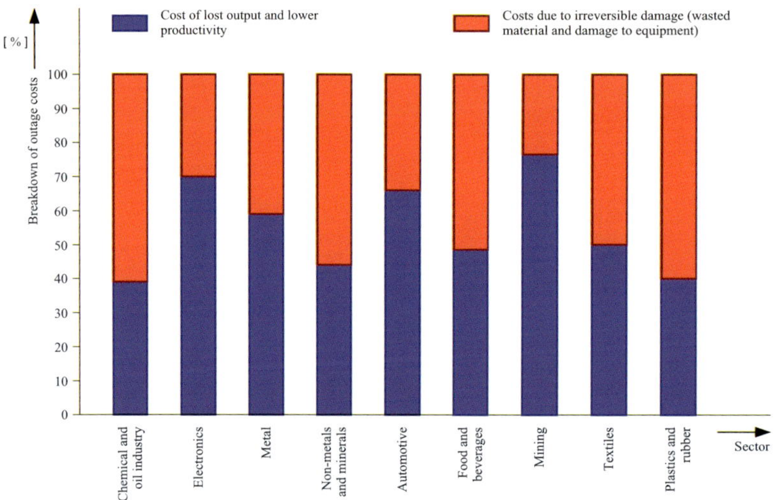

Fig. A2.7 Breakdown of the outage costs for various industries according to [2.13]

The outage costs determined according to the model [2.13] permit financial evaluation of the investment to avoid impermissible supply interruptions as compared with the risk of damage if the necessary *SR* requirements are not met. However, the outage costs stated in Figs. A2.6 and A2.7 are only guidance values. Damage costs due to interruptions are subject to great statistical variance and may differ greatly from one company to another within a single industry. To be precise and verifiable, an *SR* cost-efficiency study must apply the damage costs due to supply interruptions actually incurred by the industrial company.

Conclusion

No stipulations for supply reliability are provided by standards. Consequently, $(n-1)$-based planning decisions are standard practice. To plan the supply reliability according to the $(n-1)$ criterion, the interruption durations divided into classes provided in Table A2.5 can be used. Non-restrictive consideration of the supply reliability is only possible if the damage costs due to interruptions are known.

2.3.2.2 Voltage quality

Like the supply reliability (*SR*), the voltage quality (*VQ*) is an essential component of the quality of supply (*QS*). Indeed, the direct dependence of the quality of supply on the voltage quality has become more pronounced than in the past. This new dependency is most noticeable from the more stringent requirements that modern industrial plants make of the voltage quality in all industries. Above all, sensitive production processes and the industrial application of high technologies such as

- continuous production processes in the plastics and chemical industries (e.g. injection moulding processes),
- microprocessor-controlled automation (computer technology),
- nanotechnological processes (nanotechnology),
- semiconductor technology (e.g. wafer production) and
- photolithography and electron beam lithography (e.g. production of lithographic screens for exposing wafers)

call for solutions specially tailored to the specific *VQ* requirements.

The production processes performed in modern industrial plants make very high demands on the voltage quality but production processes performed in such plants may themselves have an adverse effect on the voltage quality because of their non-linear and fluctuating loads.

The negative effect on the voltage quality caused by the process is evaluated using continuous-time characteristic quantities. Such quality characteristics include flicker, harmonics, voltage unbalance and voltage fluctuations. Event-driven quality characteristics include voltage dips, short interruptions of the voltage and overvoltages. These discrete-time characteristic quantities are used to assess the negative impact of faults in the power system on voltage quality. The relevant characteristics of the power quality are explained in Table A2.8 and shown in Fig. A2.9.

To verify the internal compatibility of all loads of a process and the external compatibility of this process with the supply system, precisely defined limit values and compatibility levels are required for the voltage quality. The most important definitions for voltage quality are to be found in the following standards and guidelines:

- DIN EN 50160 (EN 50160): 2008-04 [2.15] (Voltage characteristics of electricity supplied by public distribution networks),
- DIN EN 61000-2-2 (VDE 0839-2-2): 2003-02 [2.16] or IEC 61000-2-2: 2002-03 [2.17] (Compatibility levels for low-frequency conducted disturbances and signalling in public low-voltage power supply systems),
- DIN EN 61000-2-4 (VDE 0839-2-4): 2003-05 [2.18] or IEC 61000-2-4: 2002-06 [2.19] (Compatibility levels in industrial plants for low-frequency conducted disturbances),
- D-A-CH-CZ Guideline 2007 [2.20] (Technical Rules for Assessing System Perturbations).

Table A2.8 Discrete-time and continuous-time characteristics of the voltage quality

Causes of the reduction in voltage quality	Quality characteristic quantities	
	discrete-time	continuous-time
Faults and similar events in the system	Voltage dips [1]	
	Short interruptions of the voltage [2]	
	Overvoltages/surges [3]	
System perturbations		Flicker [4]
		Harmonics [5]
		Voltage unbalances [6]
		Voltage fluctuations [7]

1) Short circuits in the system are not the only cause of voltage dips. Closing operations and connection of large machines (motors, transformers) and loads also cause voltage dips.

2) Voltage interruptions are a special case of voltage dips. They are typical, for example, in 110-kV overhead systems in operation with automatic reclosure.

3) A distinction is made between external and internal overvoltages. External overvoltages include, for example, lightning surges. Typical internal overvoltages include switching and fault surges.

4) Flicker is low-frequency fluctuation of the voltage amplitude ($f < 25$ Hz). It is caused by the intermittent operation of impulse loads (e. g. welding machines). The human eye perceives flicker in the form of fluctuations in luminance, which are found to be extremely irritating. Flicker can also be caused by the so-called stroboscopic effect, which resembles a slowing or stopping rotary machine.

5) Harmonics are currents or voltages whose frequencies are an integer multiple of the sinusoidal 50(60)-Hz fundamental. They are caused by non-linear loads (e. g. static converter equipment, equipment with electronic power supply units, energy-saving lamps, etc.).

6) Voltage unbalance is a state in the three-phase system in which the RMS values of the three voltages or the angles between two consecutive phases are not of equal magnitude. Voltage unbalances can occur transiently (e. g. asymmetrical fault) or temporarily (e. g. load unbalance).

7) Voltage fluctuations are a sequence of voltage changes that are caused by operation of large fluctuating loads (e. g. arc furnaces)

The standard DIN EN 50160 (EN 50160): 2008-04 [2.15] defines the quality characteristics of the voltage for the supply of electrical energy from the public supply system (Table A2.10).

The standards DIN EN 61000-2-2 (VDE 0839-2-2): 2003-02 [2.16] / IEC 61000-2-2: 2003-03 [2.17] and DIN EN 61000-2-4 (VDE 0839-2-4): 2003-05 [2.18] / IEC 61000-2-4: 2002-06 [2.19] and the D-A-CH-CZ Guideline [2.20], on the other hand, define the *VQ* requirements to be observed in the consumption of electrical energy. The essential difference between the above-stated *VQ* standards is the compliance with the limit values or the permissible compatibility level over the stipulated times. The standard DIN EN 50160 (EN 50160): 2008-04 [2.15] states a large number of limit values that only apply to 95 % of each weekly interval (see Table A2.10). However, the standard limits the validity duration to 95 %, which means that the defined limit values can be violated throughout 5 % of a weekly interval, i.e. 8.4 h.

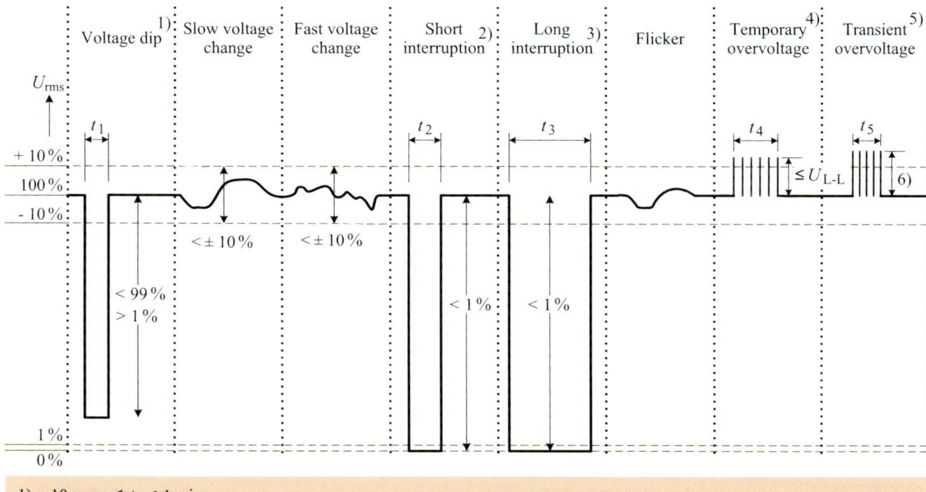

1) 10 msec $\leq t_1 < 1$ min

2) 0.3 sec $< t_2 \leq 3$ min for automatic reclosures in 110-kV overhead systems (special case of a voltage dip)

3) $t_3 > 3$ min (Long interruptions are not a characteristic quantity of the voltage quality. They are a characteristic of the supply reliability)

4) sec $\leq t_4 < 1$ min

5) 1 µsec $< t_5 \leq$ msec

6) Max. 6 kV at the LV level

Fig. A2.9 Graphical representation of the characteristics of the voltage quality [2.14]

The standards DIN EN 61000-2-2 (VDE 0839-2-2): 2003-02 [2.16] / IEC 61000-2-2: 2002-03 [2.17] and DIN EN 61000-2-4 (VDE 0839-2-4): 2003-05 [2.18] / IEC 61000-2-4: 2002-06 [2.19] do not permit such a mitigation of voltage quality. For more specific evaluation of the voltage quality for consumption of electrical energy, DIN EN 61000-2-4 (VDE 0839-2-4): 2003-05 [2.18] / IEC 61000-2-4: 2002-06 [2.19] have also introduced environment classes. In terms of compliance with the process-related *VQ* requirements, these environment classes have the significance of supply or system classes.

A total of three different supply or system classes are described in Table A2.11. For sound system planning, the process to be supplied with power must be assigned to one of the three system classes. The limit values and compatibility levels that are essential to comply with the voltage quality requirements are derived from correct assignment of the process to a certain class. Table A2.12 provides a selection of the most important limit values and compatibility levels of the voltage quality for consumption of electrical energy.

All system perturbations and faults in the system that result in violation of the compatibility levels in Table A2.12 obtained from DIN EN 61000-2-4 (VDE 0839-2-4): 2003-05 [2.18] / IEC 61000-2-4: 2002-06 [2.19] adversely affect the required voltage quality.

In the LPQI study [2.21], an analysis and evaluation of the annual financial loss in Europe due to adverse effects on the voltage quality were conducted separately for the various industries and quality characteristics. Fig. A2.13 shows the most important result of this study.

Table A2.10 Quality characteristics of the voltage for the supply of electrical energy from the public supply system according to DIN EN 50160 (EN 50160): 2008-04 [2.15]

Quality characteristic	Values or value ranges		Measurement and evaluation parameters			
	Low voltage (LV)	Medium voltage (MV)	Basic quantity	Integration interval	Period of observation	Percentage
Slow voltage changes	230 V ± 10 %	U_c ± 10 %	RMS value	10 min	1 week	95 %
Fast voltage changes	5 % max. 10 %	4 % max. 6 %	RMS value	10 msec	1 day	100 %
Flicker (definition for long-term flicker only)	$P_{lt} = 1$		Flicker algorithm	2 h	1 week	95 %
Voltage dips (10 msec $\leq t <$ 1 min)	a few 10s to 1,000 per year (below 85 % U_c)		RMS value	10 msec	1 year	100 %
Short supply interruptions ($t \leq 3$ min)	a few 10s to several 100 per year (below 1 % U_c)		RMS value	10 msec	1 year	100 %
Random long supply interruptions ($t > 3$ min)	a few 10s to 50 per year (below 1 % U_c)		RMS value	10 msec	1 year	100 %
Temporary power-frequency overvoltages (line – earth)	< 1.5 kV	1.7 to 2.0 (depending on neutral-point connection)	RMS value	10 msec	No data	100 %
Transient overvoltages (line – earth)	< 6 kV	According to the insulation coordination	Peak value	None	No data	100 %
Voltage unbalance (ratio of negative- to positive-sequence system)	Mostly 2 %, in special cases, up to 3 %		RMS value	10 min	1 week	95 %
Total harmonic distortion	THD = 8 %		RMS value	10 min	1 week	95 %
Relative harmonic voltage U_h, odd harmonics, not multiples of 3	$h = 5$: 6.0 % $h = 7$: 5.0 % $h = 11$: 3.5 % $h = 13$: 3.0 % $h = 17$: 2.0 % $h = 19$: 1.5 % $h = 23$: 1.5 %		RMS value	10 min	1 week	95 %
Relative harmonic voltage U_h, odd harmonics, multiples of 3	$h = 3$: 5.0 % $h = 9$: 1.5 % $h = 15$: 0.5 % $h = 21$: 0.5 %		RMS value	10 min	1 week	95 %
Relative harmonic voltage U_h, even harmonic	$h = 2$: 2.0 % $h = 4$: 1.0 % $6 \leq h \leq 24$: 0.5 %		RMS value	10 min	1 week	95 %

U_c Conventional voltage (U_c is equal to the nominal operating voltage U_{nN} of the distribution system)
P_{lt} Long-term flicker intensity
THD Total harmonic distortion
h Order of the harmonic voltage

As Fig. A2.13 shows, voltage dips and transient overvoltages are responsible for most of the financial losses. These high financial losses underline the importance of the immunity of processes to voltage dips and transient overvoltages.

The immunity to voltage dips and transient overvoltages is described using voltage tolerance envelopes. The best-known voltage tolerance envelope is the ITIC or CBEMA curve (Fig. A2.14). Its use is no longer restricted to just the IT industry. In other high-tech industries, too, the tolerance specifications of the ITIC curve can be used to define the immunity to voltage dips. In individual cases, especially sensitive production processes can require more severe specifications than the generally applied voltage tolerance specification of the ITIC curve. Fig. A2.15 shows an example from the chemical industry.

Barring some exceptional cases, the results of national and international measurement studies on event-driven influencing quantities [2.22, 2.23] show that, on voltage dips and overvoltages, practically no adverse affects on high-tech production processes occur within the region of uninterrupted function of the ITIC curve (Fig. A2.14).

Table A2.11 Classification of power systems for the supply of processes according to DIN EN 61000-2-4 (VDE 0839-2-4): 2003-05 [2.18] / IEC 61000-2-4: 2002-06 [2.19]

Class	Content description
1 (power systems with special VQ characteristics)	Power supply for sensitive processes requiring compatibility levels for disturbances of the voltage quality that are below the valid levels for electrical power from the public network. Class 1 applies to protected power supplies (e. g. power supply to the infrastructure of high-tech companies or data centres).
2 (public MV and LV systems)	Power supply for processes permitting compatibility levels for disturbances of the voltage quality that are the same as the valid levels and quality characteristics for electrical power from the public network. Class 2 applies to points of common coupling (PCC) with the public network and to in-plant points of coupling (IPC) with industrial and other non-public supply systems.
3 (industrial power systems with multiple voltage levels)	Power supply for more robust processes that permit compatibility levels for disturbances of the voltage quality that are, in somes cases, above the valid levels for the supply of electrical power from the public network (e. g. welding processes, production processes with large and frequently starting motors and a high static converter component in the total load). Class 3 applies exclusively to the in-plant points of coupling (IPC) of industrial power systems.

VQ Voltage quality
PCC Point of common coupling/connection
IPC In-plant point of coupling/connection

Conclusion

The information in Table A2.11, the characteristics in Table A2.12 and the voltage tolerance envelope in Fig. A2.14 can be used to determine the voltage quality requirements of a specific production process. The voltage tolerance envelope shown in Fig. A2.15 is recommended for especially sensitive processes in the chemical industry.

Table A2.12 Permissible compatibility levels for the voltage quality according to DIN EN 61000-2-4 (VDE 0839-2-4): 2003-05 [2.18] / IEC 61000-2-4: 2002-06 [2.19]

Characteristic quantities	Classes for power supply to processes		
	Class 1	Class 2	Class 3
Permissible compatibility levels for voltage and voltage unbalance			
Voltage fluctuation range $\Delta U / U_{nN}$	± 8 %	± 10 %	+ 10 % to - 15 %
Voltage unbalance $\Delta U_{negative} / \Delta U_{positive} \hat{=} k_{U\text{-}perm}$	2 %	2 %	3 %
Permissible compatibility levels for harmonics			
Harmonic order	$U_{h\text{-}perm}$ for the odd-numbered orders, non-multiples of 3		
$h = 5$	3 %	6 %	8 %
$h = 7$	3 %	5 %	7 %
$h = 11$	3 %	3.5 %	5 %
$h = 13$	3 %	3 %	4.5 %
$h = 17$	2 %	2 %	4 %
$17 < h \leq 49$	2.27 % · (17/h) – 0.27 %	2.27 % · (17/h) – 0.27 %	4.5 % · (17/h) – 0.5 %
Harmonic order	$U_{h\text{-}perm}$ for the odd-numbered orders, multiples of 3		
$h = 3$	3 %	5 %	6 %
$h = 9$	1.5 %	1.5 %	2.5 %
$h = 15$	0.3 %	0.4 %	2 %
$h = 21$	0.2 %	0.3 %	1.75 %
$21 < h \leq 45$	0.2 %	0.2 %	1 %
Harmonic order	$U_{h\text{-}perm}$ for the even-numbered orders		
$h = 2$	2 %	2 %	3 %
$h = 4$	1 %	1 %	1.5 %
$h = 6$	0.5 %	0.5 %	1 %
$h = 8$	0.5 %	0.5 %	1 %
$h = 10$	0.5 %	0.5 %	1 %
$10 < h \leq 50$	0.5 %	0.5 %	1 %
Permissible compatibility levels for the total distortion			
Total harmonic distortion THD_{perm}	5 %	8 %	10 %

U_{nN} Nominal system voltage
ΔU Absolute voltage difference
$\Delta U_{negative} / \Delta U_{positive}$ Voltage ratio of negative- to positive-sequence system
$k_{U\text{-}perm}$ Permissible unbalance factor of the voltage
THD_{perm} Permissible total harmonic distortion

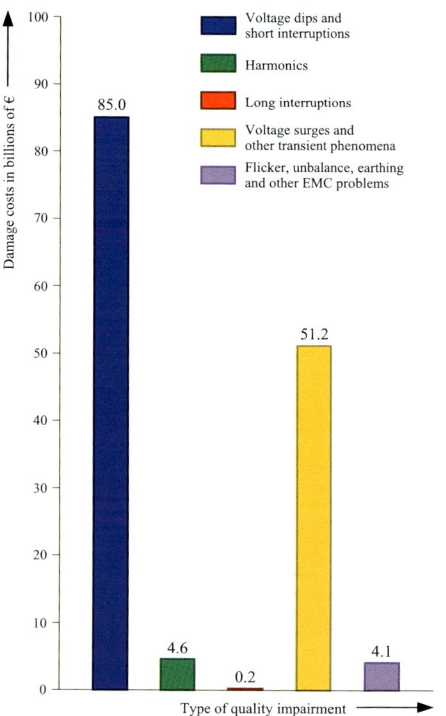

Fig. A2.13
Annual financial loss due to adversely affected voltage quality in the industries of European countries according to [2.21]

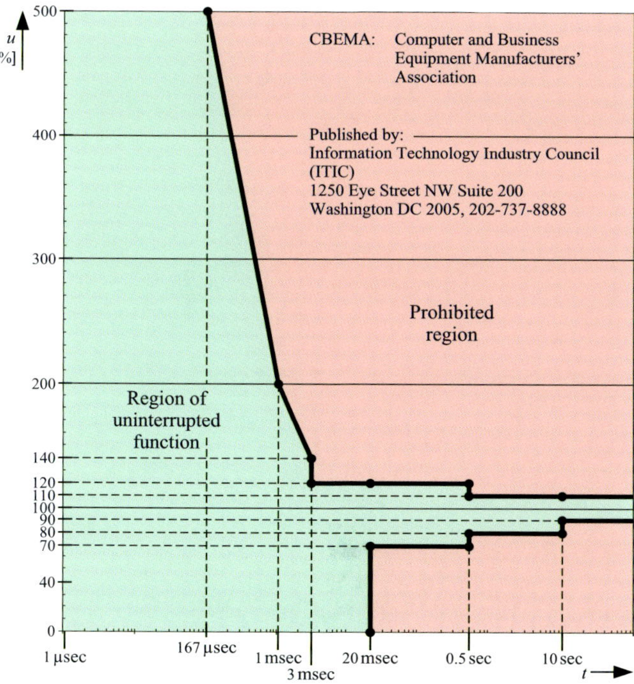

Fig. A2.14 ITIC voltage tolerance envelope for processes in the IT industry (CBEMA curve)

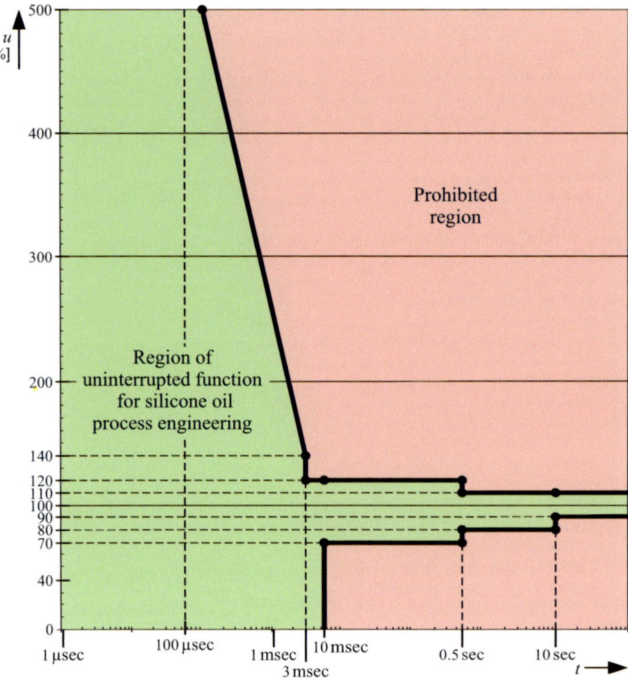

Fig. A2.15 Voltage tolerance envelope for an extremely sensitive process in the chemical industry (Dow Corning Corporation)

2.4 Determining of process-compliant power supply variants

The alternative variants for supplying a production process with power must meet the requirements, be permissible and be feasible. With this objective, possible power supply variants can be determined in the following process:

Step 1: Collating the optimization variables

The optimization variables represent the existing degrees of freedom in power system design, dimensioning and operation. Depending on the specific task to be performed, the following degrees of freedom exist for planning industrial power supply plants:

a) Design

- Type of connection to the public power system
- System configuration for the distribution and load level
- Locations for substations (supply and distribution level)
- Routes for cable installation
- Integration of the main equipment (generators, transformers, short-circuit limiting reactors, HV motors)
- Substation concepts (single-busbar and double-busbar switchgear)

b1) Dimensioning for steady-state operation

- Nominal voltage for the supply, distribution and load level

- Load current and fault current-carrying capacity of the equipment (equipment parameters)
- Fixed compensation units without reactors for PF correction in systems with a low harmonic component (< 15 %)
- Fixed compensation units with reactors for PF correction in systems with a high harmonic component (> 15 %)
- Passive tuned filters (L-C circuits) for static compensation of harmonics

b2) *Dimensioning for immunity to continuous-time and discrete-time system perturbations*

- Short-circuit power at the MV and LV level
- Motor starting connections (reactor starting, reduced-voltage starting, main-circuit-transformer starting)
- SVC systems (static var compensators) for flicker compensation and voltage stability
- DVR/DySC systems (dynamic voltage restorer/dynamic sag correctors) for compensating for temporary voltage reductions and transient system voltage dips
- Active system filter (sine wave filter with integrated IGBT modules (insulated gate bipolar transistors)) for complete compensation of harmonics
- DDUPS systems (dynamic diesel uninterrupted power supply systems) for high-quality (no continuous-time or discrete-time system perturbations) and reliable uninterruptible power supply

c) *Mode of operation*

- Method of neutral earthing of the MV system (only for electrical isolation from the public power system)
- Type of LV system earthing (TN, TT and IT systems)
- Automatic transfer gear and system decoupling equipment
- Protection and automation equipment of the power system
- Circuit state in normal operation (NOP) and operation under fault conditions (OPFC)
- Standby generating systems (emergency generating sets) to back up long-lasting power failures

Step 2: Assignment of technically meaningful implementations to the optimization variables

In this step, technically meaningful sets of values are assigned to the optimization variables that can be used for a planning decision. The set of values of a relevant optimization variable corresponds to the set of all possible discrete implementations. By way of example, Table A2.16 contains the sets of values of relevant optimization variables for planning a new 110-kV/MV industrial power system incoming supply.

Step 3: Formation of possible variants

Possible variants are formed by combining the discrete implementations of the relevant optimization variables. Table A2.17 shows the principle of variant formation. The formation of variants in Table A2.17 is based on the planning example from step 2.

Table A2.16 Example assignment of value sets to optimization variables

Optimization variable	Value set
Transformer substation connection to the public 110-kV OHL system	▪ Loop-in ▪ Double radial line
Nominal voltage of the MV power system	▪ U_{nN} = 10 kV ▪ U_{nN} = 20 kV
Neutral-point connection of the MV power system	▪ Low-impedance neutral earthing (NOSPE) ▪ Short-time low-impedance neutral earthing (KNOSPE)
Switching state of the bus sectionalizing circuit-breaker in the MV main substation	▪ Closed during normal operation ▪ Open during normal operation

Table A2.17 Example variant formation for a new 110-kV/MV industrial power system incoming supply

<table>
<tr><th rowspan="2">Variant</th><th rowspan="2">Design of the transformer substation connection to the public 110-kV OHL system</th><th rowspan="2">Rating of the supplying 110-kV/MV transformers</th><th colspan="2">Operating mode of the MV power system</th></tr>
<tr><th>Method of neutral-point connection</th><th>Switching state of the bus sectionalizing circuit-breaker in normal operation</th></tr>
<tr><td>V_{11}</td><td rowspan="8">Loop-in</td><td rowspan="4">U_{rT} = 10 kV
S_{rT} = 2 x 25 MVA</td><td rowspan="2">NOSPE</td><td>closed</td></tr>
<tr><td>V_{12}</td><td>open</td></tr>
<tr><td>V_{13}</td><td rowspan="2">KNOSPE</td><td>closed</td></tr>
<tr><td>V_{14}</td><td>open</td></tr>
<tr><td>V_{15}</td><td rowspan="4">U_{rT} = 20 kV
S_{rT} = 2 x 25 MVA</td><td rowspan="2">NOSPE</td><td>closed</td></tr>
<tr><td>V_{16}</td><td>open</td></tr>
<tr><td>V_{17}</td><td rowspan="2">KNOSPE</td><td>closed</td></tr>
<tr><td>V_{18}</td><td>open</td></tr>
<tr><td>V_{21}</td><td rowspan="8">Double radial line</td><td rowspan="4">U_{rT} = 10 kV
S_{rT} = 2 x 25 MVA</td><td rowspan="2">NOSPE</td><td>closed</td></tr>
<tr><td>V_{22}</td><td>open</td></tr>
<tr><td>V_{23}</td><td rowspan="2">KNOSPE</td><td>closed</td></tr>
<tr><td>V_{24}</td><td>open</td></tr>
<tr><td>V_{25}</td><td rowspan="4">U_{rT} = 20 kV
S_{rT} = 2 x 25 MVA</td><td rowspan="2">NOSPE</td><td>closed</td></tr>
<tr><td>V_{26}</td><td>open</td></tr>
<tr><td>V_{27}</td><td rowspan="2">KNOSPE</td><td>closed</td></tr>
<tr><td>V_{28}</td><td>open</td></tr>
<tr><td>NOSPE</td><td colspan="4">German for: low-impedance neutral earthing</td></tr>
<tr><td>KNOSPE</td><td colspan="4">German for: short-time low-impedance neutral earthing</td></tr>
<tr><td>OHL</td><td colspan="4">Overhead line</td></tr>
</table>

Step 4: Checking the variants

If it was not possible in the previous steps, the variants formed must finally be checked for compliance with the requirements, for reliability and for feasibility. As defined in Section 2.3, the requirements are based on power demand and quality of supply. Reliability is checked based on the technical constraints. Technical constraints are the current-carrying capacity and immunity conditions to be complied with (Table A2.18).

Table A2.18 Classification of technical constraints

Technical constraints			
Current-carrying capacity conditions		Disturbance/interference immunity conditions	
Load current-carrying capacity	Fault current-carrying capacity	System perturbations	Single fault
▪ Load current condition ▪ Voltage drop condition	▪ Short-circuit current conditions	▪ Voltage stability condition ▪ Flicker compatibility condition ▪ Voltage unbalance condition ▪ Harmonic compatibility conditions	▪ $(n-1)$ criterion

Compliance with the immunity conditions ensures the necessary immunity from discrete-time and continuous-time system perturbations and control of the single fault.

The current-carrying capacity conditions must be complied with to ensure reliable steady-state system operation. The following rules apply to the individual conditions:

- *Load current condition*

$$I_{\text{load-max}} \leq I_{perm} \tag{2.4}$$

$I_{\text{load-max}}$ maximum load current
I_{perm} continuous current-carrying capacity

Definitions of the current-carrying capacity of PVC, PE and XLPE power cables are found in the following standards:

- DIN VDE 0276-1000 (VDE 0276-1000): 1995-06 [2.24]
- DIN VDE 0265 (VDE 0265): 1995-12 [2.25]
- DIN VDE 0271 (VDE 0271): 2007-01 [2.26]
- DIN VDE 0276-620 (VDE 0276-620): 2009-05 [2.27] or IEC 60502-2: 2005-03 [2.28]
- DIN VDE 0298-4 (VDE 0298-4): 2003-08 [2.29]

The current-carrying capacity of the dry-type transformers preferably used in industry is governed by the standard DIN EN 60076-12 (VDE 0532-76-12): 2008-01 [2.30] or IEC 60076-12:2008-11 [2.31].

- *Voltage drop condition*

$$\Delta u_{\max} \leq \Delta u_{perm} \tag{2.5}$$

$\Delta u_{\max}$ maximum voltage dop, occurring at $I_{\text{load-max}}$
Δu_{perm} permissible steady-state voltage drop

The measure for permissible steady-state voltage drop is the voltage fluctuation range $\Delta U/U_{\text{nN}}$ at the consumer's nodes of the industrial system. This fluctuation range, which is defined in standards, is indicated in Table A2.12.

The load current and voltage drop condition must be complied with both in normal operation and in operation under fault conditions. The "Load Flow" calculation module of the interactive system analysis and system planning program PSS™SINCAL [2.32, 2.33] is available for checking these two loadability conditions. The PSS™SINCAL Load Flow module provides evaluation options (either the Newton-Raphson or the current iteration method) for the capacity utilization of lines and transformers and of voltage ranges within freely selectable limits.

- *Short-circuit current conditions*

$$I_b \leq \begin{cases} I_{\text{sc}} & \text{for MV switchgear} \\ I_{\text{cs}}(I_{\text{cu}}) & \text{for LV switchgear} \end{cases} \tag{2.6}$$

I_{b} symmetrical short-circuit breaking current
I_{sc} rated short-circuit breaking current
I_{cs} rated service short-circuit breaking capacity
I_{cu} rated ultimate short-circuit breaking capacity

$$i_{\text{p}} \leq \begin{cases} I_{\text{ma}} & \text{for MV switchgear} \\ I_{\text{cm}} & \text{for LV switchgear} \end{cases} \tag{2.7}$$

i_{p} peak short-circuit current
I_{ma} rated short-circuit making current
I_{cm} rated short-circuit making capacity

$$i_{\text{p}} \leq I_{\text{pk}} \tag{2.8}$$

I_{pk} rated peak withstand current

$$I_{\text{th}} \leq I_{\text{thp}} \tag{2.9}$$

I_{th} thermal equivalent short-circuit current
I_{thp} permissible thermal fault withstand capability

The necessary strength values in case of a short circuit are found on the right-hand side of Eqs. (2.6) to (2.9). The left-hand side states the short-circuit currents that determine the stress on the equipment. These short-circuit currents can be calculated with the "Short-Circuit" module of the PSS™SINCAL system analysis program [2.32, 2.33]. The stress values required to check the short-circuit current conditions are stored in the PSS™SINCAL database after simulation and are available for further evaluation in the form of system graphics or fault-location-related tables.

Detailed data on short-circuit-proof dimensioning of the equipment are provided in Chapters 4 (MV systems) and 9 (LV systems).

- *Voltage stability condition*

$$\Delta u' \leq \Delta u'_{\mathrm{perm}} \tag{2.10}$$

$\Delta u'$ relative voltage dip on a symmetrical or asymmetrical load change
$\Delta u'_{\mathrm{perm}}$ permissible voltage dip

The D-A-CH-CZ Guideline for Assessing System Perturbations [2.20] provides an extensive set of formulas and rules for calculating voltage dips on symmetrical and asymmetrical load changes. Specific examples of calculation of voltage dips are given in Chapter 10.

Like the steady-state voltage drop Δu_{perm}, the discrete-time voltage dip $\Delta u'_{\mathrm{perm}}$ permitted in the power system is limited by the fluctuation range of the load voltage $\Delta U/U_{\mathrm{nN}}$. At the in-plant points of coupling of the industrial power systems, this fluctuation range is $-15\,\%$ (Table A2.12). For a standardized fluctuation range of $-0.15 \cdot U_{\mathrm{nN}}$ and a limit value for steady-state load voltage of $0.95 \cdot U_{\mathrm{nN}}$, the permissible voltage change can be $\Delta u'_{\mathrm{perm}} = 10\,\%$. Limitation of the dynamic voltage change to $\Delta u'_{\mathrm{perm}} = 10\,\%$ has proven useful above all in the operation of welding power systems in the car industry. For example, incorrect welding can usually be avoided in the event of voltage dips of $\Delta u' < 10\,\%$ even without the use of dynamic compensation equipment.

In industrial systems that are mainly used to supply motor loads with power, larger voltage dips are permissible. In these power systems, the permissible voltage dip is, above all, determined by the necessity for the motor to have completed starting within the defined time. According to this requirement, values in the range $10\,\% < \Delta u' \leq 15\,\%$ are often permitted for 6 or 10-kV industrial systems. In LV systems, a sufficient minimum voltage is still applied to the terminals of a starting motor if the permissible voltage dip is $\Delta u'_{\mathrm{perm}} = 25\,\%$ [2.34].

- *Flicker compatibility condition*

$$\Delta u'_{\mathrm{fluc}} \leq \Delta u'_{perm}(r, P_{\mathrm{st}}, P_{\mathrm{lt}}) \tag{2.11}$$

$\Delta u'_{\mathrm{fluc}}$ fluctuating voltage dip in case of intermittent load
$\Delta u'_{\mathrm{perm}}(r, P_{\mathrm{st}}, P_{\mathrm{lt}})$ permissible voltage dip depending on the repetition rate r and the short-term or long-term flicker intensity P_{st} or P_{lt}

The flicker compatibility condition (2.11) can be checked using standardized flicker reference curves and flicker limit curves (Fig. A2.19).

As the figure shows, the regular square-wave voltage change amplitudes are limited depending on the repetition rate r and taking the short-term flicker intensity into account. Limitation to permissible voltage dips also depends on the time used to evaluate the flicker. Short-term intervals (10 minutes) or long-term intervals (2 hours) can be considered.

For the threshold of flicker irritability of the human eye, perception of light fluctuations is described by way of the reference short-term flicker intensity $P_{\mathrm{st\text{-}ref}} = 1$ [2.20]. This reference short-term flicker intensity must not be exceeded by interaction of all sources of disturbance in the power system.

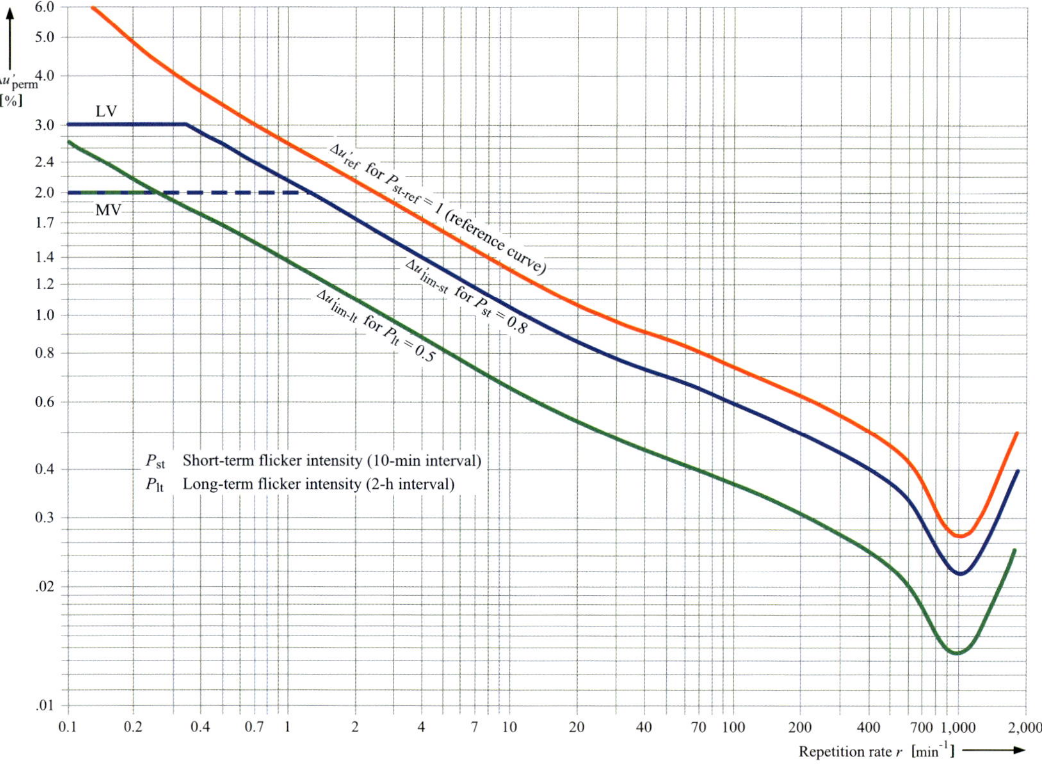

Fig. A2.19 Flicker reference and flicker limit curves $\Delta u'$ for square-wave voltage changes depending on the repetition rate r [2.20]

- *Voltage unbalance condition*

$$k_U \leq k_{U\text{-perm}} \tag{2.12}$$

k_U unbalance factor of the voltage
$k_{U\text{-perm}}$ permissible unbalance factor of the voltage

The voltage unbalance caused by two-phase loads (line-to-line connection) and single-phase loads (line-to-neutral connection) is described by the unbalance factor k_U. This is approximately calculated as follows:

$$k_U \approx \frac{S_{pr}}{S''_{k\text{-IPC}}} \cdot 100 \tag{2.13}$$

S_{pr} power rating of the single or two-phase load
$S''_{k\text{-IPC}}$ short-circuit power at the in-plant point of coupling

The unbalance factor of the voltage $k_{U\text{-perm}}$ that is permissible for steady-state system operation is given in Table A2.12.

- *Harmonic compatibility conditions*

$$THD \leq THD_{perm} \tag{2.14}$$

THD_{perm} permissible total harmonic distortion (see Tables A2.12, C12.24, C12.25)
THD total harmonic distortion factor

The total harmonic distortion factor is the ratio of the RMS value of the sum of all harmonic components up to a defined harmonic order H to the RMS value of the fundamental component. It is calculated as follows:

$$THD = \sqrt{\sum_{h=2}^{H} \left(\frac{Q_h}{Q_1}\right)^2} \tag{2.15}$$

Q current or voltage
Q_1 RMS value of the fundamental component
h order of the harmonics
Q_h RMS value of the harmonic component of order h
H final value of the summation (Generally $H = 50$. If the risk of resonances is slight at higher orders, $H = 25$ can be chosen.)

In addition to the compatibility condition for the total harmonic distortion of the voltage (Eq. 2.14), the compatibility condition for the relevant harmonic voltages must also be complied with:

$$U_h \leq U_{h\text{-perm}} \tag{2.16}$$

$U_{h\text{-perm}}$ permissible harmonic voltage of the harmonic order h in %
(see Tables A2.12, C12.24, C12.25)
U_h harmonic voltage of order h caused in %

The harmonic compatibility conditions (2.14) and (2.16) can be checked using the “Harmonics“ calculation module of the PSS™SINCAL system analysis software. The PSS™SINCAL “Harmonics“ module [2.32, 2.33] is used to calculate the harmonic distribution in the power system. The calculated harmonic currents and voltages are provided with the appropriate limit values as graphical output for all nodes and system levels.

- *(n – 1) criterion*

The $(n-1)$ criterion states that the not improbable failure of any item of equipment must not result in an impermissible supply interruption. The permissibility of a supply interruption mainly depends on the interruption duration for which a production process can still be continued without damage or outage costs. For the permissible interruption duration of a production process, compliance with the $(n-1)$ criterion can be evaluated by means of Table A2.5.

In addition to the current-carrying capacity and immunity conditions, problem-specific constraints may also have to be taken into account that result from existing special aspects of the generation, distribution and consumption of electrical energy. For example, when back-pressure turbines are used for cogeneration plants, the generation of electrical energy is directly dependent on the process steam requirement in production. The generated power is therefore only a “waste product“ of the process steam generation. The supply concept must be variably adjusted for this to cover the electrical power demand.

Along with the reliability of the power supply variants, their feasibility must be evaluated. Above all, the products and systems available on the market and the state of the art will influence the feasibility.

The feasibility can also depend on particular customer requests (e.g. preference for a certain product range, such as switchgear with silicone-free insulants, designs without fuses, etc.), and material and financial circumstances.

2.5 Search for the optimum solution

The optimum solution has to be sought within a defined optimization space. This optimization space comprises the set of requirement-compliant, permissible and feasible power supply variants. The method for searching for the optimum power supply variant is described below.

2.5.1 Decision objectives

The search for the optimum solution is a multiple-objective-oriented decision process. The following nine individual objectives affect the planning decision to be made:

① low investment costs

② low system power loss costs or high energy efficiency

③ process-related power demand coverage

④ high supply reliability

⑤ high voltage quality

⑥ low hazard for persons and equipment

⑦ low maintenance and servicing expense

⑧ uncomplicated system management

⑨ high level of environmental compatibility

The individual objectives are either neutral with respect to each other, mutually beneficial or conflict with each other [2.35]. The tradeoff between conflicting objectives is the main problem in the search for the optimum solution because improving fulfilment of one objective necessarily worsens fulfilment of other objectives. Table A2.20 shows the conflicts affecting planning decisions in industry when a number of objectives have to be pursued. According to this, the following main conflicts apply:

- Objective ① conflicts with objectives ②, ③, ④, ⑤, ⑥, ⑧, ⑨
- Objective ⑦ conflicts with objectives ②, ③, ④, ⑤

There is no conflict between objective ① and objective ⑦ that could affect the decision. The following two reasons are decisive:

- No additional investment costs are incurred, thereby reducing the maintenance and service expense through the use of maintenance-free switchgear and self-monitoring numerical system protection devices. Maintenance-free equipment is currently state of the art.
- With a smaller investment outlay and therefore fewer items of equipment, the maintenance expense is also lower.

According to this logic, objectives ① and ⑦ are mutually beneficial. The mutually beneficial relationship between objective ① and ⑦ allows the planner to concentrate on finding the optimum tradeoff between objective ① and objectives ②, ③, ④, ⑤, ⑥ and ⑧.

Table A2.20 Conflicts in case of multiple-objective-oriented planning decisions in industry

	Individual objectives	①	②	③	④	⑤	⑥	⑦	⑧	⑨
①	Low investment costs	–	X	X	X	X	X	(X)	X	X
②	Low power system losses	X	–	(X)	(X)	(X)	(X)	(X)	(X)	(X)
③	Process-related power demand coverage	X	(X)	–	(X)	(X)	(X)	X	(X)	(X)
④	High supply reliability	X	(X)	(X)	–	(X)	(X)	X	(X)	(X)
⑤	High voltage quality	X	(X)	(X)	(X)	–	(X)	X	(X)	(X)
⑥	Low hazard to persons and equipment	X	(X)	(X)	(X)	(X)	–	(X)	(X)	(X)
⑦	Low maintenance and servicing expenses	(X)	X	X	X	X	(X)	–	(X)	(X)
⑧	Uncomplicated system management	X	(X)	(X)	(X)	(X)	(X)	(X)	–	(X)
⑨	High level of environmental compatibility	X	(X)	(X)	(X)	(X)	(X)	(X)	(X)	–

X (red) Conflict; (X) (green) No or only insignificant conflict

2.5.2 Decision-making method

Decisions involving multiple conflicting objectives are necessarily compromises. Compromise decisions can be made by the combined application of the following criteria [2.35, 2.36]:

Optimality criterion

The optimality criterion means that the planner must seek the variant that meets the defined requirements at minimum cost. The defined requirements represent the degree of compliance with objectives ③, ④ and ⑤ as defined in terms of the process.

In the planning process (Fig. A2.1), definition of the degree of compliance with objectives ③, ④ and ⑤ corresponds to step 2 "Determining of the process-specific requirements".

The degree of compliance with objectives ⑥, ⑦, ⑧ and ⑨ is defined in step 3 "Determination of the possible variants".

With one possible variant, the defined degrees of compliance with objectives ③ to ⑨ can be achieved. To determine the variant that incurs the lowest cost from the set of possible variants, in step 4 "Search for the optimum solution", objectives ① and ② are combined in a calculation. This combined calculation is as follows:

$$TOTEX_{\mathrm{i}} = (CAPEX + LOSSEX)_{\mathrm{i}} \tag{2.17}$$

$$CAPEX = C_{\mathrm{Invest\text{-}t_0}} \tag{2.17.1}$$

$$LOSSEX = \sum_{\mathrm{t}=1}^{n_{\mathrm{p}}} C_{\mathrm{Loss-t}} \cdot q^{-\mathrm{t}} \tag{2.17.2}$$

$TOTEX_i$	total financial expense of variant *i* (present value of the total costs)
$CAPEX$	expense for investments (present value of the investment costs)
$LOSSEX$	expense for system power losses (present value of the system loss costs)
$C_{\text{Invest-t}_0}$	investment costs incurred completely at the beginning of the period under consideration t_0
$C_{\text{Loss-t}}$	system loss costs in year t
n_p	number of years of the period under consideration
q	interest rate factor

For the choice of the optimum variant V_{opt} the following applies:

$$V_{opt} = \left\{ V_i \Big|_{V_i \in V \wedge \min\limits_i \text{TOTEX}_i} \right\} \tag{2.18}$$

V_i	possible variant *i*
V	set of all possible variants

Expansion of the optimality criterion such that the financial expense is calculated as a function of the degree of compliance for objectives ④ and ⑤, too, has usually proven to be largely impracticable. Above all, calculation of the verifiable outage costs due to interruptions and damage costs caused by reduced voltage quality is very problematic. If the remaining conflict is taken into account after the combined calculation of objectives ① and ②, the appropriate compromises can be also be found by combined application of the decision aids and criteria given below.

Acceptability criterion

The acceptability criterion is used to define the acceptable level of events causing damage. To define this level, both the effects and the probabilities of unwanted events are considered (Fig. A2.21).

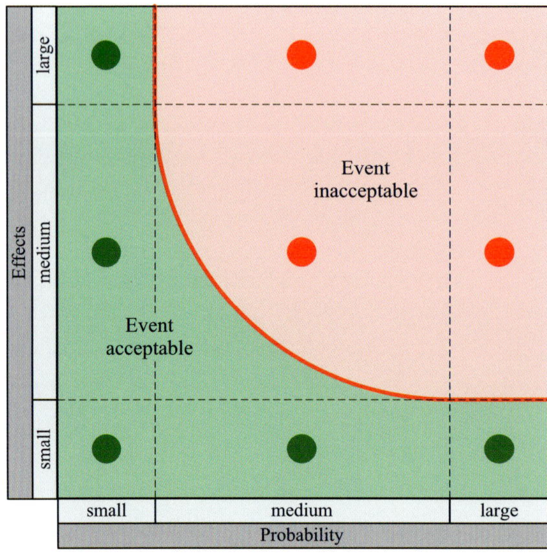

Fig. A2.21
Decision-making using the acceptability criterion

The compromises must consider not only the severity of the damage, but also the probability that the damage will occur. This method of finding a compromise is exemplified by the busbar fault. With the use of single busbar switchgear, a busbar fault often causes failure of all loads involved in the production process. A supply interruption lasting several hours or even days can result in very serious consequential damage (direct or indirect outage costs). Despite the serious consequential damage, however, installation of double busbar switchgear is not always justified. With Siemens SF_6-insulated fixed-mounted circuit-breaker switchgear of type NXPLUS C, for example, the screened single-pole solid insulation of the busbars makes a busbar short-circuit practically impossible. In view of its extremely low probability, a busbar short-circuit with the use of NXPLUS C switchgear is not an unacceptable event.

Sensitivity criterion

In line with the criterion-based analysis method, the sensitivity criterion is used to verify the sensitivity of individual objectives. Verifying the sensitivity tells the planner whether further improvement of an objective can only be achieved by disproportionate worsening of another objective. The example shown in Fig. A2.22 plots the mutual sensitivity of objectives ① and ④. Fig. A2.22 shows that the compromises only make sense if regions with only minimum improvement of objectives or clear worsening of objectives are avoided. For that reason, application of the sensitivity criterion is an important absolute condition for appropriate compromises.

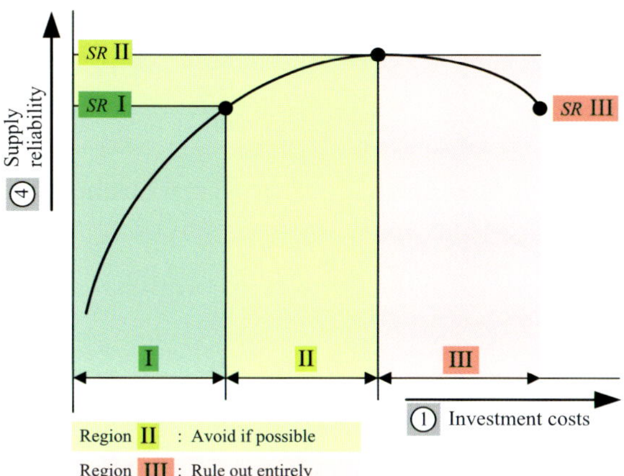

Fig. A2.22
Sensitivity verification of objectives ① and ④

Comparison criterion

In certain cases, comparisons can make it easier to define degrees of compliance with objectives. To support the findings of the compromise, the comparison criterion uses relevant comparisons with the past and with power distribution systems of allied industries at home and abroad.

Conclusion

In application of the decision criteria described, there is always some margin for discretion, which is influenced by various customer interests. This discretionary margin allows the compromise found to tend more towards low investment costs or more towards a high quality of supply.

Using the fundamentals of planning explained in Part A as a basis, Parts B and C of this book focus on providing planning recommendations for the design, dimensioning and operation of industrial medium-voltage and low-voltage power systems.

B Planning recommendations for medium-voltage systems

3 Choosing the MV system voltage

3.1 Incoming supply level

The voltages that can be chosen for the incoming supply and distribution levels are defined in the standard DIN IEC 60038 (VDE 0175): 2002-11 [3.1] or IEC 60038: 2009-06 [3.2]. The standard voltages that planners can use for the ranges 1 kV < U_{nN} ≤ 35 kV (medium voltage) and 35 kV < U_{nN} ≤ 230 kV (high voltage) are listed in Table B3.1.

From a power system engineering viewpoint, connection of industrial plants to the upstream high-voltage power system is recommended if the process-related power demand is about one third of the installed transformer power of the public medium-

Range	Highest voltage for equipment U_m [kV]	Nominal system voltage U_{nN} [kV]	
		Standard value 1	Standard value 2
1 kV < U_{nN} ≤ 35 kV	3.6 [1)]	3.3 [1)]	3 [1)]
	7.2 [1)]	6.6 [1)]	6 [1)]
	12	11	10
	(17.5) [2)]	--	(15) [2)]
	24	22	20
	36 [3)]	33 [3)]	30
	40.5 [3)]	--	35 [3)]
35 kV < U_{nN} ≤ 230 kV	(52)	(45)	--
	72.5	66	69
	123	110	115
	145	132	138
	(170)	(150)	(154)
	245	220	230

1) These values should not be used for public distribution systems.
2) The values indicated in parentheses should be considered non-preferred. It is recommended that these values should not be used for new systems to be constructed in future.
3) The unification of these values is under consideration.

Table B3.1 Standard voltages from 1 kV to 230 kV acc. to DIN IEC 60038 (VDE 0175): 2002-11 [3.1] or IEC 60038: 2009-06 [3.2]

voltage network [3.3]. Public medium-voltage networks usually have the following parameters:

- $U_{nN} = 10$ kV and $S_k'' = 250 - 350$ MVA,
- $U_{nN} = 20$ kV and $S_k'' = 500$ MVA.

Based on these power system parameters and an upper percent impedance voltage of the power transformers of $u_{rZ} = 12.5$ %, the following power demand would make connection to the 110-kV system necessary:

- $S_{max} \geq 14.5$ MVA ($P_{max} \geq 13$ MW)
  when the public MV distribution network is operated at $U_{nN} = 10$ kV,
- $S_{max} \geq 20.0$ MVA ($P_{max} \geq 18$ MW)
  when the public MV distribution network is operated at $U_{nN} = 20$ kV.

Below the limits S_{max} (10 kV) < 14.5 MVA and S_{max} (20 kV) < 20.0 MVA, industrial plants can usually be directly connected to the existing medium-voltage network of the power supply company responsible for them. This power system engineering condition applies to medium-voltage cable networks with a high consumer density (e.g. urban industrial zones).

In rural supply areas, further conditions apply to the connection of an industrial plant to the public medium-voltage network. If the short-circuit power of the network periphery is excessively low, fluctuating loads of the industrial plant can cause impermissible voltage fluctuations in the entire medium-voltage power system. Impermissible voltage fluctuations can only be avoided by increasing the short-circuit power of the system or by dynamic compensation equipment (DVR systems). Instead of incurring investment costs to bolster the MV power system or to install a DVR system (dynamic voltage restorer), it may be more practical and cost-efficient to connect the planned industrial plant to the upstream high-voltage system.

3.2 Distribution level

If the industrial plant is connected to the public medium-voltage network, this more or less already determines the choice of nominal system voltage for the distribution level. The voltage can only be freely selected if it is supplied from the upstream high-voltage system. One of the standard voltages stipulated in DIN IEC 60038 (VDE 0175): 2002-11 [3.1] / IEC 60038: 2009-06 [3.2] in the range 1 kV < $U_{nN} \leq 35$ kV can be chosen. The different nominal system voltages that have been chosen in the past from the range 1 kV < $U_{nN} \leq 35$ kV for industrial plants in Germany are shown in Fig. B3.2.

As Fig. B3.2 shows, 38 % of all industrial power systems in Germany are operated with $U_{nN} = 6$ kV or $U_{nN} = 5$ kV. The 6(5)-kV voltage level continues to be commonly used because the proportion of motor loads in the total load of industrial power systems is often very high. In the event of a short circuit, the high-voltage asynchronous motors that are in operation produce additional short-circuit current that can be about 30 % of the total short-circuit current, depending on the system structure and the fault location ($I_{kM}'' = 0.3 \cdot I_{k\Sigma}''$) [3.5, 3.6]. The additional short-circuit current from the motors places additional thermal and dynamic stress on the equipment. This additional short-circuit current stress can prove problematic at the standard voltage of $U_{nN} = 6$ kV because a sufficient short-circuit power is necessary from the outset if asynchronous motors are to start and restart correctly. At the nominal system voltage $U_{nN} = 6$ kV, the short-circuit currents are 67 % higher than at the nominal system voltage $U_{nN} = 10$ kV. In power systems with a high motor load density, the 10-kV voltage level has the following advantages over the 6-kV voltage level for the same short-circuit power:

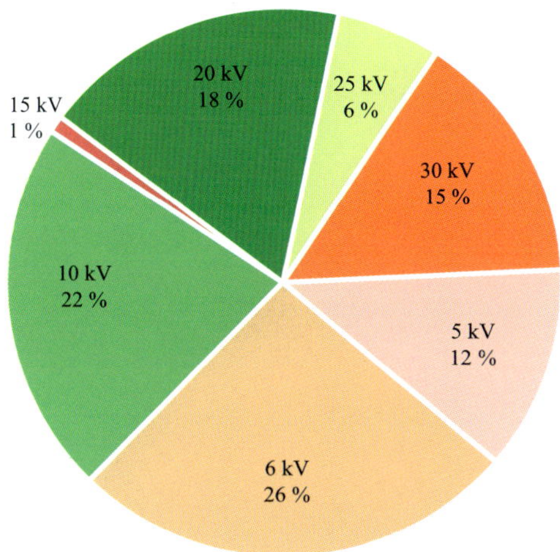

Fig. B3.2
Breakdown of MV voltage levels in various German industries [3.4]

- reduction of the rated short-circuit breaking current I_{sc} and the rated short-circuit making current I_{ma} by one or two steps,
- reduction of the cable cross-sectional areas by two to three steps,
- increase in energy efficiency due to lower system power losses,
- sufficient grading times ensured to permit use of DTL as the main and back-up protection.

Installation and system loss costs are saved due to the reduction of the short-circuit making/breaking current ratings and the increased energy efficiency. In 10-kV industrial power systems with a high proportion of motor loads relative to the total load, the savings in installation and system loss costs are always greater than the additional costs of 10-kV motors as compared with 6-kV motors. When new power systems with a high motor load density are planned (e.g. chemical industry, cement factories), the distribution voltage U_{nN} = 10 kV is preferable. [3.7] also favours the use of the distribution voltage U_{nN} = 10 kV in such power systems.

A further criterion in choosing the MV distribution voltage is a low cost for direct transformation to the low-voltage level. This criterion is a primary consideration in planning industrial power systems with a high low-voltage power demand (few HV motors, a large number of LV motor drives and LV loads with different power ratings and duty ratios). For a high LV power demand, the 20-kV voltage level proves a flexible and efficient MV distribution level [3.8]. The high flexibility and efficiency of U_{nN} = 20 kV is expressed as follows:

- low-cost direction transformation to either
  6 kV (additional connection of HV motors) and
  0.69 kV or 0.4 kV (priority connection of LV loads),
- better conditions for parallel connection of transformers with a high rating in terms of power system engineering,
- higher instantaneous reserve for cable and transformer faults,
- improved load current and fault current-carrying capacity,
- lower power system losses.

Conclusion

The decision as to the number of voltage levels and the nominal value of the distribution voltages depends on many constraints. Generally, there should be only a few voltage levels between the supplying high-voltage network and the low-voltage power system close to the process. If the industrial plant is supplied with power from the upstream high-voltage network, the following voltages can be recommended for the MV distribution level:

- 10 (11) kV
  Industrial power systems with a large number of HV drives and/or a comparatively low LV power demand
- 20 (22) kV
  Industrial power systems without HV drives and with a high LV power demand
- 20 (22) kV/6 (6.6) kV
  Industrial power systems with a small number of HV drives and a high LV power demand

4 Determining short-circuit stress and the necessary short-circuit withstand capability

4.1 Choosing the short-circuit power

One important prerequisite for safe power system operation is the presence of a sufficiently high initial symmetrical short-circuit power (also known simply as short-circuit power) [4.1].

According to DIN EN 60909-0 (VDE 0102): 2002-07 [4.2] / IEC 60909-0: 2001-07 [4.3], the short-circuit power can be calculated as follows:

$$S_k'' = \sqrt{3} \cdot U_{nN} \cdot I_k'' = \frac{c \cdot U_{nN}^2}{K \cdot Z_k} \quad (4.1)$$

U_{nN} nominal voltage of the power system
I_k'' initial symmetrical short-circuit current
c voltage factor ($0.95 \le c \le 1.10$)
K impedance correction factor for network transformers, synchronous generators and power station units (see [4.2] and [4.3])
Z_k short-circuit impedance of the equipment

For supply from the public medium-voltage network, reliable operation of the industrial plant depends on the short-circuit power at the point of connection (PC) or point of common coupling (PCC). This short-circuit power is invariable. The magnitude of the MV short-circuit power can only be influenced if the industrial plant is connected to the upstream high-voltage power system. In this case, the short-circuit power can be influenced by the choice of the impedance in the HV/MV incoming supply. The impedance of the HV/MV incoming supply is determined by the rated power S_{rT} and the percent impedance voltage u_{rZ} of the power transformers used.

The impedance in the incoming supply must be defined, taking technical and economic criteria into account. One important economic criterion are the costs incurred for the short-circuit-proof dimensioning of the MV switchgear and MV cables. The main factor influencing these costs is the choice of the nominal system voltage. With the following short-circuit powers, which refer to the chosen nominal system voltage, cost-efficient use of equipment is possible:

- $250 \text{ MVA} \le S_k'' < 350 \text{ MVA}$ if $U_{nN} = 6 \text{ kV}$
- $350 \text{ MVA} \le S_k'' < 500 \text{ MVA}$ if $U_{nN} = 10 \text{ kV}$
- $500 \text{ MVA} \le S_k'' \le 750 \text{ MVA}$ if $U_{nN} = 20 \text{ kV}$

The technical criteria are particularly relevant to the choice of short-circuit power. One important technical criterion is the voltage stability under impulsive and fluctuating loads. The voltage changes due to impulsive or fluctuating loads depend directly on the short-circuit power at the system nodes. Eq. (4.2) expresses this dependency.

$$\Delta u' = \left[\frac{\Delta S}{S_{k\text{-PC/PCC}}''} \cdot \cos(\varphi_{PC/PCC} - \varphi_{\Delta S}) \right] \cdot 100\,\% \quad (4.2)$$

$$\varphi_{PC/PCC} = \text{arc}\cot(R_{PC/PCC}/X_{PC/PCC}) \quad (4.2.1)$$

ΔS	load change (change in apparent power)
$S''_{k\text{-}PC/PCC}$	short-circuit power at the point of connection or point of common coupling
$\varphi_{PC/PCC}$	impedance angle at the point of connection or point of common coupling
$\varphi_{\Delta S}$	angle of the load change
$R_{PC/PCC}/X_{PC/PCC}$	ratio of resistance to reactance at the point of connection or point of common coupling

Table B4.1 provides an overview of the network points at which a sufficient short-circuit power is required to ensure reliable uninterrupted functioning of loads.

Voltage change	Functional reliability of the	
	disturbing load	influenced load
Voltage dip	PC	PCC
Voltage fluctuation	PC	PCC
PC Point of connection PCC Point of common coupling		

Table B4.1 Classification of nodes with respect to the reliable functioning of loads

Table B4.2 A comparison of methods for calculating the short-circuit power at points of connection power or points of common coupling

Performance indicator	Calculation method	
	DIN EN 60909-0 (VDE 0102):2002-07 [4.2]/ IEC 60909-0: 2001-07 [4.3]	D-A-CH-CZ Guideline for Assessing System Perturbation [4.4]
Source voltage	Voltage factor $0.95 \leq c \leq 1.1$ and nominal system voltage U_{nN}	Line-to-line voltage at the PC or PCC
Type of short circuit	Three-phase short circuit, line-to-line short circuit with earth connection and line-to-line short circuit clear of earth, line-to-earth short circuit	Three-phase short circuit
Line impedance	MV: Temperature at the end of the short-circuit duration LV: 80 °C	MV: No data LV cable: 70 °C
Impedance correction factors	To be applied to network transformers, synchronous generators and power station units	No correction factors
Switching state of the power system	Unambiguous conditions for calculation of the minimum short-circuit current	Convenient operation, that is, system impedance is to be maximized at the PC or PCC
Frequency influence	No influence of the frequency on the impedances	Frequency influencies of the impedances may have to be considered
Short-circuit quantities	Initial symmetrical short-circuit current I''_k steady-state short-circuit current I_k or short-circuit powers S_k and S''_k	Short-circuit power at the PC or PCC $S''_{k\text{-}PC/PCC}$
Contribution to short-circuit currents by motors	Omitted in calculation of the minimum short-circuit current	Convenient operation, that is, system impedance is to be maximized at the PC or PCC
Short-circuit power at the PC or PCC	$S''_{k-PC/PCC} = \frac{c \cdot U^2_{nN}}{Z_{k-PC/PCC}}$	$S''_{k-PC/PCC} = \frac{U^2_{PC/PCC}}{Z_{k-PC/PCC}}$

The short-circuit power at the points of connection power and points of common coupling can be calculated using both the method according to standard DIN EN 60909-0 (VDE 0102): 2002-07 [4.2] or IEC 60909-0: 2001-07 [4.3] and the method stated in the D-A-CH-CZ Guideline [4.4]. Table B4.2 provides a comparison of the two calculation methods. As the comparison shows, there are considerable differences in the way the short-circuit power is calculated. Example calculations of the short-circuit power have shown that results based on the D-A-CH-CZ Guideline may be more than 10 % off the true value [4.5]. For that reason, the short-circuit power to assess system perturbations of industrial plants should be calculated according to DIN EN 60909-0 (VDE 0102): 2002-07 [4.2] or IEC 60909-0: 2001-07 [4.3].

The short-circuit power at the connecting points or points of common coupling is not only a measure of the voltage stability under impulsive and fluctuating loads. It is also a measure of the hazards posed to people and equipment in the event of a short circuit. Damage to equipment caused by arcing faults directly depends on the product I^2t. The destructive power of the arc is therefore twice as large at $S_k'' = 500$ MVA than at $S_k'' = 350$ MVA. Because of the destructive power of the arc, which depends on the short-circuit current, fast fault clearance is especially crucial in power systems with a high short-circuit power.

Conclusion

In defining the short-circuit power, a tradeoff must be found between high voltage stability and a low hazard to people and equipment.

4.2 Short-circuit withstand capability of the equipment

The standards DIN VDE 0101 (VDE 0101): 2000-01 [4.6] and IEC 61936-1: 2002-10 [4.7] stipulate that installations must reliably withstand the mechanical and thermal effects of a short-circuit current. To comply with this safety requirement, all equipment must have short-circuit-proof ratings.

4.2.1 MV switchgear

The following short-circuit current conditions must be met for short-circuit-proof rating of MV switchgear:

$$I_b \leq I_{sc} \tag{4.3}$$

$$I_b = I_{kN}'' + \mu_M \cdot q_M \cdot I_{kM}'' \tag{4.3.1}$$

I_b	symmetrical short-circuit breaking current
I_{sc}	rated short-circuit breaking current
I_{kN}''	initial symmetrical short-circuit current of the supplying system
$\mu_M \cdot q_M \cdot I_{kM}''$	contribution of the motors to the symmetrical short-circuit breaking current
μ_M	decay factor that depends on the ratio I_{kM}''/I_{rM} and on the minimum time delay t_{min}
q_M	decay factor that depends on the ratio P_{rM}/p_M and on the minimum time delay t_{min}
I_{kM}''	initial symmetrical short-circuit current contributed by the motors
I_{kM}''/I_{rM}	ratio of the initial symmetrical short-circuit current and rated current of the motors contributing to the short-circuit current

P_{rM}/p_M ratio of the rated power and number of pole pairs ($p_M = f \cdot 60/n_{syn}$) of the motors contributing to the short-circuit current

n_{syn} synchronous motor speed (e.g. n_{syn} = 1,500 rpm)

$$i_p \leq \begin{cases} I_{ma} & \text{for circuit-breakers and switches} \\ I_{pk} & \text{for switchgear} \end{cases} \quad (4.4)$$

$$i_p = i_{pN} + i_{pM} \quad (4.4.1)$$

i_p peak short-circuit current
I_{ma} rated short-circuit making current
I_{pk} rated peak withstand current
i_{pN} peak short-circuit current of the supplying system
i_{pM} peak short-circuit current contributed by the motors

$$I_{th} \leq \begin{cases} I_{thr} & \text{for } t_k \leq t_{thr} \\ I_{thr}\sqrt{\dfrac{t_{thr}}{t_k}} & \text{for } t_k > t_{thr} \end{cases} \quad (4.5)$$

$$I_{th} = I_k'' \cdot \sqrt{m+n} \quad (4.5.1)$$

$$m = \frac{1}{2 \cdot f \cdot t_k \cdot \ln(\kappa - 1)} \cdot [e^{4 \cdot f \cdot t_k \cdot \ln(\kappa - 1)} - 1] \quad (4.5.1.1)$$

$$n = \begin{cases} 1 & \text{for } I_k''/I_k = 1 \\ f(I_k''/I_k, t_k) & \text{for } I_k''/I_k > 1.25 \end{cases} \quad (4.5.1.2)$$

I_{th} thermal equivalent short-circuit current
I_{thr} rated short-time withstand current
t_{thr} rated short time
t_k maximum short-circuit duration
I_k'' initial symmetrical short-circuit current ($I_k'' = I_{kN}'' + I_{kM}''$)
I_k steady-state short-circuit current
m factor for the heat effect of the DC component of the short-circuit current
n factor for the heat effect of the AC component of the short-circuit current (for distribution systems (far-from-generator short circuit), n = 1 can usually be applied)
f system frequency
κ asymmetrical current peak factor

The short-circuit currents I_b, i_p and I_{th} that determine the mechanical and thermal stress on the MV switchgear must be calculated according to DIN EN 60909-0 (VDE 0102): 2002-07 [4.2] / IEC 60909-0: 2001-07 [4.3] and DIN EN 60865-1 (VDE 0103): 1994-11 [4.8] / IEC 60865-1: 1993-09 [4.9]. The necessary short-circuit withstand capability values are obtained by calculating the maximum short-circuit stress quantities occurring during a short circuit. The short-circuit ratings of MV switchgear complementing the stress quantities are standardized. They are listed in Table B4.3.

Table B4.3 Standardized short-circuit ratings of MV switchgear

Ratings	Standardized stepping [1] of the ratings (currents in kA)									
Rated short-circuit breaking current I_{sc}	8	10	12.5	16	20	25	31.5	40	50	63
Rated short-circuit making current I_{ma}	20	25	31.5	40	50	63	80	100	125	160
Rated peak withstand current I_{pk}	20	25	31.5	40	50	63	80	100	125	160
Rated short-time withstand current I_{thr}	8	10	12.5	16	20	25	31.5	40	50	63
Rated short-time t_{thr}	1 sec or 3 sec									

1) Normally, only MV switchgear systems with I_{sc} (I_{thr}) $\geq$ 16 kA and I_{ma} (I_{pk}) $\geq$ 40 kA are available on the market.

Tables B4.4 and B4.5 provide a simplified way of determining the necessary short-circuit withstand capability of MV main switchgear. The ratings are determined depending on the system parameters of the 110-kV/MV supply. Appropriate rating margins are stated for motors contributing short-circuit current to the mechanical and thermal stresses in case of a short circuit.

Example B1

Short-circuit-proof dimensioning of the MV main switchgear with parallel supply through two 31.5-MVA transformers according to Table B4.5, column $U_{nN} = 10$ kV.

Data of the incoming supply:

$U_{rT} = 110$ kV/10 kV

$S_{rT} = 2 \times 31.5$ MVA

$u_{rZ} = 10$ %

Short-circuit stress:

$S_k'' = 629.1$ MVA

$I_b = \sum I_{kN}'' = 36.3$ kA

$i_p = \sum i_{pN} = 102.7$ kA

Required short-circuit withstand capability:

$I_{sc} = 50$ kA

$I_{ma} = 125$ kA

$I_{pk} = 125$ kA

$I_{thr} = 50$ kA (1 sec)

Because the maximum peak short-circuit current $i_{p\text{-}max} = 102.7$ kA exceeds the rating $I_{ma} = 100$ kA or $I_{pk} = 100$ kA (see Table B4.3), 50-kA switchgear must be chosen. This switchgear ensures a dynamic short-circuit capability margin of $i_{pM} = 125$ kA $-$ 102.7 kA $= 22.3$ kA for HV motors contributing to the short-circuit stress. According to the short-circuit capability margin of $i_{pM} = 22.3$ kA, the apparent power for the 10-kV-side motor contribution to the short-circuit stress can be at least $S_M = 25$ MVA.

Table B4.4 Short-circuit current carrying capacity for MV switchgear supplied through a 110-kV/MV transformer

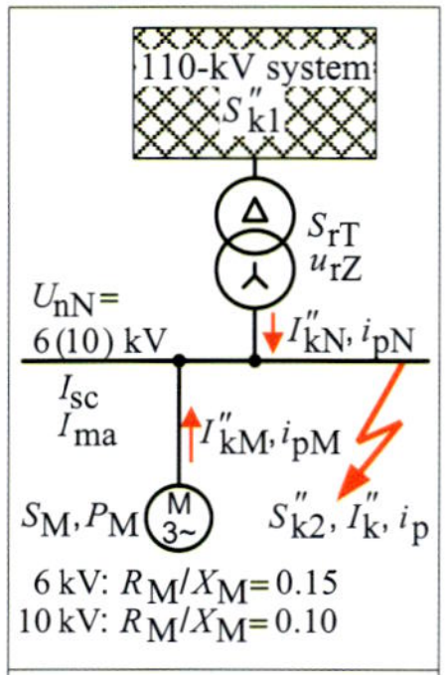

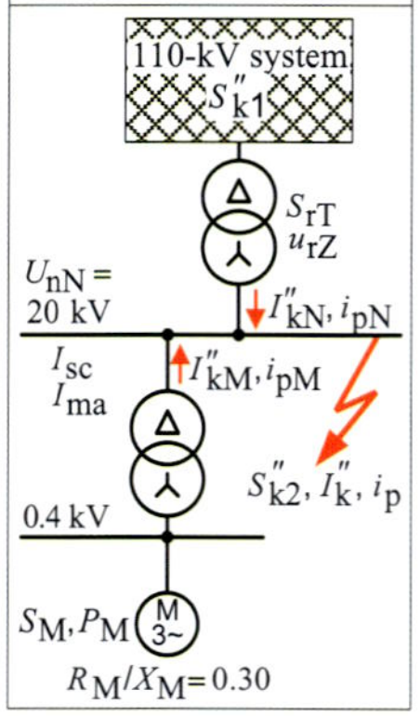

	S''_{k1}	Short-circuit power of the 110-kV system acc. to DIN EN 60076-5 (VDE 0532-76-5):2007-01 [4.10] or IEC 60076-5: 2006-02 [4.11] in GVA																	
		6																	
Data of the 110-kV/MV supply	U_{nN}	Nominal system voltage in kV																	
		6						10						20					
	S_{rT}	Rated transformer power in MVA																	
		16	20	25	31.5	40	63	16	20	25	31.5	40	63	16	20	25	31.5	40	63
	u_{rZ}	Minimum value of the percent impedance voltage acc. to IEC 60076-5: 2006-02 [4.11] in %																	
		8			10		11	8			10		11	8			10		11
	I_{rT}	Rated transformer current in A																	
		1,540	1,925	2,406	3,031	3,849	6,062	924	1,155	1,443	1,819	2,309	3,637	462	577	722	909	1,155	1,819
Maximum short-circuit stress without contribution to short-circuit currents by motors	S''_{k2}	Short-circuit power of the MV network in MVA																	
		212.8	263.7	326.1	332.0	415.4	580.6	212.8	263.7	326.1	332.0	415.4	580.6	212.8	263.7	326.1	332.0	415.4	580.6
	$I''_k = I''_{kN}$	Initial symmetrical short-circuit current in kA																	
		20.5	25.4	31.4	31.9	40.0	55.9	12.3	15.2	18.8	19.2	24.0	33.5	6.1	7.6	9.4	9.6	12.0	16.8
	$i_p = i_{pN}$	Peak short-circuit current in kA ($\kappa \leq 2{,}0$)																	
		≤ 58.0	≤ 71.8	≤ 88.8	≤ 90.2	≤ 113.1	≤ 158.1	≤ 34.8	≤ 43.0	≤ 53.2	≤ 54.3	≤ 67.9	≤ 94.8	≤ 17.2	≤ 21.5	≤ 26.6	≤ 27.2	≤ 33.9	≤ 47.5
Dimensioning – Required strength	I_{sc}	Rated short-circuit breaking current in kA																	
		25	31.5	40	40	50	63	16 [1]	20	25	25	31.5	40	16 [1]	16 [1]	16 [1]	16 [1]	16 [1]	20
	I_{ma}	Rated short-circuit making current in kA																	
		63	80	100	100	125	160	40	50	63	63	80	100	40	40	40	40	40	50
Dimensioning – Rated margin for contribution to short-circuit currents by motors	i_{pM}	Dynamic short-circuit strength margin [3] for contribution to peak short-circuit current by motors in kA																	
		5.0	8.2	11.2	9.8	11.9	1.9	5.2	7.0	9.8	8.7	12.1	5.2	22.8	18.5	13.4	12.8	6.1	2.5
	S_M	Guaranteed permissible apparent power for contribution to short-circuit currents by motors in MVA																	
		3.7	5.9	8.1	7.1	8.6	1.35	5.9	7.9	11.2	9.8	13.8	5.9	16 [2]	20 [2]	25 [2]	31.5 [2]	15.8	5.6
	P_M	Guaranteed permissible active power for contribution to short-circuit currents by motors in MW																	
		3.2	5.1	7.0	6.1	7.4	1.2	5.1	6.8	9.7	8.5	11.9	5.1	12.8	16.0	20.0	25.1	12.6	4.5

1) Smallest rated value for short-circuit current for commercially available MV switchgear

2) Limit rating, taking into account the available rated power of the transformer S_{rT}. Theoretically, the dynamic short-circuit strength margin i_{pM} would permit a higher power S_M.

3) Guaranteed minimum value

Table B4.5 Short-circuit current carrying capacity for MV switchgear with a parallel supply through two 110-kV/MV transformers

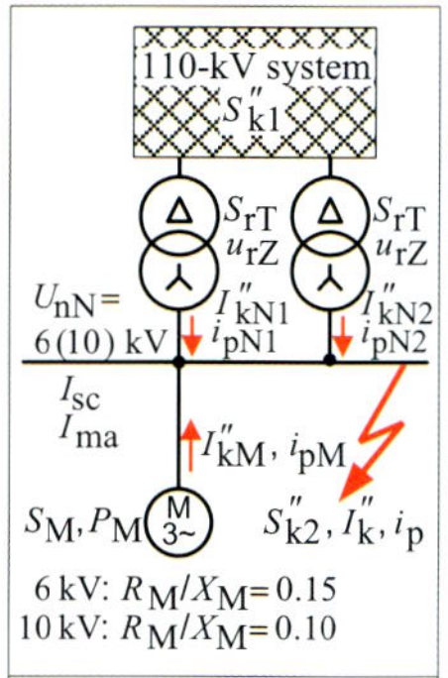

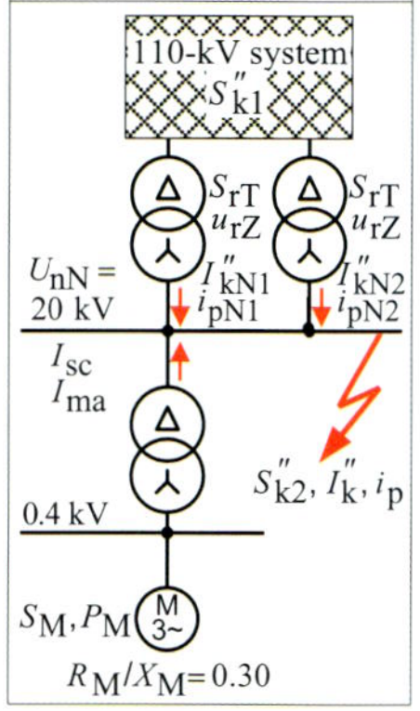

Data of the 110-kV/MV supply	S''_{k1}	Short-circuit power of the 110-kV system acc. to DIN EN 60076-5 (VDE 0532-76-5):2007-01 [4.10] or IEC 60076-5: 2006-12 [4.11] in GVA																	
		6																	
	U_{nN}	Nominal system voltage in kV																	
		6						10						20					
	S_{rT}	Rated transformer power in MVA																	
		2x16	2x20	2x25	2x31.5	2x40	2x63	2x16	2x20	2x25	2x31.5	2x40	2x63	2x16	2x20	2x25	2x31.5	2x40	2x63
	u_{rZ}	Minimum value of the percent impedance voltage acc. to IEC 60076-5: 2006-12 [4.11] in %																	
		8			10		11	8			10		11	8			10		11
	I_{rT}	Rated transformer current in A																	
		2x1,540	2x1,925	2x2,406	2x3,031	2x3,849	2x6,062	2x924	2x1,155	2x1,443	2x1,819	2x2,309	2x3,637	2x462	2x577	2x722	2x909	2x1,155	2x1,819
Maximum short-circuit stress without contribution to short-circuit currents by motors	S''_{k2}	Short-circuit power of the MV network in MVA																	
		411.1	505.2	618.5	629.1	776.9	1,058.8	411.1	505.2	618.5	629.1	776.9	1,058.8	411.1	505.2	618.5	629.1	776.9	1,058.8
	$I''_k = \Sigma I''_{kN}$	Initial symmetrical short-circuit current in kA																	
		39.6	48.6	59.5	60.5	74.8	101.9	23.7	29.2	35.7	36.3	44.8	61.1	11.9	14.6	17.9	18.2	22.4	30.6
	$i_p = \Sigma i_{pN}$	Peak short-circuit current in kA ($\kappa \leq 2.0$)																	
		≤ 112.0	≤ 137.5	≤ 168.3	≤ 171.1	≤ 211.6	≤ 288.2	≤ 67.0	≤ 82.6	≤ 101.0	≤ 102.7	≤ 126.7	≤ 172.8	≤ 33.6	≤ 41.3	≤ 50.6	≤ 51.5	≤ 63.4	≤ 86.5
Dimensioning: Required strength	I_{sc}	Rated short-circuit breaking current in kA																	
		50	63					31.5	40	50	50	63		16[1]	20	25	25	31.5	40
	I_{ma}	Rated short-circuit making current in kA																	
		125	160					80	100	125	125	160		40	50	63	63	80	100
Dimensioning: Rated margin for contribution to short-circuit currents by motors	i_{pM}	Dynamic short-circuit strength margin [3] for contribution to peak short-circuit current by motors in kA																	
		13.0	22.5					13.0	17.4	24.0	22.3	33.3		6.4	8.7	12.4	11.5	16.6	13.5
	S_M	Guaranteed permissible apparent power for contribution to short-circuit currents by motors in MVA																	
		9.4	16.4					14.9	19.8	25[2]	25.4	38.0		16[2]	20[2]	25[2]	31.5[2]	40[2]	37.7
	P_M	Guaranteed permissible active power for contribution to short-circuit currents by motors in MW																	
		8.1	14.2					12.9	17.1	21.6	21.9	32.8		12.8	16.0	20.0	25.1	31.9	30.1

1) Smallest rated value for short-circuit current for commercially available MV switchgear

2) Limit power considering the (*n*-1) principle. The dynamic short-circuit strength margin i_{pM} would permit a higher power S_M.

3) Guaranteed minimum value

4.2.2 MV cables

In a short circuit, the heat generated by the short-circuit current is mainly stored in the conductor. Under this condition the conductor must not be heated beyond the permissible short-circuit temperature ϑ_e. Thus, the conductor temperature ϑ_a at the beginning of the short circuit as well as the maximum short-circuit duration t_k must be considered [4.13]. The following short-circuit condition must therefore be met:

$$I_{th} \leq I_{thp} \tag{4.6}$$

$$I_{thp} = A_n \cdot J_{thr} \cdot \sqrt{t_{thr}/t_k} \tag{4.6.1}$$

I_{th} thermal equivalent short-circuit current (see Eq. 4.5.1)
I_{thp} thermal short-circuit current-carrying capacity of the MV cable
A_n nominal cross-sectional area of the conductor (Table B4.6)
J_{thr} rated short-time current density (Table B4.7)
t_{thr} rated short time (t_{thr} = 1 sec)

Table B4.6 Standardized cross-sectional areas of conductors of MV cables

Cross-section	Standardized steps												
A_n in mm²	25	35	50	70	95	120	150	185	240	300	400	500	630

Table B4.7 Permissible short-circuit temperatures and rated short-time current densities J_{thr} for power cables [4.13]

Construction	Permissible operating temperature ϑ_{Lr} in °C	Permissible short-circuit temperature ϑ_e in °C	Conductor temperature ϑ_a at the beginning of the short circuit: 90 °C	80 °C	70 °C	65 °C	60 °C	50 °C	40 °C	30 °C	20 °C
			Rated short-time current density J_{thr} in A/mm² for t_{thr} = 1 sec								
Cable with copper conductors											
XLPE cable, EPR cable	90	250	143	148	154	157	159	165	170	176	181
PE cable	70	150	--	--	109	113	117	124	131	138	145
PVC cable ≤ 300 mm²	70	160	--	--	115	119	122	129	136	143	150
PVC cable > 300 mm²	70	140	--	--	103	107	111	118	126	133	140
Cable with aluminium conductors											
XLPE cable, EPR cable	90	250	94	98	102	104	105	109	113	116	120
PE cable	70	150	--	--	72	75	77	82	87	91	96
PVC cable ≤ 300 mm²	70	160	--	--	76	78	81	85	90	95	99
PVC cable > 300 mm²	70	140	--	--	68	71	73	78	83	88	93

XLPE cable: Insulation made of crosslinked polyethylene (2X)
EPR cable: Insulation made of ethylene propylene rubber
PE cable: Insulation made of polyethylene (2Y)
PVC cable: Insulation made of polyvinyl chloride (Y)

Example B2

For the transformer feeder shown in Fig. B4.8, the short-circuit withstand capability of the Protothen-X cable used (XLPE insulation, flame-retardant PVC sheath) must be verified.

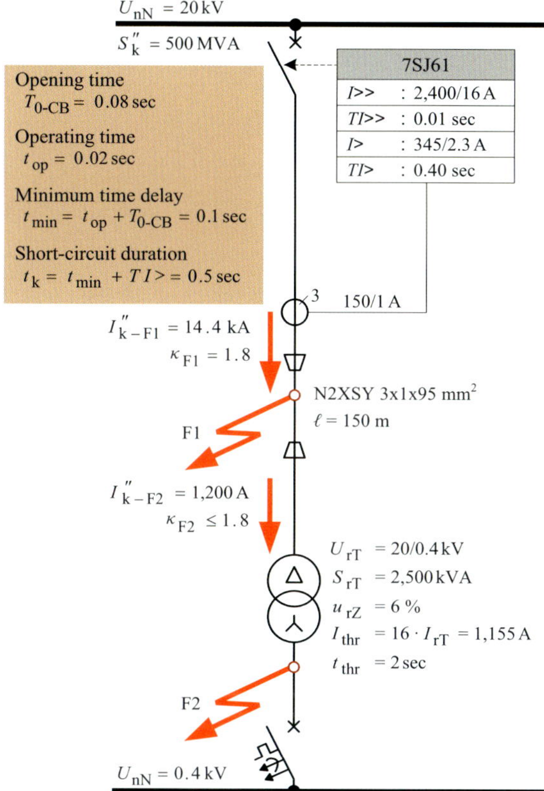

Fig. B4.8
Short-circuit-proof rating of a transformer feeder

Using the values from Fig. B4.8 for the far-from-generator fault F1 and applying equations (4.5.1), (4.5.1.1) and (4.5.1.2), a thermal equivalent short-circuit current of

$$I_{th\text{-}F1} = I''_{k\text{-}F1} \cdot \sqrt{m+n} = 14.4\ \text{kA} \cdot \sqrt{0{,}09+1} = 15.0\ \text{kA}$$

can be calculated. The MV cable of type N2XSY 3 × 1 × 95 mm² used must withstand the effects of this thermal equivalent short-circuit current. According to Table B4.7, the rated short-time current density for this XLPE cable with a copper conductor at an initial temperature $\vartheta_a = 90\,°C$ and a final temperature $\vartheta_e = 250\,°C$ is $J_{thr} = 143\ A/mm^2$. For the rated short-time current density $J_{thr} = 143\ A/mm^2$ and a short-circuit duration of $t_k = 0.5$ sec, Eq. (4.6.1) yields the following short-circuit capacity:

$$I_{thp} = A_n \cdot J_{thr} \cdot \sqrt{t_{thr}/t_k} = 95\ \text{mm}^2 \cdot 143\ \frac{A}{\text{mm}^2} \cdot \sqrt{\frac{1\ \text{sec}}{0.5\ \text{sec}}} = 19.2\ \text{kA}$$

The thermal equivalent short-circuit current of $I_{th\text{-}F1} = 15.0$ kA is less than this short-circuit capacity; that is, if the transformer connecting cable N2XSY $3 \times 1 \times 95$ mm^2 is used, the short-circuit condition (4.6) must be reliably fulfilled. The cable is short-circuit-proof.

4.2.3 MV distribution transformers

The requirements and constraints by which transformers have to be dimensioned and built to withstand the thermal and mechanical effects of external short circuits are defined in DIN EN 60076-5 (VDE 0532-76-5): 2007-01 [4.10] or IEC 60076-5: 2006-02 [4.11]. For transformers to be able to withstand the thermal effects of an external short circuit without damage, the following short-circuit condition must be met:

$$I_{th} \leq I_{thp} \tag{4.7}$$

$$I_{thp} = I_{thr} \cdot \sqrt{t_{thr}/t_k} \tag{4.7.1}$$

I_{th} thermal equivalent short-circuit current (see Eq. 4.5.1)
I_{thp} thermal short-circuit capacity of the transformer
I_{thr} rated short-time withstand current
t_{thr} rated short time ($t_{thr} = 2$ sec)

The rated short-circuit current I_{thr} corresponds to the RMS value of the three-phase short-circuit current of three-phase transformers with two separate windings to be calculated according to DIN EN 60076-5 (VDE 0532-76-5): 2007-01 [4.10] / IEC 60076-5: 2006-02 [4.11]. Rated short-time withstand currents I_{thr} for MV distribution transformers calculated according to this standard are given in Table B4.9.

Table B4.9 Rated short-time withstand currents I_{thr} for MV distribution transformers

Rated power S_{rT} in kVA	Relative impedance voltage at rated current [1] u_{rZ} in %	Rated current I_{rT} in A		Rated short-time withstand current I_{thr} at $S_k'' = 500$ MVA [2] for $t_{thr} = 2$ sec [3]
		$U_{rT} = 10$ kV	$U_{rT} = 20$ kV	
400	4	23.1	11.5	$24 \cdot I_{rT} \leq I_{thr} < 25 \cdot I_{rT}$
500	4	28.9	14.4	
630	4	36.4	18.2	
800	5	46.2	23.1	$19 \cdot I_{rT} \leq I_{thr} < 20 \cdot I_{rT}$
1,000	5	57.7	28.9	
1,250	5	72.2	36.1	
1,600	6	92.4	46.2	$15.5 \cdot I_{rT} \leq I_{thr} < 16.5 \cdot I_{rT}$
2,000	6	115.5	57.7	
2,500	6	144.3	72.2	

1) Minimum values acc. to DIN EN 60076-5 (VDE 0532-76-5): 2007-01 [4.10] / IEC 60076-5: 2006-02 [4.11]

2) Value of the short-circuit power of the 10(20)-kV system that can be used according to DIN EN 60076-5 (VDE 0532-76-5): 2007-01 [4.10] / IEC 60076-5: 2006-02 [4.11] to prove the thermal short-circuit strength if data are not specified.

3) Rated short time for MV distribution transformers according to DIN EN 60076-5 (VDE 0532-76-5): 2007-01 [4.10] / IEC 60076-5: 2006-02 [4.11]

Example B3

The short-circuit withstand capability is to be verified for the 2,500-kVA GEAFOL cast-resin transformer shown in Fig. B4.8.

The thermal short-circuit stress of the 2,500-kVA GEAFOL cast-resin transformer is determined by the thermal equivalent short-circuit current $I_{\text{th-F2}}$ of the external fault F2. With the values from Fig. B4.8, the result is

$$I_{\text{th-F2}} = I''_{\text{k-F2}} \cdot \sqrt{m+n} = 1{,}200\ \text{A} \cdot \sqrt{0.09+1} = 1{,}253\ \text{A}$$

Damage caused by the thermal effects of the external short circuit F2 to the transformer used is ruled out if the thermal short-circuit capacity I_{thp} is greater than the calculated fault current $I_{\text{th-F2}} = 1{,}253$ A.

According to Eq. (4.7.1), the following thermal short-circuit capacity results from the values stated in Fig. B4.8:

$$I_{\text{thp}} = I_{\text{thr}} \cdot \sqrt{t_{\text{thr}}/t_{\text{k}}} = 1{,}155\ \text{A} \cdot \sqrt{2\ \text{sec}/0.5\ \text{sec}} = 2{,}310\ \text{A}$$

The thermal short-circuit capacity $I_{\text{thp}} = 2{,}310$ A is greater than the thermal equivalent short-circuit current $I_{\text{k-F2}} = 1{,}253$ A.

The 2,500-kVA GEAFOL cast-resin transformer withstands the thermal effects during external faults because the short-circuit condition (4.7) is reliably fulfilled.

5 Defining optimum system configurations for industrial power supplies

5.1 MV load structure in the metal-processing industry

In the metal-processing industry, there are only a few individual large loads that have to be connected directly to the MV system. These include, for example, large compressor drives, arc furnaces and large engine test benches. Due to of the small number of large loads, in plants of the metal-processing industry the focus of attention is on configuration of the supply from the MV into the LV system [5.1].

5.2 Best MV/LV incoming supply variant in terms of power system engineering

In configuration of the MV/LV incoming supply, a distinction is made between centralized and decentralized multiple incoming supply [5.2, 5.3]. Whereas the transformers of the centralized multiple incoming supply are installed in an enclosed electrical operating room at the edge of the hall, the transformers of a decentralized multiple incoming supply are containerized and installed in the load centres of the production and function areas.

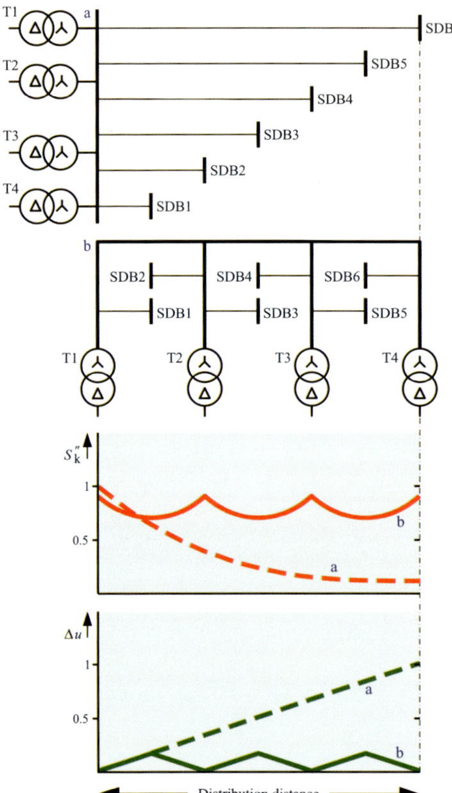

Fig. B5.1
Curves for the short-circuit power S_k'' and the voltage drop Δu with a centralized (a) and decentralized (b) multiple incoming supply

With the hall dimensions that are usual today, a decentralized multiple incoming supply has the following advantages (Fig. B5.1):

- lower system losses and voltage drops because power is mainly transmitted on cables at the medium-voltage level,
- better voltage stability and reduced system perturbations because of the short-circuit power that is of equal magnitude at all system nodes,
- greater flexibility in response to load shifts caused by changing production processes.

Due to these advantages, a decentralized multiple incoming supply is usually selected for large-area production halls. The transformer load-centre substation (for load-centre substations, see Section 11.2.5) has proven the most efficient component for setting up decentralized MV/LV incoming supplies.

5.3 Optimum system configuration for connecting transformer load-centre substations

In principle, decentralized load-centre substations can be connected on the MV side by means of ring-cable or radial-cable configurations. Fig. B5.2 shows a typical example of a ring-cable and radial-cable connection of decentralized load-centre substations.

Using Fig. B5.2, it is necessary to decide which of the two MV-side connection types is preferable for constructing a decentralized multiple incoming supply. A sound planning decision can by made by a multiple-objective-oriented evaluation of the ring system and load-centre system.

The multiple-objective-oriented evaluation of the two system configurations is performed based on the decision objectives discussed in Section 2.4.1. These decision objectives are met as follows:

- *Investment costs* ↓

  The magnitude of the investment costs is primarily determined by the expense for cable installation and switchgear.

  The expense for cable installation is approximately equal for the two system configurations. Although a shorter cable length is required for connecting the load-centre substations in a ring system, a larger cable cross-sectional area is required for the $(n-1)$ redundancy of the powerline transmission and short-circuit-proof rating than in a load-centre system.

  The expense of the switchgear required to build the two types of system is always different. To implement a load-centre system with four decentralized distribution transformers, only four switch-disconnector panels with HV HRC fuses are required.

  A suitable ring-main connection, on the other hand, can be constructed with only two additional circuit-breaker panels equipped with definite-time overcurrent relays and a further eight switch-disconnector panels (Fig. B5.2). As a result, up to 60 % of switchgear costs can be saved with a load-centre system. A comparison of the space requirement for setting up the load-centre substations in the production area is also favourable for the load-centre system.

- *System losses* ↓

  Losses in a ring system do not differ greatly from those in a load-centre system. The difference in system losses ΔP_V during operation of the two networks is not relevant to the decision.

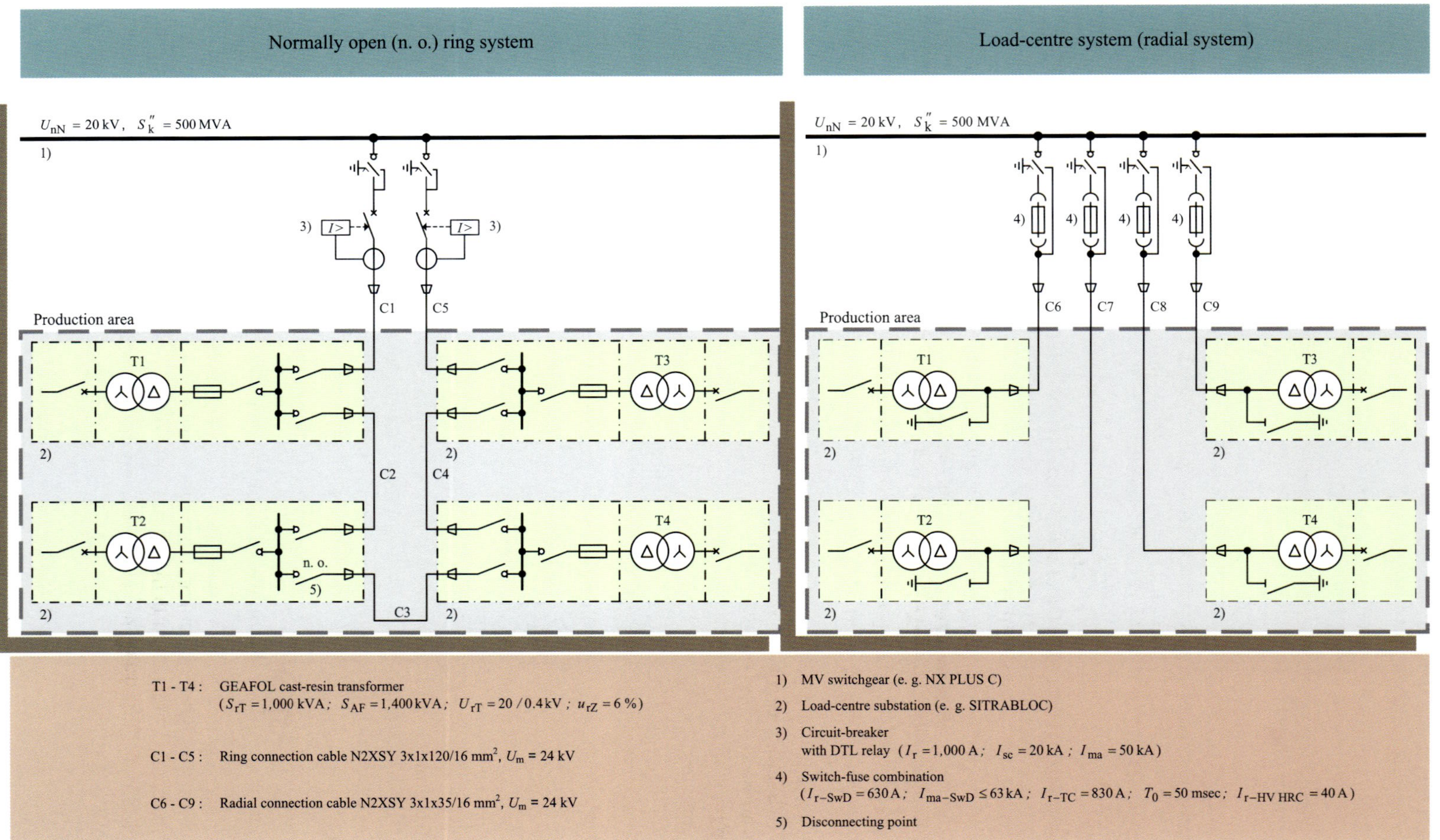

Fig. B5.2 MV-side system configurations for connecting decentralized load-centre substations

- *Coverage of power demand* ↑

  The process-related power demand for factory halls in industrial plants can be distributed both on a ring system and on a load-centre system without any problem. Only when the power demand reaches P_{max} = 10...15 MW or more does the grouping together of all load-centre substation feeders into one central transfer substation result in unmanageably large MV main switchgear. Moreover, in MV switchgear of such large dimensions, keeping spare panels for the final expansion stage can result in problems along the cable routes. The alternative in this case is to erect substations for the separate configuration of load-centre systems (Section 5.4.3).

- *Supply reliability* ↑

  Evaluation of the supply reliability of ring and load-centre systems is based on the ability to handle an ($n-1$) fault by isolation of the fault location by means of protection equipment. In the case of a transformer fault, the ($n-1$) fault is handled in both types of system without interruption because the necessary power demand can be provided by the instantaneous reserve in AF operation ($S_{AF} = 1.4 \cdot S_{rT}$ with fans switched on) of $n-1$ GEAFOL cast-resin transformers.

  In the case of an MV-side cable fault in a normally open (n.o.) ring system, the number of simultaneously failing transformers or load-centre substations of a half-ring depends on the location of the n.o. disconnecting point.

  According to the example in Fig. B5.2, two load-centre substations will fail simultaneously in the n.o. ring system on a cable fault. Because all four load-centre substations of the decentralized multiple incoming supply are meshed on the LV side and thus supply the production process together in an interconnected system, failure of one half-ring would have the effect of a double fault. Failure of one half-ring can therefore result in an overload and disconnection of the transformers that are still operating.

  This problem does not occur in a load-centre system. A cable fault in a load-centre system only ever causes failure of one load-centre substation. A comparable supply reliability level can be achieved with the ring system only in normally closed (n.c.) operation.

  For n.c. operation, all ring-main panels of the load-centre substations would have to be equipped with circuit-breakers and line differential protection or directional protection. For approximately equal supply reliability, the investment costs for the ring system would be considerably higher than for a suitable load-centre system.

- *Voltage quality* ↑

  The quantities that adversely affect the voltage quality include short-time voltage dips that occur during short circuits in the power system. In the event of short circuits in the ring system, the best fault clearance time that can be achieved with the circuit-breaker-protection-relay combination is t_a = 70...120 msec. The voltage dip occurring during this fault duration can greatly interfere with sensitive production processes (e.g. wafer production in the semiconductor industry) without the use of a dynamic voltage restorer (DVR) and cause irreversible production damage. With the switch-fuse combination of the load-centre system, it is possible to reduce the fault clearance time to $t_a \leq 5$ msec in the event of a short circuit. Because of this fault clearance time and the current-limiting effect of the HV HRC fuses, the voltage dips caused by short circuits in a load-centre system have more or less negligible impact on the voltage quality.

- *Hazard to persons and equipment* ↓

  One measure of the hazard is the arc energy released during a short circuit. The following guidance values apply to the hazard posed [5.1]:

  - $W_{arc} \leq 250$ kWsec → harmless
  - $W_{arc} \leq 500$ kWsec → non-hazardous
  - $W_{arc} > 500$ kWsec → hazardous

  The arc energy released W_{arc} directly depends on the clearing time in case of a fault. In the case of a 20-kV-side short-circuit power of $S_k'' = 500$ MVA, until fault clearance by the circuit-breaker-protection-relay combination in the ring system ($t_a = 70...120$ msec), an arc energy of $W_{arc} = 1300...4500$ kWsec is released. In the vicinity of cable racks and load-centre substations installed exposed in the production area, arc energy of this magnitude poses a substantial hazard to people in the event of a fault. Moreover, the destructive power of the arc during short circuits on cable racks can also cause failure of systems not directly involved in the fault.

  In the 20-kV load-centre system, the destructive power of the arc is much smaller for the same short-circuit power. Due to the short clearing time of $t_a \leq 5$ msec and the current-limiting effect of the HV HRC fuse, the arc energy released is only $W_{arc} = 10...30$ kWsec. There is no hazard to people. Arc damage remains limited to the part of the installation in which the fault occurs.

- *Maintenance and servicing expense* ↓

  Because switchgear is increasingly designed to be maintenance-free and system protection devices have become self-monitoring with the advent of digital technology, the difference between the maintenance and servicing expenses incurred by operation of a ring or load-centre system is negligible.

- *Ease of system management* ↑

  Decentralized switching operations are always necessary in operation of a n.o. ring system. In particular, fault-confining switching operations and switching actions for creating a damaged-condition configuration make system management more complicated than in a load-centre system. In a load-centre system, all switching operations can be performed from the centrally located MV switchgear. Moreover, the load-centre system affords the operator a better overview of the circuit state on the MV side. A better overview of the circuit state reduces the danger of spurious switching operations.

- *Environmental compatibility* ↑

  SF_6-insulated MV switchgear is also widely used in industry owing to the special properties that sulphur hexafluoride has for switching and insulating. SF_6 is an inert, non-combustible, non-toxic, non-ozone-depleting insulant with a high global warming potential. Because it is known that SF_6 is a persistent and very effective greenhouse gas in the atmosphere, use of SF_6-insulated MV switchgear in industrial power systems complies with the "Voluntary commitment of the SF_6 producers, manufacturers and users of electrical installations > 1 kV for the transmission and distribution of electrical energy in the Federal Republic of Germany to the use of SF_6 as an insulating and arc extinguishing gas" [5.4]. This voluntary commitment ensures environmentally sound use of SF_6-insulated MV switchgear in all industrial power systems.

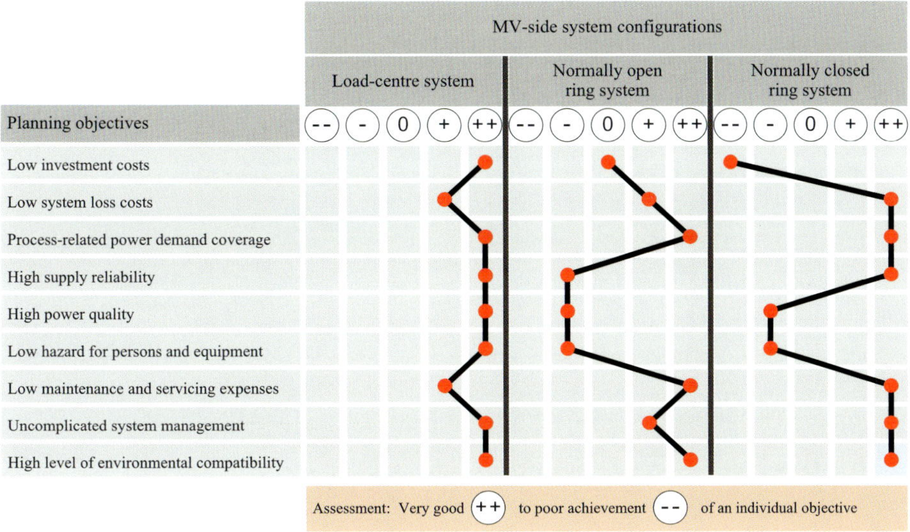

Fig. B5.3 Multiple-objective-oriented evaluation of MV-side system configurations for decentralized power supply

A high level of environmental compatibility is achieved not only with air-insulated but also with SF_6-insulated MV switchgear in all system configurations.

In addition to the verbal description, an overview of the multiple-objective-oriented evaluation of ring and load-centre systems is provided in Fig. B5.3. The overview clearly shows that the load-centre system is the configuration with the best total score.

Conclusion

The load-centre system is the preferred configuration for efficient decentralized multiple incoming supplies of power to large-area factory halls in industrial plants.

5.4 System structures and concepts meeting the requirements for industrial plants

Preferred system structures and concepts for small, medium-sized and large industrial plants and high-technology businesses that meet the different requirements for power demand and supply quality are presented below. Part A of this guide provides the theoretical basis for the planning recommendations for power system design. Practical experience from industrial power system planning is also considered.

5.4.1 Small industrial plants

Small industrial plants ($P_{max} \leq 3$ MW at $U_{nN} = 10$ kV and $P_{max} \leq 5$ MW at $U_{nN} = 20$ kV) can be supplied with power directly from the public MV network. Public MV networks are usually operated as normally open (n.o.) ring systems.

The following guidance values apply to the maximum transmission capacity of an n.o. cable ring:

- $S_{max} = 7$ MVA at $U_{nN} = 10$ kV,
- $S_{max} = 12$ MVA at $U_{nN} = 20$ kV.

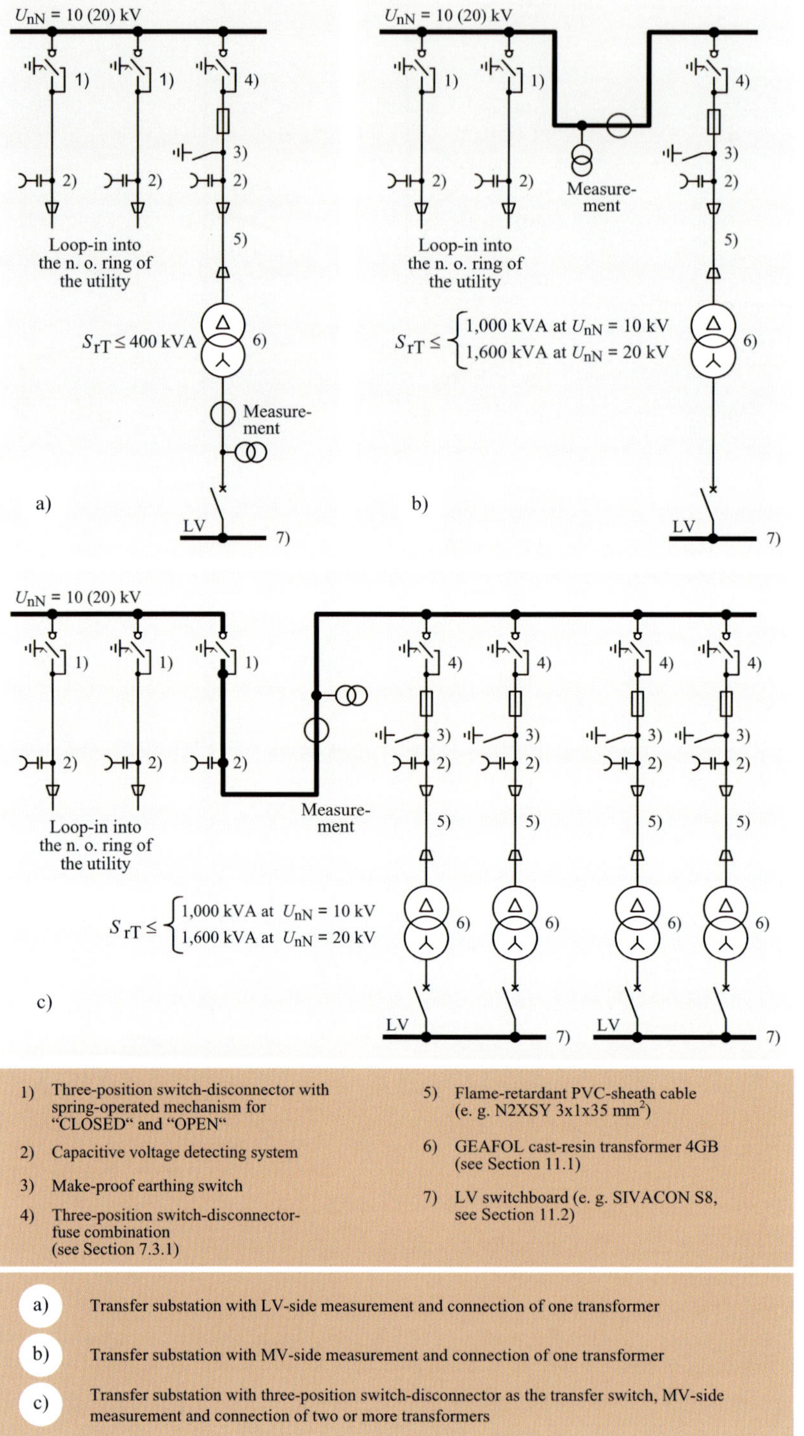

Fig. B5.4 Power supply concept for a small industrial plant

Based on these values, the transfer substation for small industrial plants can almost always be looped into an existing MV cable ring. The basic circuits depicted in Fig. B5.4 have proven convenient.

The incoming feeder panels of the MV substation are normally switch-disconnector panels with a spring-operated mechanism for closing and opening. Only when the operating conditions of the industrial customer or the power system engineering conditions require this must each of the incoming feeder panels be equipped with a circuit-breaker and the appropriate protection devices. A circuit-breaker is only used as a transfer switch in exceptional circumstances. The transfer switch is usually a switch-disconnector. For operational reasons, the responsible operating company of the public power supply system may impose the following additional requirements on the switchgear design [5.5]:

- possibility of connecting devices for fault location,
- possibility of mounting short-circuit indicators,
- possibility of measuring the residual current in case of an earth fault, if applicable, by the installation of core balance current transformers.

The following Siemens switchgear is recommended for the transfer substation of small industrial plants:

- 8DH10 switchgear for indoor installation with the features [5.6, 5.7]: SF_6-insulated, type-tested, three-pole metal-enclosed, metal-clad, single busbar,
- SIMOSEC switchgear for indoor installation with the features [5.8, 5.9]: air-insulated, type-tested, three-pole metal-enclosed, metal-clad, single busbar.

5.4.2 Medium-sized industrial plants

For individual consumers looped into an MV cable ring, an upper reference power is usually available that corresponds to hardly more than 50 % of the maximum transmission capacity of the ring main. For annual maximum demands in the range

- $3\ \text{MW} < P_{max} \leq 13\ \text{MW}$ at $U_{nN} = 10\ \text{kV}$ or
- $5\ \text{MW} < P_{max} \leq 18\ \text{MW}$ at $U_{nN} = 20\ \text{kV}$

a basic circuit arrangement according to Fig. B5.5 is recommended. According to Fig. B5.5, the consumer transfer substation is supplied with power through two parallel radial cable systems directly from the transformer substation or from a central switching substation of the public MV network. With this type of supply, the incoming feeder panels of the MV substation must be equipped with circuit-breakers and protection devices.

The protection devices for the double-radial-line connection in the transfer substation must always be coordinated with the existing protection devices of the MV public distribution network. They must be selected and set in consultation with the responsible utility company or public network operating company. Directional time-overcurrent protection is normally deployed with double-radial-line connection of transfer substations. For directional time-overcurrent protection devices (e.g. SIPROTEC relays 7SJ62, 7SJ63 and 7SJ64), both current and voltage transformers are provided.

With the basic circuit arrangement shown in Fig. B5.5, it is possible to implement a very flexible power system and installation concept. Unlike the circuit arrangement in Fig. B5.4, in this case separation of the system into two independent sections is possible by means of bus sectionalizing between the two incoming feeders. This enables part-load operation to continue even during expansion and maintenance work.

The full advantage of this circuit arrangement is only really obtained if the distribution transformers of each LV power system are distributed evenly to both sections of the

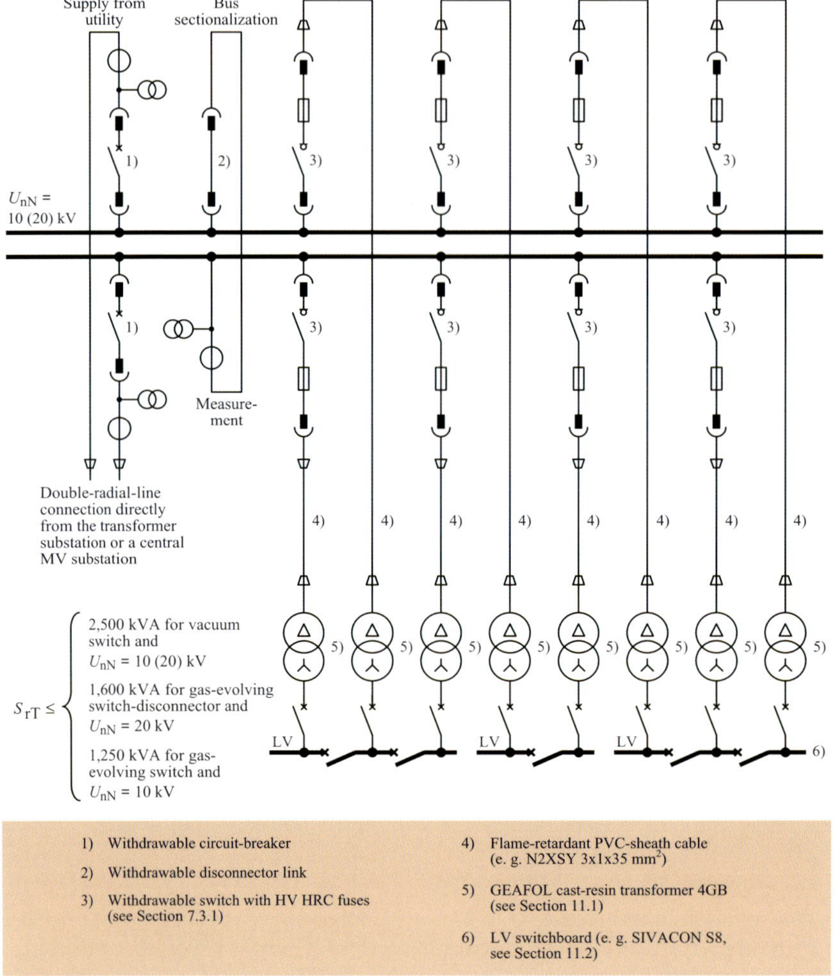

Fig. B5.5 Power supply concept for a medium-sized industrial plant

busbar. If this is implemented, partial power is available in all supplied areas when one busbar section is completely disconnected.

During normal operation, the two busbar sections can be considered as a single busbar. The busbar must therefore not be sectionalized using a circuit-breaker with selective protection because this would inevitably result in spurious tripping and overloads in the LV power system. However, non-availability of one busbar section due to a fault can be all but ruled out with the modern switchgear available today. The following Siemens switchgear can therefore be used for the transfer substations of medium-sized industrial plants:

- NXPLUS C fixed-mounted circuit-breaker switchgear for indoor installation with the features [5.10, 5.11]: SF_6-insulated, type-tested, three-pole metal-enclosed, metal-clad, single busbar,
- NXAIR (M) withdrawable circuit-breaker switchgear for indoor installation with the features [5.12, 5.13]: air-insulated, type-tested, three-pole metal-enclosed, metal-clad, single busbar.

The use of air-insulated switchgear of type NXAIR (M) is absolutely necessary if the technical specifications of DIN EN 62271-105 (VDE 0671-105): 2003-12 [5.16] or IEC 62271-105: 2002-08 [5.17] for transformer protection with a switch-fuse combination can only be met by means of a vacuum-switch or vacuum-contactor-fuse combination (see Section 7.3.1).

5.4.3 Large industrial plants

Large industrial plants with annual maximum demands

- $P_{max} > 13$ MW at $U_{nN} = 10$ kV and
- $P_{max} > 18$ MW at $U_{nN} = 20$ kV

must be supplied with power from the public high-voltage network through dedicated 110/20(10)-kV power transformers. In power supply concepts according to Fig. B5.6 and B5.7, the main switchgear of the transfer substation consists of a combination of a large number of circuit-breaker and switch panels. Therefore 110-kV/MV transfer substations should preferably be equipped with the following Siemens switchgear types:

- NXPLUS C fixed-mounted circuit-breaker switchgear for indoor installation with the features [5.10, 5.11]: SF_6-insulated, type-tested, three-pole metal-enclosed, metal-clad, single busbar,
- NXAIR (M) withdrawable circuit-breaker switchgear for indoor installation with the features [5.12, 5.13]: air-insulated, type-tested, three-pole metal-enclosed, metal-clad, single busbar.

In addition to the central transfer substation, MV distribution substations are usually constructed on the site of large industrial plants. Before additional MV distribution substations are built, the following economic and technical considerations must be examined:

a) To evaluate cost efficiency, above all, the investment costs for construction of a distribution substation must be calculated. A double-radial-line connection according to Fig. B5.6 requires, for each distribution substation, four circuit-breaker panels, one busbar sectionalizer in the distribution substation and the installation of two cable systems (dimensioned according to the necessary load current or fault current-carrying capability).

   For direct radial-line connection of the load-centre substations from the main substation, by comparison, only the additional expense for the longer cable runs has to be calculated. An overall comparison of the individual cost factors shows that 4 radial cables can be installed for the 2 main cables for the incoming supply to the distribution substation. For the additional expense for switchgear, a further 8 to 10 radial cables could be installed. From this it follows that cost efficiency advantages for a substation only begin after connection of more than 10 to 14 load-centre substations. Looping in two distribution substations in a ring main (Fig. B5.7) only makes a slight difference to the costs.

b) From a technical viewpoint, the installation of distribution substations necessarily results in an increase in grading time for the definite-time overcurrent-time protection in the main switchgear. Due to the increase in grading time by at least 1 to 2 stages, the I^2t-dependent arc energy in case of a fault also increases. High arc energy is also released in MV-side short circuits in factory halls because the distribution substations are connected to the main substation using circuit-breaker panels.

As regards minimization of damage in case of a short circuit in a large-area supply, there would initially seem to be many advantages favouring central switch-HV-HRC-fuse connection of all load-centre substations to the main switchgear. However, this is outweighed by the fact that, in practice, as from an annual maximum demand of

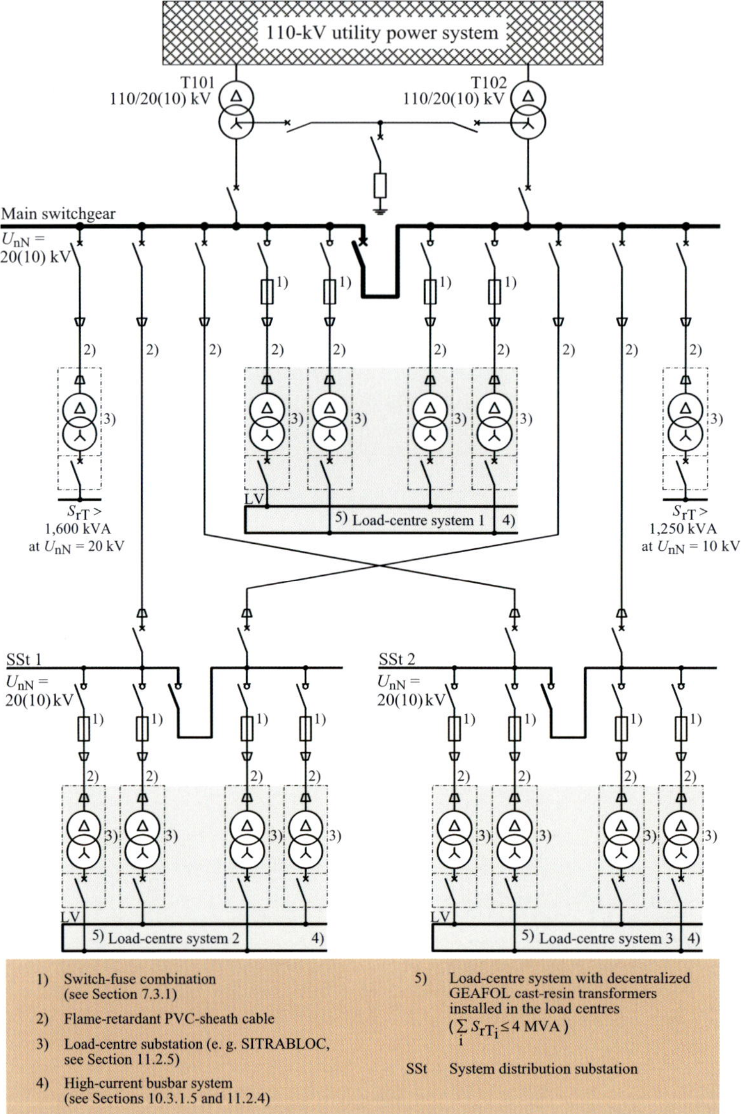

Fig. B5.6 Power supply concept for a large industrial plant with double-radial-line connection of the system distribution substations

P_{max} = 10...15 MW, grouping together all substation feeders usually results in unmanageably large main switchgear, and that keeping spare panels for the final expansion stage results in problems along the cable routes [5.1]. For this reason, installation of distribution substations is technically advisable even at the lower cost efficiency limit, i.e.

- P_{max} = 4.5...5.5 MW at U_{nN} = 10 kV and
- P_{max} = 6.0...7.5 MW at U_{nN} = 20 kV

Installation of distribution substations provides marginally higher supply reliability if one busbar section fails in the transfer substation. According to the power supply con-

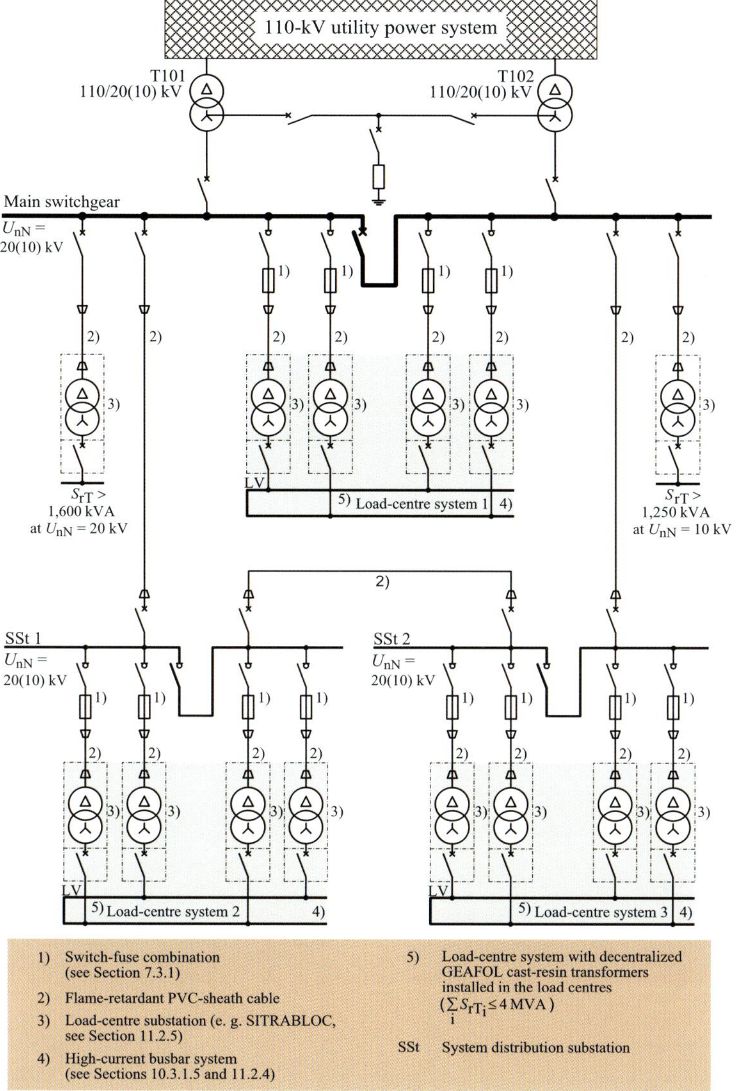

Fig. B5.7 Power supply concept for a large industrial plant with looping in of the system distribution substations into an MV main ring

cept shown in Figs. B5.6 and B5.7, the load-centre systems connected to the distribution substations are not affected by this failure. Only in the event of a fault in one busbar section in the distribution substation do half of the connected load-centre substations fail.

Fig. B5.8 shows a power supply concept that ensures the $(n-1)$ instantaneous reserve for the full power demand of the production process both in the event of a failure of one busbar section in the transfer substation and in the event of failure of one busbar section in a distribution substation.

The ability to handle a busbar fault in the transfer substation requires that each supplying 110/20(10)-kV power transformer is rated according to DIN IEC 60076-7 (VDE 0532-

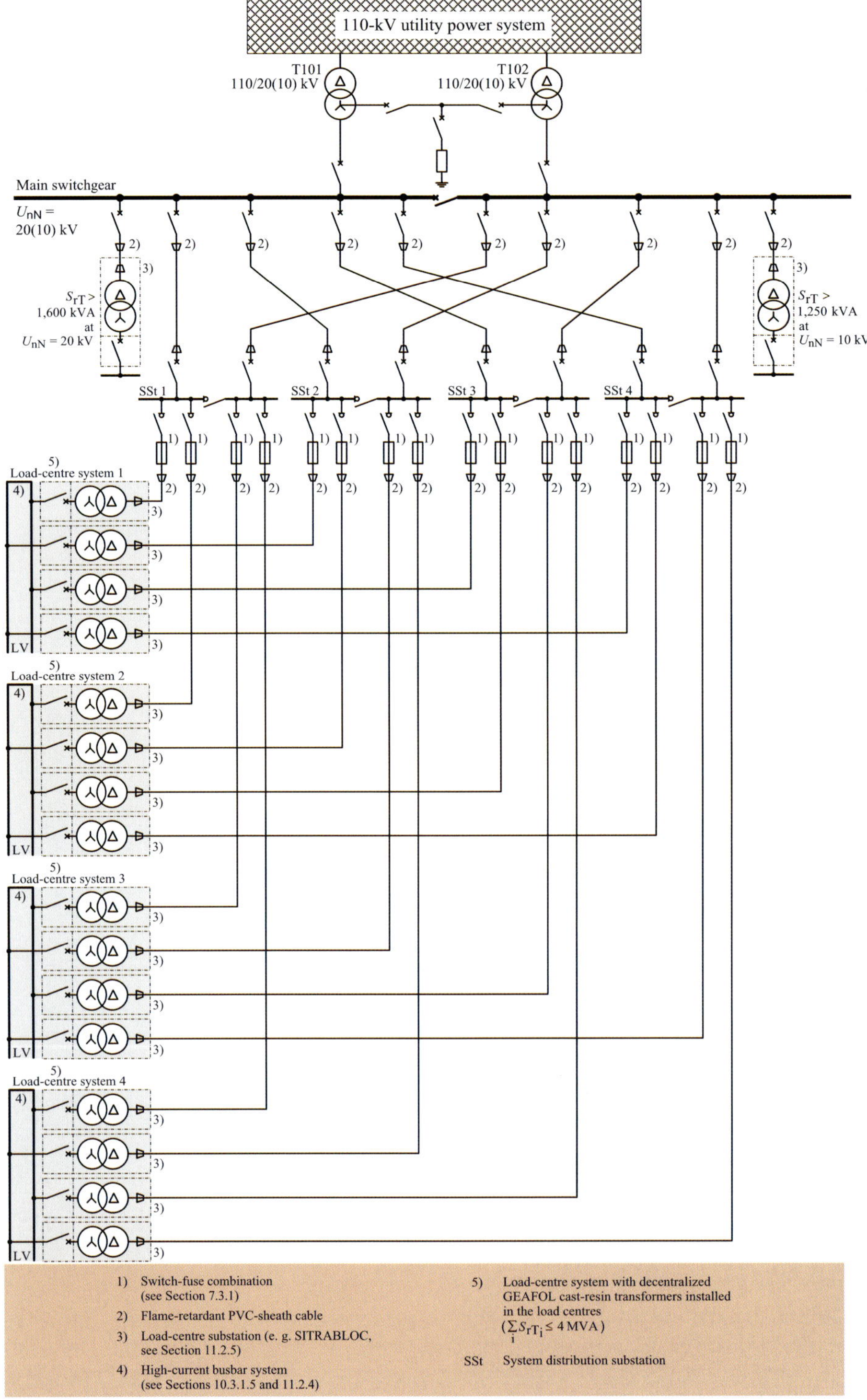

Fig. B5.8 Power supply concept for a large industrial plant with complete $(n-1)$ instantaneous reserve of the normal power supply

76-7): 2008-02 [5.18] or IEC 60076-7: 2005-12 [5.19] for the expected annual maximum demand of the industrial plant (Eq. 2.1).

The power supply concept according to Fig. B5.8 can be implemented with the following Siemens switchgear types:

a) Main switchgear in the transfer substation

- NXPLUS C fixed-mounted circuit-breaker switchgear for indoor installation with the features [5.10, 5.11]: SF_6-insulated, type-tested, three-pole metal-enclosed, metal-clad, single busbar,
- NXAIR (M) withdrawable circuit-breaker switchgear for indoor installation with the features [5.12, 5.13]: air-insulated, type-tested, three-pole metal-enclosed, metal-clad, single busbar,
- 8DA fixed-mounted circuit-breaker switchgear for indoor installation with the features [5.14, 5.15]: SF_6-insulated, type-tested, single-pole metal-enclosed, metal-clad, single busbar.

b) Switchgear in the distribution substations SSt1 to SSt4

- 8DH10 switchgear for indoor installation with the features [5.6, 5.7]: SF_6-insulated, type-tested, three-pole metal-enclosed, metal-clad, single busbar,
- SIMOSEC switchgear for indoor installation with the features [5.8, 5.9]: air-insulated, type-tested, three-pole metal-enclosed, metal-clad, single busbar.

The switchgear of the substations SSt1 to SSt4 is connected to both busbar sections of the main switchgear by a double-radial-line. Radial cables are alternately installed from the distribution substations to the individual load-centre substations of the LV power system. With this structure, no more than one load-centre substation fails per LV power system irrespective of the fault location (transformer, bus section or cable route). Assuming that the incoming supply of the LV systems is dimensioned according to the $(n-1)$ criterion, an instantaneous reserve can be guaranteed for any fault in the entire plant with the power supply concept according to Fig. B5.8.

Power system concepts of industrial plants with in-plant generation and imported power

In some industries, the quantity of steam required for the production process is so large that it is cost-efficient to run in-plant generation using backpressure turbines [5.20]. Among other considerations, a combination of in-plant generation and imported power can be advisable to ensure a high quality of supply.

Fig. B5.9 shows a power supply concept of a large industrial plant with in-plant generation and imported power. In this concept, the imported power system and the in-plant generation mutually independently supply power to two separate MV networks, the imported-power network and the in-plant generation network. The two networks are interconnected through a bus coupler equipped with a tripping device for ensuring steady-state and dynamic stability. The stability behaviour and system dynamics for the relevant operating states must be examined in the context of planning supply systems with in-plant generation and imported power. This examination must cover the behaviour during

- interconnected operation,
- islanding and
- island operation.

Knowledge of the stability limits is an important prerequisite for dimensioning and setting the tripping device for network splitting and its correct coordination with the power system protection and generator protection devices.

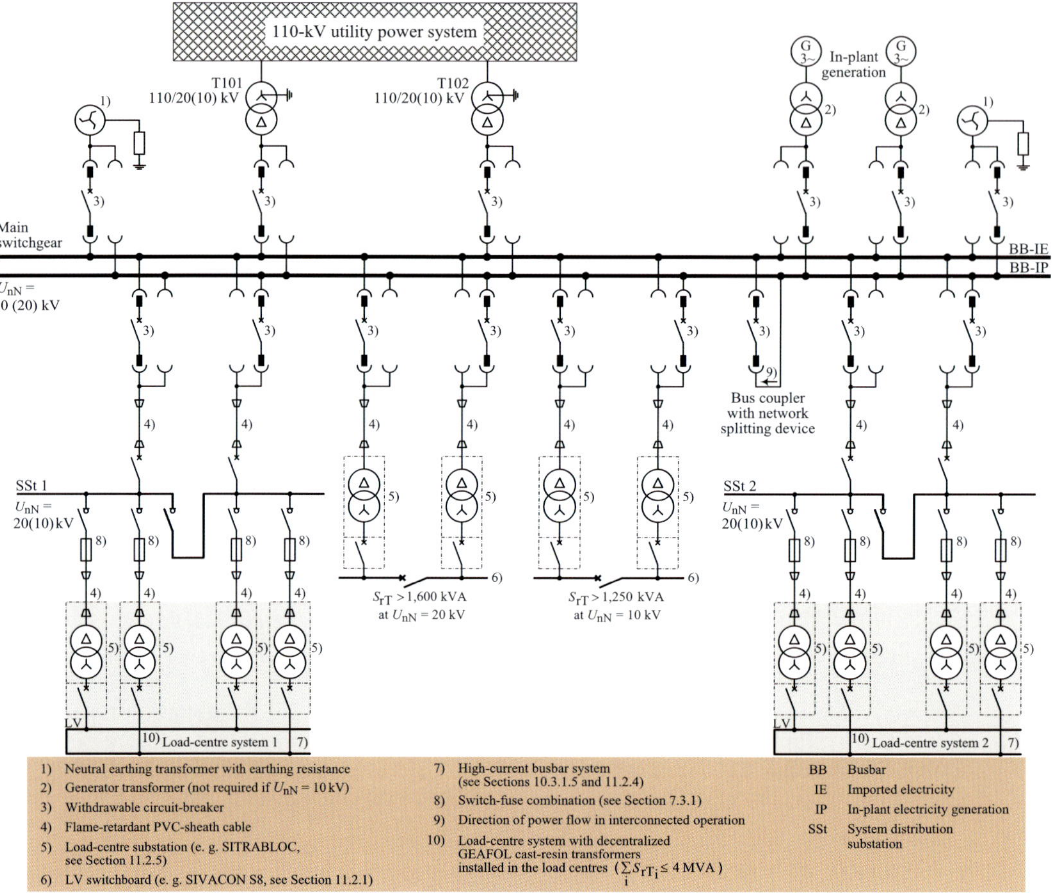

Fig. B5.9 Power supply concept of a large industrial plant with in-plant generation and imported power (main switchgear with double-busbar system, e.g. NXAIR P or 8DB10)

The simulation calculations required for this can be performed using the PSS™NETOMAC power system planning software [5.21, 5.22]. The following criteria are suitable for simple assessment of the stability behaviour of industrial power systems with in-plant generation and imported power:

a) The power flow in interconnected operation must always be from the busbar of the in-plant generation (BB-IP) to the busbar for imported electricity (BB-IE). If this is ensured, there will be no lack of power in the island network remaining in operation immediately after the network splitting due to a fault. The island network remains stable and permits successful resynchronisation with the imported power network thanks to the available turbine output.

b) Only the priority loads of the industrial plant (e.g. essential systems to maintain or prevent damage to the production process) must be connected to the in-plant generation busbar (BB-IP). The power of these loads must be limited to 70 % of the installed generator output.

c) The installed generator output should be at least twice as large as the power rating of all drives that are important to operation, in order to minimize the probability of spurious tripping of DTL protection devices in the incoming feeders due to excessive motor currents. The speed of the motors already slows during short-term ($t < 300$ msec) fault-induced voltage dips ($U < 0.8\ U_{nN}$), so that a sort of restart is performed after the full supply voltage has recovered. The motors absorb larger currents. A fault-induced voltage dip already causes a rise in current that can be twice the nominal current of the motor on recovery of the full supply voltage. Almost the full starting current, that is, three to five times the nominal current of the motor, already flows after voltage dips lasting $t = 0.5$ sec.

d) In island operation, the backpressure control of the turbines must be switched over to speed control (frequency control) because network splitting in the event of a fault offloads active power from the generators. Using speed control, the synchronous generators can be adapted to changes in the active power during island operation.

e) To prevent the frequency from leaving the range 49.6 Hz $< f <$ 50.5 Hz, transition from interconnected to island operation should be performed as instantaneously as possible [5.23].

The protection of synchronous generators always has priority over the selectivity of the power system protection. In most cases, however, selectivity, reliable generator protection and dynamic system stability can all be achieved by association and logical combination of islanding criteria with the settings of the DTL protection, differential protection and distance protection devices. Fig. B5.10 shows the relevant tripping criteria for the transition of an industrial plant with in-plant generation and imported power to island operation.

The double busbar switchgear shown in Fig. B5.9, in which the in-plant generation (BB-IP) and imported electricity busbars (BB-IE) are connected through a bus-coupler equipped with circuit-breaker and network splitting device, can be implemented, if $U_{nN} = 10$ kV, reliably and cost-efficiently with the NXAIR P air-insulated, metal-enclosed and metal-clad withdrawable circuit-breaker switchgear in a back-to-back or face-to-face

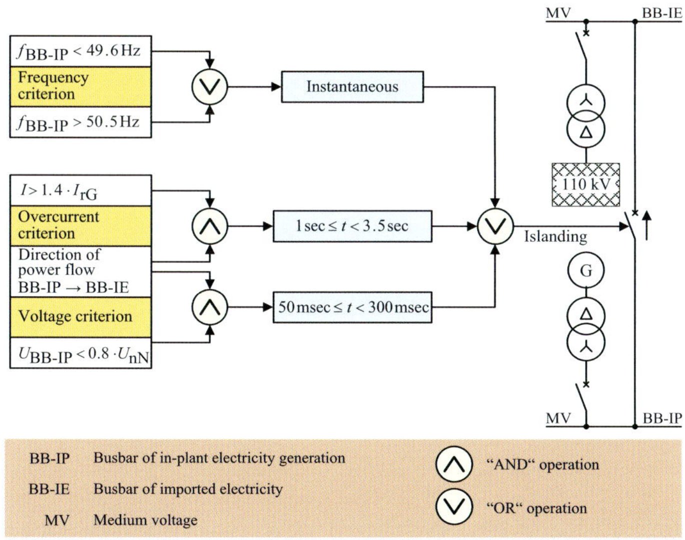

Fig. B5.10 Tripping criteria for islanding for an industrial plant with in-plant generation and imported electricity

arrangement [5.12, 5.13]. If gas-insulated equipment is preferred and/or for a nominal system voltage of $U_{nN} = 20$ kV, use of the 8DB10 SF_6-insulated, metal-enclosed and metal-clad fixed-mounted circuit-breaker switchgear for double-busbar application and indoor installation [5.14, 5.15] is recommended.

5.4.4 Production facilities of high-technology businesses

Supplying the infrastructure of high-technology businesses with power poses the greatest challenge for quality of supply *QS*. Due to the preferred application of high technologies and the wide use of computer-controlled automation systems, processes in modern production plants respond especially sensitively to voltage dips even in the millisecond range. Power supply concepts for production plants of high-technology businesses must provide not only a high supply reliability *SR* but also, above all, a high voltage quality *VQ*. The *VQ* requirements usually correspond to the class 1 from DIN EN 61000-2-4 (VDE 0839-2-4): 2003-05 [5.24] or IEC 61000-2-4: 2002-06 [5.25]. If a process is assigned to supply class 1, the compatibility levels for the influencing quantities of the *VQ* must be below the valid level for the supply of electrical energy from the public network. The solutions so far usually deployed by industrial consumers for connection to the public network are not sufficient to fulfil these high *VQ* requirements.

If the requirements of a sensitive production process with respect to the power demand and *SR* can be met by a system design with $(n-1)$ redundancy, the combination of an MV system connection designed for the *SR* requirements and a dynamic voltage restorer (DVR) provides a convenient solution for the power supply [5.26]. Fig. B5.11 shows such a solution. The DVR connected serially between the voltage source and load ensures that fault-induced voltage dips (e.g. on short circuits in the power system) or short-time voltage dips (e.g. due to automatic reclosure in the 110-kV overhead line system or automatic switchover in the MV system) do not have any effect on the process.

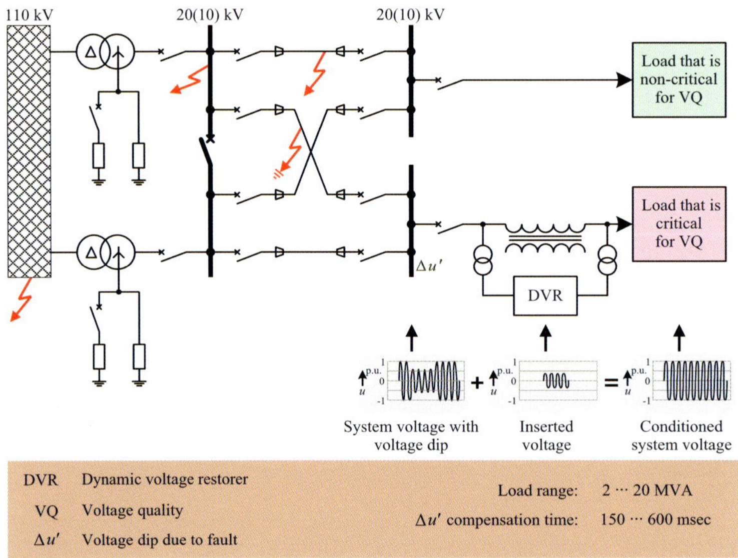

Fig. B5.11 Power supply concept for a high-technology business with integration of a dynamic voltage restorer (DVR) into the MV network

The energy required for compensation of a voltage dip or a short-time voltage interruption is drawn from an energy store of the DVR consisting of a large DC capacitor bank and fed into the power system for the duration of the fault or switchover ($t = 150...600$ msec) [5.27]. Energy supply concepts with integrated DVR systems therefore provide sensitive processes with reliable protection from critical voltage dips and short interruptions. To protect sensitive processes from long-time outages of the normal power supply with high *VQ*, dynamic diesel UPS systems (DDUPS) [5.28 to 5.30] are required. Dynamic diesel UPS systems from Hitec [5.28] (Fig. B5.12a to B5.12d) consist of a diesel engine (1) and an electrical machine unit that comprises an induction coupling (3) and synchronous generator (4). The induction coupling is used as a kinetic energy store to back up during the starting time of the diesel engine in case of a power failure. The mechanical connection between the diesel engine and electrical machine

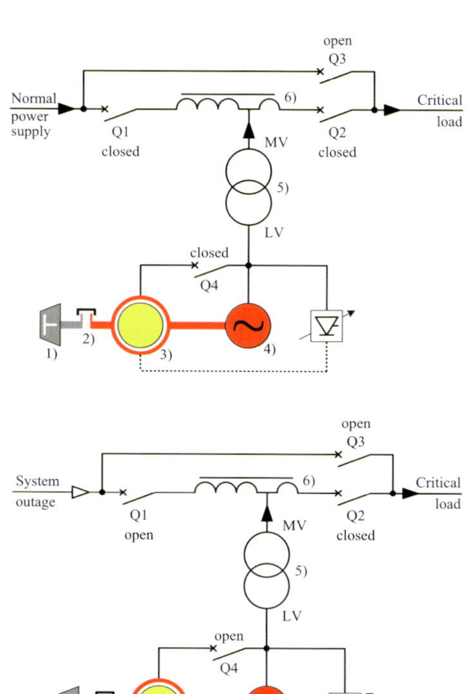

Fig. B5.12a
DDUPS in normal mode

Fig. B5.12b
DDUPS in change-over to diesel mode

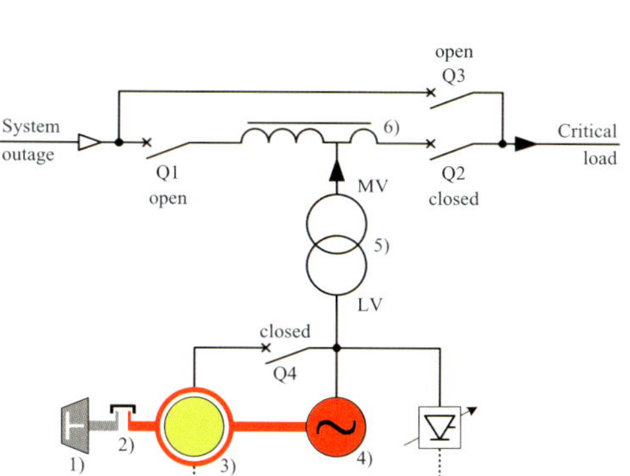

Fig. B5.12c
DDUPS in diesel mode

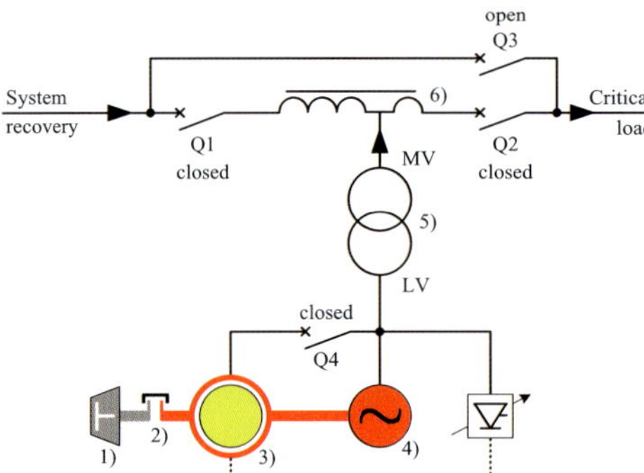

Fig. B5.12d
DDUPS on return to normal mode

unit is formed by a free-wheeling clutch (2). The connection of the DDUPS system to the MV power system is effected through a step-up transformer (5) and a reactor (6).

Up to a nominal system voltage of U_{nN} = 11 kV, DDUPS systems can also be implemented as pure MV systems [5.28, 5.29]. The normal MV power supply is decoupled by means of a reactor from all the loads that have especially stringent requirements for the *VQ* (class 1 of standard DIN EN 61000-2-4 (VDE 0839-2-4): 2003-05 [5.24] or IEC 6100-2-4: 2002-06 [5.25]). Operation of a Hitec DDUPS system is characterized by the following four system modes [5.28]:

- *Normal mode*

By the combination of a synchronous generator (4) with a reactor (6), the DDUPS system acts as an active filter in normal mode (Fig. B5.12a). This active filter protects the sensitive production process from all continuous-time and discrete-time disturbing quantities (e.g. flicker, harmonics, voltage unbalance and fluctuations, voltage dips, short interruptions in the voltage). The synchronous generator (4) runs as a motor and drives the outer rotor of the induction coupling (3) at a speed of n = 1,500 rpm (f = 50 Hz) or n = 1,800 rpm (f = 60 Hz). Through excitation of the two-pole three-phase winding of the outer rotor, the inner rotor reaches a speed of n = 3,000 rpm (f = 50 Hz) or n = 3,600 rpm (f = 60 Hz). As a result, kinetic energy is stored in the inner rotor. The free-wheeling clutch (2) ensures that the outer rotor of the induction coupling and the diesel engine are mechanically decoupled in normal mode.

- *Change-over to diesel mode*

On failure of the normal power supply, circuit-breaker Q1 opens. Circuit-breaker Q1 opening causes the DDUPS system to go into diesel operation (Fig. B5.12b). In the first phase of this transition, the kinetic energy stored in the inner rotor is transferred to the outer rotor of the induction coupling (3). At the same time, the diesel engine (1) starts. Due to the energy transfer from the inner to the outer rotor, the generator speed remains constant throughout engine starting at n = 1,500 rpm (f = 50 Hz) or n = 1,800 rpm (f = 60 Hz). After the diesel engine has started in less than two seconds, the free-wheeling clutch (2) engages. In the phase following automatic engagement of the free-wheeling clutch, the diesel engine (1) and the induction coupling (3) together drive the generator (4) for a short time. After about 5 to 10 seconds, the energy to drive the generator (4) is supplied by the diesel engine (1) alone.

- *Diesel mode*

Diesel operation (see Fig. B5.12c) is used to back up a long-time outage of the normal power supply while maintaining the voltage quality of normal mode. The duration of diesel mode to back up a long-time outage depends entirely on the amount of fuel stored. In diesel mode, the winding of the outer rotor is re-energized, causing the inner rotor of the induction coupling (3) to ramp up to a speed of $n = 3{,}000$ rpm ($f = 50$ Hz) or $n = 3{,}600$ rpm ($f = 60$ Hz) again.

The speed of the diesel engine is monitored and digitally controlled to ensure a constant output frequency in the range $49.6\text{ Hz} < f < 50.5\text{ Hz}$ or $59.5\text{ Hz} < f < 60.6\text{ Hz}$. The *VQ*-compliant power supply of sensitive loads is additionally supported in diesel mode by the induction coupling (3).

- *Return to normal mode*

After termination of the long-time outage, the UPS system synchronizes with the recovered power system of the normal power supply and closes the circuit-breaker Q1. After the circuit-breaker Q1 has been closed again, the speed of the diesel engine (1) is ramped down to $n = 1{,}450$ rpm ($f = 50$ Hz) or $n = 1{,}750$ rpm ($f = 60$ Hz).

The speed limitation to $n = 1{,}450$ (1,750) rpm causes mechnical decoupling of the diesel engine (1) from the electrical machine unit (3), (4) by the free-wheeling clutch (2). After mechanical decoupling, the synchronous generator (4) continues to run as a motor so that the outer rotor of the induction coupling (3) can resume its nominal speed of $n = 1{,}500$ (1,800) rpm. To cool the diesel engine, it is again operated at no load for a short time. After the diesel engine has completed its cool-down run it will shut down and return to standby mode.

For a reliable power supply to sensitive processes, Table B5.13 provides a selection of DDUPS systems that can be integrated into the MV distribution systems of high-technology businesses.

Figs. B5.14 and B5.15 show two power supply concepts that have proven successful with a DDUPS system integrated into the MV system as general single-line diagrams. The general single-line diagram in Fig. B5.14 shows an individual customer solution for the supply of a semiconductor factory with a total power demand of $S_{max} = 50$ MVA from the public 110-kV network. The power required for wafer production in this factory is supplied uninterruptibly and with the highest possible voltage quality by a DDUPS system with $(n-1)$ redundancy.

Table B5.13 Commercially available DDUPS systems for integration into the MV power system

Technical features	DDUPS system		
	Hitec [5.28]	Hitzinger [5.29]	Piller [5.30]
Maximum rated power S_r per unit	$S_r \le 2{,}000$ kVA		$S_r \le 2{,}500$ kVA
Configurations for connection to the MV power system	○ LV-UPS module, step-up transformer and MV reactor ○ MV-UPS module ($U_{nN} \le 11$ kV) and MV reactor		○ LV-UPS module, step-up transformer and MV reactor
Ensuring voltage quality	Generator-reactor combination		Power electronics and generator-reactor combination
Generator short-circuit current I_{kG}	$I_{kG} \le 14 \cdot I_{rG}$	$14 \cdot I_{rG} \le I_{kG} \le 17 \cdot I_{rG}$	$I_{kG} \le 14 \cdot I_{rG}$
Subtransient reactance X''_d	$X''_d = 10\,\%$	$5\,\% \le X''_d \le 8\,\%$	$X''_d = 10\,\%$

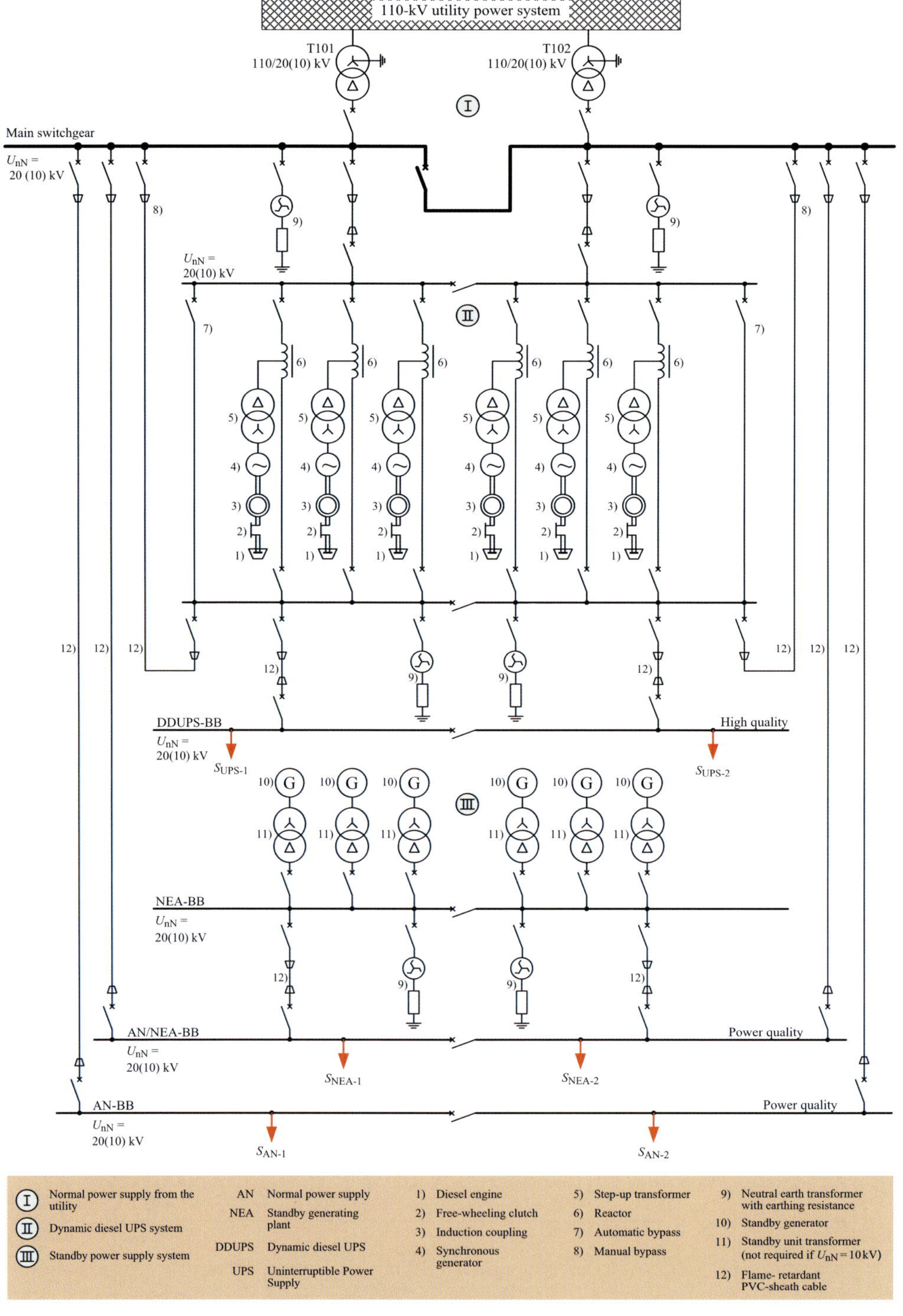

Fig. B5.14 Power supply concept for a high-technology business with integration of standby generating sets and dynamic diesel UPS systems into the in-plant MV power system (main supply from the 110-kV utility power system)

With the general single-line diagram shown in Fig. B5.15, it is possible to implement a similar power supply concept for a high-technology business whose power demand can be provided in compliance with the requirements through the public MV power system.

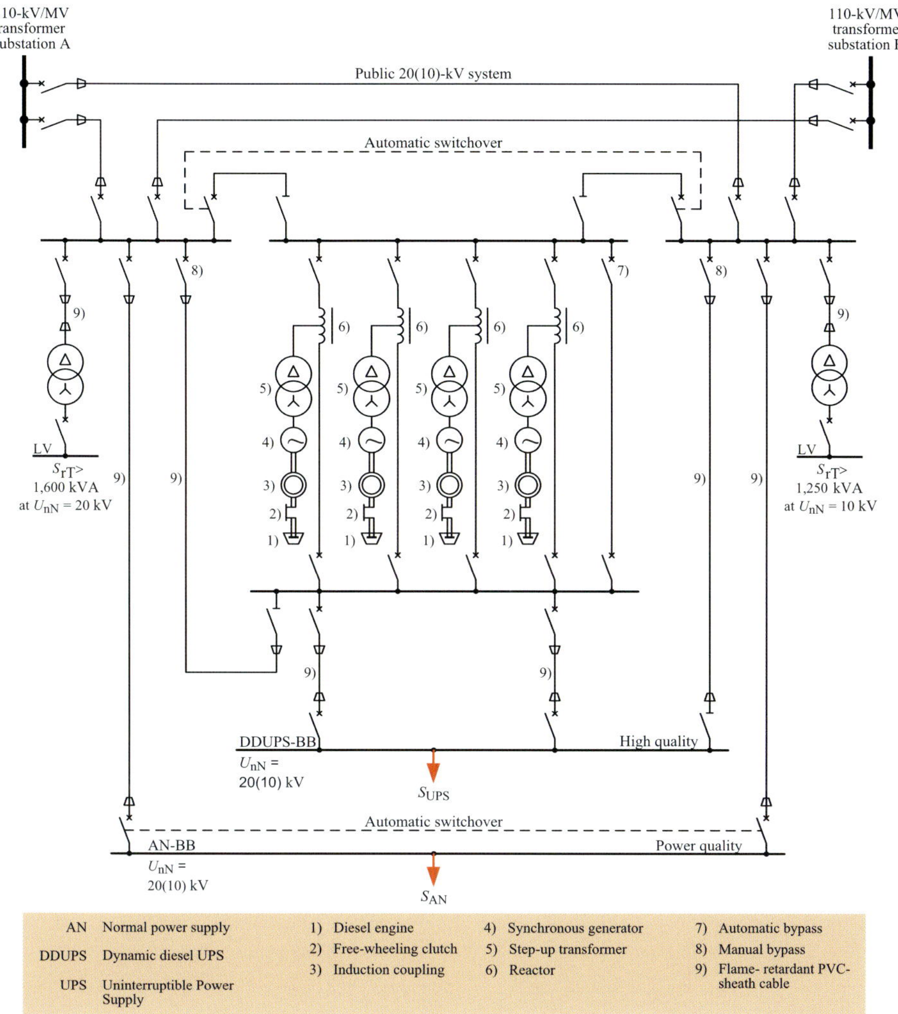

Fig. B5.15 Power supply concept for a high-technology business with integration of dynamic diesel UPS systems into the MV distribution power system (main incoming supply from the MV utility power system)

5.5 Switchgear classification for implementing the MV power system concepts

To implement the power system concepts explained in Sections 5.4.1 to 5.4.4, MV switchgear according to DIN VDE 0671-1 (VDE 0671-1): 2009-08 [5.31] or IEC 62271-1: 2007-10 [5.32] is used for primary or secondary power distribution according to its task and importance in the power system.

MV switchgear for primary power distribution mainly functions as a node between the upstream 110-kV power system and the supplying MV power system. MV switchgear for secondary power distribution, on the other hand, constitutes nodes that are used exclusively to supply LV systems from the upstream MV power system.

The main switchgear for establishing power systems for the supply of large (or medium-sized) industrial plants is typical switchgear for primary power distribution. The transfer substations of small industrial plants and the distribution substations of large industrial plants are preferably equipped with switchgear for secondary power distribution. The gas-insulated and air-insulated MV switchgear offered by Siemens for primary and secondary power distribution [5.6 to 5.15] is listed with its maximum ratings in Tables B5.16 to B5.18.

Cost-optimized switchgear for implementation of MV-side power system concepts must be dimensioned such that, taking into account the load current condition (Eq. 2.4) and short-circuit current conditions (Eqs. 4.3 to 4.5) to be complied with, the smallest possible standardized switchgear rating step (for standardized rating steps for short-circuit-proof dimensioning of MV switchgear, see Table B4.3) is always selected.

In industrial power systems with large load densities, it may not be possible to meet the short-circuit conditions (Eqs. 4.3 to 4.5) with the short-circuit current ratings of MV switchgear for secondary power distribution (e.g. SIMOSEC or 8DH10) owing to the short cable connections between the substations. In such cases, the distribution substations of the industrial power systems must also be equipped with MV switchgear for primary power distribution (e.g. NXAIR, NXAIR M or NXPLUS C).

Table B5.16 Air-insulated Siemens switchgear for primary power distribution

Ratings		Air-insulated switchgear type								
		NXAIR			NXAIR M			NXAIR P		
Rated operational voltage U_m [kV]		7.2	12	--	--	--	24	7.2	12	--
Rated short-circuit breaking current I_{sc} [kA]		≤ 31.5	≤ 31.5	--	--	--	≤ 25	≤ 50	≤ 50	--
Rated short-time withstand current I_{thr} [kA]		≤ 31.5	≤ 31.5	--	--	--	≤ 25	≤ 50	≤ 50	--
Rated short time t_{thr} [sec]		3	3	--	--	--	3	3	3	--
Rated short-circuit making current I_{ma} [kA]		≤ 80	≤ 80	--	--	--	≤ 63	≤ 125	≤ 125	--
Rated peak withstand current I_{pk} [kA]		≤ 80	≤ 80	--	--	--	≤ 63	≤ 125	≤ 125	--
Rated normal current I_r [A]	Busbar	≤ 2,500	≤ 2,500	--	--	--	≤ 2,500	≤ 4,000	≤ 4,000	--
	Feeders	≤ 2,500	≤ 2,500	--	--	--	≤ 2,500	≤ 4,000	≤ 4,000	--
Single-busbar application		X			X			X		
Double-busbar application		--			X			X		

Table B5.17 Gas-insulated Siemens switchgear for primary power distribution

Ratings		Gas-insulated switchgear type								
		NXPLUS C			8DA			8DB		
Rated operational voltage U_m [kV]		7.2	12	24	--	12	24	--	12	24
Rated short-circuit breaking current I_{sc} [kA]		≤ 31.5	≤ 31.5	≤ 25	--	≤ 40	≤ 40	--	≤ 40	≤ 40
Rated short-time withstand current I_{thr} [kA]		≤ 31.5	≤ 31.5	≤ 25	--	≤ 40	≤ 40	--	≤ 40	≤ 40
Rated short time t_{thr} [sec]		3	3	3	--	3	3	--	3	3
Rated short-circuit making current I_{ma} [kA]		≤ 80	≤ 80	≤ 63	--	≤ 100	≤ 100	--	≤ 100	≤ 100
Rated peak withstand current I_{pk} [kA]		≤ 80	≤ 80	≤ 63	--	≤ 100	≤ 100	--	≤ 100	≤ 100
Rated normal current I_r [A]	Busbar	≤ 2,500	≤ 2,500	≤ 2,500	--	≤ 5,000	≤ 5,000	--	≤ 5,000	≤ 5,000
	Feeders	≤ 2,500	≤ 2,500	≤ 2,000	--	≤ 2,500	≤ 2,500	--	≤ 2,500	≤ 2.500
Single-busbar application		X			X			--		
Double-busbar application		X			--			X		

Table B5.18 Air-insulated and gas-insulated Siemens switchgear for secondary power distribution

Ratings		Switchgear type										
		Air-insulated						Gas-insulated				
		SIMOSEC						8DH10				
Rated operational voltage U_m [kV]		7.2		12		24		7.2		12		24
Rated short-time withstand current I_{thr} [A]	t_{thr} = 1 sec	≤ 20	≤ 25	≤ 20	≤ 25	16	≤ 20	≤ 20	≤ 25	≤ 20	≤ 25	≤ 20
	t_{thr} = 3 sec	≤ 20	--	≤ 20	--	--	≤ 20	--	--	≤ 20	--	≤ 20
Rated short-circuit breaking current I_{sc} [kA]		≤ 20	≤ 25	≤ 20	≤ 25	16	≤ 20	≤ 20	≤ 25	≤ 20	≤ 25	≤ 20
Rated short-circuit making current I_{ma} [kA]		≤ 50	≤ 63	≤ 50	≤ 63	40	≤ 50	≤ 50	≤ 63	≤ 50	≤ 63	≤ 50
Rated peak withstand current I_{pk} [kA]		≤ 50	≤ 63	≤ 50	≤ 63	40	≤ 50	≤ 50	≤ 63	≤ 50	≤ 63	≤ 50
Rated normal current I_r [A]	Busbar	630 [1] 1,250 [2]		630 [1] 1,250 [2]		630 [1] 1,250 [2]		630 [1] 1,250 [2]		630 [1] 1,250 [2]		630 [1] 1,250 [2]
	Feeders	≤ 1,250		≤ 1,250		≤ 1,250		≤ 630		≤ 630		≤ 630
Single-busbar application		X						X				
Double-busbar application		--						--				
1) Standard	2) Option											

6 Choosing the neutral earthing

6.1 Importance of neutral earthing

The most common fault in all distribution systems is a line-to-earth fault. Fault statistics collected over many years show that 70 % to 90 % of all power system disturbances start as a line-to-earth fault, that is, as a puncture in the insulation of a conductor to earth [6.1 to 6.5]. On occurrence of a line-to-earth fault, three processes occur simultaneously within a very short time (μs < t < ms) [6.6]:

- *The line-to-earth voltage* of the line affected by the earth fault collapses. In the case of a dead short circuit to earth, this voltage drops to zero.
- *The instantaneous values of the line-to-earth voltages* of the two lines without earth faults jump to the instantaneous value of the assigned line-to-line voltage (delta voltage).
- *The instantaneous value of the displacement voltage* suddenly increases from the normal level (less than 10 % of the nominal voltage) to a value that, in the case of a dead short circuit to earth, corresponds to the instantaneous value of the line-to-neutral voltage (star voltage).

The sudden change in the voltages in the three-phase three-wire system is accompanied by short high-frequency oscilllation (transient condition, igniting oscillation) that adds up to the fundamental of the displacement voltage. This short transient condition can produce high voltage surges if damping is not provided. The effects of an earth fault on system operation are considerably influenced by whether and how the neutral point of the system supply is connected to earth. Practically, the fault currents occurring on faults to earth and voltage stress of the equipment depend only on the method of neutral earthing. Because of the high percentage that earth faults represent in the total number of faults, the neutral earthing also has a major influence on the supply reliability in the event of disturbances in the distribution system. For reliable operation of MV industrial power systems, choosing the most advantageous neutral earthing is therefore very important.

6.2 Methods of neutral earthing

The mode of operation of an MV industrial power system that is galvanically isolated from the public power supply is determined by the method of neutral earthing on the secondary side of the transfer transformer. According to DIN VDE 0101 (VDE 0101): 2000-01 [6.7], the following methods of neutral earthing can be used to earth secondary-side neutral points of transfer transformers:

- isolated neutral (OSPE),
- earth-fault compensation or resonant neutral earthing (RESPE),
- low-impedance neutral earthing (NOSPE).

Power systems with low-impedance neutral earthing also include those with isolated neutral or with resonant neutral earthing whose neutral point is earthed temporarily on each occurrence of an earth fault. Solid neutral earthing, which is not mentioned, is not particulary important in MV industrial power systems due to of the high line-to-earth short-circuit currents ($0.45 \cdot I''_{k3} \leq I''_{k1} < 1.5 \cdot I''_{k3}$) and the resulting interference (EMC), earthing (permissible touch voltage) and dimensioning problems (required rated short-time withstand current of the cable screen). The main aspects and features of the methods of neutral earthing are summarized in Table B6.1.

Table B6.1 Methods and features of neutral earthing in MV distribution systems

Features	Method of neutral earthing		
	Isolated neutral (OSPE)	Resonant neutral earthing (RESPE)	Low-impedance neutral earthing (NOSPE)
Target	Self-extinguishing of the earthing fault arc and time-limited continued operation on failed arc extinction		Selective clearance of the line-to-earth fault
Dimensioning	—	Arc-suppression coil $X_{ASC} = \frac{U_{nN}}{\sqrt{3}\cdot I_{rASC}} \approx \frac{1}{3\cdot\omega\cdot C_E}$	Neutral earthing resistance $R_E \approx \frac{U_{nN}}{\sqrt{3}\cdot I''_{k1}} << \frac{1}{3\cdot\omega\cdot C_E}$
Ratio of the zero-sequence to the positive-sequence impedance Z_0/Z_1	$\frac{Z_0}{Z_1} = \left\lvert\frac{1/j\cdot\omega\cdot C_E}{Z_1}\right\rvert$	$\lim\limits_{Z_0\to\infty} \frac{Z_0}{Z_1} = \infty$	$\frac{Z_0}{Z_1} = \begin{cases} 30\cdots150 \text{ for } 10\text{ kV and } 350\text{ MVA} \\ 20\cdots100 \text{ for } 20\text{ kV and } 500\text{ MVA} \end{cases}$
Current at the fault location	$I_{CE} \approx \omega\cdot\sqrt{3}\cdot C_E\cdot U_{nN}$	$I_{resi} \approx \omega\cdot C_E\cdot\lvert d+jv\rvert\cdot\sqrt{3}\cdot U_{nN}$	$I''_{k1} = \frac{c\cdot\sqrt{3}\cdot U_{nN}}{\lvert 2\cdot \underline{Z}_1 + \underline{Z}_0\rvert}$
	$10\text{ A} < I_{CE} \leq 30\text{ A}$ [1)]	$I_{resi} \leq 60\text{ A}$ [1)]	$I''_{k1} \leq 2\text{ kA}$ [2)]
Selective earth-fault detection	sin φ measurement	cos φ measurement	Line-to-earth short-circuit protection
Current connection of the SIPROTEC relay for selective earth-fault detection	o Holmgreen circuit if $I_{CE\text{-secondary}} > 0.05\cdot I_{N2}$ o Core balance CT if $I_{CE\text{-secondary}} < 0.05\cdot I_{N2}$	Core balance current transformer	o Holmgreen circuit if $I''_{k1} > 0.1\cdot I_{N1}$ o Core balance CT if $I''_{k1} < 0.1\cdot I_{N1}$
Voltage connection of the SIPROTEC relay for selective earth-fault detection	Three line-to-earth connections or line-to-earth connection with open-delta winding	Line-to-earth connection with open-delta winding	Line-to-earth connection for directional earth-fault protection only
Transient overvoltage at earth-fault entry	$3\cdots3.5\cdot U_{LN}$	$2.5\cdot U_{LN}$	$1.4\cdots1.8\cdot U_{LN}$
Voltage rise in the faultless lines	$\sqrt{3}\cdot U_{LN}$	$\sqrt{3}\cdot U_{LN}$	for a short period $\leq\sqrt{3}\cdot U_{LN}$
Risk of fault propagation (double earth fault, short-circuit)	exists		almost ruled out
Influence	insignificant		low if $I''_{k1} \leq 2\text{ kA}$
Fault duration	$t \leq 3$ h in the case of a non-self-extinguishing earth fault arc		$t \leq 3$ sec
System extent	very limited	limited by extinction limit $I_{resi} = 60$ A	not limited

- C_E Line-to-earth capacitance
- I_{CE} Capacitive earth-fault current
- I''_{k1} Initial line-to-earth short-circuit current
- I_{N1} Nominal primary transformer current
- I_{N2} Nominal secondary transformer current
- I_{rASC} Nominal current of the arc-suppression coil (Petersen coil)
- I_{resi} Residual earth-fault current
- R_E Resistance of the neutral earthing resistor (NOSPE resistance)
- U_{nN} Nominal system voltage
- U_{LN} Line-to-neutral voltage
- X_{ASC} Reactance of the arc-suppression coil (Petersen coil)
- Z_0 Impedance in the zero-sequence system
- Z_1 Impedance in the positive-sequence system
- c Voltage factor
- d Damping factor
- v Detuning
- ω Angular frequency ($\omega = 2\cdot\pi\cdot f$)
- 1) see Fig. B6.25
- 2) see principle i) of low-impedance neutral earthing (NOSPE)

The following explanations of the methods of neutral earthing complement the summary in the table.

Power system with isolated neutral [6.2, 6.5, 6.7 to 6.18]

A power sytem with isolated neutral (Fig. B6.2) is defined as a power system in which the neutral points of transformers and generators are either not connected to earth or only connected to earth through measurement and protective devices with a very high impedance or through an overvoltage protector. Operation with isolated neutral is the simplest form of earth-fault-oriented neutral-point connection. In the event of an earth fault, voltage displacement to earth occurs. This voltage displacement which is shown in Fig. B6.3 as a vector diagram, is characterized by the fact that the neutral point of the system (transformer neutral point) accepts the full star voltage to earth and the lines without faults increase their voltages to earth from the star voltage to the delta voltage. The line-to-line voltages of the earth-faulted power system, on the other hand, do not change. Because of this, no reaction arises that is disadvantageous for the loads connected to the power system and operation can be maintained even in the event of an earth fault.

However, the voltage displacement to earth causes the capacitive earth-fault current I_{CE} of the entire power system to flow through the earth fault location. This earth-fault current, which is largely determined by the earth capacitance of the lines C_E, has the magnitude $I_{CE} \approx 3 \cdot \omega \cdot C_E \cdot U_{LE}$.

For reliable operation of industrial cable networks with isolated neutral, a fault current range of $10\ A < I_{CE} \leq 30\ A$ is recommended. In this range, it can be expected that both the risk of intermittent earth faults with high transient overvoltages and the thermal effect of the earth-fault arc are relatively minor. Due to the comparatively low thermal stress at $I_{CE} < 30\ A$, the earth-fault arc can burn for longer without destroying the insulation of the intact conductors. Destruction of this insulation would entail a serious danger of the earth fault developing into a double earth fault or short circuit. Danger of fault propagation and the risk of a double earth fault can only be prevented with

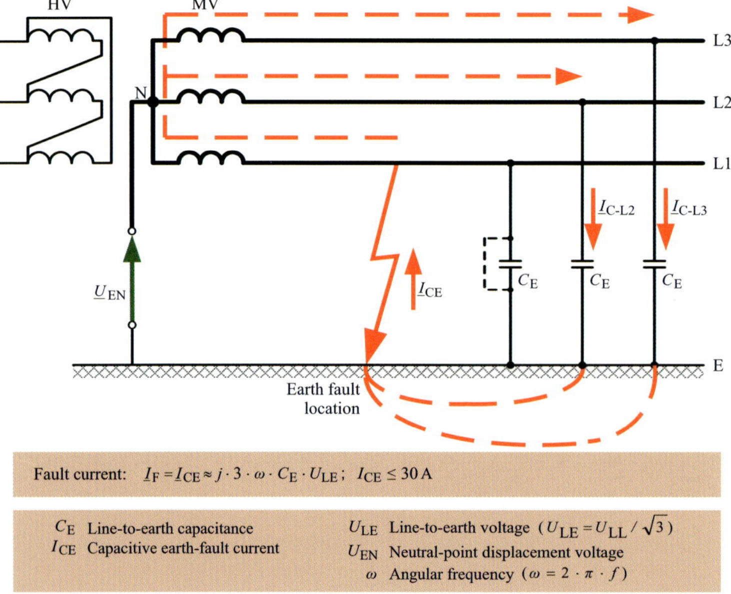

Fig. B6.2 MV system with isolated neutral during an earth fault of line L1

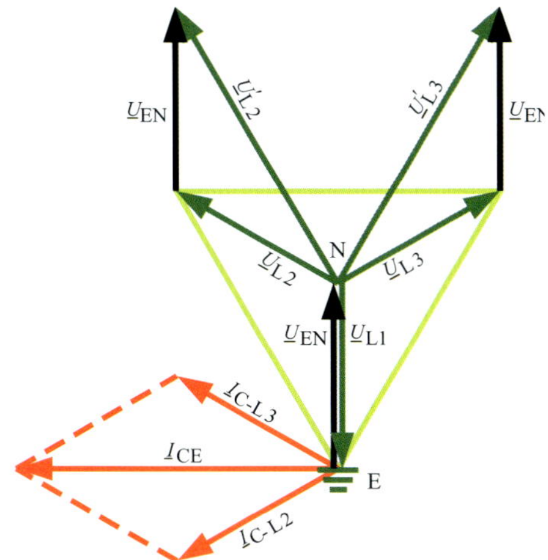

Fig. B6.3
Vector diagram of the voltages and currents during an earth fault of line L1 in a system with isolated neutral

$\underline{U}_{L1}$, $\underline{U}_{L2}$, $\underline{U}_{L3}$	Star voltage of lines L1, L2, L3
$\underline{U}'_{L2}$, $\underline{U}'_{L3}$	Star voltage on the faultless phases L2 and L3 raised by the factor $\sqrt{3}$ (power-frequency overvoltage)
$\underline{U}_{EN}$ or $\underline{U}_{en}$	Neutral-point displacement voltage
$\underline{I}_{C\text{-}L2}$ or $\underline{I}_{C\text{-}L3}$	Capacitive charging current of line L2 or L3
$\underline{I}_{CE}$	Capacitive earth-fault current at the fault location

small earth-fault currents if the earth fault is detected selectively and operation is continued only for a limited time. Generally, a time limitation of 3 h for continued operation should be sufficient to create the necessary conditions for earth-fault clearance without any adverse effect on the production process.

To detect the earth-fault location, SIPROTEC relays with sensitive earth-fault detection (Chapter 7) can be used. They measure the capacitive residual currents. The residual currents of the faultless and earth-faulted feeder differ in their magnitude and direc-

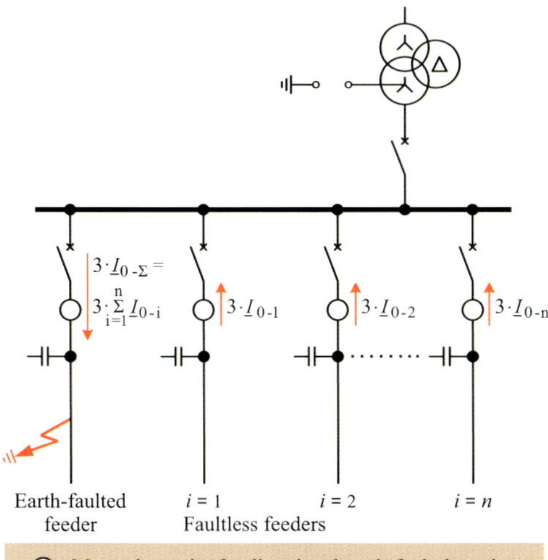

Fig. B6.4
Distribution of the residual currents on an earth fault in a system with isolated neutral

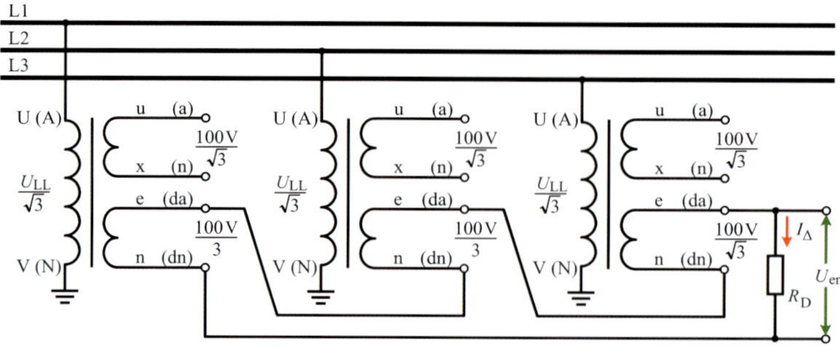

Fig. B6.5 Connection of single-pole-earthed inductive voltage transformers to measure the neutral-point displacement voltage U_{en}

Table B6.6 Standard values for damping resistors R_D

e-n or da-dn winding of the voltage transformer		Damping resistor R_D
Rated thermal limiting output $S_{r\,sec}$	Rated long-duration current I_Δ	
75 W	4 A	25 Ω / 500 W
100 W	6 A	25 Ω / 500 W
150 W	8 A	12.5 Ω / 1,000 W

tion (Fig. B6.4). In faultless feeders, all residual currents flow in the same direction. Their magnitude depends on the magnitude of the relevant capacitive charging current. The residual current of the earth-faulted feeder is the sum of the charging currents of all faultless feeders flowing in the opposite direction. By means of the measurement with reference to the neutral-point displacement voltage U_{en} of this capacitive residual current (sin φ measurement), the SIPROTEC relay detects the feeder with the earth fault.

The neutral-point displacement voltage U_{en} required for earth-fault direction detection is obtained by connecting single-pole-earthed inductive voltage transformers (Fig. B6.5). If this is used in a system with isolated neutral, there is a risk of relaxation oscillations (ferroresonance).

Relaxation oscillations are caused by interaction of the non-linear no-load inductance of the voltage transformers connected to earth with the earth capacitance of the network. The relaxation oscillations, which mainly occur when an earth fault arc is extinguished or on an energizing operation, put the iron core in the saturated state and cause high core losses. As a consequence of these high core losses, the voltage transformer can be thermally overloaded and finally destroyed.

The simplest and safest way of avoiding relaxation oscillations is to insert an ohmic damping resistor R_D into the earth-fault windings of the three voltage transformer units interconnected in an open delta (Fig. B6.5). The damping resistor R_D is rated so that neither it nor the voltage transformer is thermally overloaded. Table B6.6 contains standard values for damping resistors that have proven convenient in practice.

The choice of R_D is based on the thermal limiting output (rated long-duration current) of the e-n or da-dn winding of the voltage transformer. If these standard values cannot be used, other values can be calculated. The calculation must be performed as follows:

$$R_{\text{D-req}} = \sqrt{3} \cdot \frac{U_{\text{r}_{\text{sec}}}^2}{S_{\text{r}_{\text{sec}}}} \tag{6.1}$$

$$R_{\text{D-select}} \geq R_{\text{D-req}} \tag{6.1.1}$$

$$P_{\text{V-req}} = \frac{(3.3 \cdot U_{\text{r}_{\text{sec}}})^2}{R_{\text{D-select}}} \tag{6.2}$$

$$P_{\text{V-select}} \geq P_{\text{V-req}} \tag{6.2.1}$$

$R_{\text{D-req}}$ required damping resistor (minimum value)
$R_{\text{D-select}}$ selected damping resistor
$P_{\text{V-req}}$ required thermal load capacity (minimum value)
$P_{\text{V-select}}$ selected thermal load capacity
$U_{\text{r}_{\text{sec}}}$ secondary rated voltage of the earth-fault winding
$S_{\text{r}_{\text{sec}}}$ thermal rated limiting output

Example B4

Example of calculation of the damping resistor dimensioning according to Eqs. (6.1) and (6.2): see Table B6.7.

Table B6.7 Calculation of the damping resistor dimensioning (Example B4)

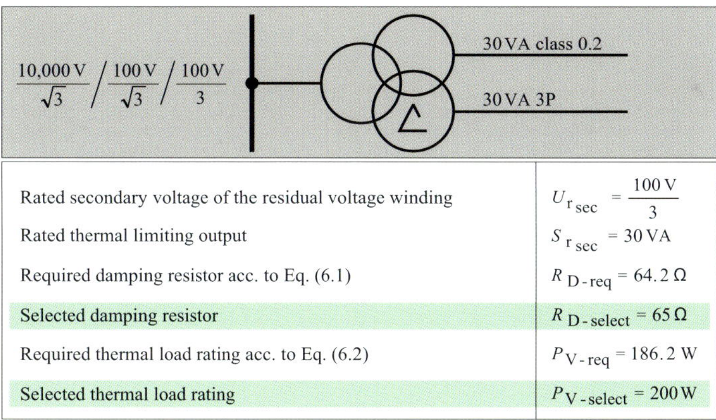

Rated secondary voltage of the residual voltage winding	$U_{\text{r sec}} = \frac{100\text{ V}}{3}$
Rated thermal limiting output	$S_{\text{r sec}} = 30\text{ VA}$
Required damping resistor acc. to Eq. (6.1)	$R_{\text{D-req}} = 64.2\ \Omega$
Selected damping resistor	$R_{\text{D-select}} = 65\ \Omega$
Required thermal load rating acc. to Eq. (6.2)	$P_{\text{V-req}} = 186.2\text{ W}$
Selected thermal load rating	$P_{\text{V-select}} = 200\text{ W}$

A further negative phenomenon in isolated-neutral systems are overvoltage levels that result in an excessive voltage stress on the equipment insulation. On occurrence of an earth fault, the capacitive charge reversal of the faultless lines is brought about by a transient. This transient occurs as a medium-frequency oscillation that briefly causes an overvoltage. The transient overvoltage on occurrence of an earth fault can be 3 to 3.5 times the star voltage. This transient overvoltage is brought under control by the insulation level that is standardized and assigned to the specific rated voltage of the item of equipment (Table B6.8).

Table B6.8 Standardized insulation levels in the range 1 kV < U_m ≤ 36 kV according to DIN EN 60071-1 (VDE 0111-1): 2006-11 [6.17] or IEC 60071-1: 2006-01 [6.18]

Nominal system voltage U_{nN} [kV] RMS value	Highest voltage for equipment U_m [kV] RMS value	Rated short-duration power-frequency withstand voltage U_{rd} 1) [kV] RMS value	Rated lightning impulse voltage $\hat{U}_{rp}$ 2) [kV] Peak value	Rated switching impulse withstand voltage $\hat{U}_{r\,SIL}$ 2) [kV] Peak value
6	7.2	20	40 60	32 48
10	12	28	60 75 95	48 60 76
20	24	50	95 125 145	76 100 116
30	36	70	145 170	116 136

1) Alternating voltage with a frequency between 48 Hz and 62 Hz and a duration of 60 sec

2) Voltage pulse with a rise time of 1.2 µsec and a time to half-value of 50 µsec

3) For the MV range 1 kV < U_m ≤ 36 kV, no rated switching impulse level is defined in a standard. It is considered in the insulation coordination with a absolute value of $0.8 \cdot \hat{U}_{rp}$. Because of the longer time to half-value of the transient switching overvoltage, $\hat{U}_{r\,SIL}$ is less than $\hat{U}_{rp}$.

Handling of long-lasting power-frequency overvoltages by the equipment rated voltage of the cable network (e.g. U_m = 24 kV at U_{nN} = 20 kV) is ensured if the individual earth fault does not exist for longer than 8 h and the sum of all earth-fault times over a year does not exceed about 125 h [6.19, 6.20]. Compliance with these time limits must be ensured in the operation of cable networks with isolated neutral.

Despite its negative side-effects (relaxation oscillations, high transient and long-lasting power-frequency overvoltages), operation with isolated neutral is very important for industrial power supplies. This method of neutral earthing is preferred in small cable networks and those without $(n-1)$ redundancy.

Power system with resonant neutral earthing [6.2 to 6.5, 6.7, 6.21 to 6.46]

A system with resonant neutral earthing or earth-fault compensation (Fig. B6.9) is a power system in which the neutral point of one or more transformers is earthed through an arc-suppression coil (Petersen coil) with inductance L_{ASC}. The resulting inductance of the arc-suppression coil(s) is essentially coordinated with the line-to-earth capacitances of the power system. Coordination is undertaken in such a way that the power-frequency inductive alternating current $\underline{I}_{ASC}$ that flows through the arc-suppression coils during a line-to-earth fault largely compensates for the power-frequency capacitive component $\underline{I}_{CE}$ of the earth-fault current $\underline{I}_F$.

The residual earth-fault current $\underline{I}_{resi}$ that remains after compensation of $\underline{I}_{CE}$ reduces the thermal load at the earth-fault location.

Its magnitude $|\underline{I}_{resi}| = I_{resi}$ can be calculated as follows:

$$I_{resi} = \sqrt{(I_{CE} - I_{ASC})^2 + I_R^2 + \sum_h I_h^2} \tag{6.3}$$

$$I_R = I_{R\text{-}ASC} + I_{R\text{-}CE} \tag{6.3.1}$$

I_{resi} residual earth-fault current
I_{CE} capacitive earth-fault current of the system
I_{ASC} inductive arc-suppression coil current
I_R residual active current (I_R = (0.02...0.05) · I_{CE} in MV cable networks)
I_h harmonic currents, caused by magnetization currents of the transformers
$I_{R\text{-}ASC}$ residual active current due to ohmic losses of the arc-suppression coil
$I_{R\text{-}CE}$ residual active current due to leakage and dielectric losses of the lines

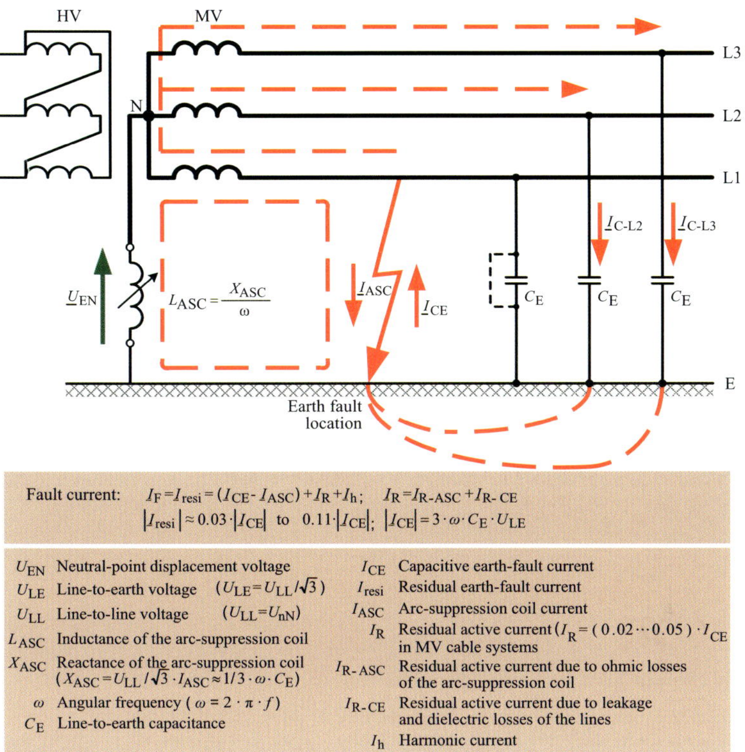

Fig. B6.9 MV power system with earth-fault compensation or resonant neutral earthing (RESPE)

In a system ideally tuned to resonance (detuning factor $v = 1 - I_{ASC}/I_{CE} = 0$), only the power-frequency residual active current $\underline{I}_R$ (50-Hz currrent) and a higher-frequency harmonic residual current $\underline{I}_h$ (e.g. 250-Hz current) flow through the fault location. The tuning to resonance has no direct influence on the voltage conditions in case of an earth fault. Practically, the same voltage conditions to earth occur in a power system with resonant neutral earthing as in a power system with isolated neutral. Fig. B6.10 shows the vector diagram of the power-frequency voltages and currents during an earth fault of line L1 in a resonant neutral earthed system.

The necessary resonance tuning in power systems with resonant neutral earthing is best achieved with the use of plunger-core arc-suppression coils. Standardized nominal powers Q_{rASC} with the corresponding reactive current ranges $I_{min\text{-}ASC} \leq I_{ASC} \leq I_{rASC}$ are listed in Table B6.11. Plunger-core arc-suppression coils are used in conjunction with a

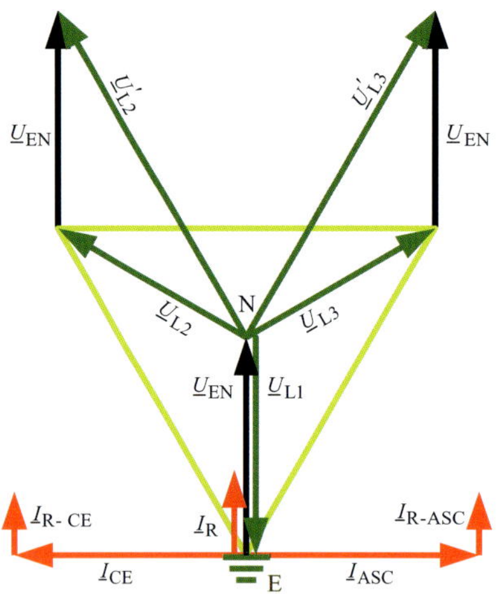

Fig. B6.10
Vector diagram of the voltages and currents during an earth fault of line conductor L1 in a system with earth-fault compensation or resonant neutral earthing (RESPE)

$\underline{U}_{L1}$, $\underline{U}_{L2}$, $\underline{U}_{L3}$	star voltage of lines L1, L2, L3
$\underline{U}'_{L2}$, $\underline{U}'_{L3}$	star voltage on the faultless phases L2 and L3 raised by the factor $\sqrt{3}$ (power-frequency overvoltage)
$\underline{U}_{EN}$ or $\underline{U}_{en}$	neutral-point displacement voltage
$\underline{I}_{CE}$	capacitive earth-fault current of the system
$\underline{I}_{ASC}$	inductive arc-suppression coil current
$\underline{I}_{R}$	resultant residual active current
$\underline{I}_{R\text{-}CE}$	active component of $\underline{I}_{CE}$
$\underline{I}_{R\text{-}ASC}$	active component of $\underline{I}_{ASC}$

resonant neutral earthing controller that automatically tunes the inductive arc-suppression coil current I_{ASC} for the prevailing system conditions. The maximum displacement voltage (voltage between the neutral point and earth) functions as the closed-loop control setpoint during fault-free operation. The displacement voltage during fault-free operation is caused by the unequal line-to-earth capacitance of the individual phases.

With precise resonance tuning ($I_{ASC} = I_{CE}$), this voltage reaches its maximum value. In resonant-earthed cable networks, the maximum displacement voltage is only small. Due to the great symmetry of the line-to-earth capacitances, cable networks have only relatively small displacement voltages and flat resonance curves.

If resonance curves are too flat, controllability of the plunger-core arc-suppression coils used for the resonant neutral earthing must be questioned. The only remedy may be to increase the displacement voltage using artificially produced capacitive system unbalance. One good way of adjusting the setpoint of the resonant neutral earthing controller based on measurements is an artificial increase in the capacitive system unbalance by 3-pole disconnection and 1-pole reclosing of selected cable routes.

If resonance tuning is correct, the heat generated at the fault location will be very slight. It is largely determined by the residual active current $\underline{I}_R$ flowing through the fault location. Because the residual active current in cable networks is relatively small, the time that elapses while an earth fault becomes a short circuit can usually be considerably prolonged. The actual advantage of resonant neutral earthing in the operation of cable networks is derived from this extra time. For example, if $(n-1)$ redundancy is limited, it proves advantageous that disconnection of the earth-faulted cable can be deferred until a time that is more favourable for operation even with large capacitive earth-fault currents ($\underline{I}_{CE} > 30\text{A}$).

Table B6.11 Standardized power ratings and inductive reactive current ranges of plunger-core arc-suppression coils

Rated power Q_{rASC} in kvar	Plunger-core arc-suppression coil (ASC) for resonant neutral earthing					
	U_{nN}=10 kV(U_{LE}=5.774 kV)		U_{nN}=20 kV(U_{LE}=11.547 kV)		U_{nN}=30 kV(U_{LE}=17.320 kV)	
	Rated current I_{rASC} in A	Range of [4] adjustment 10 ··· 100 %	Rated current I_{rASC} in A	Range of [4] adjustment 10 ··· 100 %	Rated current I_{rASC} in A	Range of [4] adjustment 10 ··· 100 %
200	35	3[1] ··· 35	--	--	--	--
250	43	4 ··· 43	--	--	--	--
315	54	5 ··· 54	--	--	--	--
400	69	7 ··· 69	--	--	--	--
630	109	11 ··· 109	54	5 [2]··· 54	--	--
800	139	14 ··· 139	69	7 ··· 69	--	--
1,250	216	22 ··· 216	108	11 ··· 108	--	--
1,600	277	28 ··· 277	139	14 ··· 139	--	--
2,000	346	35 ··· 346	173	17 ··· 173	115	11[3]··· 115
2,500	433	43 ··· 433	216	22 ··· 216	144	14 ··· 144
3,150	545	54 ··· 545	273	27 ··· 273	182	18 ··· 182
4,000	693	69 ··· 693	346	34 ··· 346	231	23 ··· 231
6,300	1,091	109 ··· 1,091	545	54 ··· 545	364	36 ··· 364
8,000	--	--	693	69 ··· 693	462	46 ··· 462
10,000	--	--	866	86 ··· 866	577	57 ··· 577
12,500	--	--	1,082	108 ··· 1,082	722	72 ··· 722
16,000	--	--	--	--	932	93 ··· 932

1) Minimum coil current I_{ASC} at U_{nN} = 10 kV
2) Minimum coil current I_{ASC} at U_{nN} = 20 kV
3) Minimum coil current I_{ASC} at U_{nN} = 30 kV
4) A slightly overcompensated adjustment is preferred ($I_{CE} < I_{ASC} \leq 1.15 \cdot I_{CE}$)

In cable networks, the extensive compensation of the capacitive earth-fault current also has a significant disadvantage: In the case of small residual earth-fault currents $\underline{I}_{resi}$, the earth faults may suppress themselves. Because XLPE cables do not have a self-healing ability, locations with impaired insulation remain after suppression of the earth fault. These weak points are then not easy to find. Unlocated weak points can later result in chains of multiple earth faults and serious system disturbances in the event of sustained earth faults.

System disturbances due to unlocated weak points can only be prevented by reliable earth-fault detection. Such reliable earth-fault detection in resonant-earthed systems is made more difficult by the fact that the direction of the residual current resulting from the superimposition of capacitive charging currents with the inductive arc-suppression current does not bear any fixed relation to the fault location. For that reason, the relay in the faulted feeder can also measure a residual current that matches the direction of the capacitive residual current in the faultless feeders. In this way, the earth-fault direction can only be accurately determined by a measurement of the residual active current $\underline{I}_R$ (cos φ measurement) referred to the neutral-point displacement voltage $\underline{U}_{EN}$. Measurement of the active component of the residual earth-fault current $I_R = I_{resi} \cdot \cos\varphi$ requires a very high level of precision. Reliable directional earth-fault detection can therefore only be expected if the relay is connected to the core balance current transformer.

For use in power systems with resonant neutral earthing, SIPROTEC relays of type 7SJ62-7SJ64 are suitable, which must be connected to the current and voltage transformers according to Fig. B6.12. The open-delta winding of the voltage transformer connection shown in Fig. B6.12 does not have to be provided with a damping resistor. Unlike in isolated-neutral systems, in resonant-earthed systems no relaxation oscillations occur. Moreover, the transient overvoltages are smaller than in the isolated-neutral configuration. They reach values that can be up to 2.5 times the star voltage. Furthermore, the processes that occur at the fault location are much smoother. With such a smooth process, the slow rise in the recovering voltage considerably reduces the risk of arc-backs and intermittent earth faults.

However, in the operation of resonant-earthed MV cable networks with $(n-1)$ redundancy, the disadvantages explained previously outweigh the advantages stated above.

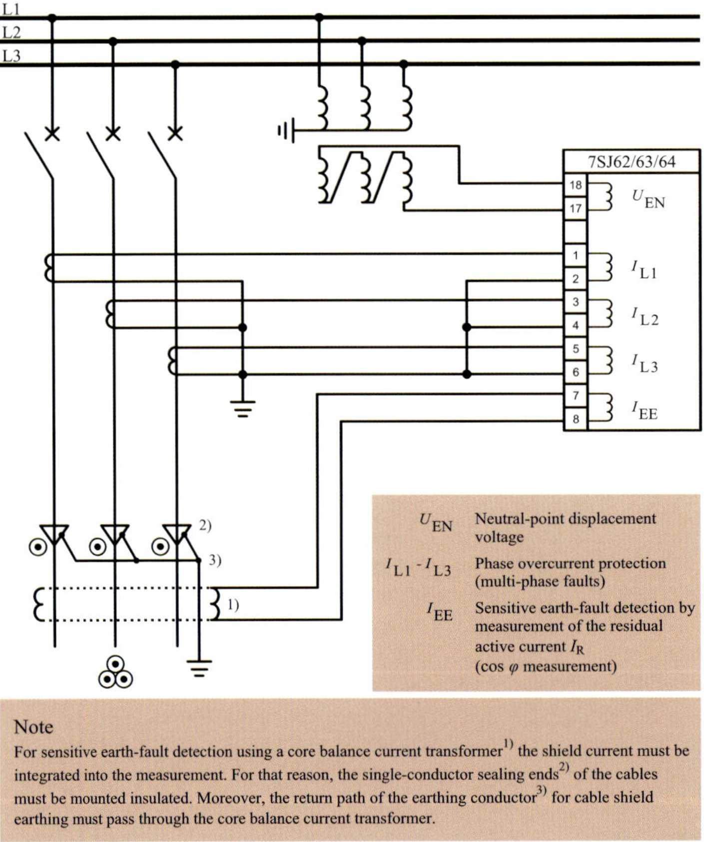

Fig. B6.12 SIPROTEC relay connection for directional earth-fault detection in MV cable networks with resonant neutral earthing

Power system with low-impedance neutral earthing [6.1 to 6.3, 6.5, 6.7, 6.47 to 6.59]

A power system with low-impedance neutral earthing (Fig. B6.13) is one in which the neutral point of one or more power transformers, neutral earthing transformers or generators is earthed through impedance $\underline{Z}_E$ that limits the line-to-earth short-circuit current. The purpose of earthing the transformer neutral point through an impedance $\underline{Z}_E$ is to clear any earth fault occurring in the system selectively and in the shortest possible time ($t_a < 3$ sec). To achieve this aim, the earthing impedance must be dimensioned in such a way that even a minimal line-to-earth short-circuit current in a cable feeder will reliably cause the power system protection equipment to trip. This minimum line-to-earth short-circuit current of a cable feeder can be calculated based on the equivalent circuit of the symmetrical component systems shown in Fig. B6.14.

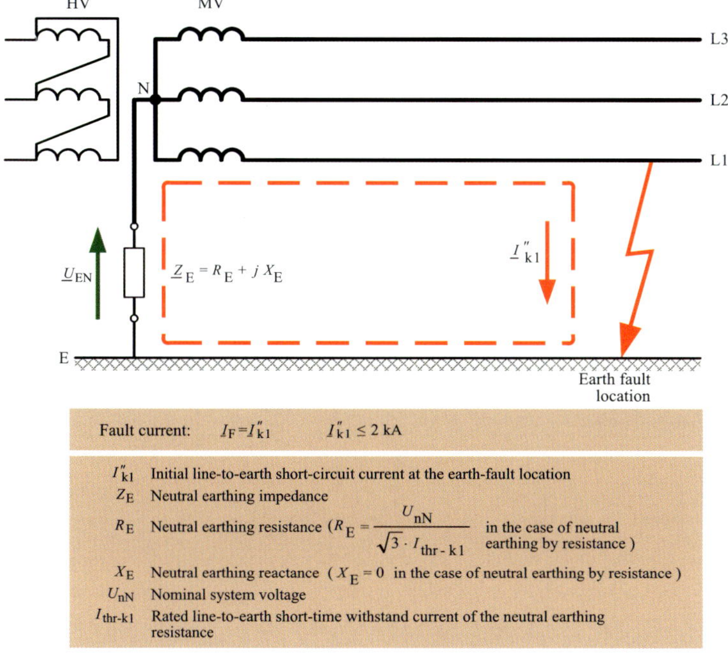

Fig. B6.13 MV power system with low-impedance neutral earthing (NOSPE)

For calculation of the line-to-earth short-circuit current $\underline{I}''_{k1}$ according to Fig. B6.14, the following applies:

$$\underline{I}''_{k1} = \frac{c \cdot \sqrt{3} \cdot \underline{U}_{nN}}{2 \cdot (\underline{Z}_{1N} + \underline{Z}_{1T} + \underline{Z}_{1C}) + \underline{Z}_{0T} + \underline{Z}_{0C} + 3 \cdot \underline{Z}_E} \tag{6.4}$$

In the case of resistance earthing ($\underline{Z}_E = R_E$), the modulus of the complex line-to-earth short-circuit current $\underline{I}_{k1}$ can be calculated as follows:

$$\left|\underline{I}''_{k1}\right| = \frac{c \cdot \sqrt{3} \cdot U_{nN}}{\sqrt{[2 \cdot (R_{1N} + R_{1T} + R_{1C}) + R_{0T} + R_{0C} + 3 \cdot R_E]^2 + [2 \cdot (X_{1N} + X_{1T} + X_{1C}) + X_{0T} + X_{0C}]^2}} \tag{6.5}$$

$$I''_{\text{k1-min}} = \frac{1}{f_D} \cdot \left| \underline{I}''_{\text{k1}} \right| \tag{6.5.1}$$

$I''_{\text{k1-min}}$	minimum initial line-to-earth symmetrical short-circuit current of the cable feeder
U_{nN}	nominal system voltage
f_D	reduction factor for damping by the earth contact resistance and arc resistance ($f_D = 1.25...1.5$)
c	voltage factor ($1.0 \le c \le 1.1$ at $U_{\text{nN}} > 1$ kV)
R_{1N}, X_{1N}	resistance or reactance of the upstream network in the positive-sequence system
R_{1T}, X_{1T}	resistance or reactance of the transfer transformer in the positive-sequence system
R_{1C}, X_{1C}	resistance or reactance of the cable route in the positive-sequence system
R_{0T}, X_{0T}	resistance or reactance of the transfer transformer in the zero-sequence system
R_{0C}, X_{0C}	resistance or reactance of the cable route in the zero-sequence system
R_E	neutral-point resistance

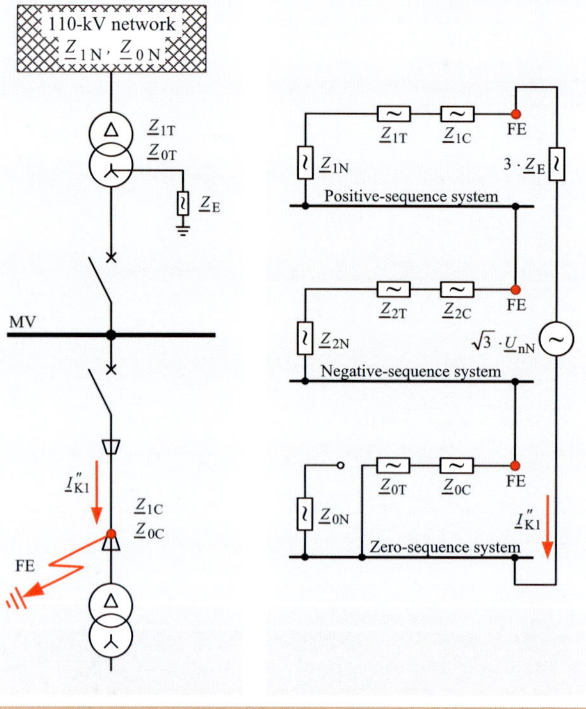

$\underline{Z}_{1N}$	Positive-sequence impedance of the upstream network ($\underline{Z}_{1N} = R_{1N} + j\,X_{1N}$)
$\underline{Z}_{0N}$	Zero-sequence impedance of the upstream network ($\underline{Z}_{0N} = R_{0N} + j\,X_{0N}$)
$\underline{Z}_{1T}$	Positive-sequence impedance of the transfer power transformer ($\underline{Z}_{1T} = R_{1T} + j\,X_{1T}$)
$\underline{Z}_{0T}$	Zero-sequence impedance of the transfer power transformer ($\underline{Z}_{0T} = R_{0T} + j\,X_{0T}$)
$\underline{Z}_{1C}$	Positive-sequence impedance of the cable ($\underline{Z}_{1C} = R_{1C} + j\,X_{1C}$)
$\underline{Z}_{0C}$	Zero-sequence impedance of the cable ($\underline{Z}_{0C} = R_{0C} + j\,X_{0C}$)
$\underline{Z}_{E}$	Neutral earthing impedance ($\underline{Z}_E = R_E + j\,X_E$, $X_E = 0$ in the case of neutral earthing by resistance)

Fig. B6.14 Connection of the symmetrical component systems in the event of an earth fault in the cable feeder of an MV power system with low-impedance neutral earthing

To operate MV power systems with low-impedance neutral earthing, a number of specific power system and installation engineering principles must be adhered to. The following principles must be heeded or complied with:

a) The $(n-1)$ principle is also relevant for more frequently occurring line-to-earth faults. When operating a low-impedance neutral-earthed MV industrial power system, it is important to pay attention to the redundancy of the electrical equipment intervening in the production process (e.g. electromotive drives).

b) The neutral earthing impedance must be implemented as a resistor to limit the line-to-earth short-circuit current. The evaluation criteria for safe and reliable low-impedance neutral earthing operation are better met with resistance than with reactance earthing (Table B6.15).

Table B6.15 Evaluation of resistance and reactance earthing

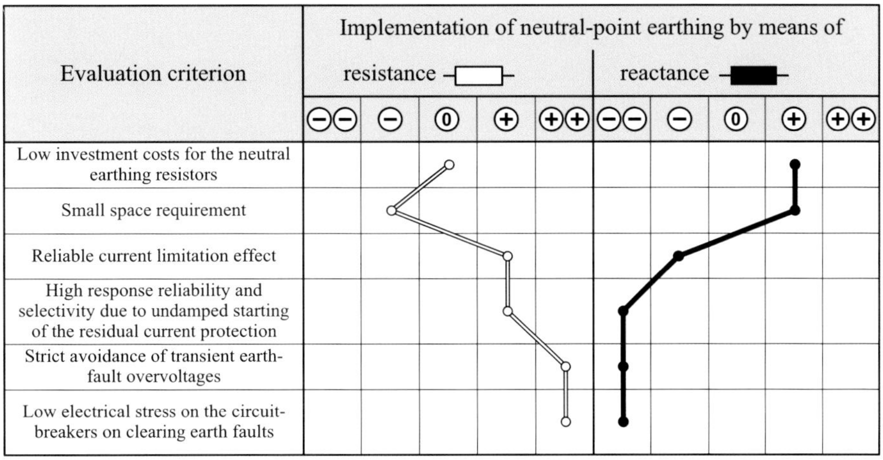

Evaluation criterion	Implementation of neutral-point earthing by means of resistance					reactance				
	−−	−	0	+	++	−−	−	0	+	++
Low investment costs for the neutral earthing resistors			○						●	
Small space requirement		○							●	
Reliable current limitation effect				○			●			
High response reliability and selectivity due to undamped starting of the residual current protection				○		●				
Strict avoidance of transient earth-fault overvoltages					○	●				
Low electrical stress on the circuit-breakers on clearing earth faults					○	●				

c) The current rating of the neutral earthing impedance is determined by contrary demands: to ensure high response reliability of the protection devices, the line-to-earth short-circuit currents should be as large as possible, but to avoid high touch voltages and impermissible interference with information equipment, the line-to-earth short-circuit currents should be as small as possible.

d) The largest line-to-earth short-circuit current $I''_{k1\text{-max}}$ that still meets all contrary demands must be used as the rated short-time withstand current of neutral earthing impedance $I_{thr\text{-}k1}$, i. e. $I_{thr\text{-}k1} \geq I''_{k1\text{-max}}$.

e) The neutral resistors must be rated for a load duration (rated short time) of $t_{thr} = 5...10$ sec. During this time, they must reliably withstand the thermal loads due to the line-to-earth short-circuit current that is flowing.

f) If there is a neutral point on the secondary side, the neutral earthing resistor R_E can be connected directly to the transfer transformer. The precondition for this is that the zero-sequence impedance of the supplying transformer is sufficiently low. With transformers having delta-connected primary winding or stabilizing winding, this precondition is fulfilled. In this case, the necessary earthing resistance R_E can be determined with sufficient precision from the star voltage of the power system and the largest line-to-earth short-circuit current. For simplified calculation, the following applies:

$$R_E \approx \frac{U_{LN}}{I''_{k1\text{-max}}} = \frac{U_{LL}}{\sqrt{3} \cdot I''_{k1\text{-max}}} \tag{6.6}$$

U_{LN} line-to-neutral voltage (star voltage)
U_{LL} line-to-line voltage (delta voltage)
$I''_{k1\text{-max}}$ largest line-to-earth short-circuit current that was defined for low-impedance neutral earthing operation

g) Transfer power transformers with a delta-connected secondary winding do not have a neutral point on the secondary-side. In this case, neutral or other earthing transformers must be used for low-impedance neutral earthing. Fig. B6.16 shows a system with low-impedance earthing through a ZN neutral earthing transformer and the corresponding distribution of the residual currents on an earth fault.

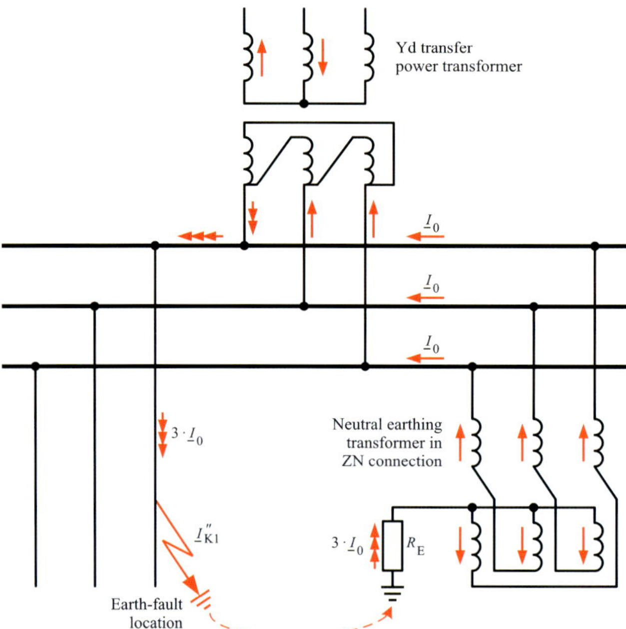

Fig. B6.16 System with low-impedance earthing through a ZN neutral earthing transformer and the current distribution on an earth fault

h) The current rating of the neutral earthing transformer and neutral earthing resistor has been defined according to the largest line-to-earth short-circuit current that was defined for operation of the low-impedance neutral-earthed system. For a defined zero-sequence impedance of the neutral earthing transformer (NTr), the necessary neutral earthing resistor can be calculated as follows:

$$R_E = \frac{1}{3} \cdot \left[\sqrt{\left(\frac{\sqrt{3} \cdot U_{nN}}{I''_{k1\text{-max}}} \right)^2 - X^2_{0\text{-NTr}}} - R_{0\text{-NTr}} \right] \tag{6.7}$$

R_E	neutral earthing resistor for limitation of the line-to-earth short-circuit current to its largest value defined for low-impedance neutral earthing operation
U_{nN}	nominal system voltage
$I''_{k1\text{-max}}$	largest line-to-earth short-circuit current that was defined for low-impedance neutral earthing operation
$X_{0\text{-NTr}}$	zero-sequence reactance of the neutral earthing transformer
$R_{0\text{-NTr}}$	zero-sequence resistance of the neutral earthing transformer

If the neutral earthing resistor is connected to a neutral earthing transformer with a zigzag reactor (ZN connection), sufficient damping of possible overvoltages must be ensured. Possible overvoltages are sufficiently damped if two conditions are fulfilled. These conditions are defined as follows:

$$\frac{R_0}{X_0} = \frac{R_E + R_{0\text{-NTr}}}{X_{0\text{-NTr}}} \geq 2 \tag{6.8}$$

$$\frac{X_{0\text{-NTr}}}{X_{1\text{-N}}} \leq 10 \tag{6.9}$$

The damping conditions (6.8) and (6.9) can only be met if the zero-sequence reactance of the neutral earthing transformer $X_{0\text{-NTr}}$ is not too large.

Example B5

Fulfilment of the damping conditions (6.8) and (6.9) if a ZN neutral earthing transformer and a neutral earthing resistor are used for limitation of the line-to-earth short-circuit current in a 20-kV system to I''_{k1} = 1,000 A (Fig. B6.17).

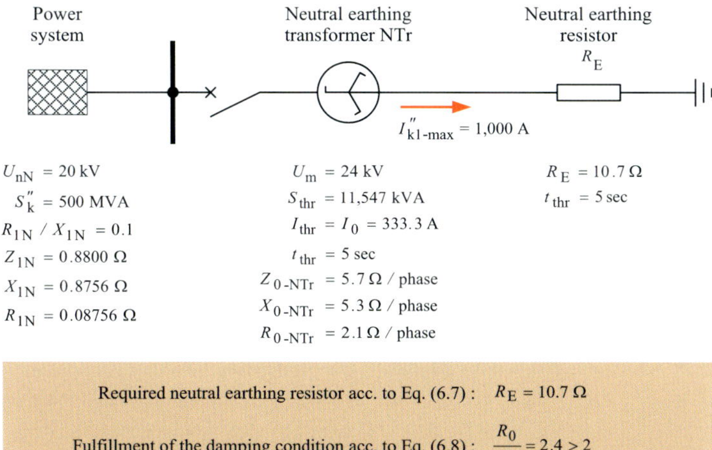

Fig. B6.17 Calculation example for low-impedance neutral earthing using a neutral earthing transformer and neutral earthing resistor (Example B5)

i) The permissible touch voltage U_{Tp} according to DIN VDE 0101 (VDE 0101): 2000-01 [6.7] must be complied with. The characteristic of the permissible touch voltage U_{Tp} defined in the standards as a function of the duration of current flow t_F is shown in the chart in Fig. B6.18.

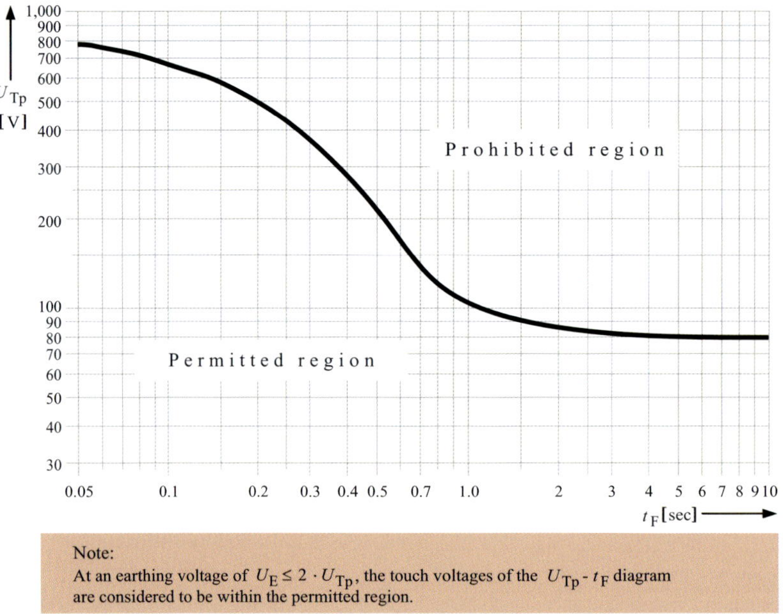

Fig. B6.18 Permissible touch voltage U_{Tp} as a function of the duration of current flow t_F

When operating low-impedance-earthed MV industrial power systems, no hazard need to be expected due to excessive touch voltages if the line-to-earth short-circuit current is limited to $I''_{k1} \leq 2{,}000$ A. In MV industrial power systems, the earth-electrode effect of cables, extensive building earthing along the cable routes and the galvanic connection of HV and LV system earth electrodes provide very favourable earthing conditions.

Favourable earthing conditions are indicated by controlled characteristics of the earth-surface voltage φ in the factory halls and a slight rise in potential at the earth-fault location. Fig. B6.19 shows that impermissible step and touch voltages can be ruled out with a controlled characteristic of the earth-surface voltage φ and small earthing voltages U_E. It is therefore usually possible to dispense with verification of the step and touch voltage in MV industrial power systems.

j) In a low-impedance neutral-earthed MV power system, the earth fault also changes the positive-sequence voltage. This causes temporary changes in the line-to-earth and line-to-line voltage in the LV system. For that reason, the line-to-earth short-circuit current must be limited in such a way that, on an earth fault in the MV system, the LV voltage does not exceed or fall below its permissible fluctuation band (e.g. $\Delta U/U_{nN} = \pm 10\,\%$).

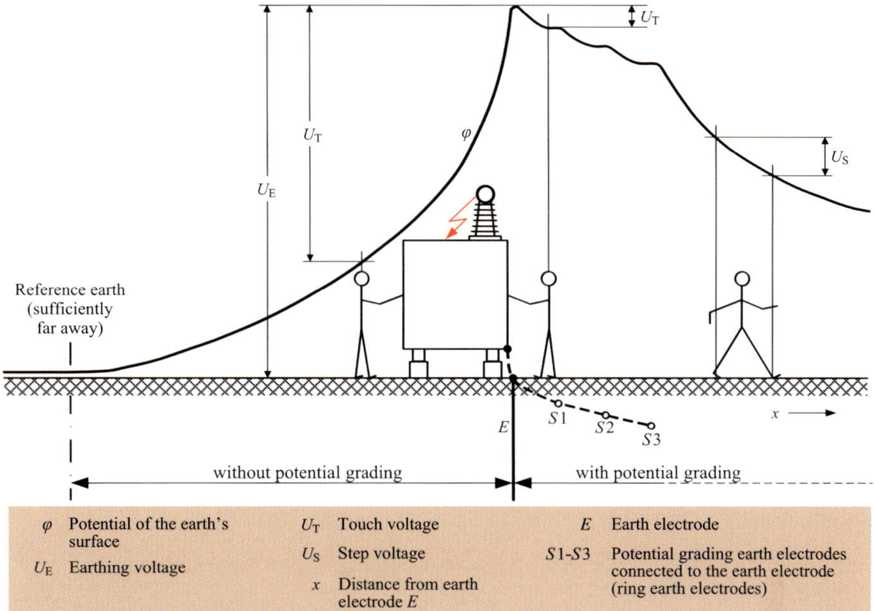

Fig. B6.19 Characteristics of the earth-surface voltage and the voltages for a current-carrying earthing electrode

Fig. B6.20 shows the voltage phasors on the LV side of a 20/0.4-kV transformer for limiting the line-to-earth short-circuit current in a real industrial power system to I''_{k1} = 1,000 A or I''_{k1} = 2,000 A. From Fig. B6.20 it can be seen that the LV voltage falls below its permissible fluctuation band at I''_{k1} = 2,000 A. Only if the line-to-earth short-circuit current is limited to I''_{k1} = 1,000 A will the short-time voltage changes remain in the permissible range of $\Delta U/U_{nN} = \pm 10\ \%$.

k) In the case of low-impedance neutral earthing of MV power systems with high-voltage motors (1 kV < U_{rM} < 10(11) kV), the line-to-earth short-circuit current must be limited to values $I''_{k1} \leq 200$ A to avoid core burning. Protection tripping must be instantaneous.

l) To detect line-to-earth short circuits in all three lines, each MV switchgear panel must be equipped with three phase-current transformers. For connection of the protection relays to the three phase-current transformers, the Holmgreen circuit must be used (Fig. B6.21).

Using the Holmgreen circuit, it is possible to obtain the residual current $3 \cdot \underline{I}_0$ for the residual current starting I_E> from the three phase currents. Residual currents occur neither during normal operation nor during an overload. For that reason, the residual current starting I_E> can be set to be more sensitive than phase current starting I>.

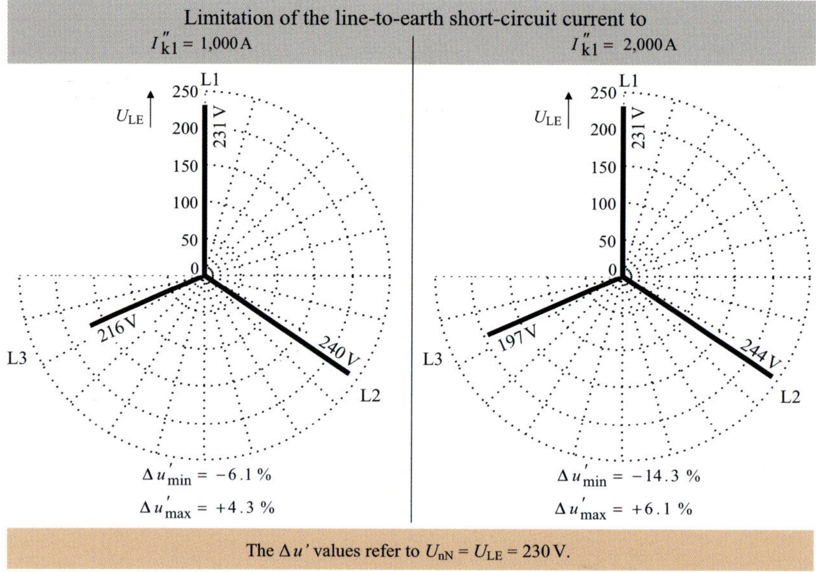

Fig. B6.20 LV-side voltage changes during line-to-earth short circuits in a real 20-kV industrial power system with low-impedance neutral earthing

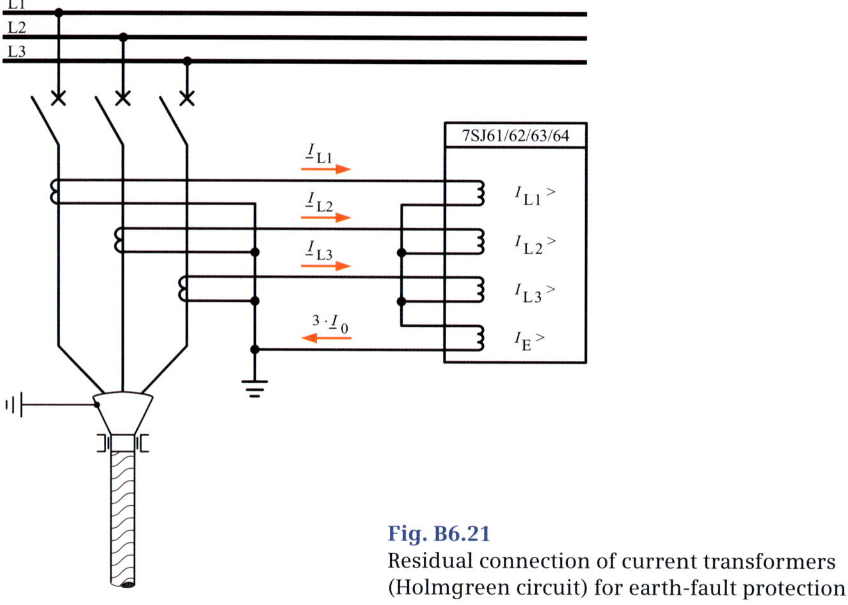

Fig. B6.21
Residual connection of current transformers (Holmgreen circuit) for earth-fault protection

Reliable detection of line-to-earth short circuits using the Holmgreen circuit is only possible if the following condition has been met:

$$I''_{k1\text{-min}} = 3 \cdot I_0 > 0.1 \cdot I_{N1} \tag{6.10}$$

$I''_{k1\text{-min}}$ minimum initial line-to-earth symmetrical short-circuit current (see Eq. 6.5)
I_0 residual current
I_{N1} primary nominal current of the phase current transformer

m) In the case of small line-to-earth short-circuit currents ($I''_{k1\text{-min}} < 0.1 \cdot I_{N1}$), earth faults have to be detected using core balance current transformers. Core balance current transformers can only be used in combination with protection relays featuring sensitive residual current starting I_{EE}> (Fig. B6.22). Obtaining the residual current by magnetic summation requires very high measurement precision. To meet this requirement, the shield current of the cable must be integrated into the measurement. As shown in Fig. B6.22, this is done by returning the earth conductor for cable shield earthing through the core balance current transformer. Moreover, the cable sealing end must be mounted insulated against earth.

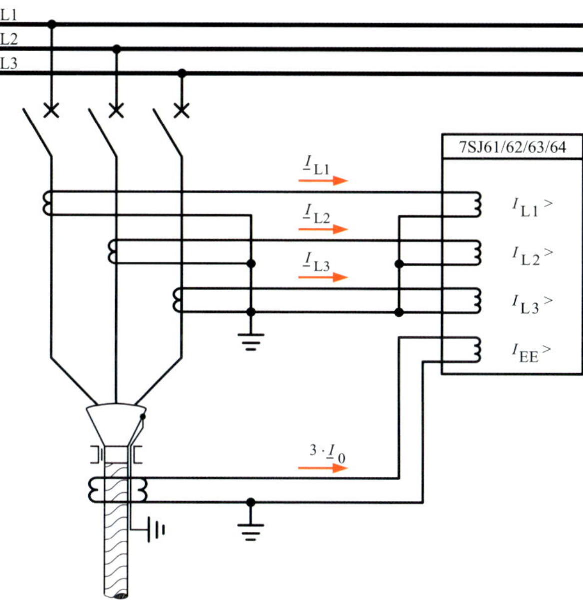

Fig. B6.22 Residual connection of a core balance current transformer (CBCT) for earth-fault protection

n) For measurement of residual currents using the Holmgreen circuit (Fig. B6.21) or using core balance current transformers (Fig. B6.22), SIPROTEC relays with four current inputs are required.

The SIPROTEC 7SJ61-64 relays are equipped with four current inputs. The SIPROTEC 7SJ61 relay does not have a voltage input. With this relay, therefore, neither directional earth-fault protection $\overrightarrow{I_E}$> (67N) nor sensitive directional earth-fault detection $\overrightarrow{I_{EE}}$> (67Ns) is possible.

o) In low-impedance-earthed MV load-centre systems with alternating current switch-fuse combinations, the minimum, initial line-to-earth symmetrical short-circuit current $I''_{k1\text{-min}}$ of the feeder must be greater than the minimum breaking current $I_{b\text{-HV HRC-min}}$ of the HV HRC fuse (see Section 7.3.1).

If principles a) to o) are adhered to, the low-impedance neutral earthing ensures safe and reliable operation of MV industrial power systems.

Power system with short-time low-impedance neutral earthing [6.5, 6.7, 6.57, 6.58, 6.60]

Short-time low-impedance neutral earthing (KNOSPE) is a combination of resonant neutral earthing (RESPE) and low-impedance neutral earthing (NOSPE). In case of a sustained earth fault, an earthing resistor is connected parallel with the arc-suppression coil by means of a single-pole circuit-breaker that earths the MV system with low impedance for a short time. The parallel connection of the neutral earthing resistor changes the earth fault to a short circuit to earth that can be easily located and cleared (Fig. B6.23). Short-time low-impedance neutral earthing is therefore an attempt to associate the advantages of resonant neutral earthing and low-impedance neutral earthing and use both at the same time. The short-time low-impedance neutral earthing principle has proven useful, above all, in locating sustained earth faults in public MV networks with resonant neutral earthing.

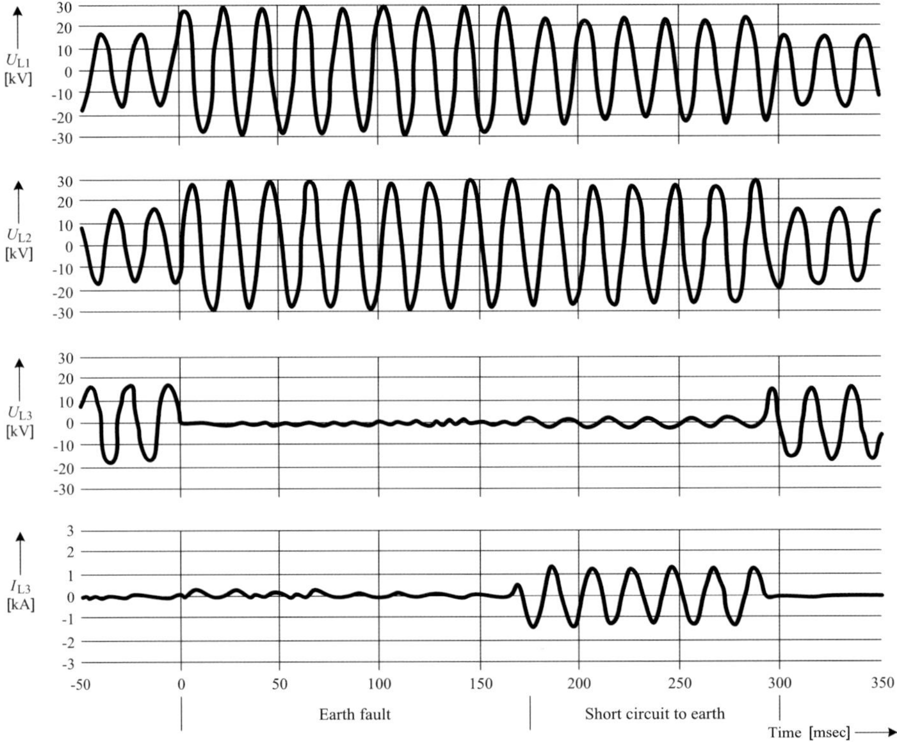

Fig. B6.23 Current and voltages in operation with short-time low-impedance neutral earthing [6.58]

6.3 Selection criterion and decision aid

To be able to recommend a method of neutral-point connection based on the traditional optimality criterion, we require a cost function to minimize the total financial expense, taking the expected value for earth-fault-induced damage costs into account. The following must apply to this cost function:

$$SPE_{\mathrm{min}} = \left\{ SPE_{\mathrm{i}} \;\middle|_{SPE_{\mathrm{i}} \in SPE \wedge \min\limits_{\mathrm{i}} TOTEX_{\mathrm{i}}} \right\} \tag{6.11}$$

with $TOTEX_{\mathrm{i}} = (CAPEX + OPEX + FAILEX)_{\mathrm{i}}$ (6.11.1)

SPE_{min}	neutral earthing variant with the minimum total expense
$TOTEX_{\mathrm{i}}$	total financial expense of neutral earthing variant i
$CAPEX$	investment costs (expense)
$OPEX$	operating costs (expense)
$FAILEX$	costs (expense) of the expected earth-fault-induced outage
i	incrementing index for neutral earthing variants

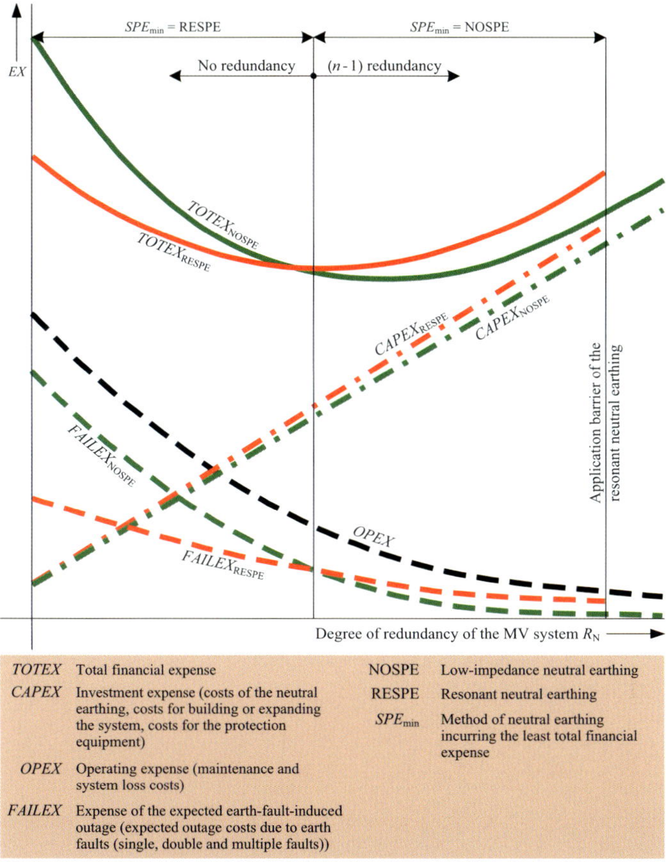

Fig. B6.24 Basic curves for the financial expense for the example comparison between the variants NOSPE and RESPE [6.63]

The expense component *CAPEX* contains the investment costs for the actual neutral earthing (*SPE*), general power system layout and protection equipment. Component *OPEX* includes the maintenance and servicing and system loss costs. The expense component *FAILEX*, which is subject to relatively large uncertainties, records the process-related damage due to earth-fault-induced supply interruptions. This damage depends on the level of redundancy of the MV system and the method of neutral earthing (*SPE*). The level of redundancy is a measure of the supply reliability level of a power system.

The method of neutral earthing influences the frequency and duration of earth-fault-induced single and consequential double and multiple faults. [6.61] presents the probabilistic calculation model that considers not only the investment and operating costs in the expense minimization but also the expected outage costs due to earth-fault-induced supply interruptions. Using the example calculations performed with the model, the results were generalized [6.62, 6.63]. The generalized result of the expense minimization for the variant comparison of low-impedance neutral earthing (NOSPE) and resonant neutral earthing (RESPE) is shown in Fig. B6.24. This figure shows that, in MV systems with $(n-1)$ redundancy, low-impedance neutral earthing is preferred.

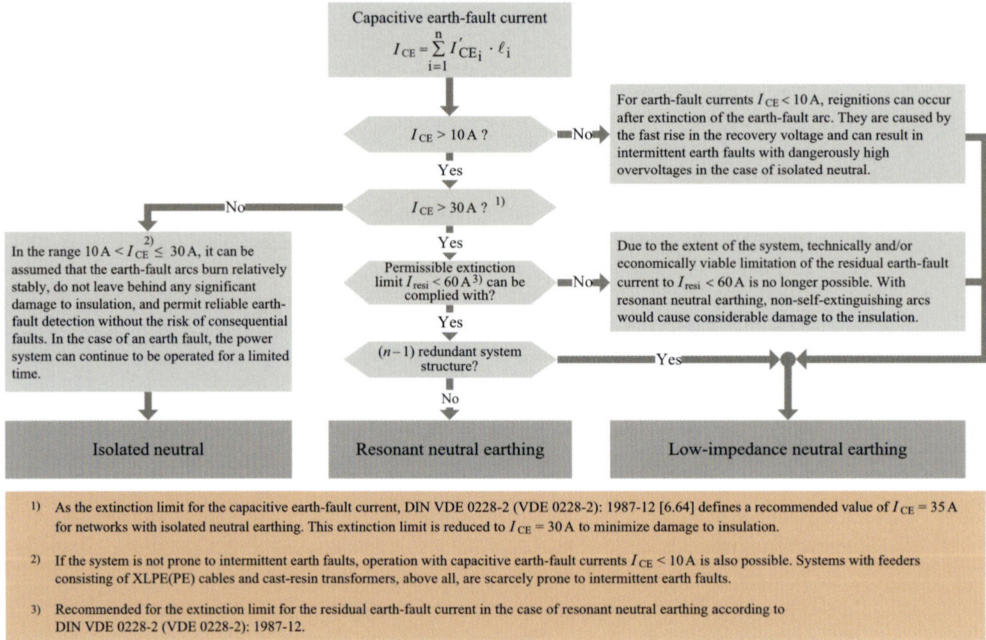

Fig. B6.25 Simplified decision aid for choosing the neutral earthing in MV systems

Table B6.26 Capacitive earth-fault current per unit lenght in MV cable networks (guidance values)

Capacitive earth-fault current per unit length	Nominal system voltage U_{nN}		
	6 kV	10 kV	20 kV
I'_{CE}	0.5 ··· 1 A/km	1 ··· 3 A/km	1.5 ··· 4 A/km

The choice of neutral earthing depends not only on the total financial expense and level of redundancy of the power system but also on the application barriers of the isolated neutral and the resonant neutral earthing. Taking these application barriers into account, Fig. B6.25 shows a simplified decision aid for choosing the neutral earthing in MV power systems.

To use this decision aid, the capacitive earth-fault current I_{CE} of the power system must be known. For rough calculations, the guidance values given in Table B6.26 can be used. The precise type-specific I'_{CE} values for MV cables can be found in the Siemens cable manual [6.65].

6.4 Selection recommendation for operation of MV cable networks in industry

Statistical surveys [6.66 to 6.68] show that resonant neutral earthing is the most frequently used method of neutral earthing in Germany. In particular for the operation of OHL distribution systems, resonant neutral earthing has proven a safe and reliable method. OHL systems are characterized by the fact that at the earth fault location after extinction of the arc, intact insulation is restored. After extinction of earth fault arcs in cable systems, by contrast, weak points with reduced insulation strength remain. XLPE, PE and PVC cables are preferred for use in industrial systems. The partial self-healing property of paper-insulated mass-impregnated cable is not provided by XLPE, PE and PVC cables. The weak points that necessarily remain without self-healing cable insulation in industrial systems with resonant neutral earthing can later result in chains of multiple earth faults and serious disturbances of the production process in the event of sustained earth faults. For that reason, resonant neutral earthing for operation of pure cable systems is regarded, with some exaggeration, as an engineering error in [6.69]. All fields of industry are indeed switching from resonant neutral earthing to low-impedance neutral earthing. In the automotive industry, above all, resonant neutral earthing is being widely replaced by low-impedance neutral earthing [6.70]. This is being driven by the following reasons and advantages:

- $(n-1)$ redundancy in the network design permits selective clearance of line-to-earth faults,
- disconnection of the line-to-earth fault location by the protection equipment is performed without any supply interruption,
- clearly defined protection tripping and switching state changes make integrated power system automation possible,
- operation with low-impedance neutral earthing (neutral earthing resistor) prevents high transient and long-lasting power-frequency overvoltages,
- the risk of fault propagation and double earth faults is eliminated,
- short clearing times limit the consequential damage at the fault location due to earth faults,
- no equipment wear nor loss of service life is caused by earth-fault-induced overvoltages,
- the frequency of faults is permanently reduced.

Because of its advantages, low-impedance neutral earthing has proven its value as the optimum neutral earthing for operation of MV cable systems. Thoroughly positive experience has been obtained, above all, in operation of low-impedance earthed cable systems in the automotive industry [6.70, 6.71]. Low-impedance earthing has also proven valuable in the public power supply for the operation of urban underground cable systems [6.59, 6.72].

Conclusion

Low-impedance neutral earthing is the solution of choice for operation of cable systems with ($n-1$) redundancy in industry.

6.5 Neutral earthing on both sides of transfer transformers in operation of MV industrial power systems

For certain reasons of power system and protection engineering (selectivity and response reliability of the 110-kV-side line-to-earth short-circuit protection, compliance with the extinction limit in extensive resonant-earthed 110-kV systems), it can be desirable to earth the neutral points on both sides of the transfer transformers. Earthing the neutral points on both sides of a transformer between two systems with different voltage levels couples their zero-sequence systems.

Coupling the zero-sequence voltage from the upstream 110-kV to the downstream MV power system causes the voltage in the downstream system in the event of an earth fault in the upstream system to fall more in the phase affected by the fault than if the zero-sequence systems were not coupled [6.73, 6.74].

The voltage decrease associated with transfer of the zero-sequence voltage from the upstream HV to the downstream MV system can cause the undervoltage tripping of motor protection relays or spurious drop-out of motor contactors.

Table B6.27 technically evaluates the connection variants of the neutral earthing on both sides of 110-kV/MV transformers with respect to their feasibility. The basis for the technical evaluation are per-unit values for the coupling of the power-frequency steady-state and transient zero-sequence voltage into the MV system. These values were calculated in [6.74] as examples of different ways of connecting the neutral point of transfer transformers on both sides. According to Table B6.27, for the earthing of transfer transformers on both sides in the operation of MV industrial power systems, only the connection variants c) (transformer vector group YNynd0, solid neutral earthing on the primary side and low-impedance neutral earthing on the secondary side) and h) (transformer vector group YNynd0, resonant neutral earthing on the primary side and low-impedance neutral earthing on the secondary side) can be considered.

To meet the ($n-1$) criterion by handling of a transformer fault in the 110-kV/MV incoming supply without interruption, two transfer transformers must be constantly connected in parallel during normal operation.

In the case of parallel connection and neutral earthing on both sides of the transfer transformers, earth-fault-induced circulating currents occur. These circulating currents are non-critical if the two neutral points of the transformers operated in parallel are earthed through an arc-suppression coil on the primary side (Table B6.27, connection variant h). In the case of solid earthing of both neutral points on the primary side, on the other hand, a line-to-earth short circuit in the upstream 110-kV power system can cause the occurrence of high circulating currents in the downstream MV power system.

Table B6.27 Connection variants for neutral earthing on both sides of 110-kV/MV transfer transformers

Connection variants	Coupling of the zero-sequence voltage from the upstream HV to the downstream MV power system acc. to [6.74]		Evaluation
	Per-unit steady-state value u_0	Per-unit transient value $u_{0\omega}$	
a) 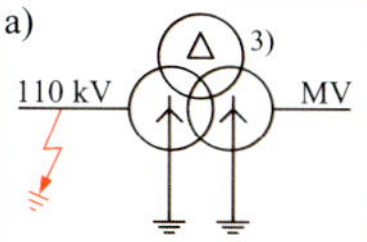 110 kV, MV, Δ 3)	0.25	0.25	- relatively low interference due to zero-sequence voltage - due to high line-to-earth short-circuit currents in the MV system $(0.45 \cdot I''_{k3} \le I''_{k1} < 1.5 \cdot I''_{k3})$ this connection variant is not recommended
b) 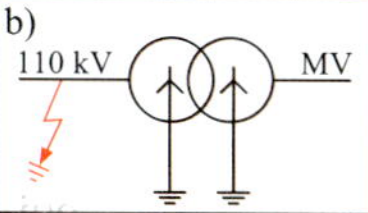 110 kV, MV	1.0	1.0	- high coupling of the zero-sequence voltage into the MV system - connection variant is impermissible
c) 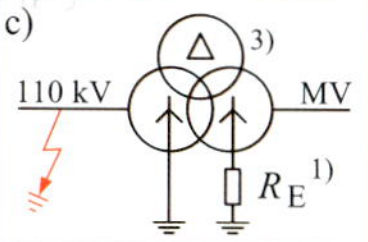 110 kV, MV, Δ 3), R_E 1)	0.24	0	- low coupling of the power-frequency zero-sequence voltage into the MV system - connection variant is permissible
d) 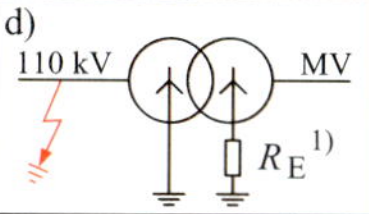 110 kV, MV, R_E 1)	1.0	0	- high interference due to the power-frequency zero-sequence voltage - connection variant is impermissible
e) 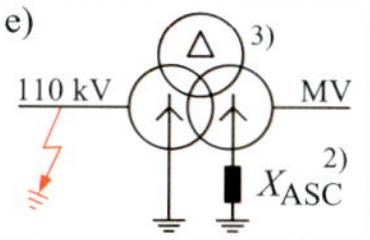 110 kV, MV, Δ 3), X_{ASC} 2)	≈ 7	≈ 7	- high interference due to the power-frequency and transient zero-sequence voltage - connection variant is impermissible
f) 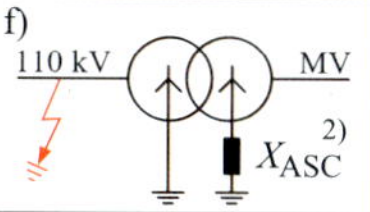 110 kV, MV, X_{ASC} 2)	» 7	» 7	- very high interference due to the power-frequency and transient zero-sequence voltage - connection variant is impermissible
g) 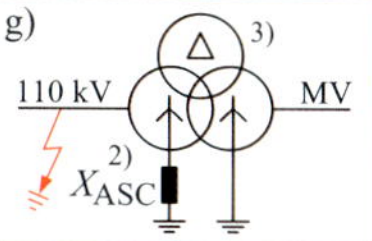 110 kV, MV, Δ 3), X_{ASC} 2)	0.02	0.02	- low interference due to the zero-sequence voltage - due to high line-to-earth short-circuit currents in the MV system $(0.45 \cdot I''_{k3} \le I''_{k1} < 1.5 \cdot I''_{k3})$ this connection variant is not recommended
h) 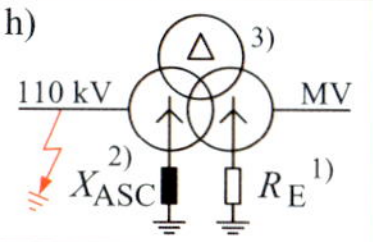 110 kV, MV, Δ 3), X_{ASC} 2), R_E 1)	0.02	0	- low interference due to the power-frequency zero-sequence voltage - connection variant is permissible
i) 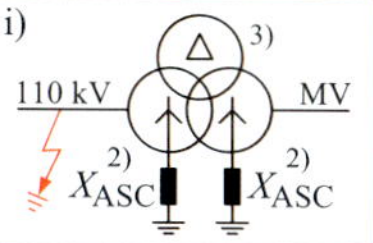 110 kV, MV, Δ 3), X_{ASC} 2), X_{ASC} 2)	≈ 0.4	≈ 0.4	- significant interference due to the zero-sequence voltage and switching-state-dependent effectiveness of the earth-fault compensation - connection variant is not recommended
j) 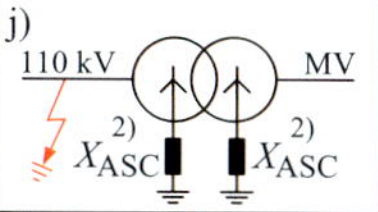 110 kV, MV, X_{ASC} 2), X_{ASC} 2)	Quality of the earthing of the 110-kV power system depends on the operating state of the MV power system. The effect of this earthing is slight and unpredictable.		- connection variant must always be avoided

1) Resistance of the neutral earthing resistor (R_E is dimensioned for $500\,A \le I''_{k1} \le 2{,}000\,A$)

2) Reactance of the arc-suppression coil ($X_{ASC} \approx |-j/3 \cdot \omega \cdot C_E|$)

3) Stabilizing winding ($S_{r\Delta} = 0.33 \cdot S_{rT}$)

Table B6.28 Connection variants of transfer transformers operated in parallel in the case of solid neutral earthing on the primary side and low-impedance neutral earthing on the secondary side

Connection variant	Operating problem [6.74]	Evaluation
k) 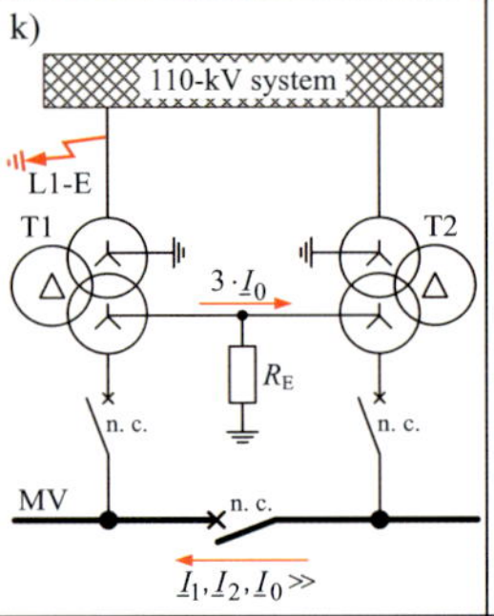	Circulating currents occur in the downstream MV system. The phase currents $\underline{I}_{L1} = 2 \cdot \underline{I}_1 + \underline{I}_0$ and $\underline{I}_{L2} = \underline{I}_{L3} = \underline{I}_0 - \underline{I}_1$ flow on the busbar if a line-to-earth short circuit occurs in line L1 in the 110-kV power system. If the ohmic components are neglected, the phase currents have the same direction. They can be of considerable magnitude ($1.0 \cdot I_{rT} < \lvert \underline{I}_0 \rvert \leq 3 \cdot I_{rT}$; $0.5 \cdot I_{rT} < \lvert \underline{I}_1 \rvert < 1.0 \cdot I_{rT}$) and result in spurious tripping of protection equipment.	 If the primary side is solidly earthed, it is impermissible to interconnect the neutral points of the two transformers on the secondary side and to earth them through a common resistor R_E.
l) 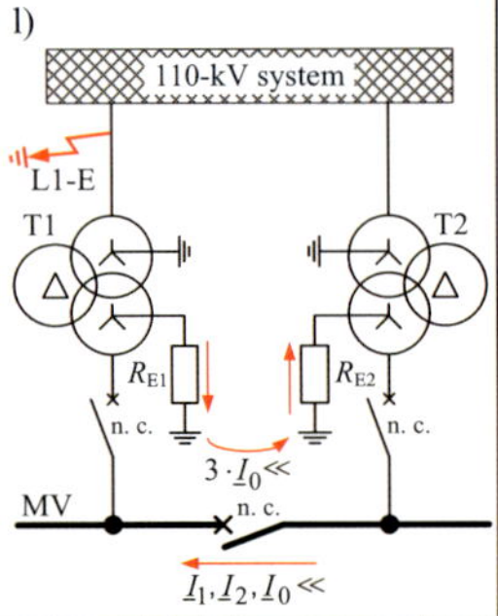	The circulating currents occurring in the downstream MV power system are negligible. In the case of line-to-earth short circuit in the upstream 110-kV system, strong limitation of the zero-sequence current is achieved by the series connection $R_{E1} + R_{E2}$. Spurious tripping of protection equipment due to excessive zero-sequence currents can usually be excluded.	 Parallel operation of transformers in which the two neutral points are solidly connected on the primary side and the two neutral points on the secondary side are each earthed through a resistor is permissible with the rules of power system design.

$\underline{I}_{L1}, \underline{I}_{L2}, \underline{I}_{L3}$	Phase currents	$\underline{I}_1$	Current in the positive-sequence system
I_{rT}	Absolute value of the rated current of the transformer	$\underline{I}_2$	Current in the negative-sequence system
		$\underline{I}_0$	Current in the zero-sequence system

Table B6.28 compares the connection variants of transfer transformers operated in parallel with solid neutral earthing on the primary side and low-impedance neutral earthing on the secondary side with respect to circulating currents. According to this table, only connection variant l) is permissible from a power system engineering viewpoint. In this variant, both neutral points on the primary side are solidly earthed while the two neutral points on the secondary side are earthed through one resistance each. It is basically impermissible to interconnect the neutral points of the two transfer transformers on the secondary side and to earth them together through a common resistance (Table B6.28, connection variant k). Due to the high circulating currents in the zero-sequence system occurring in connection variant k) ($1.0 \cdot I_{rT} \leq \lvert \underline{I}_0 \rvert \leq 3 \cdot I_{rT}$), spurious protection tripping can occur that makes reliable parallel operation of transfer transformers impossible.

Conclusion

Earthing of the transformer neutral points on both sides results in mutual influence between the upstream HV and the downstream MV system in the event of earth faults. The influence due to coupling of the zero-sequence voltage is non-critical in the case of

- solid neutral earthing on the primary side and low-impedance neutral earthing on the secondary side
- resonant neutral earthing on the primary side and low-impedance neutral earthing on the secondary side.

As the transfer transformers, transformers with stabilizing windings (vector group YNynd0, $S_{r\Delta} = 0.33 \cdot S_{rT}$) should always be used. Owing to the risk of circulating currents, the neutral points of transfer transformers operated in parallel should not be interconnected.

7 Design of the MV power system protection

7.1 Fundamentals of protection engineering and equipment

The power system protection devices used on the MV side are primarily intended to provide reliable and selective short-circuit and earth-fault protection. The occurrence of short circuits and earth faults can never be entirely ruled out even in strategically planned industrial power systems. In the protection of strategically planned power systems, it is not absolutely necessary to monitor compliance with the load current-carrying capacity by the phase current starting element. Because of this, the setting values of the phase-current starting element $I>$ must be chosen such that not only is the protection device not started at the maximum load current $I_{load\text{-}max}$ but actually drops out, while the circuit-breaker reliably trips at the minimum short-circuit current $I_{k\text{-}min}$. The starting condition for short-circuit protection results from this:

$$f_{oper} \cdot I_{load\text{-}max} \leq I> \leq \frac{I_{k\text{-}min}}{f_{arc}} \tag{7.1}$$

f_{oper}	safety factor that considers operating conditions such as load development, operation under fault conditions, transient phenomena, measuring errors of CTs and resetting ratio ($f_{oper} = 1.7$ for cables and $f_{oper} = 2.0$ for transformers)
f_{arc}	safety factor for damping by arc resistance ($f_{arc} = 1.25...2.0$)
$I_{load\text{-}max}$	maximum load current (for $I_{load\text{-}max}$, the current permissible according to the $(n-1)$ criterion in normal operation must be used)
$I_{k\text{-}min}$	minimum short-circuit current (calculation of $I_{k\text{-}min}$ must be based on a line-to-line short circuit and the most unfavourable fault location)

The starting condition for the short-circuit protection (Eq. 7.1) must be complied with in all MV power systems, irrespective of the method of neutral earthing. For earth-fault protection, on the other hand, the starting condition to be met depends on the method of neutral earthing of the MV power system. The starting conditions to be met depending on the type of neutral-point connection (Eqs. 7.2 to 7.8) are listed in Table B7.1.

In addition to the phase current and the earth-current starting conditions, the clearing time condition must also be met. The following must apply to each protection trip:

$$t_{a\text{-total}} \leq t_{a\text{-total-perm}} \tag{7.9}$$

$$t_{a\text{-total-perm}} = \begin{cases} \left(\dfrac{I_{thr}}{I_{th}}\right)^2 \cdot t_{thr} & \text{for switchgear and transformers} \\ \left(\dfrac{A_n \cdot J_{thr}}{I_{th}}\right)^2 \cdot t_{thr} & \text{for cables} \end{cases} \tag{7.9.1}$$

$t_{a\text{-total}}$	total clearing time (equivalent to the maximum short-circuit duration dsee example B2 in Section 4.2.2)
$t_{a\text{-total-perm}}$	permissible total clearing time
I_{th}	thermal equivalent short-circuit current (see Eq. 4.5.1)
I_{thr}	rated short-time withstand current of the equipment (see Tables B4.3, B4.9 and B5.16 to B5.18)
t_{thr}	rated short time of the equipment (see Tables B4.3, B4.9 and B5.16 to B5.18)
A_n	standardized conductor cross-sectional area of the cable (see Table B4.6)
J_{thr}	rated short-time current density of the cable (see Table B4.7)

Table B7.1 Starting conditions and SIPROTEC relay connection for the earth-fault or line-to-earth short-circuit protection

Type of neutral-point connection	Relevant earth fault current	Protection function	Starting condition	Eq. No.	SIPROTEC relay connection: Current connection	SIPROTEC relay connection: Voltage connection
Low-impedance neutral earthing	$0.1 \cdot I_{N1} < I''_{k1} \leq 2{,}000$ A	Line-to-earth short-circuit protection $I_E>$, 51N	$I_{CE\text{-max-feeder}} < I_E> < I''_{k1\text{-min}}$	(7.2)	Holmgreen circuit with 3 phase current transformers	—
Low-impedance neutral earthing	$0.1 \cdot I_{N1} < I''_{k1} \leq 2{,}000$ A	Line-to-earth short-circuit prot., directional $\overrightarrow{I_E}>$, 67N	$I_{CE\text{-max-feeder}} < \overrightarrow{I_E}> < I''_{k1\text{-min}}$	(7.3)	Holmgreen circuit with 3 phase current transformers	Line-to-earth connection
Low-impedance neutral earthing	$I''_{k1} < 0.1 \cdot I_{N1}$	Sensitive earth-fault prot. for high-resistance line-to-earth short circuits $I_{EE}>$, 51Ns	$0.003 \cdot I_{N1\text{-CBCT}} < I_{EE}> < I_{k1\text{-min}}$	(7.4)	Core balance current transformer (CBCT)	—
Low-impedance neutral earthing	$I''_{k1} < 0.1 \cdot I_{N1}$	Sensitive earth-fault prot. for high-resistance line-to-earth short circuits, directional $\overrightarrow{I_{EE}}>$, 67Ns	$0.003 \cdot I_{N1\text{-CBCT}} < \overrightarrow{I_{EE}}> < I_{k1\text{-min}}$	(7.5)	Core balance current transformer (CBCT)	Line-to-earth connection
Isolated neutral	$I_{CE\text{-sec}} > 0.05 \cdot I_{N2}$	Sensitive earth-fault direction detection $\overrightarrow{I_{EE}}>$, 67Ns (sin φ measurement)	$0.05 \cdot I_{N1} < \overrightarrow{I_{EE}}> < 0.5 \cdot I_{CE}$	(7.6)	Holmgreen circuit with 3 phase current transformers	3 line-to-earth connections or line-to-earth connection with open delta winding
Isolated neutral	$I_{CE\text{-sec}} < 0.05 \cdot I_{N2}$	Sensitive earth-fault direction detection $\overrightarrow{I_{EE}}>$, 67Ns (sin φ measurement)	$0.003 \cdot I_{N1\text{-CBCT}} < \overrightarrow{I_{EE}}> < 0.5 \cdot I_{CE}$	(7.7)	Core balance current transformer (CBCT)	3 line-to-earth connections or line-to-earth connection with open delta winding
Resonant neutral earthing	$I_{resi\text{-sec}} < 0.05 \cdot I_{N2}$ $I_{resi} = I_R$ $I_R = (0.02 \cdots 0.05) \cdot I_{CE}$ in MV cable systems	Sensitive earth-fault direction detection $\overrightarrow{I_{EE}}>$, 67Ns (cos φ measurement)	$0.003 \cdot I_{N1\text{-CBCT}} < \overrightarrow{I_{EE}}> < 0.5 \cdot I_{resi}$	(7.8)	Core balance current transformer (CBCT)	Line-to-earth connection with open delta winding

I_{N1} Nominal primary current of the phase current transformer (standardized values: 50 A; 75 A; 100 A; 150 A; 200 A; 300 A; 400 A; 600 A; 800 A; 1,000 A; 1,250 A ··· 4,000 A)

I_{N2} Nominal secondary current of the phase current transformer (standardized values: 1 A; 5 A; I_{N2} = 1 A is preferred)

I''_{k1} Initial line-to-earth short-circuit current (see Eq. (6.5))

$I''_{k1\text{-min}}$ Minimum, initial line-to-earth short-circuit current of the cable feeder (see Eq. (6.5.1))

$I_{CE\text{-max-feeder}}$ Maximum, capacitive earth-fault current of the feeder ($I_{CE\text{-max-feeder}} = I'_{CE} \cdot \ell_{cable}$)

$I_{N1\text{-CBCT}}$ Nominal primary current of the core balance current transformer (standardized values: 50 A; 60 A; 100 A; $I_{N1\text{-CBCT}}$ = 60 A is preferred)

$I_{N2\text{-CBCT}}$ Nominal secondary current of the core balance current transformer (standardized value: 1 A)

I_{CE} Capacitive earth-fault current ($I_{CE} \approx 3 \cdot \omega \cdot C_E \cdot U_{LE}$)

I_{resi} Residual earth-fault current (see Eq. (6.3))

I_R Residual active current (see Eq. (6.3.1))

The clearing time condition (Eq. 7.9) corresponds to the thermal short-circuit conditions (Eqs. 4.5 to 4.7). If the thermal short-circuit current conditions are fulfilled, the clearing time condition is automatically also fulfilled.

Given the high short-circuit power in industrial power systems, a short total clearing time of $t_{a\text{-total}}$ must be complied with. To keep arcing damage to equipment to a minimum during short circuits, the total clearing time should, if possible, be limited to $t_{a\text{-total}} \leq t_{thr}$.

Selectivity is one of the most important line protection criteria alongside starting reliability and speed. In line with the selectivity criterion, the protection settings must be chosen such that only the protection device closest to the fault is tripped without tripping the neighbouring protection device that is closer to the incoming feeder. In addition, back-up protection that is as independent as possible must be active in case the main protection fails.

The extensive SIPROTEC equipment range [7.1] is available for implementation of all necessary main and back-up protection functions (overcurrent, directional overcurrent, differential and distance protection). In industrial power systems, in particular, the use of numerical SIPROTEC devices has some important advantages. These advantages include [7.2]:

a) *Optimum filtering of the measured quantities*

The digital filtering and numerical measurement methods ensure a high level of measuring accuracy and short pick-up and drop-out times even if the measured quantities are subject to distortion and hunting. With the usual 50(60)-Hz fundamental measurement, harmonic currents of converter equipment or current distortions caused by arc furnaces, for example, only have a slight influence on the measuring accuracy.

b) *Small transient overreach*

The DC component of the short-circuit current causes transient overreaches. With large transient overreaches, there is a risk of spurious tripping. This risk can be eliminated using Fourier filters that eliminate the DC component of the short-circuit current.

c) *Reduced influence of inrush currents*

When a transformer is energized, high overcurrents that may be transient (lasting a few 10 milliseconds) to quasi-steady (lasting several seconds) must be expected. These are called inrush currents. Because the inrush currents are many times larger than the nominal current and contain not only the 2nd-order harmonic (100(120) Hz) but also a considerable fundamental component (50(60) Hz), the power system protection may fail when transformers are energized. Such malfunctions are reliably prevented using inrush restraint, which is based on evaluation of the 2nd-order harmonic. The inrush restraint blocks the inrush-induced response of the phase current elements (I>/t>, I_p/t_p) and earth current starting elements (I_E>/t_E>, I_{Ep}/t_{Ep}) of the time-delay time-overcurrent protection.

Because the I>/t> and I_p/t_p phase current starting element is insensitive to the magnetizing inrush, sensitive back-up protection can be provided for faults on the secondary side of the transformer.

d) *Improved short-circuit protection for motors*

On motor starting, a transient inrush current is superimposed on the starting current. The inrush peaks and the direction current component of the starting current are eliminated by digital filtering of the inrush current superimposition. Because of this, the phase current starting element can be set to be much more sensitive. The phase current starting element can also be set in two stages (I>> and I>/t> stage).

With the aid of the instantaneous I>> stage, for short-circuits near to the terminals, a clearing time of $t_a \leq 100$ msec (operating time $t_{op} \leq 20$ msec and circuit-breaker opening time $T_{0\text{-CB}} \leq 80$ msec) can be achieved.

e) *Additionally implementable overload protection*

In addition to short-circuit protection, overload protection can be implemented for all equipment (generators, motors, transformers and cables) in the form of a thermal replica. Using the overload protection, it is possible to monitor the load current condition (Eq. 2.4) using measurement equipment. Above all, in industrial power systems that have arisen historically and are planned to meet immediate operative requirements, monitoring of the load current with measurement equipment provides the necessary protection against overloading of cables and transformers.
In a methodically correctly planned power system, overload protection for cables and transformers offers an additional safety measure.

f) *Flexible protection coordination*

Numerical SIPROTEC relays have separate setting ranges and tripping characteristics for short-circuit and line-to-earth fault protection. This convenient equipment permits extremely flexible protection time grading in low-impedance-earthed systems. For use in power systems with isolated neutral or with resonant neutral earthing, a sensitive directional earth-fault detection function is integrated into the SIPROTEC relay. The high function integration in a relay provides protection coordination that is very flexible and efficient.

g) *Short grading times*

The short overtravel time of numerical SIPROTEC relays of about 30 msec (up to 150 msec for electromechanical relays) and the high precision of the zone times of about 10 msec (60 msec for electromechanical relays) mean that the grading times can be shortened from $\Delta T = 400...500$ msec to $\Delta T = 250...300$ msec. This both reduces the total clearing time $t_{a\text{-total}}$ for a given number of protection sections and increases the selectivity if more protection sections are added. Reducing the total clearing time $t_{a\text{-total}}$ makes it easier to meet the clearing time condition (Eq. 7.9) with high system short-circuit powers S_k'' and small rated short times t_{thr} of the equipment.

h) *Integrated breaker failure protection*

If the circuit-breaker that is connected directly upstream of the short circuit fails, this protection trips the circuit-breaker or circuit-breakers in the incoming supply. Owing to the speed of the numerical SIPROTEC protection equipment, a short-circuit current interruption is reached within $t_{a\text{-total}} \leq 250$ msec on a breaker failure. This assumes a typical mechanical delay of $T_{0\text{-CB}} = 80$ msec for the MV circuit-breakers. With its very fast back-up clearance of short circuits, the breaker failure protection makes a considerable contribution to increasing the safety of people and equipment in the operation of industrial power systems.

i) *Absolute directional selectivity*

If numerical SIPROTEC relays are used, the short-circuit direction can also be determined by measuring external voltages or using voltage stores. In this way, SIPROTEC relays ensure absolute directional selectivity in the case of faults both at the end of the line and on the busbar.

j) *Numerical measurement of the short-circuit impedance*

The minimum protection distance that can be set with numerical distance protection relays with numerical resistance measurement is five times shorter than for electromechanical distance protection relays. The setting of the numerical distance protection can therefore be adapted to very short line lengths ($l \geq 200$ m).

k) *Reduced requirements for current transformers (CTs)*

Due to the extremely low device burdens ($S_{relay} \leq 0.1$ VA) and the vastly improved stability against CT saturation (e.g. by means of integrated saturation detectors), numerical SIPROTEC relays make lower demands of the transient response of current transformers. The high stability during CT saturation also ensures reliable and selective protection tripping in industrial power systems with large power system time constants (50 msec $\leq \tau <$ 500 msec).

l) *Adaptive setting group change and additional signal comparison*

Considering switching state changes and compliance with the permissible total clearing time $t_{a\text{-total-perm}}$, it is possible to limit the selective coordination of the power system protection by means of protection time grading. In this case, it is possible to use the option that numerical relays provide for setting group change and signal comparison.

In MV industrial power systems, the significant advantages of numerical protection are implemented by the coordinated use of the protection devices described below.

Time-overcurrent protection devices [7.1 to 7.13]

Time-overcurrent equipment is used in MV industrial power systems both as standard main protection and as additional back-up protection. Use as the main protection is possible depending on the

- system structure (supply through feeder, line, radial or ring cable),
- method of neutral earthing (isolated neutral, resonant neutral earthing or low-impedance neutral earthing) and
- type and size of the equipment to be protected.

With time-overcurrent protection devices as the only main protection, HV motors up to $P_{rM} \leq 2$ MW and transformers up to $S_{rT} \leq 10$ MVA can be protected. For larger rated powers P_{rM} and S_{rT}, time-overcurrent protection devices should only be used as back-up protection. In that case, differential protection devices are used for the main protection.

For the short-circuit protection of radial cables and normally open cable rings, simple time-overcurrent protection with a DTL characteristic (definite time-lag overcurrent) or IDMTL characteristic (inverse definite minimum time-lag overcurrent) can be used (Fig. B7.2).

SIPROTEC time-overcurrent protection devices permit selection between DTL and IDMTL tripping characteristics.

Four different characteristics with IDMTL tripping can be selected. It is possible to choose between the following types of characteristics:

- NORMAL INVERSE (IEC type A),
- VERY INVERSE (IEC type B),
- EXTREMELY INVERSE (IEC type C),
- LONG INVERSE (IEC type B).

In Fig. B7.2, the set of curves for the usually preferred IDMTL tripping characteristic “NORMAL INVERSE“ is shown. By selecting a different IDMTL tripping characteristic (e.g. “VERY INVERSE“, IEC type B), it may be possible to achieve a shorter tripping time in the event of short-line high-current faults.

For parallel operation of feeder and line cables, directional DTL or IDMTL protection must be provided due to reverse flow of fault current in the event of a short circuit.

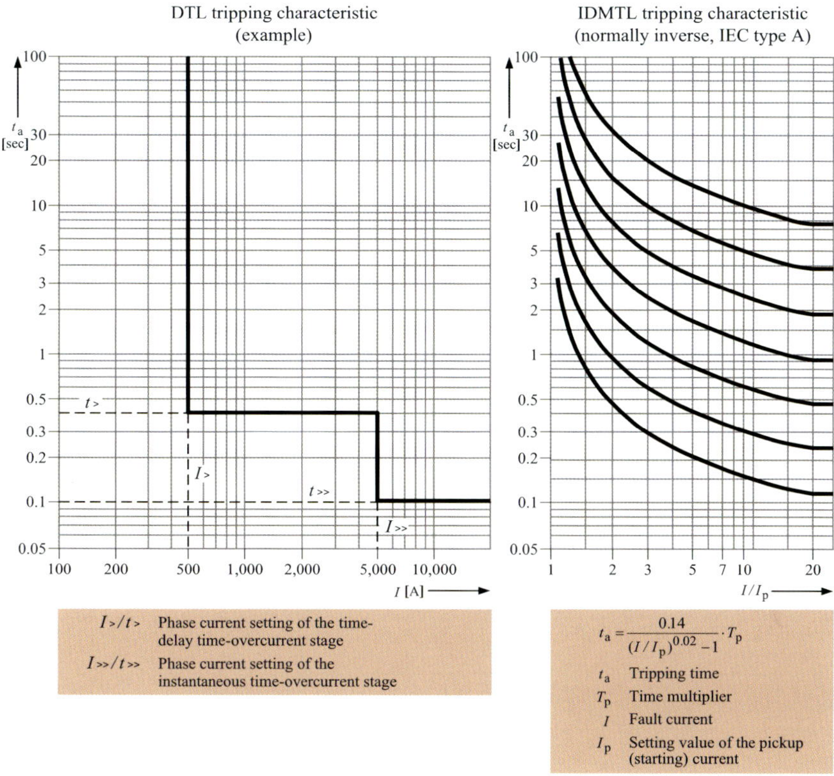

Fig. B7.2 Tripping characteristics of the definite-time and inverse-time overcurrent protection

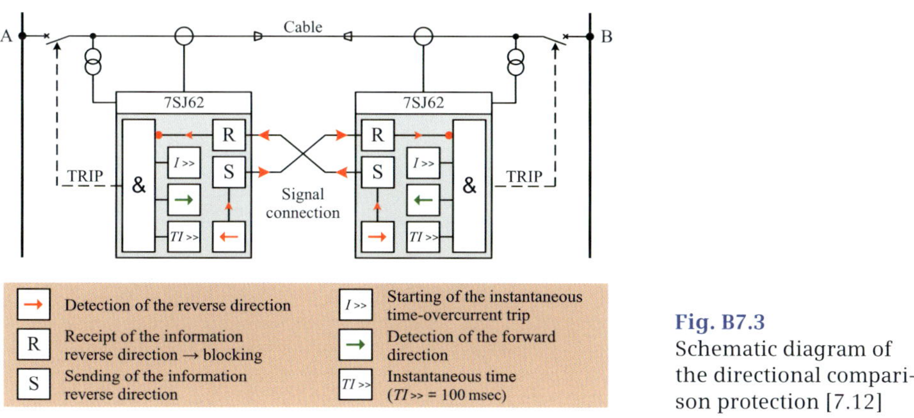

Fig. B7.3 Schematic diagram of the directional comparison protection [7.12]

When a system distribution substation is supplied through parallel feeder and line cables, the directional time-overcurrent protection can function as the main protection because SIPROTEC time-overcurrent protection devices ensure absolute selectivity by direction sensing and keep the total clearing time $t_{a\text{-total}}$ as small as possible.

Table B7.4 Protection functions of SIPROTEC time-overcurrent protection devices (7SJ600, 7SJ61–64)

Scope of functions	Abbreviations	ANSI device function number	SIPROTEC time-overcurrent protection relays				
			7SJ600	7SJ61	7SJ62	7SJ63	7SJ64
• Current inputs	--	--	3	4	4	4	4
• Voltage inputs	--	--	0	0	3 – 4	3	4
• Instantaneous time-overcurrent protection (phase)	$I>>$	50	X	X	X	X	X
• Instantaneous time-overcurrent protection (earth)	$I_E>>$	50N	X	X	X	X	X
• Time-delay time-overcurrent protection (phase)	$I>>/t>>$, $I>/t>$, I_p/t_p	51	X	X	X	X	X
• Time-delay time-overcurrent protection (earth)	$I_E>>/t_E>>$, $I_E>/t_E>$, I_{Ep}/t_{Ep}	51N	X	X	X	X	X
• Voltage-controlled time-overcurrent protection	$t=f(I)+U<$	51V	--	--	X	--	X
• Directional time-overcurrent protection (phase)	$\overrightarrow{I>>}/t>>$ $\overrightarrow{I>}/t>$, $\overrightarrow{I_p}/t_p$	67	--	--	(X)	(X)	(X)
• Directional time-overcurrent protection (earth)	$\overrightarrow{I_E>>}/t_E>>$, $\overrightarrow{I_E>}/t_E>$, $\overrightarrow{I_{Ep}}/t_{Ep}$	67N	--	--	(X)	(X)	(X)
• Sensitive earth-fault detection (instantaneous)	$I_{EE}>>$	50Ns	--	(X)	(X)	(X)	(X)
• Sensitive earth-fault detection (time delay)	$I_{EE}>>/t_{EE}>>$, $I_{EE}>/t_{EE}>$, I_{EEp}/t_{EEp}	51Ns	--	(X)	(X)	(X)	(X)
• Sensitive earth-fault direction detection	$\overrightarrow{I_{EE}>>}/t_{EE}>>$, $\overrightarrow{I_{EE}>}/t_{EE}>$, $\overrightarrow{I_{EEp}}/t_{EEp}$	67Ns	--	--	(X)	(X)	(X)
• Displacement voltage	$U_{en}>$	64	--	--	(X)	(X)	(X)
• Intermittent earth fault	--	--	--	(X)	(X)	(X)	(X)
• Circuit-breaker failure protection	--	50BF	--	X	X	X	X
• Undervoltage protection	$U<$	27	--	--	(X)	(X)	(X)
• Overvoltage protection	$U>$	59	--	--	(X)	(X)	(X)
• Frequency protection	$f>$, $f<$	81 o/u	--	--	(X)	(X)	(X)
• Lock-out	--	86	--	X	X	X	X
• Motor protection	--	--	(X)	(X)	(X)	(X)	(X)
– Overload protection	$I^2t>$	49	(X)	(X)	(X)	(X)	(X)
– Negative-sequence overcurrent (unbalanced-load) protection	$I_2>$, $t=f(I_2)$	46	(X)	(X)	(X)	(X)	(X)
– Starting time supervision	$I^2_{start}\cdot t$	48	(X)	(X)	(X)	(X)	(X)
– Restart inhibit	I^2t	66, 49R	--	(X)	(X)	(X)	(X)
– Temperature monitoring	ϑ (RTD box)	38	--	(X)	(X)	(X)	(X)
– Load jam detection	ΔP	51M	--	(X)	(X)	--	(X)
– Undercurrent monitoring	$I<$	37	--	X	X	X	X
• Flexible protection functions	--	--	--	--	X	--	X
– Directional power monitoring	$\overrightarrow{P>}$, $\overrightarrow{P<}$	32	--	--	(X)	--	(X)
– Power factor monitoring	$\cos\varphi>$, $\cos\varphi<$	55	--	--	(X)	--	(X)
– Phase-sequence monitoring	L1, L2, L3	47	--	--	(X)	--	(X)
– Frequency change	$df/dt>$, $df/dt<$	81R	--	--	(X)	--	(X)

X Standard function (X) Optional function

Table B7.5 Current transformer requirements when using SIPROTEC time-overcurrent protection devices (7SJ600, 7SJ61–64)

Accuracy limiting factor	Equation / condition
effective	$K'_{ALF} = K_{ALF} \cdot \frac{R_{ct} + R_{rb}}{R_{ct} + R_{b}}$ (7.10)
required	$K'_{ALF} \geq I{>>}/I_{N1}$ but at least 20 (7.11)

K'_{ALF}	Accuracy limiting factor (effective)
K_{ALF}	Rated accuracy limiting factor (for example: CT class 5P10 → $K_{ALF} = 10$)
R_{ct}	Secondary winding resistance at 75 °C (inherent burden of the current transformer)
R_{rb}	Rated burden ($R_{rb} = S_{rb} / I_{N2}^2$)
S_{rb}	Rated power of the current transformer (S_{rb} = 5 VA; 10 VA; 15 VA; 20 VA; 30 VA; 60 VA)
I_{N2}	Nominal secondary transformer current (I_{N2} = 1 A or 5 A, I_{N2} = 1 A is preferred)
R_b	Connected burden ($R_b = R_{relay} + R_L$)
R_{relay}	Relay burden ($R_{relay} = S_{relay} / I_{N2}^2$)
S_{relay}	Power of the relay (S_{relay} = 0.1 VA for SIPROTEC time-overcurrent protection relays)
R_L	Burden of the connection or measurement leads ($R_L = \frac{2 \cdot \varsigma \cdot \ell}{A}$)
ℓ	Single length between current transformer and device
ς	Resistivity ($\varsigma = 0.0175\ \Omega \cdot m^2/m$ for copper at 20 °C)
A	Conductor cross-section
$I{>>}$	Setting value of the instantaneous short-circuit release
I_{N1}	Nominal primary transformer current (I_{N1} = 50 A; 75 A; 100 A; 200 A; 300 A; 400 A; 600 A; 800 A; 1,000 A; 1,250 A; .... 4,000 A)
Note:	Condition (7.11) is usually met if 5-VA current transformers, class 5P10, are used

To achieve selectivity in normally closed cable rings, the directional time-overcurrent protection must be time-graded. The time grading of the directional time-overcurrent protection devices can result in an unacceptably long total clearing time $t_{a\text{-total}}$. For that reason, SIPROTEC time-overcurrent protection devices enable the provision of alternative directional comparison protection by means of a signal connection (Fig. B7.3). By providing directional comparison protection, cable connections of normally closed ring systems can be protected by instantaneous tripping.

Tables B7.4 and B7.5 provide a general view of the protection functions that can be used and the current transformer requirements to be met when using SIPROTEC time-overcurrent protection devices in MV industrial power systems.

Current differential protection devices [7.1 to 7.9, 7.14 to 7.19]

Current differential protection devices are part of the main protection of MV industrial power systems by virtue of their functions. With differential protection devices as the main protection, optimum protection of feeder and ring cables, large motors and generators ($P_{rM,G}$ > 2 MW), power transformers (S_{rT} > 10 MVA) and important busbar systems (in particular, double and multiple busbars) can be implemented. For example, for very short feeder cable and ring cable distances in industrial power systems, differential protection devices provide ideal main protection for zone-selective instantaneous clearance (pickup time ≤ 15 msec) of short circuits on 100 % of the cable distance.

Unlike directional time-overcurrent and directional comparison protection, differential protection is not reliant on voltage transformers. SIPROTEC 7SD610 numerical two-end line differential protection relays can be used for differential protection in industrial power systems. With SIPROTEC 7SD610 relays, it is possible to implement data ex-

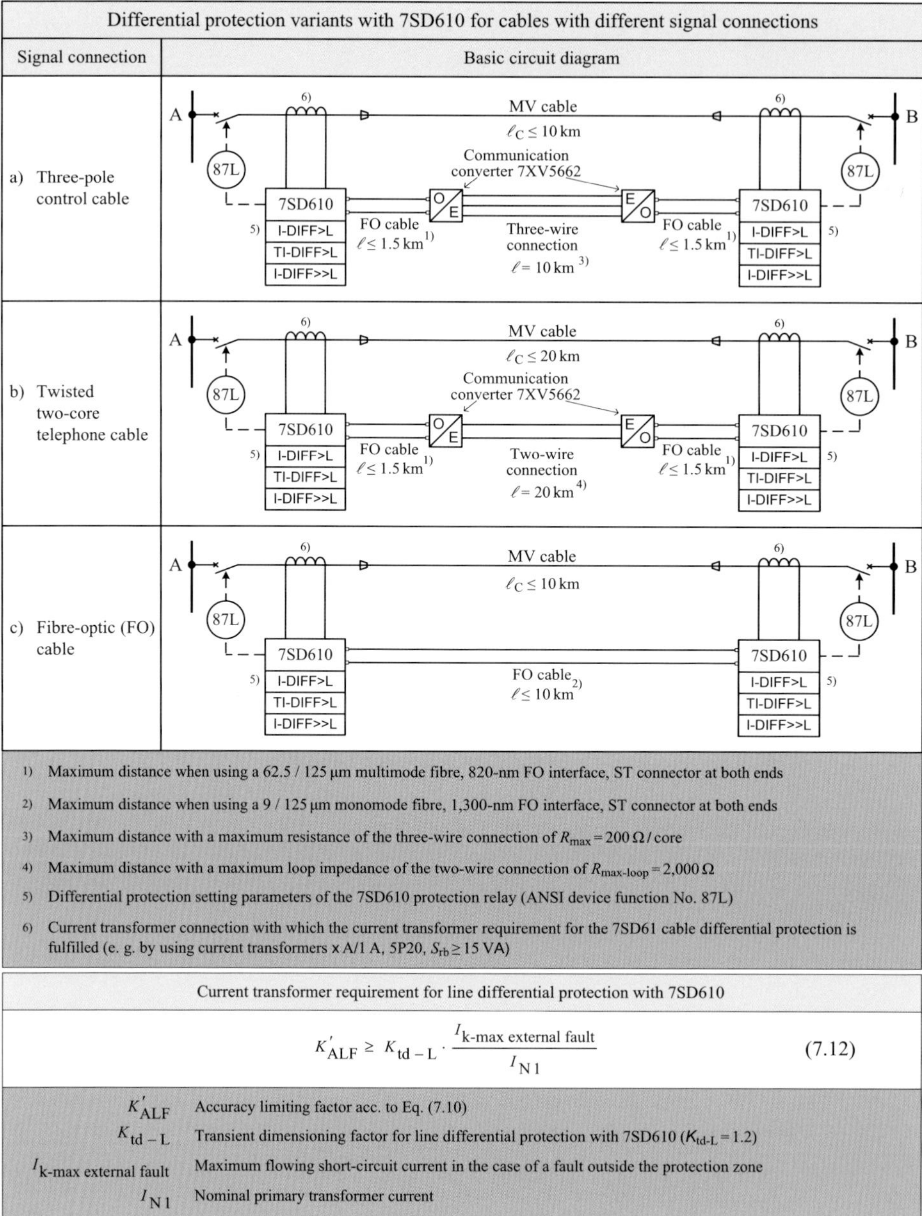

Fig. B7.6 7SD610 numerical differential protection relay for cables

change for measuring differential current by means of fibre-optic (FO) cables or existing telephone and control cables. To permit digital data exchange by means of conventional copper pilot wires, special converters (communication converters from FO to copper conductors, e.g. Siemens type 7XV5662) are required. Further details regarding the conditions of use (design of signal connections, fulfilment of current transformer requirements) of the 7SD610 numerical differential protection in industrial power systems are given in Fig. B7.6.

The SIPROTEC 7UT612 differential protection relay can be used as an independent differential protection device for power transformers. The differential protection additionally used for high-voltage motors of power rating class $P_{rM} > 2$ MW can be implemented with SIPROTEC 7UM62 differential protection relays. These two numerical differential protection devices have the following characteristics:

- current restraint tripping characteristic,
- restraint feature against inrush currents with 2nd-order harmonic,
- restraint feature against transient and steady-state fault currents with 3rd or 5th-order harmonic,
- insensitivity to DC components and current transformer saturation,
- high level of stability even with different degrees of current transformer saturation,
- high-speed instantaneous trip in case of high-current faults,
- integrated matching of the transformer vector group and transformation ratio (saving of current matching transformers).

Fig. B7.7 provides details of transformer differential protection with 7UT612 and motor differential protection with 7UM62.

The SIPROTEC 7SS52 numerical busbar protection system proves a suitable differential protection device for important substations with double and multiple busbars. This protection system consists of a central unit and distributed bay units. Communication is performed digitally using fibre-optic cables. The measured values and disconnector positions detected in the switchgear panel are transmitted to the central unit where they are evaluated according to the differential protection principle. In the reverse direction, the tripping commands are transmitted serially to the bay units that forward them to the circuit-breakers.

The digital communication principle of the 7SS52 busbar protection system is shown graphically in Fig. B7.8. Fig. B7.8 also contains the information needed to meet the current transformer requirement for busbar differential protection with 7SS52.

Distance protection devices [7.1 to 7.9, 7.20 to 7.25]

Distance protection is universal short-circuit protection whose mode of operation is based on the measurement and evaluation of short-circuit impedance. The short-circuit impedance is a measured quantity that is proportional to the distance between the relay installation location and the fault location. The distance protection function can be used with the following starting methods:

- overcurrent starting,
- voltage and current-dependent starting (underimpedance starting),
- voltage, current and phase angle-dependent starting,
- impedance starting.

These starting methods permit very good matching to different power system structures and modes of operation that are dependent on the method of neutral earthing. Industrial power systems have a high density of switchgear, that is, relatively short cable connections between substations. Because of this peculiarity, conventional dis-

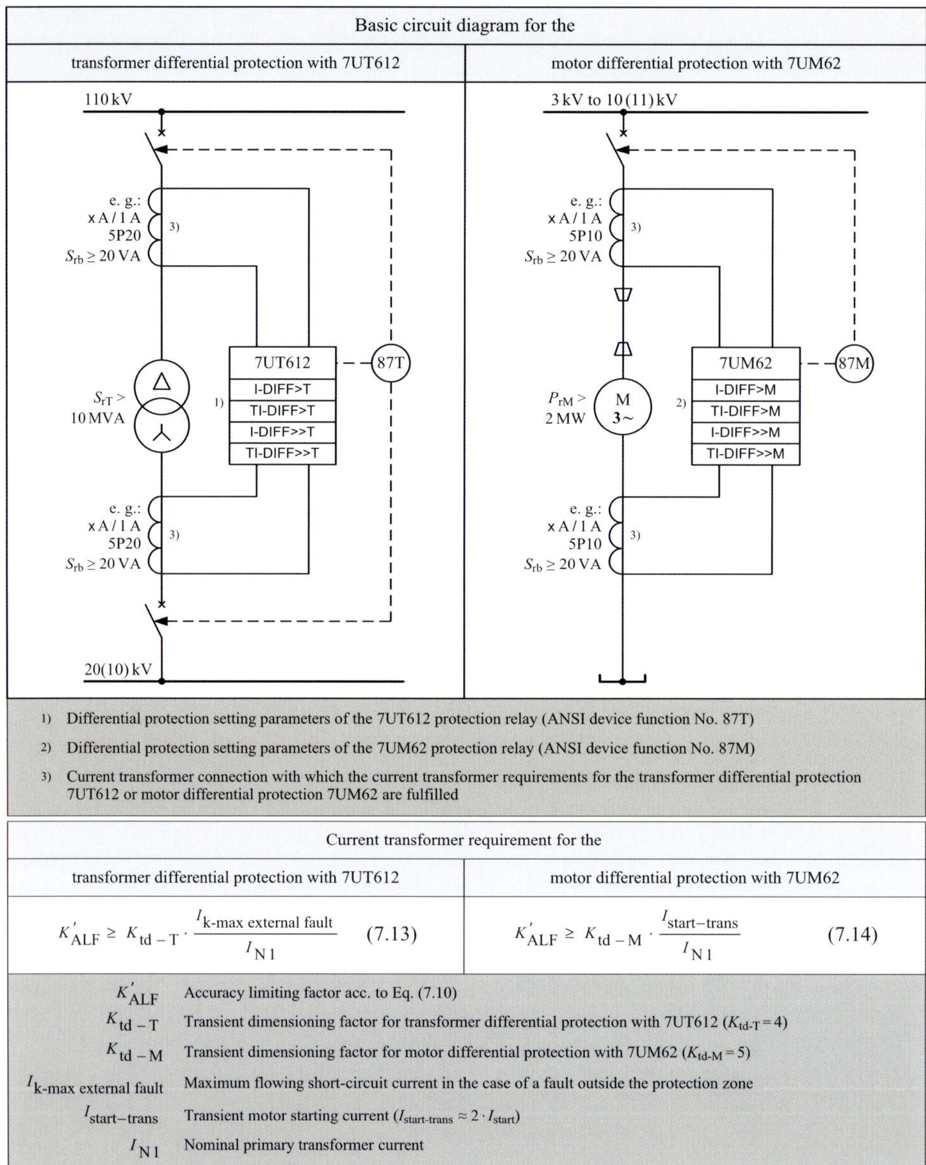

Fig. B7.7 7UT612 numerical differential protection relay for transformers and 7UM62 for motors

tance protection has in the past often only been used as back-up protection for line differential protection. Unlike conventional distance protection relays, the SIPROTEC 7SA6 numerical distance protection relay can also be used for main protection.

The numerical distance measurement of the 7SA6 is characterized by the fact that the resistance R and the reactance X can be set completely independently of one another for the zone limits. The advantage of independent setting of R and X is that even over short cable distances there is sufficient arcing reserve for distance measurement. The secondary smallest setting of the first distance zone is $X_{1\text{-sec}} = 0.05\ \Omega$ at $I_{N2} = 1$ A.

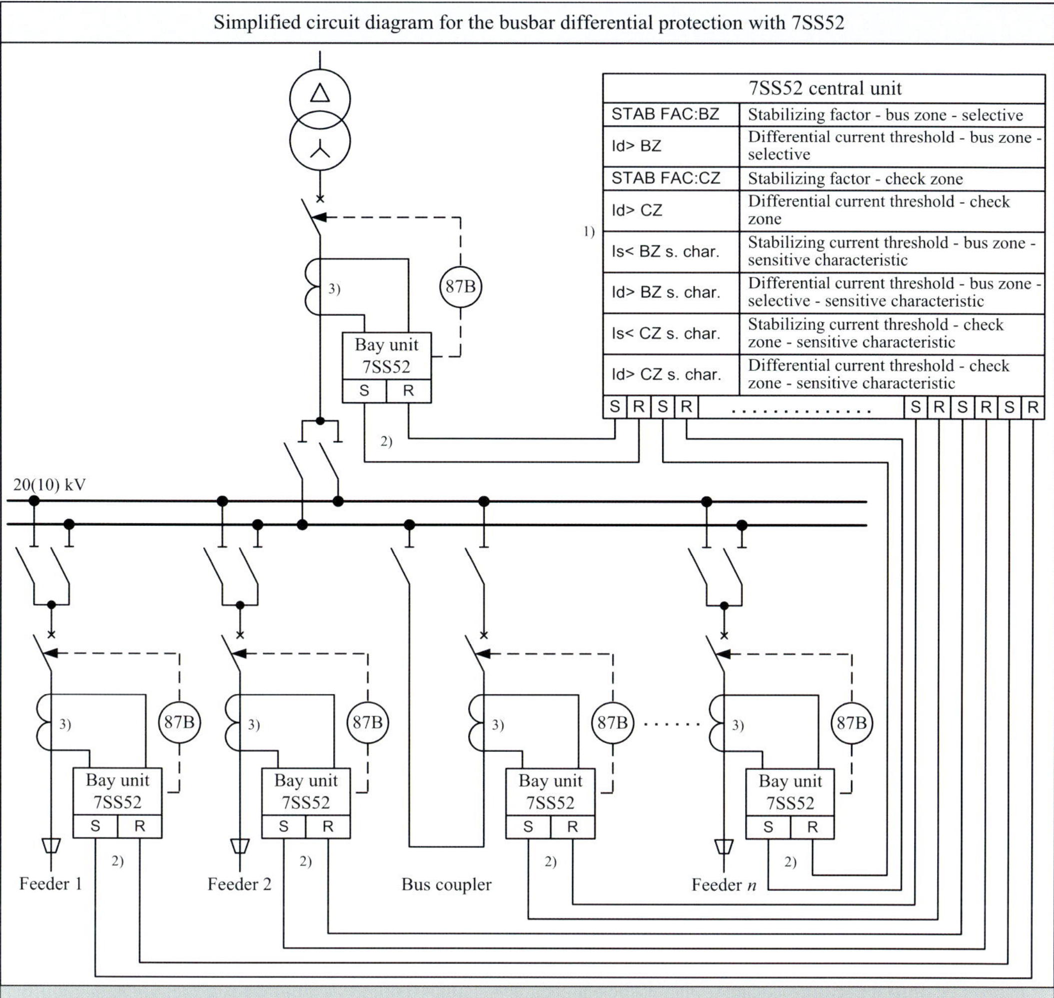

1) Example of selected differential protection setting parameters of the 7SS52 protection relay (ANSI device function No. 87B)

2) Fibre-optic (FO) cable with a transmission rate of 1.2 Mbaud. (Transmission is performed synchronously with high reliability (HDLC protocol (High-level Data Link Control) and a Hamming distance $h = 8$). FO lengths of $\ell_{max} \leq 1.5$ km are permissible.)

3) Current transformer connection with which the current transformer requirement for the 7SS52 busbar differential protection is fulfilled (e. g. by using current transformers x A/1 A, 5P10, $S_{rb} \geq 15$ VA)

Current transformer requirement for busbar differential protection with 7SS52

$$K_{td\text{-}BB} \cdot \frac{I_{k\text{-max external fault}}}{I_{N1}} \leq K'_{ALF} \leq 100 \qquad (7.15)$$

K'_{ALF}	Accuracy limiting factor acc. to Eq. (7.10)
$K_{td\text{-}BB}$	Transient dimensioning factor for busbar differential protection with 7SS52 ($K_{td\text{-}BB} = 0.5$)
$I_{k\text{-max external fault}}$	Maximum flowing short-circuit current in the case of a fault outside the protection zone
I_{N1}	Nominal primary transformer current

Fig. B7.8 7SS52 numerical differential protection relay for busbars (protection of a double-busbar system)

For this secondary minimum reactance, the resulting cable distance, depending on the conductor cross-sectional area and transformation ratios of the instrument transformers, is $l_C \approx 200$ m, which is the shortest cable distance that the 7SA6 can still protect without a signal connection.

The reach of numerical distance protection relays with a polygonal (quadrilateral) characteristic is determined by the reactance X. For distance protection grading with the reactance X, five independent zones (Z1 to Z5) can be set on the 7SA6.

- Distance protection zone 1:

  The first zone is normally set to 85% of the cable distance to be protected. The following applies to the setting:

$$X_{1\text{-sec}} = \frac{k_I}{k_U} \cdot X_{1\text{-prim}} \quad (7.16)$$

$$X_{1\text{-prim}} = 0.85 \cdot X_C' \cdot l_C \quad (7.16.1)$$

k_I	CT transformation ratio (e.g. 400 A/1 A)
k_U	VT transformation ratio (e.g. 20,000 V/100 V)
X_C'	reactance per unit length of the cable in Ω/km
l_C	length of the feeder cable in km

  Tripping on faults in zone 1 is instantaneous ($T_1 \leq 100$ msec).

- Distance protection zone 2:

  The downstream busbar is protected by the second zone. Moreover, zone 2 is back-up protection for the cable feeders connected to the busbar. These outgoing feeders must be taken into account in the distance protection setting $X_{2\text{-sec}}$ with their reactance in parallel operation. The second zone should, if possible, extend 10 to 20% beyond the busbar. $\Delta T = 300$ msec is selected as the grading distance for the tripping time. The tripping time of the second protection stage is therefore $T_2 = 400$ msec.

- Distance protection zones 3 to 5:

  Zones Z3 to Z5 are usually not required for distance protection grading in industrial power systems. Owing to of the high switchgear density, no suitable protection distances can be defined for these zones. Moreover, setting the distance zones Z3 to Z5 would make it harder to meet the clearing time condition (Eq. 7.9).

  To meet the clearing time condition, only non-directional or directional back-up tripping time grading is performed in industrial power systems. The directional or non-directional back-up tripping time of the distance protection acts as the last "emergency brake" for the short-circuit protection of the equipment.

Fig. B7.9 shows an example of distance protection grading with parallel supply of a system distribution substation. Fig. B7.9 also shows the current transformer requirements that have to be met using numerical 7SA6 SIPROTEC distance protection devices in industrial power systems.

Because of the high short-circuit currents, special attention is paid to fast and safe fault clearing in industrial power systems. Current-limiting protection devices are therefore also preferred along with time-overcurrent, differential and distance protection equipment. The current-limiting protection device explained below has been successfully used for decades.

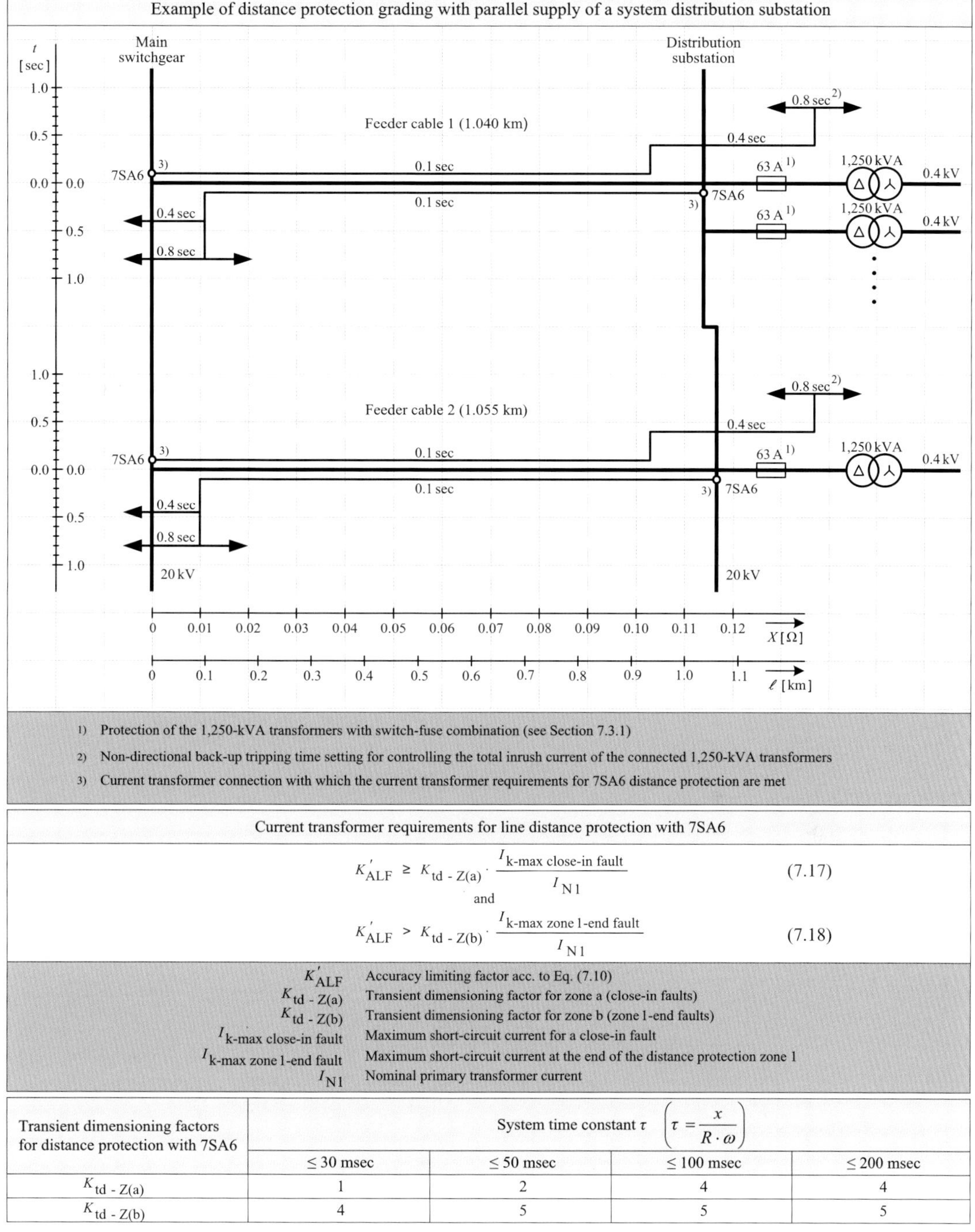

1) Protection of the 1,250-kVA transformers with switch-fuse combination (see Section 7.3.1)

2) Non-directional back-up tripping time setting for controlling the total inrush current of the connected 1,250-kVA transformers

3) Current transformer connection with which the current transformer requirements for 7SA6 distance protection are met

Current transformer requirements for line distance protection with 7SA6

$$K'_{ALF} \geq K_{td\text{ - }Z(a)} \cdot \frac{I_{k\text{-max close-in fault}}}{I_{N1}} \quad (7.17)$$

and

$$K'_{ALF} > K_{td\text{ - }Z(b)} \cdot \frac{I_{k\text{-max zone 1-end fault}}}{I_{N1}} \quad (7.18)$$

K'_{ALF} Accuracy limiting factor acc. to Eq. (7.10)
$K_{td\text{ - }Z(a)}$ Transient dimensioning factor for zone a (close-in faults)
$K_{td\text{ - }Z(b)}$ Transient dimensioning factor for zone b (zone 1-end faults)
$I_{k\text{-max close-in fault}}$ Maximum short-circuit current for a close-in fault
$I_{k\text{-max zone 1-end fault}}$ Maximum short-circuit current at the end of the distance protection zone 1
I_{N1} Nominal primary transformer current

Transient dimensioning factors for distance protection with 7SA6	System time constant τ $\left(\tau = \frac{x}{R \cdot \omega}\right)$			
	≤ 30 msec	≤ 50 msec	≤ 100 msec	≤ 200 msec
$K_{td\text{ - }Z(a)}$	1	2	4	4
$K_{td\text{ - }Z(b)}$	4	5	5	5

Fig. B7.9 7SA6 numerical distance protection relay for cable systems

Table B7.10 Electrical parameters of the 3GD2 Siemens high-voltage high-rupturing-capacity fuse-links

Rated current of fuse	3GD2 (SIBA) fuse parameters								
	Power dissipation at rated current P_{rV} in W if			Rated breaking current I_{rb} in kA if			Minimum breaking current (operating current) $I_{b\text{-HV HRC-min}}$ in A if		
$I_{r\text{-HV HRC}}$ in A	U_m = 3.6/7.2 kV	U_m = 12 kV	U_m = 24 kV	U_m = 3.6/7.2 kV	U_m = 12 kV	U_m = 24 kV	U_m = 3.6/7.2 kV	U_m = 12 kV	U_m = 24 kV
6,3	10	16	29	63	63	63	22	22	22
10	17	28	52	63	63	63	34	34	34
16	17	28	59	63	63	63	56	56	56
20	13	23	46	63	63	63	70	70	70
25	16	29	56	63	63	63	90	90	90
31.5	21	38	72	63	63	63	110	110	110
40	27	50	106	63	63	63	140	140	140
50	30	56	108	63	63	63	170	170	170
63	38	63	132	63	63	63	210	210	210
80	47	76	174	63	63	63	280	280	280
100	64	104	234	63	63	63	320	320	320
125	85	127	--	63	63	--	390	390	--
160	98	172	--	63	63	--	600	600	--
200	121	--	--	50	--	--	800	--	--
225 (250)	145	--	--	50	--	--	1,000	--	--
250 (315)	143	--	--	50	--	--	1,260	--	--

High-voltage high-rupturing-capacity fuses [7.26 to 7.33, 7.39, 7.40]

High-voltage high-rupturing-capacity fuses (HV HRC fuses) are used in the voltage rating range 3.6 kV ≤ U_m ≤ 36 kV. They protect electrical equipment from the dynamic and thermal effect of high short-circuit currents by interrupting them in the millisecond range already while the current is rising. HV HRC fuses are mainly used in combination with

- switch-disconnectors,
- vacuum contactors and
- switches

to protect transformers (Section 7.3.1), motors (Section 7.4.1) and capacitors (Section 7.4.2). Each HV HRC fuse-link has a thermal striker pin. This thermal striker pin has the following tasks:

- to trip as soon as all main fuse-elements have melted,
- to trip before the fuse-link is thermally overloaded and bursts (e.g. on fault currents that are greater than the rated fuse current $I_{r\text{-HV HRC}}$ and smaller than the minimum breaking current $I_{b\text{-HV HRC-min}}$),
- to indicate when an HV HRC fuse-link has blown.

The thermal striker pin can operate the releases of the above switching devices both mechanically and electrically. If the correct HV HRC fuse and switching device are chosen, a switch-fuse combination is able to clear both small and large fault currents reliably. Table B7.10 provides important electrical parameters for short-circuit protection with Siemens 3GD2 high-voltage high-rupturing-capacity fuses.

7.2 Protection of supplying 110-kV/MV transformers

The incoming supply of industrial plants from the 110-kV power system usually has to be configured with $(n-1)$ redundancy. In an $(n-1)$ redundant incoming supply with two parallel 110-kV/MV transformers, failure of one of the transfer transformers must be handled without interruption by disconnection of the fault location by means of protection devices. Fig. B7.11 shows a protection concept that meets this requirement for two parallel supplying 110-kV/MV two-winding transformers with low-impedance-earthed neutral points.

The 7UT612 SIPROTEC numerical differential protection device is used as the main protection in the concept shown in Fig. B7.11. With this device, it is possible to implement earth-fault differential protection I-EDS> in addition to the usual differential protection I-DIFF>. The earth-fault differential protection I-EDS> implemented with an additional current transformer in the neutral point of the transformers increases the responsiveness of the main protection to insulation faults to earth in a winding. In this way, the earth-fault differential protection I-EDS> can detect single-phase fault currents of 10 % of the nominal current of the transformer.

In addition to the Buchholz protector BH, the protection concept includes a two-stage time-overcurrent and overload protection for the two parallel supplying 110-kV/MV transformers as separate back-up protection. For this back-up protection on the primary side of the transformers, the SIPROTEC 7SJ600 time-overcurrent protection device is used.

The SIPROTEC 7SJ62 time-overcurrent protection relay is used on the secondary side of the transformers. In addition to its four current inputs, at least three voltage inputs are provided on each SIPROTEC 7SJ62 time-overcurrent protection relay (Table B7.4). In the current and voltage inputs available on the 7SJ62, directional time-overcurrent

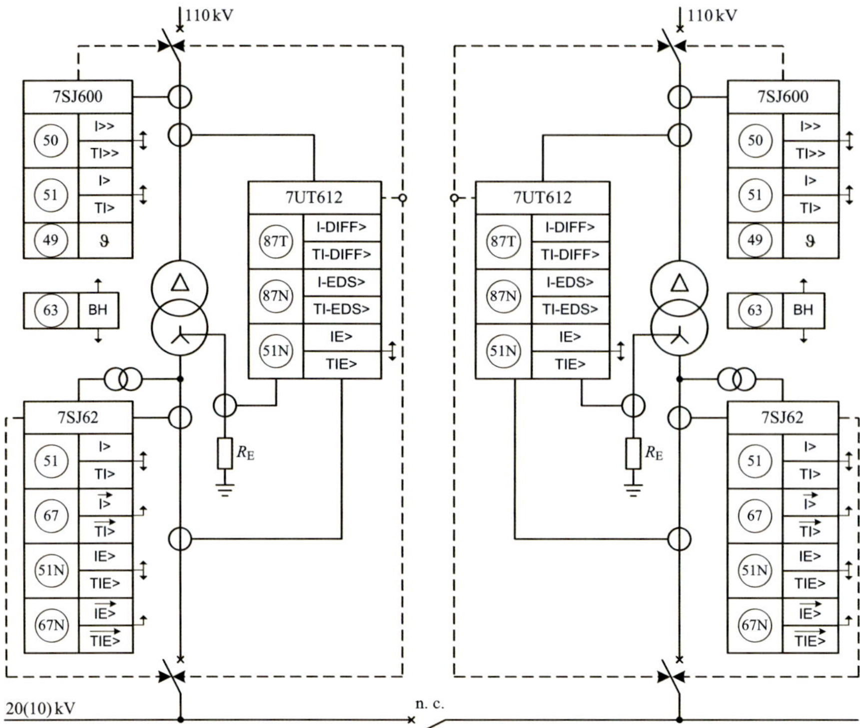

Fig. B7.11 Protection concept for two parallel supplying 110-kV/MV two-winding transformers with low-impedance-earthed neutral point

protection can be implemented on the secondary side of the transformers. This directional time-overcurrent protection is essential for selective disconnection of a faulty transformer in parallel operation.

Recommended setting parameters for protection of parallel supplying 110-kV/MV transfer transformers are listed in Tables B7.12a to B7.12c.

7.3 Protection of MV distribution transformers

The short-circuit protection of MV distribution transformers can be implemented with or without current limiting. The classic method of short-circuit protection without current limiting is a circuit-breaker-relay combination. For short-circuit protection with current limiting, a switch-fuse combination is used. When a switch-fuse combination is used, the HV HRC fuses strongly limit not only the duration but also the peak value of the short-circuit current.

Because HV HRC fuses have a strongly current-limiting effect, thermal and dynamic stress on the transformer caused by the short-circuit current remains low. This is especially important if the object being protected is an oil transformer. Short-circuit tests conducted by the German FGH Engineering&Test Corporation [7.34] on liquid-filled 20(10)-kV distribution transformers (power rating class $S_{rT} \leq 1{,}000$ kVA) have shown that, even with a three-phase fault with a short-circuit current of $I_k'' = 10$ kA and a clearing time of $t_a = 20$ msec, tank damage and oil leakage are inevitable. The greatest con-

Table B7.12a Setting parameters for protection of parallel supplying 110-kV/MV transfer transformers (transformer differential protection as the main protection)

Transformer differential protection with SIPROTEC 7UT612 unit			
ANSI device function No.	Setting parameter	Setting range	Setting
87T	I-DIFF> Current pickup (starting) value of the differential protection trip stage	$0.05 \cdots 2.00 \cdot I / I_{nO}$	I-DIFF> = $0.20 \cdot I / I_{nO}$ The setting value is referred to the nominal current of the protection object (transformer). For transformers, the sensitive setting $0.20 \cdot I / I_{nO}$ can be selected.
87T	TI-DIFF> Time setting value of the differential protection trip stage	0.00 ··· 60.00 sec	TI-DIFF> = 0.00 sec On detection of an internal transformer fault, the differential protection should normally trip instantaneously.
87N	I-EDS> Current pickup (starting) value of the earth fault differential protection trip stage	$0.05 \cdots 2.00 \cdot I / I_{nO}$	I-EDS> = $0.10 \cdot I / I_{nO}$ The setting value is referred to the nominal current of the protection object (transformer). A value of $0.10 \cdot I / I_{nO}$ has proven practicable for low-impedance neutral earthing (500 A $\le I''_{k1} \le$ 2,000 A).
87N	TI-EDS> Time setting value of the earth fault differential protection trip stage	0.00 ··· 60.00 sec	TI-EDS> = 0.00 sec On detection of an insulation fault of a winding to earth, the earth fault differential protection should normally trip instantaneously.
51N	IE> Pickup (starting) value of the DTL earth-fault current trip stage	0.05 ··· 35.00 A	IE> $\le \frac{I''_{k1\text{-}min}}{f_{arc}}$
51N	TIE> Time setting value of the DTL earth-fault current trip stage	0.00 ··· 60.00 sec	TIE> = TIE>(7SJ62) + 0.30 sec and TIE> $\le t_{thr\text{-}RE}$ The time delay must be coordinated with the grading plan of the MV system and be smaller than the rated short-time withstand current of the neutral earthing resistor.

$I''_{k1\text{-}min}$ Minimum initial line-to-earth short-circuit current

f_{arc} Safety factor for damping by the arc and earth contact resistance ($f_{arc} = 1.25 \cdots 2.00$)

$t_{thr\text{-}RE}$ Rated short time of the neutral earthing resistor ($t_{thr\text{-}RE} = 5 \cdots 10$ sec)

sequential damage, the explosion of the transformer tank, was caused by a short-circuit current flowing with the same magnitude for several cycles ($t_a \ge 100$ msec).

Because of its operating response, the HV HRC fuse interrupts fault currents in the range of its rated breaking current within the first half-wave ($t_a < 10$ msec). This reliably prevents tank explosions [7.35, 7.36]. In short-circuit protection without current limiting with a circuit-breaker and time-overcurrent relay, the total clearing time (pick-up time of the relay + operating time + opening time of the circuit-breaker) is much longer than for an HV HRC fuse. Even with the minimum opening time of $T_{0\text{-}CB} \le 80$ msec that can be achieved with today‘s vacuum circuit-breakers, many times more short-circuit energy is released than with the HV HRC fuse.

Table B7.12b Setting parameters for protection of parallel supplying 110-kV/MV transfer transformers (DTL protection on the primary side as the back-up protection)

Primary-side DTL back-up protection with SIPROTEC 7SJ600 device				
ANSI device function No.	Setting parameter		Setting range	Setting
50	I>> Pickup (starting) current of the DTL instantaneous trip stage		$0.1 \cdots 25.00 \cdot I_N$	$f_S \cdot \frac{I_{rT\text{-}pri}}{u_{rZ}/\%} \cdot 100 < I\!>> \le \frac{I_{kmin\text{-}pri}}{f_{arc}}$ $\sqrt{2} \cdot I\!>> < \hat{I}_{E1}$
	TI>> Time delay of the DTL instantaneous trip stage		$0.00 \cdots 60.00$ sec	TI>> = 0.10 sec
51	I> Pickup (starting) current of the time-delay DTL trip stage		$0.1 \cdots 25.00 \cdot I_N$	$f_O \cdot I_{rT\text{-}HV} < I\!> \le \frac{I_{kmin\text{-}pri/sec}}{f_{arc}}$
	TI> Time setting value of the time-delay DTL trip stage		$0.00 \cdots 60.00$ sec	TI> $\le t_{thr\text{-}T}$
49	ϑ Thermal overload protection	K factor	$0.40 \cdots 2.00$	$K = \frac{I_{max}}{I_{rT}}$ K = 1 in the method with hot-spot calculation
		Thermal time constant τ_{th}	$1.0 \cdots 999.9$ min	$\tau_{th} = \frac{t_{thr\text{-}T}}{60} \cdot \left(\frac{I_{thr\text{-}T}}{I_{rT}}\right)^2$

$I_{rT\text{-}pri}$	Primary-side nominal current of the transformer
$I_{kmin\text{-}pri}$	Minimum short-circuit current in a short circuit on the primary side
$I_{kmin\text{-}pri/sec}$	Minimum short-circuit current on the primary side in a short circuit in the busbar zone on the secondary side
$\hat{I}_{E1}$	First peak value of the inrush current
$I_{thr\text{-}T}$	Rated short-time withstand current of the transformer
$t_{thr\text{-}T}$	Rated short time of the transformer ($t_{thr\text{-}T}$ = 2 sec as standard)
u_{rZ}	Percent impedance voltage of the transformer
f_O	Safety factor that considers operating conditions, such as operation under fault conditions, measuring errors of CTs and resetting ratio ($f_O = 2.0$ for transformers)
f_{arc}	Safety factor for damping due to arc resistance ($f_{arc} = 1.25 \cdots 2.00$)
f_S	Safety factor for avoiding a short-circuit instantaneous trip on faults on the secondary side of the transformer ($f_S = 1.2 \cdots 1.5$)

7.3.1 Protection with a switch-fuse combination

Switch-fuse combinations are a functional unit comprising

- a switch according to DIN EN 60265-1 (VDE 0670-301):1999-05 [7.37] / IEC 60265-1: 1998-01 [7.38] and
- a current-limiting fuse according to DIN EN 60282-1 (VDE 0670-4): 2006-12 [7.39] / IEC 60282-1:2005-11 [7.40].

The HV HRC fuse of the switch-fuse combination for the protection of MV distribution transformers in industrial power systems must be selected according to the following criteria:

Table B7.12c Setting parameters for protection of parallel supplying 110-kV/MV transfer transformers (DTL protection on the secondary side as the complementary protection)

Secondary-side DTL protection with SIPROTEC 7SJ62 device			
ANSI device function No.	Setting parameter	Setting range	Setting
51	I> Pickup (starting) current of the time-delay DTL trip stage	0.10 ··· 35.00 A	$f_{O} \cdot I_{rT\text{-sec}} < I> \leq \frac{I_{kmin\text{-sec}}}{f_{arc}}$
	TI> Time setting value of the time-delay DTL trip stage	0.00 ··· 60.00 sec	TI> = TI>(feeder) + 0.3 sec The time delay must be coordinated with the grading plan of the MV system for phase current tripping.
67	$\overrightarrow{I>}$ Pickup (starting) current of the directional DTL trip stage	0.10 ··· 35.00 A	$\overrightarrow{I>} \leq \frac{I_{kmin\text{-rev}}}{f_{arc}}$
	$\overrightarrow{TI>}$ Time setting value of the directional DTL trip stage	0.00 ··· 60.00 sec	$\overrightarrow{TI>}$ = 0.00 sec The directional protection in the incoming feeder should normally trip instantaneously.
51N	IE> Pickup (starting) value of the DTL earth-fault current trip stage	0.05 ··· 35.00 A	$IE> \leq \frac{I''_{k1\text{-min}}}{f_{arc}}$
	TIE> Time setting value of the DTL earth-fault current trip stage	0.00 ··· 60.00 sec	TIE> = TIE>(feeder) + 0.3 sec The time delay must be coordinated with the grading plan of the MV system for earth-fault current tripping.
67N	$\overrightarrow{IE>}$ Pickup (starting) value of the directional DTL earth-fault current trip stage	0.05 ··· 35.00 A	$\overrightarrow{IE>} \leq \frac{I''_{k1\text{-min-rev}}}{f_{arc}}$
	$\overrightarrow{TIE>}$ Time setting value of the directional DTL earth-fault current trip stage	0.00 ··· 60.00 sec	$\overrightarrow{TIE>}$ = 0.00 sec Earth-fault directional protection in the incoming feeder should normally trip instantaneously.

$I_{rT\text{-sec}}$	Secondary-side nominal current of the transformer
$I_{kmin\text{-sec}}$	Minimum short-circuit current of the secondary-side protection zone
$I_{kmin\text{-rev}}$	Minimum short-circuit current in the reverse direction
$I''_{k1\text{-min}}$	Minimum initial line-to-earth short-circuit current of the secondary-side protection zone
$I''_{k1\text{-min-rev}}$	Minimum initial line-to-earth short-circuit current in the reverse direction
f_{O}	Safety factor that considers operating conditions, such as operation under fault conditions, measuring errors of CTs and resetting ratio ($f_{O} = 2.0$ for transformers)
f_{arc}	Safety factor for damping by the arc and earth contact resistance ($f_{arc} = 1.25 \cdots 2.00$)

- continuous permissible overload current of the transformer,
- heat effect of the inrush current,
- operating on single-phase and multi-phase terminal short circuits on the secondary side,
- selectivity to upstream and downstream protection devices,
- permissible power dissipation when fuse is installed in a narrow enclosure (moulded-plastic housing),
- earth-fault clearing condition in case of low-impedance neutral earthing.

In addition to these aspects of HV HRC fuse selection, the standard DIN EN 62271-105 (VDE 0671-105): 2003-12 [7.41] or IEC 62271-105: 2002-08 [7.42] poses specific conditions for the interaction of switches and fuses. Because switches only have a limited breaking capacity, it must be ensured that no impermissibly high fault currents have to be interrupted. This would not be a problem if the time-current characteristics of the HV HRC fuses were absolutely identical and without tolerances. However, this is not possible with the manufacturing process used. As the example in Fig. B7.13 shows for protection of a 20/0.4-kV transformer (S_{rT} = 1,250 kVA, u_{rZ} = 6 %) with a switch-fuse combination, the HV HRC fuses can interrupt the fault current of a short-circuit on the secondary side at different speeds in the three phases. The striker pin of the fastest fuse trips the switch of the switch-fuse combination.

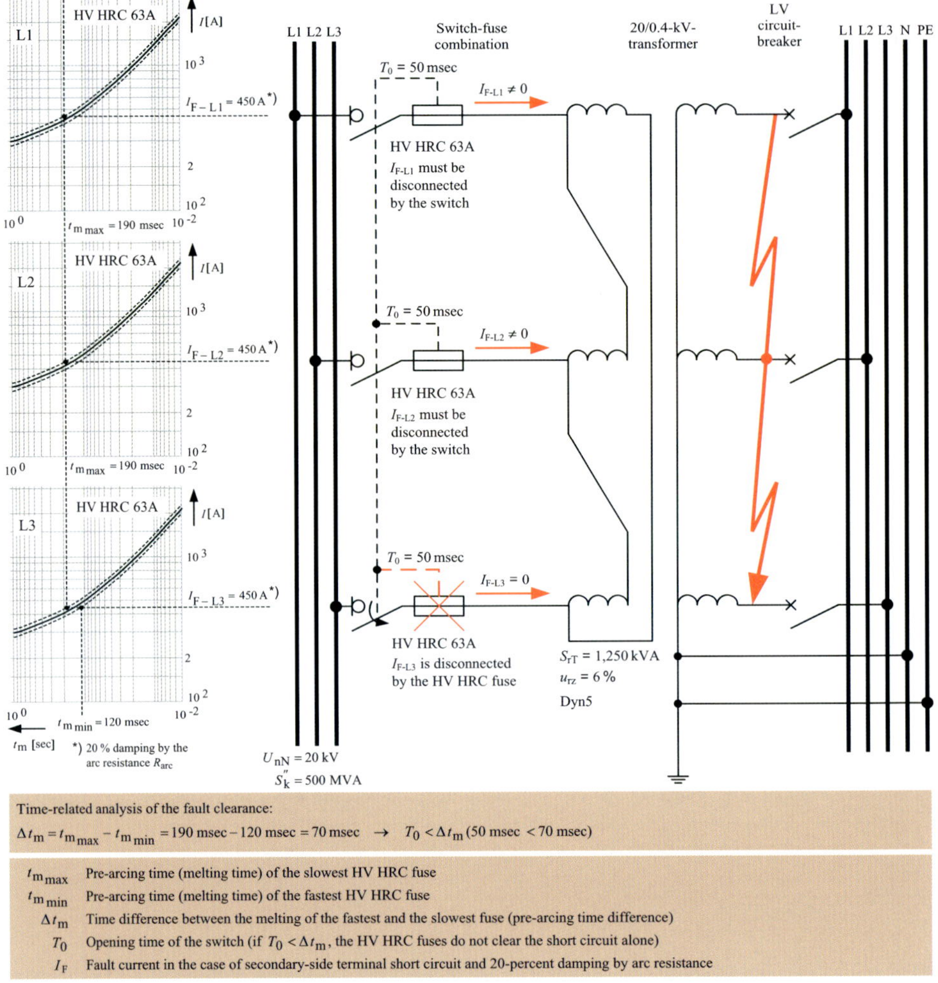

Fig. B7.13 Example of clearing of a secondary-side transformer terminal short circuit with a switch-fuse combination

The operating response of the fastest fuse is determined by the lower and that of the slowest fuse by the upper pre-arcing characteristic. The fastest 63-A HV HRC fuse in the protection of the 1,250-kVA transformer is therefore the one in line L3 and the slowest are those in lines L1 and L2.

For a fault current of $I_F = 450$ A, the pre-arcing time difference of this over the HV HRC fuse used for protection of the 1,250-kVA transformer is $\Delta t_m = 70$ msec (Fig. B7.13). The pre-arcing time difference $\Delta t_m = 70$ msec is greater than the opening time of the

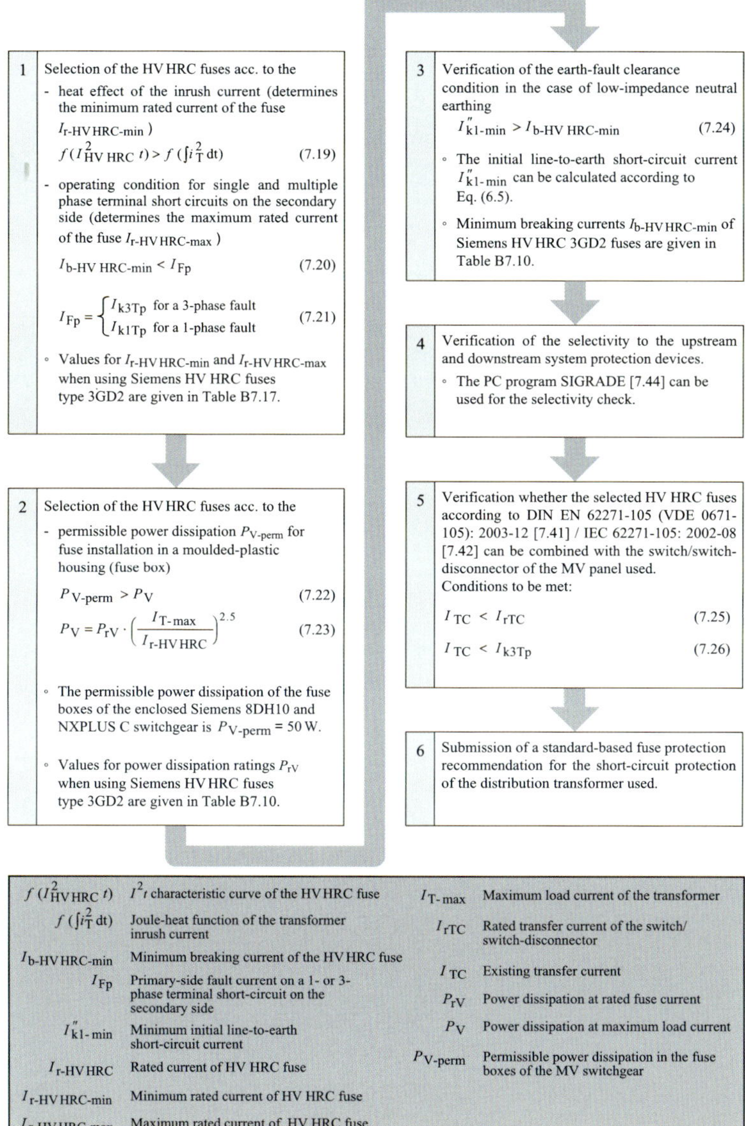

Fig. B7.14 Procedure for selecting and dimensioning a switch-fuse or switch-disconnector-fuse combination

switch of $T_0 = 50$ msec, i.e. in the phases L1 and L2, the fault current is not interrupted by the HV HRC fuses. At $T_0 < \Delta t_m$, the switching duty is split. Accordingly, the protection of distribution transformers requires a full-range switching device that can interrupt both overload and short-circuit currents reliably [7.43]. Such a full-range switching device is the switch-fuse combination selected and dimensioned according to DIN EN 62271-105 (VDE 0671-105): 2003-12 [7.41] or IEC 62271-105: 2002-08 [7.42]. Selection and dimensioning of the switch-fuse combination according to the standard as a full-range switching device can be performed step by step as shown in Fig. B7.14. The procedure according to Fig. B7.14 is explained below.

Example B6

Current-limiting short-circuit protection for a 800-kVA GEAFOL cast-resin transformer (Fig. B7.15)

The 800-kVA GEAFOL cast-resin transformer is to be connected to the main switchgear of a low-impedance-earthed 20-kV industrial power system and protected by means of a switch-disconnector-fuse combination.

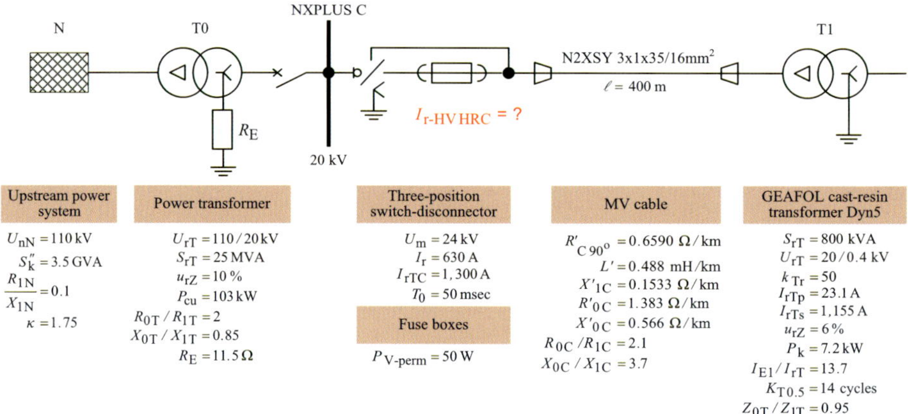

Fig. B7.15 Example of dimensioning the current-limiting short-circuit protection of a 800-kVA GEAFOL cast-resin transformer (Example B6)

The main switchgear is the SF_6-insulated fixed-mounted circuit-breaker switchgear of type NXPLUS C (Fig. B7.15). According to the standard, the Siemens 3GD2 HV HRC fuse must be used in combination with the three-position switch-disconnector of the NXPLUS C. According to Fig. B7.14, this is done as follows:

Step 1: Selection of HV HRC fuses according to the heat effect of the inrush current and operating on single-phase and multi-phase terminal short circuits on the secondary side of the transformer

Inrush-current-dependent specification of the Siemens 3GD2 HV HRC fuse required to protect the 800-kVA GEAFOL cast-resin transformer at $U_{nN} = 20$ kV is shown in Fig. B7.16. The Joule heat function $f\left(\int i_T^2 dt\right)$ of the transformer inrush current shown in the figure can be calculated from the ratio of the maximum inrush current to the rated transformer current ($I_{E1}/I_{rT} = 13.7$) and the time to half value ($K_{T0.5} = 14$ cycles). As com-

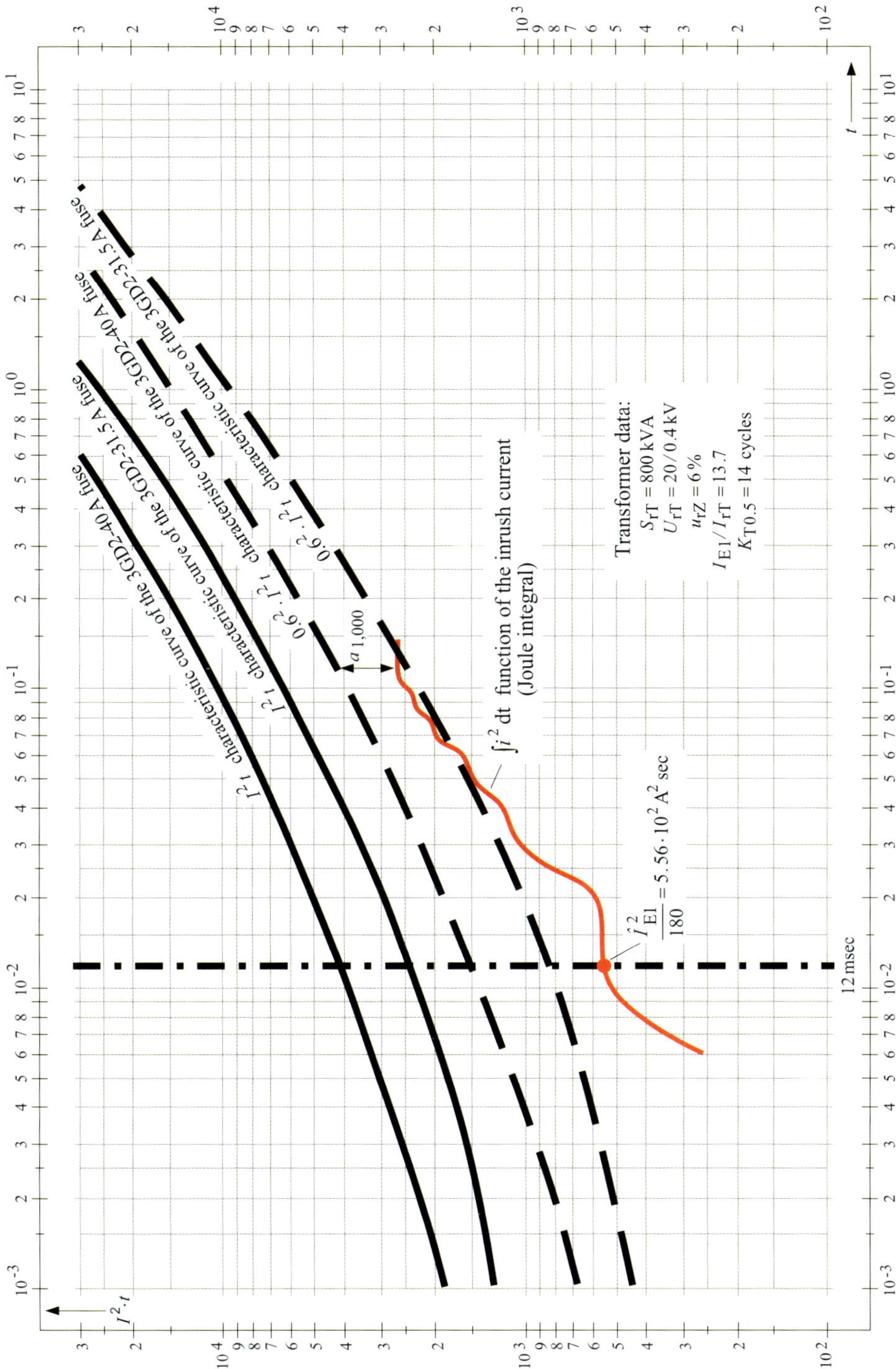

Fig. B7.16 Inrush-current-dependent specification of the Siemens 3GD2 HV HRC fuse-link required to protect the 800-kVA GEAFOL cast-resin transformer

parison of the $\int i_T^2 dt$ function (Joule integral) with the $I^2 t$ characteristics shows, the rated current of the Siemens 3GD2 HV HRC fuses for protecting the 800-kVA GEAFOL cast-resin transformer must be at least $I_{r\text{-HV HRC-min}} = 40$ A.

A Siemens 3GD2 HV HRC fuse with a rated current of $I_{r\text{-HV HRC-min}} = 40$ A reliably withstands the heat effect of the inrush current on switch-on, even taking its ageing deterioration into account.

In addition to the minimum rated fuse current, a maximum rated fuse current must also be defined for protection of the 800-kVA GEAFOL cast-resin transformer. Definition of the maximum rated fuse current $I_{r\text{-HV HRC-max}}$ depends on the operating response of the HV HRC fuse on single-phase and multi-phase terminal short circuits on the secondary side of the transformer. To ensure clearance of the secondary-side terminal short circuits on the transformer, the primary-side flowing fault current I_{Fp} must be larger than the minimum breaking current $I_{b\text{-HV HRC-min}}$ of the HV HRC fuse (Eq. 7.20). The minimum breaking currents $I_{b\text{-HV HRC-min}}$ of Siemens 3GD2 HV HRC fuses are given in Table B7.10. Using the transformer data shown in Fig. B7.15, the following primary-side fault currents I_{Fp} flow in the case of solid terminal short circuits on the secondary side:

- 3-phase solid terminal short circuit (zero fault impedance)

$$I_{Fp} = I_{k3Tp} = \frac{1}{k_{Tr}} \cdot I_{k3Ts} = \left(\frac{1}{k_{Tr}} \cdot \frac{I_{rTs}}{u_{rZ}} \cdot 100\right) \tag{7.27.1}$$

$$I_{Fp} = I_{k3Tp} = \frac{1}{50} \cdot \left(\frac{1{,}155 \text{ A}}{6\,\%} \cdot 100\right) = 385 \text{ A}$$

- 1-phase solid terminal short circuit (zero fault impedance)

$$I_{Fp} = I_{k1Tp} = \frac{1}{\sqrt{3} \cdot k_{Tr}} \cdot I_{k1Ts} = \frac{1}{\sqrt{3} \cdot k_{Tr}} \cdot \left(\frac{3}{2 + \dfrac{Z_{0T}}{Z_{1T}}} \cdot I_{k3Ts}\right) \tag{7.27.2}$$

$$I_{Fp} = I_{k1Tp} = \frac{1}{\sqrt{3} \cdot 50} \cdot \left(\frac{3}{2 + 0.95} \cdot 19{,}250 \text{ A}\right) = 226 \text{ A}$$

In a 2-phase terminal short circuit on the secondary side, the magnitude of the 3-phase short-circuit current is always effective in one phase on the primary side of Dyn transformers. The worst case for HV HRC fuse operating is therefore a 1-phase secondary terminal short circuit. To handle the worst case of Example B6, the minimum breaking current $I_{b\text{-HV HRC-min}}$ of the 3GD2 HV HRC fuse must be selected according to the primary-side fault current $I_{Fp} = I_{k1Tp} = 226$ A. At $U_m = 24$ kV, this choice results in a maximum rated fuse current of $I_{r\text{-HV HRC-max}} = 63$ A (Table B7.10). The minimum breaking current $I_{b\text{-HV HRC-min}} = 210$ A of the 63-A 3GD2 HV HRC fuse is smaller than the primary-side fault current $I_{k1Tp} = 226$ A, that is, the operating condition (7.20) for 1-phase terminal short circuits on the secondary side is complied with (210 A < 226 A). The following Siemens HV HRC fuses can therefore be used to protect the 800-kVA GEAFOL cast-resin transformer:

- 3GD2-40A
  ($I_{b\text{-HV HRC-min}} = 140$ A permits 60 % damping of the fault current in case of a 1-phase terminal short circuit),

- 3GD2-50A
  ($I_{b\text{-HV HRC-min}}$ = 170 A permits 30 % damping of the fault current in case of a 1-phase terminal short circuit),
- 3GD2-63A
  ($I_{b\text{-HV HRC-min}}$ = 210 A permits 7 % damping of the fault current in case of a 1-phase terminal short circuit).

The transformer fuse protection Table B7.17 simplifies the selection of fuses according to the heat effect of the inrush current and the operation response in case of single-phase and multi-phase terminal short circuits on the secondary side. With this selection aid it is possible to assign the rated currents of Siemens 3GD2 HV HRC fuses to transformer power ratings without additional calculation.

Table B7.17 Assignment of the rated currents of Siemens 3GD2 HV HRC fuse-links to transformer power ratings (fuse protection recommendations)

Rated voltage for		Rated power of the transformer [1]	Rated primary current of the transformer [1]	Impedance voltage at rated current	Rated current for the	
Transformer [1]	HV HRC fuse				smallest HV HRC fuse to be connected line-side [2]	largest HV HRC fuse to be connected line-side [3]
U_{rT} [kV]	U_m [kV]	S_{rT} [kVA]	I_{rT} [A]	u_{rZ} [%]	$I_{r\text{-HV HRC-min}}$ [A]	$I_{r\text{-HV HRC-max}}$ [A]
6	7.2	250	24.1	4	40	80
		315	30.3	4	50	100
		400	38.5	4	63	100
		500	48.1	4	80	125
		630	60.6	4	100	160
		800	77.0	6	125	160
		1,000	96.2	6	160	200
		1,250	120.3	6	250	250
		1,600	154.0	6	2 x 160	
10	12	250	14.4	4	25	40
		315	18.2	4	32	50
		400	23.1	4	40	63
		500	28.9	4	50	80
		630	36.4	4	63	100
		800	46.2	6	80	125
		1,000	57.7	6	100	160
		1,250	72.2	6	160	160
		1,600	92.4	6	200	200
		2,000	115.5	6	250	250
20	24	250	7.2	4	16	25
		315	9.1	4	16	25
		400	11.6	4	20	40
		500	14.4	4	25	50
		630	18.2	4	31.5	63
		800	23.1	6	40	63
		1,000	28.9	6	50	80
		1,250	36.1	6	80	80
		1,600	46.2	6	100	100
		2,000	57.7	6	125	125

1) Transformers with the preferred standard vector group Dy5

2) Smallest possible rated current to control the heat effect of the inrush current on switch-on without damaging the fuse

3) Largest possible rated current to ensure operating in response to a secondary-side terminal short circuit

Step 2: Selection of HV HRC fuses according to the permissible power dissipation $P_{V\text{-perm}}$ if the fuses are installed in moulded-plastic housings (fuse boxes)

If HV HRC fuses are freely installed in open and enclosed MV switchgear, sufficient volumes of air and natural circulation of air provide the necessary dissipation of heat losses. Installing the HV HRC fuses in a moulded-plastic housing (fuse box) changes the thermal behaviour of the HV HRC fuse as compared with conventional installation. To prevent thermal destruction of the moulded-plastic housing, the power dissipation of the HV HRC fuses must be limited. The limit value applicable to the power dissipation of HV HRC fuses that can be installed in the fuse boxes of the SF_6-insulated, metal-enclosed and metal-clad Siemens NXPLUS C switchgear is $P_{V\text{-perm}} = 50$ W.

According to the transformer fuse protection Table B7.17, 3GD2 HV HRC fuses with a rated current in the range $40\text{ A} \leq I_{r\text{-HV HRC}} \leq 63$ A can be used for protection of the 800-kVA GEAFOL cast-resin transformer ($U_{rT} = 20$ kV). For fuse selection according to the power dissipation that is permissible in the fuse boxes of the NXPLUS C ($P_{V\text{-perm}} =$ 50 W), the dissipated power loss of these fuses must be calculated at maximum load current. The results of the power dissipation calculation are given in Table B7.18. As the calculation results in Table B7.18 show, all suitable Siemens 3GD2 HV HRC fuses meet the condition $P_{V\text{-perm}} > P_V$ (Eq. 7.22). Of all HV HRC fuses that meet the condition (7.29), the HV HRC fuse with the smallest rated current shall be selected.

The HV HRC fuse with the smallest rated current offers the best protection in case of secondary-side terminal short circuits with a high contact and arc resistance. For combination with the three-position switch-disconnector, the 3GD2-40A HV HRC fuse is therefore preferred.

Table B7.18 Dissipated power losses of Siemens 3GD2 HV HRC fuses used in the protection of an 800-kVA GEAFOL cast-resin transformer

HV HRC fuse	P_{rV} 1)	Dissipated power loss at maximum load current	
		Calculation according to equation (7.23)	Power dissipation P_V that must be controlled
3GD2-40A	106 W	$P_V = P_{rV} \cdot \left(\frac{I_{T-max}\ ^{2)}}{I_{r\text{-HV HRC}}}\right)^{2.5}$	$106\text{ W} \cdot \left(\frac{28.9\text{ A}}{40\text{ A}}\right)^{2.5} = 47.0\text{ W}$
3GD2-50A	108 W		$108\text{ W} \cdot \left(\frac{28.9\text{ A}}{50\text{ A}}\right)^{2.5} = 27.4\text{ W}$
3GD2-63A	132 W		$132\text{ W} \cdot \left(\frac{28.9\text{ A}}{63\text{ A}}\right)^{2.5} = 18.8\text{ W}$

1) Power dissipation at rated current $I_{r\text{-HV HRC}}$ (see Table B7.10)

2) Maximum load current of the 800-kVA GEAFOL cast-resin transformer
(The 800-kVA GEAFOL cast-resin transformer is used in the Siemens SITRABLOC transformer load-centre substation. The SITRABLOC substation has separate ventilation. While the separate ventilation (AF mode) is running, the transformer can be permanently loaded with up to 140% of its rated power ($k_{AF} \leq 1.4$). The SITRABLOC fans should only switch on to ensure the necessary instantaneous reserve capacity in operation under fault conditions (OPFC) if a transformer fails. During normal operation (NOP), they must remain switched off for noise abatement reasons. The continuously permitted overload capability is therefore reduced from $k_{AF} = 1.4$ to $k_{AF} = n/n-1$, if the total power for $n-1$ transformers in AF mode (fans running) is greater than for n transformers in AN mode (fans switched off). Example B6 (Fig. B7.15) applies to a decentralized multiple incoming supply with $n = 5$ interconnected 800-kVA GEAFOL cast-resin transformers. If $n = 5$ AF-ventilated 800-kVA GEAFOL cast-resin transformers the following maximum load current can occur in operation under fault conditions:

$$I_{T-max} = I_{T-OPFC} = k_{AF-OPFC} \cdot I_{rTp} = \frac{n}{n-1} \cdot I_{rTp} = \frac{5}{4} \cdot 23.1\text{ A} = 28.9\text{ A})$$

Step 3: Verification of the earth-fault clearing condition in the case of low-impedance neutral earthing

Step 3 entails checking whether the selected 3GD2-40A HV HRC fuse will be able to clear earth faults quickly ($t_m < 100$ msec) and reliably. Fast and reliable clearance of an earth fault in the case of low-impedance earthing is ensured if the minimum initial line-to-earth symmetrical short-circuit current $I''_{k1\text{-min}}$ of the feeder is greater than the minimum breaking current $I_{b\text{-HV HRC-min}}$ of the HV HRC fuse used (Eq. 7.24).

The mininum initial line-to-earth symmetrical short-circuit current for protection of the 800-kVA GEAFOL cast-resin transformer with a switch-disconnector-fuse combination is $I''_{k1\text{-min}} = 637.3$ A (Fig. B7.19). This earth-fault current is greater than the mini-

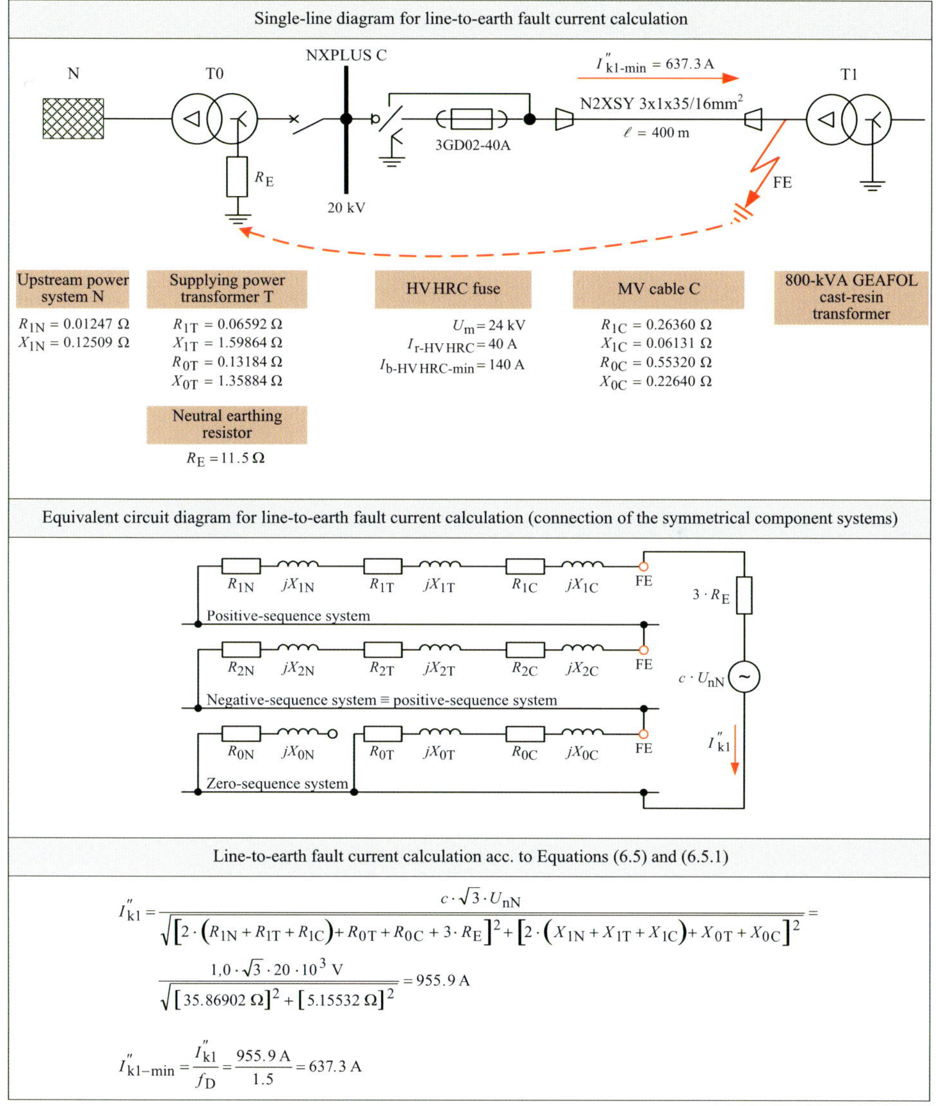

Fig. B7.19 Minimum initial line-to-earth symmetrical short-circuit current of the transformer feeder

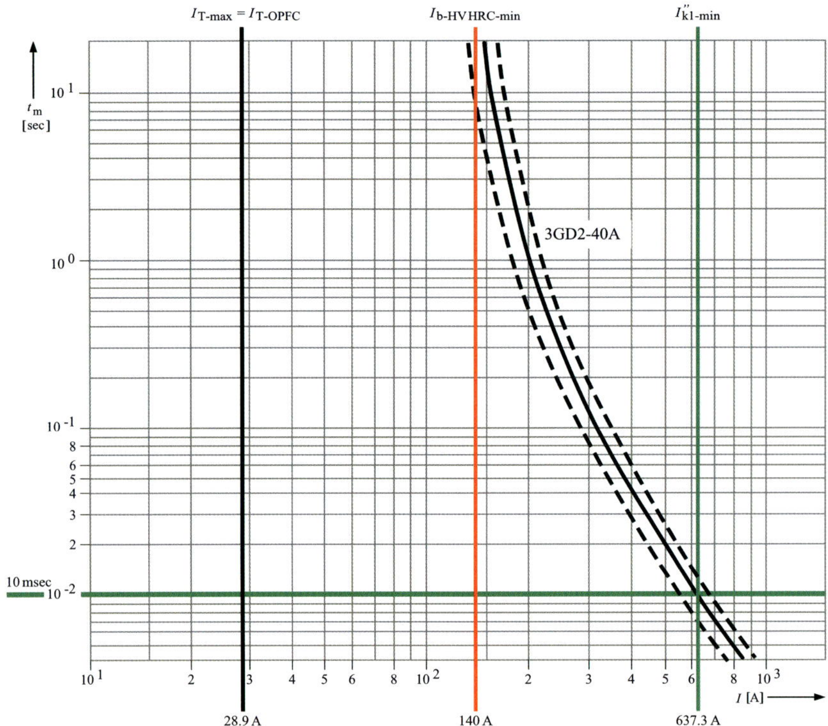

Fig. B7.20 Time-current characteristic of the 3GD2-40A HV HRC fuse

mum breaking current $I_{b\text{-HV HRC-min}} = 140$ A of the selected 3GD2-40A HV HRC fuse. The earth-fault clearing condition (7.31) is met without any problem (637.3 A > 140 A).

As Fig. B7.20 shows, the 3GD2-40A HV HRC fuse quickly (7 msec ≤ t_m ≤ 14 msec) and reliably clears earth faults occuring in its protection zone.

Step 4: Verification of the selectivity to the upstream and downstream system protection devices

Step 4 involves verification of the selectivity to the upstream and downstream system protection devices at the current-time level. This verification can be performed with computer assistance using the SIGRADE PC program [7.44].

Fig. B7.21 shows the selectivity verification obtained with SIGRADE for the example 20/0.4-kV load-centre system with $n = 5$ 800-kVA GEAFOL cast-resin transformers in a decentralized installation that feed in parallel into the SIVACON 8PS low-voltage high-current busbar system (Section 11.2.4). The time grading diagrams in Fig. B7.21 verify that all faults occurring in the 20/0.4-kV load-centre system can be cleared selectively. According to this, use of the 3GD2-40A HV HRC fuse is also permissible in terms of selective fault clearance.

After that, it is necessary to check whether the 3GD2-40A HV HRC fuse can be combined with the three-position switch-disconnector of the NXPLUS C.

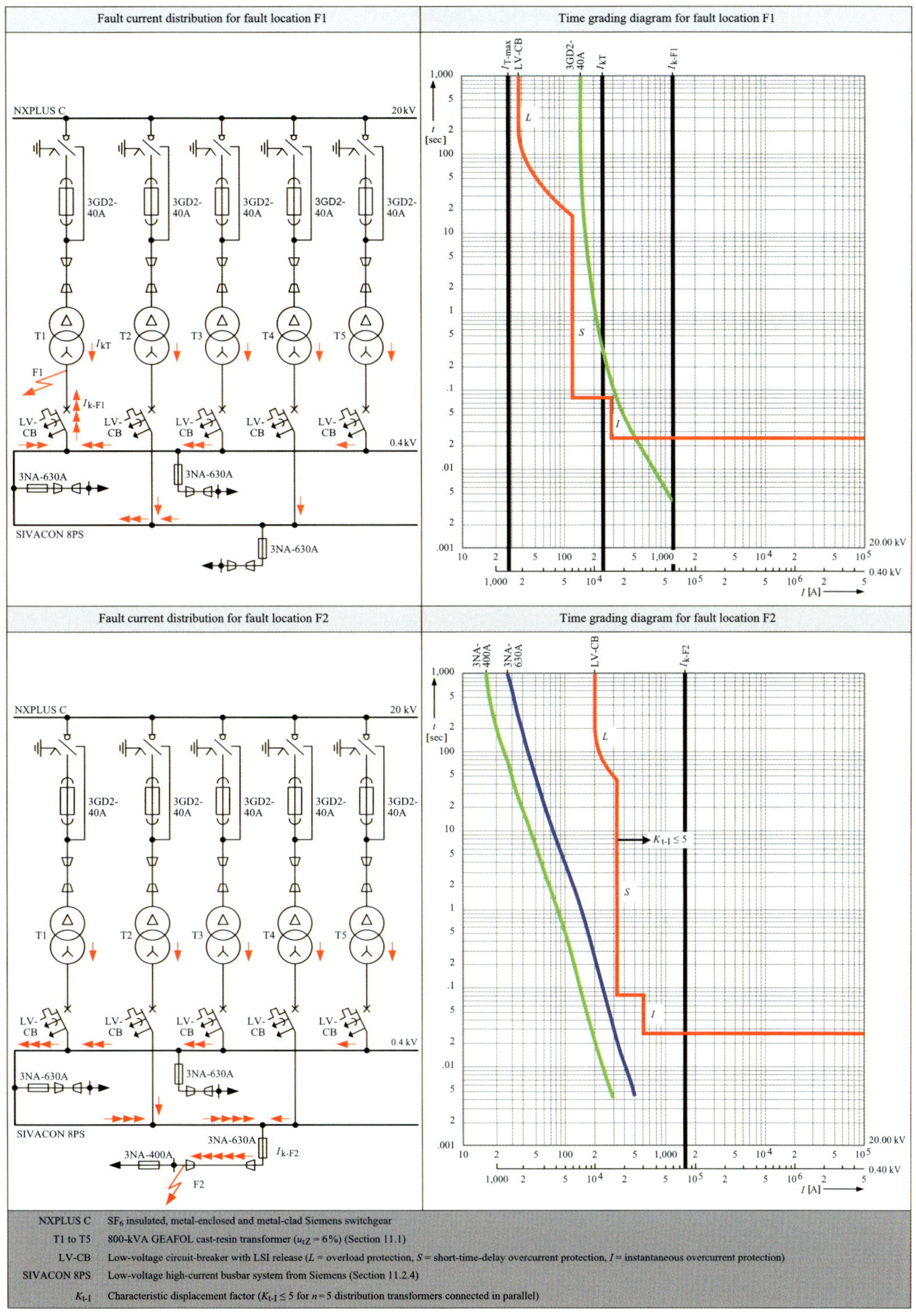

Fig. B7.21 Selectivity verification for a 20/0.4-kV load-centre power system with $n = 5$ parallel supplying 800-kVA distribution transformers

Step 5: Verification whether the selected HV HRC fuse can be combined with the switch or switch-disconnector of the MV switchgear used

Verification of correct functional interaction between the three-position switch-disconnector and the selected 3GD2-40A HV HRC fuse is defined in the standard DIN EN 62271-105 (VDE 0671-105): 2003-12 [7.41] or IEC 62271-105: 2002-08 [7.42].

This standard defines the switching duty for the switch or switch-disconnector and the HV HRC fuse depending on the fault current that is flowing. A split of the switching duty takes place with the transfer current I_{TC} of the combination.

To determine the transfer current, a line must be drawn parallel with the current axis through point $0.9 \cdot T_0$ on the pre-arcing time axis. The point of intersection with the lower time-current characteristic of the HV HRC fuse-link proposed to protect the distribution transformer yields the transfer current I_{TC} of the combination. The transfer current I_{TC} determined for the combination must be smaller than the rated transfer current I_{rTC} of the switch or switch-disconnector (Eq. 7.25).

The standard DIN EN 62271-105 (VDE 0671-105): 2003-12 [7.41] or IEC 62271-105: 2002-08 [7.42] also stipulates that the transfer current I_{TC} must be smaller than the short-circuit current I_{k3Tp} that flows on the primary side in case of a solid (non-arcing) terminal short circuit on the secondary side of the transformer. The primary side fault condition caused by a solid short circuit on the transformer secondary terminals corresponds to very high transient recovery voltages (TRV values), which the switch in a combination may not be able to cope with. For that reason, the selected HV HRC fuses must be able to deal with such a fault condition without throwing any of the breaking duty onto the switch or switch-disconnector. This demand made of the HV HRC fuses of the combination involves the condition $I_{TC} < I_{k3Tp}$ which must additionally be met (Eq. 7.26). If the condition according to Eq. (7.26) is met, the transfer currents I_{TC} on striker pin tripping also correspond to faults for which arc resistance or fault line impedance reduce both the short-circuit current and the TRV values, and increase the power factor [7.41, 7.42].

Fig. B7.22 shows verification of the conditions for functionally correct interaction between the three-position switch disconnector of the NXPLUS C with 3GD2-40A HV HRC fuses to be met according to DIN EN 62271-105 (VDE 0671-105): 2003-12 [7.41] or IEC 62271-105: 2002-08 [7.42].

As the verification shows, the three-position switch-disconnector of the NXPLUS C combined with 3GD2-40A HV HRC fuses constitutes a standard-compliant full-range switching device for protection of the 800-kVA GEAFOL cast-resin transformer.

Step 6: Submission of a standard-based fuse protection recommendation for short-circuit protection of the distribution transformer used

Finally, a binding (i. e. standard-based) recommendation for the fuse protection of the distribution transformer must be submitted.

For fuse protection of the distribution transformer used in the example, the 800-kVA GEAFOL cast-resin transformer (Fig. B7.15), 3GD2-40A HV HRC fuses ($I_{r\text{-HV HRC}} = 40$ A, $I_{b\text{-HV HRC-min}} = 140$ A) must be used. In combination with the three-position switch-disconnector of the NXPLUS C ($I_{rTC} = 1{,}300$ A, $T_0 = 50$ msec), all conditions of the standard DIN EN 62271-105 (VDE 0671-105): 2003-12 [7.41] or IEC 62271-105: 2002-08 [7.42] can be reliably met with these HV HRC fuses.

After thorough calculation, it may be found that the switching device conditions (Eqs. 7.25 and 7.26) cannot be met by combining the HV HRC fuses available for selection with the three-position switch-disconnector or gas-evolving switch. For example, large impedance voltages at rated current u_{rZ}, low values of the rated transfer current

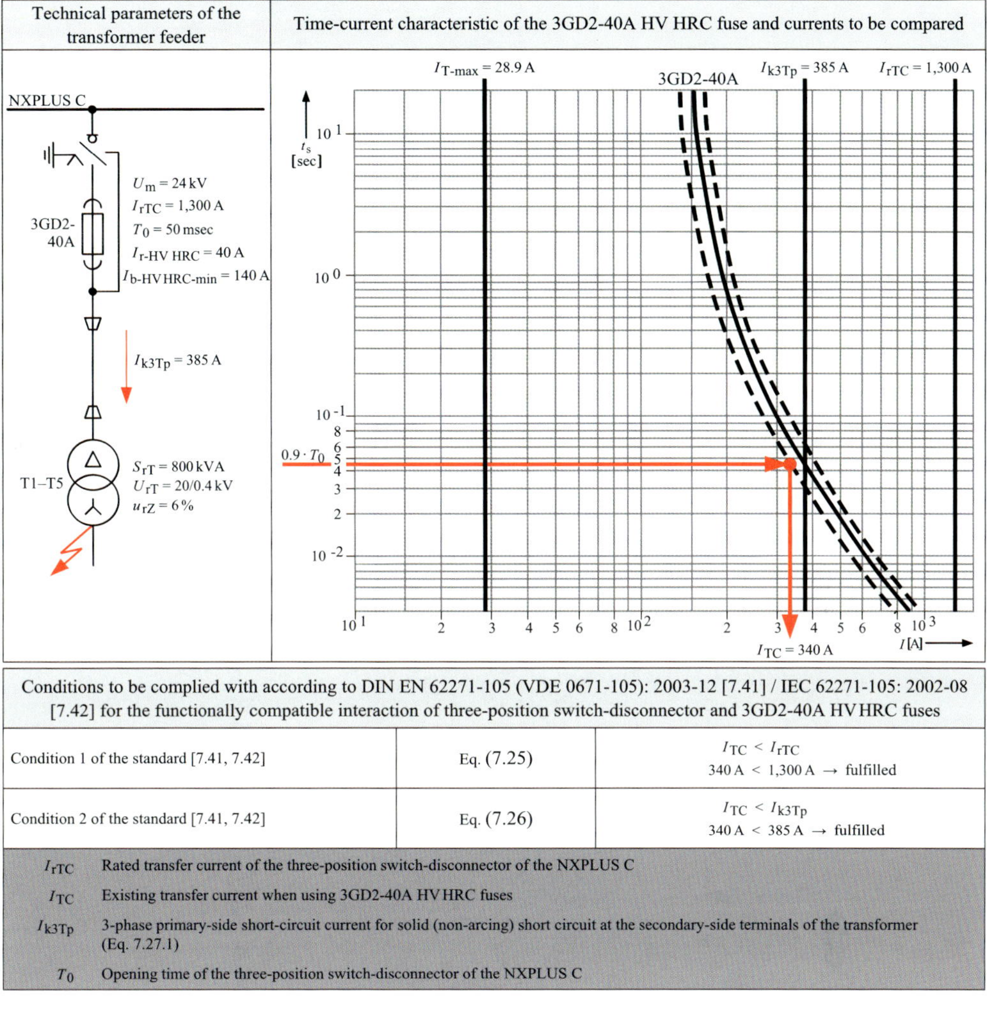

Fig. B7.22 Verification of the conditions as defined in the standards for the interaction of three-position switch-disconnector and 3GD2-40A HV HRC fuses

I_{rTC} or short opening times T_0 can result in the switching conditions not being met [7.45].

Use of a standard-compliant, current-limiting short-circuit protection for MV distribution transformers is subject to physical limits, above all due to the restricted breaking capacity of the three-position switch-disconnector or gas-evolving switch. Current-limiting short-circuit protection of distribution transformers can only be implemented without gaps using the withdrawable air-insulated vacuum-switch-fuse combination described in [7.46]. By using a vacuum switch with a breaking capacity type-tested according to DIN EN 62271-100 (VDE 0671-100): 2004-04 [7.47] or IEC 62271-100: 2008-04 [7.48] (test duty: "short-circuit current – make and break test"), MV distribution transformers can also be protected with current limiting in the power rating range 1,000 kVA ≤ S_{rT} ≤ 2,500 kVA (Table B7.23).

Table B7.23 Fuse protection recommendation for short-circuit protection of distribution transformers with a vacuum-switch-fuse combination [7.46]

Rated voltage U_m [kV]	Rated power of the transformer S_{rT} [kVA]	Rated current of the HV HRC fuse $I_{r\text{-HV HRC}}$ [A]
7.2	1,000	200
	1,250	250
	1,600	2 x 160
	2,000	2 x 200
	2,500	2 x 250
12	1,000	125
	1,250	160
	1,600	200
	2,000	2 x 125
	2,500	2 x 160
24	1,000	63
	1,250	80
	1,600	100
	2,000	100
	2,500	125

The transient recovery voltages (TRV values) occurring on the primary-side fault clearance of solid secondary-side terminal short circuits are reliably handled by type-tested vacuum switches. For vacuum switches type-tested according to DIN EN 62271-100 (VDE 0671-100): 2004-04 [7.47] or IEC 62271-100: 2008-04 [7.48], it is immaterial whether the switching device condition $I_{TC} < I_{k3Tp}$ (Eq. 7.26) is met. For that reason, in the combination of type-tested vacuum switches, HV HRC fuses with large rated currents (e.g. $I_{r\text{-HV HRC}} = 2 \times 250$ A at $U_m = 7.2$ kV) are also used. The use of HV HRC fuses with large rated currents increases the range of applications of the switch-fuse combination in the protection of distribution transformers.

7.3.2 Protection with a circuit-breaker-relay combination

Before the decision in favour of a circuit-breaker-relay combination, it should first be ascertained whether protection of the MV distribution transformer would be better achieved with HV HRC fuses in combination with a vacuum switch. The circuit-breaker-relay combination may only be used for transformer protection if there is no alternative to the use of fully enclosed SF_6-insulated MV switchgear (e.g. NXPLUS C, 8DH10) and the defined power rating of the transformer S_{rT} forces the decision. When used for transformer protection in low-impedance-earthed MV systems, the circuit-breaker-relay combination must meet the following protection requirements:

- reliable and selective clearance of multi-phase faults located between the cable sealing end of the feeder panel and the primary-side transformer terminals by DTL instantaneous tripping I>>,
- reliable and selective clearance of single-phase faults located between the cable sealing end of the feeder panel and the primary-side transformer terminals by DTL earth-fault tripping I_E>,
- reliable and selective clearance of single-phase and multi-phase faults located between the secondary-side transformer terminals and the LV incoming feeder circuit-breaker by time-delay DTL tripping I>,

- reliable back-up clearance of single-phase and multi-phase faults with high arc resistance located in the region of the LV busbar by time-delay DTL tripping $I>$,
- reliable prevention of spurious inrush-current-induced protection tripping,
- high operating response by an unsaturated current signal profile of the current transformers in case of a fault.

Example B7

Short-circuit protection for a 2,500-kVA GEAFOL cast-resin transformer

The short-circuit protection for a 2,500-kVA GEAFOL cast-resin transformer shown in Fig. B7.24 by way of example meets these requirements. The SIPROTEC 7SJ61 numerical time-overcurrent relay is selected as the relay for the short circuit protection. The 7SJ61 is set in accordance with the starting conditions for short circuit (Eq. 7.1), line-to-earth short-circuit protection (Eq. 7.2) and the clearing time condition (Eq. 7.9). The time-delay DTL trip stage of the 7SJ61 must be set in such a way that both multi-phase and single-phase faults on the secondary side of the transformer (fault location F3) can be cleared. For that reason, when setting the starting current $I>$ of the time-delay DTL trip stage, it is important to ensure that the conversion of the single-phase fault current is performed from the secondary to the primary side of a Dyn transformer using the $I_{k1Ts}/(\sqrt{3} \cdot k_{Tr})$ quotient. The minimum single-phase fault current of the 2,500-kVA transformer converted from the 0.4-kV to the 20-kV side for the protection setting is $I_{k1\text{-}min\text{-}F3} = 630$ A. By setting time-delay DTL tripping to $I> = 345$ A/2.3 A and $TI> = 0.3$ sec, this fault is reliably and selectively cleared (Fig. B7.24). A time-delay tripping time in the range $0.3 \text{ sec} \leq TI> \leq 0.5$ sec is usually sufficient to achieve selectivity as far as the downstream protective devices in the LV system.

Primary-side faults (fault locations F1 and F2) must be cleared instantaneously ($TI>> \leq 0.1$ sec). Instantaneous clearance is only possible if faults on the low-voltage side do not cause the DTL high-set stage $I>>$ to trip. Because MV distribution transformers ensure current selectivity between the MV and LV voltage levels, this requirement is almost always met. This can be seen in the time grading diagram in Fig. B7.24 from the fact that the short-circuit bands for faults on the 20-kV side (fault locations F1 and F2) and the 0.4-kV side (fault location F3) do not overlap.

Accordingly, the tripping characteristic TR-MV-CB of the SIPROTEC 7SJ61 time-overcurrent relay for short-circuit protection of the 2,500-kVA GEAFOL cast-resin transformer shown green in Fig. B7.24 ensures reliable and selective clearance of all possible faults. With the actual accuracy limiting factor $K'_{ALF} = 37$ calculated for the circuit-breaker-relay combination TR-MV-CB, the current transformer stability conditions (Eq. 7.11) for the implemented SIPROTEC 7SJ61 time-overcurrent relay are reliably met. The SIPROTEC 7SJ61 time-overcurrent protection relay features inrush restraint. As comparison of the DTL tripping characteristic of the 7SJ61 with the inrush current of the 2,500-kVA GEAFOL cast-resin transformer in Fig. B7.25 shows, spurious inrush-current-induced protection tripping is reliably prevented if the inrush restraint is defective or deactivated.

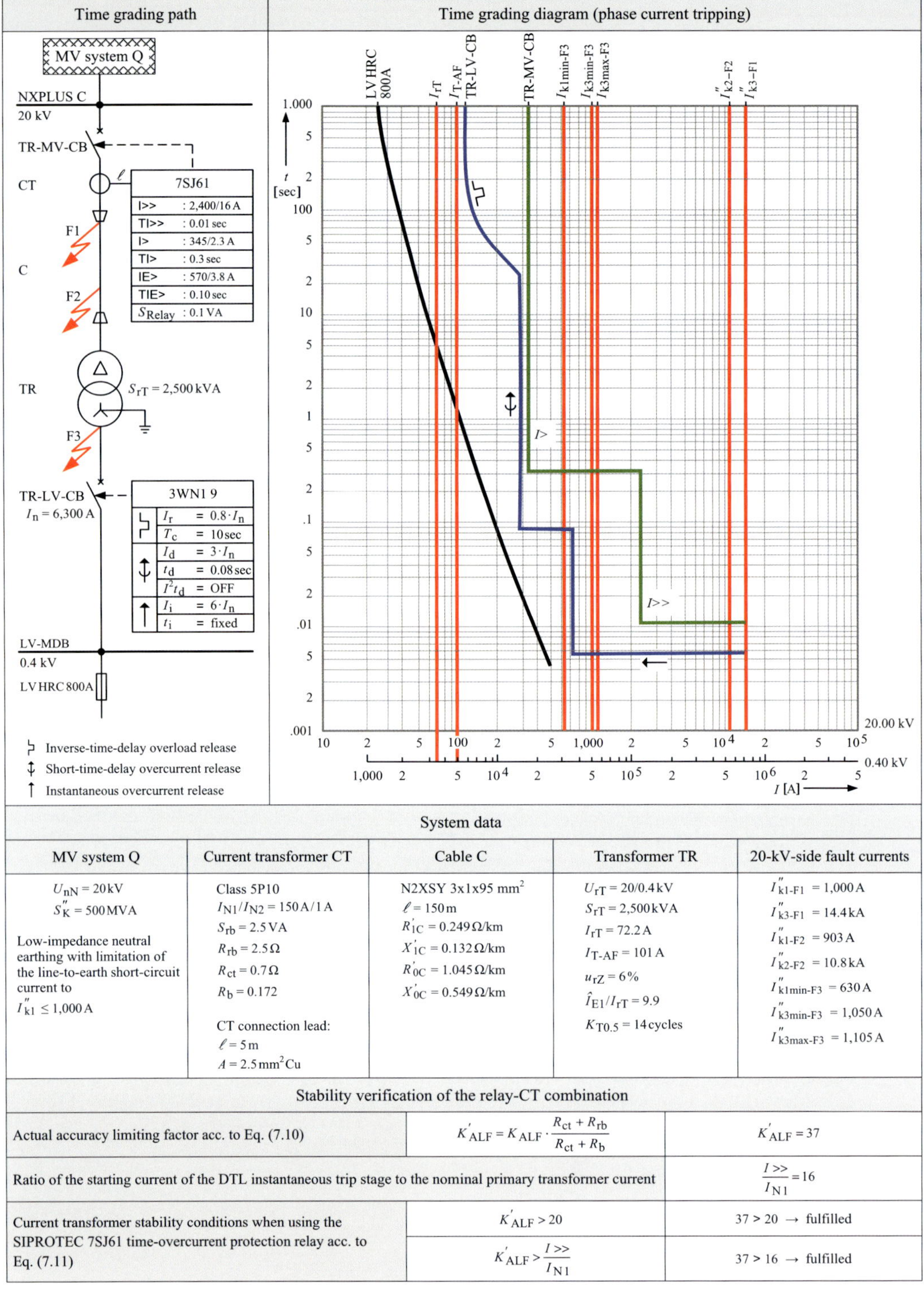

System data				
MV system Q	**Current transformer CT**	**Cable C**	**Transformer TR**	**20-kV-side fault currents**
$U_{nN} = 20\,kV$ $S''_K = 500\,MVA$ Low-impedance neutral earthing with limitation of the line-to-earth short-circuit current to $I''_{k1} \leq 1{,}000\,A$	Class 5P10 $I_{N1}/I_{N2} = 150\,A/1\,A$ $S_{rb} = 2.5\,VA$ $R_{rb} = 2.5\,\Omega$ $R_{ct} = 0.7\,\Omega$ $R_b = 0.172$ CT connection lead: $\ell = 5\,m$ $A = 2.5\,mm^2\,Cu$	N2XSY 3x1x95 mm² $\ell = 150\,m$ $R'_{1C} = 0.249\,\Omega/km$ $X'_{1C} = 0.132\,\Omega/km$ $R'_{0C} = 1.045\,\Omega/km$ $X'_{0C} = 0.549\,\Omega/km$	$U_{rT} = 20/0.4\,kV$ $S_{rT} = 2{,}500\,kVA$ $I_{rT} = 72.2\,A$ $I_{T\text{-}AF} = 101\,A$ $u_{rZ} = 6\%$ $\hat{I}_{E1}/I_{rT} = 9.9$ $K_{T0.5} = 14\,cycles$	$I''_{k1\text{-}F1} = 1{,}000\,A$ $I''_{k3\text{-}F1} = 14.4\,kA$ $I''_{k1\text{-}F2} = 903\,A$ $I''_{k2\text{-}F2} = 10.8\,kA$ $I''_{k1min\text{-}F3} = 630\,A$ $I''_{k3min\text{-}F3} = 1{,}050\,A$ $I''_{k3max\text{-}F3} = 1{,}105\,A$

Stability verification of the relay-CT combination		
Actual accuracy limiting factor acc. to Eq. (7.10)	$K'_{ALF} = K_{ALF} \cdot \frac{R_{ct} + R_{rb}}{R_{ct} + R_b}$	$K'_{ALF} = 37$
Ratio of the starting current of the DTL instantaneous trip stage to the nominal primary transformer current		$\frac{I>>}{I_{N1}} = 16$
Current transformer stability conditions when using the SIPROTEC 7SJ61 time-overcurrent protection relay acc. to Eq. (7.11)	$K'_{ALF} > 20$	37 > 20 → fulfilled
	$K'_{ALF} > \frac{I>>}{I_{N1}}$	37 > 16 → fulfilled

Fig. B7.24 Example of short-circuit protection of a 2,500-kVA GEAFOL cast-resin transformer with a circuit-breaker-relay combination (Example B7)

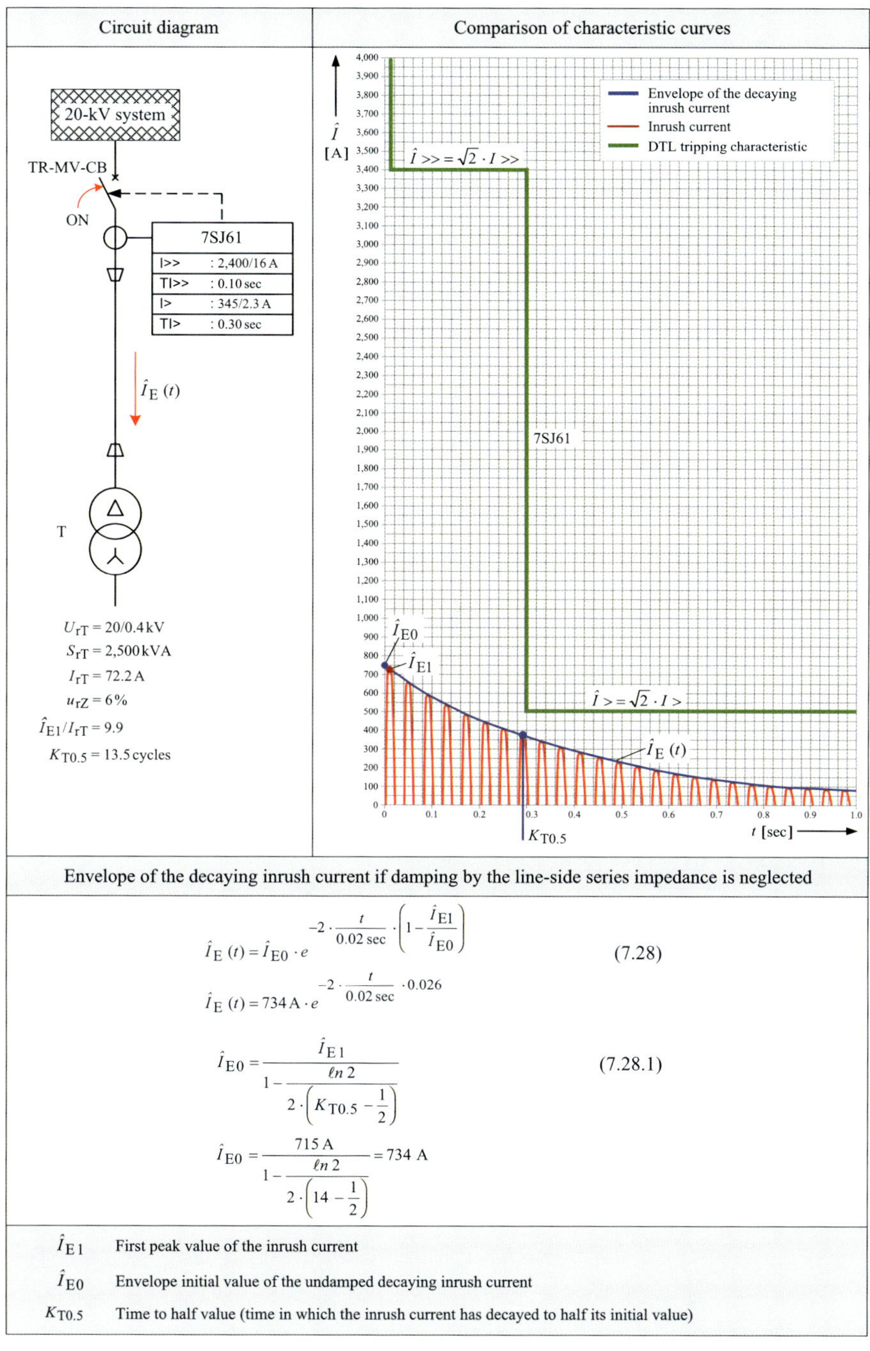

Fig. B7.25 Comparison of the DTL tripping characteristic of the SIPROTEC 7SJ61 time-overcurrent protection relay with the inrush current of the 2,500-kVA GEAFOL cast-resin transformer

7.4 Current-limiting short-circuit protection of motors and capacitors

Due to the high short-circuit powers in MV industrial systems, HV HRC fuses also provide excellent short-circuit protection for motors and capacitors. The rules and principles to be followed in the high-voltage-side fuse protection of motors and capacitors are explained below.

7.4.1 Fuse protection of HV motors

HV HRC fuses for protecting HV motors are preferably used in combination with vacuum contactors for rated voltages in the range 7.2 kV $\leq U_m \leq$ 12 kV. Vacuum contactors are switching devices used to switch currents of approximately the magnitude of their rated load current. When switching load currents, they are characterized by extremely high switching rates (1,200 operating cycles per hour when the Siemens 3TL vacuum contactor is used). The end of the mechanical life is not reached until after approximately 10^6 operating cycles at rated load current.

Table B7.26 Fuse protection of HV motors with 3GD2 HV HRC fuses for rated operational voltage U_m = 7.2 kV [7.31]

Operating data of the motor		Maximum permissible starting current $I_{start\text{-}perm}$ in A at fuse rated current $I_{r\text{-}HV\,HRC}$									
Starting time t_{start}	Number of starts n per h	50 A	63 A	80 A	100 A	125 A	160 A	250 A	315 A	2 x 160 A	2 x 250 A
≤ 6 sec	2	125	155	200	290	400	525	1,200	1,580	1,120	2,480
	4	115	140	185	260	360	480	1,070	1,440	1,110	2,230
	8	105	130	170	240	330	440	990	1,300	910	2,010
	16	-	120	155	220	305	410	910	1,220	820	1,810
≤ 15 sec	2	110	135	170	240	340	440	990	1,250	975	2,120
	4	100	120	155	220	310	400	895	1,150	870	1,910
	8	-	110	140	200	280	370	820	1,050	790	1,730
	16	-	100	130	185	260	340	755	975	710	1,550
≤ 30 sec	2	-	125	155	220	300	390	850	1,100	825	1,770
	4	-	110	140	200	270	360	775	1,000	740	1,590
	8	-	105	130	180	250	325	705	900	670	1,440
	16	-	-	120	170	230	300	655	850	600	1,290

Table B7.27 Fuse protection of HV motors with 3GD2 HV HRC fuses for rated operational voltage U_m = 12 kV [7.31]

Operating data of the motor		Maximum permissible starting current $I_{start\text{-}perm}$ in A at fuse rated current $I_{r\text{-}HV\,HRC}$			
Starting time t_{start}	Number of starts n per h	50 A	63 A	100 A	160 A
≤ 6 sec	2	125	155	290	525
	4	110	140	260	480
	8	100	130	240	440
	16	-	120	220	105
≤ 15 sec	2	110	135	240	440
	4	-	120	220	400
	8	-	110	205	370
	16	-	100	190	340
≤ 30 sec	2	-	125	220	390
	4	-	115	200	355
	8	-	105	180	315
	16	-	-	170	300

Vacuum contactors are therefore especially suitable for frequent switching of HV motors [7.49]. Owing to the relatively small breaking capacity (rated breaking current I_{rb} = 3,200 A when 7.2-kV 3TL vacuum contactors are used), they cannot interrupt high short-circuit currents. This duty must be performed by the current-limiting HV HRC fuses. The HV HRC fuses used exclusively for short-circuit protection of the motor circuit must be selected according to

- rated operational voltage U_m,
- starting current I_{start},
- starting time t_{start} and
- starting frequency.

If these data are taken into account, the following fuse protection condition for HV motors results:

$$I_{start\text{-}perm}|_{I_{r\text{-}HV\,HRC}} > I_{start} \tag{7.29}$$

$$I_{start\text{-}perm}|_{I_{r\text{-}HV\,HRC}} = f(t_{start}, n) \tag{7.29.1}$$

I_{start} starting current of the HV motor to be protected

$I_{start\text{-}perm}|_{I_{r\text{-}HV\,HRC}}$ maximum permissible motor starting current at fuse rated current $I_{r\text{-}HV\,HRC}$

The maximum permissible starting current handled by an HV HRC fuse without predamage depends on the starting time t_{start} and the number n of motor starts per hour. Taking this dependency into account, Tables B7.26 (U_m = 7.2 kV) and B7.27 (U_m = 12 kV) state the fuse protection recommendations for HV motors with Siemens 3GD2 HV HRC fuses.

The fuse rated currents contained in Tables B7.26 and B7.27 correspond to the rated currents for the smallest line-side connected HV HRC fuse in each case. The time-current characteristic of this HV HRC fuse must be coordinated with the relay tripping characteristic for the overload protection of the HV motor.

Example B8

Example of coordination of the current-limiting short-circuit protection with the overload protection for a 6-kV motor (Fig. B7.28)

The following principles apply to the coordination of the HV HRC fuse with the other components of the motor circuit:

- The fuse rated current $I_{r\text{-}HV\,HRC}$ must be larger than the rated current I_{rM} of the motor ($I_{r\text{-}HV\,HRC} > 2.5 \cdot I_{rM}$).
- The time-current characteristic of the HV HRC fuse must be located on the right of the motor starting current I_{start} (point A in the diagram of Fig. B7.28).
- The current I_B that results from the point of intersection of the t-I characteristic of the HV HRC fuse with the relay tripping curve for the overload protection (point B in the diagram in Fig. B7.28) must be greater than the minimum breaking current $I_{b\text{-}HV\,HRC\text{-}min}$ of the HV HRC fuse ($I_B > I_{b\text{-}HV\,HRC\text{-}min}$). If the condition $I_B > I_{b\text{-}HV\,HRC\text{-}min}$ cannot be met, the vacuum contactor must interrupt the overload currents not detected by the overload protection relay via the thermal striker of the HV HRC fuse-link.
- Correct interaction of the vacuum contactor with the HV HRC fuses in clearing short circuits can be checked using the coordination method for switch-fuse combinations according to DIN EN 62271-105 (VDE 0671-105): 2003-12 [7.41] or IEC 62271-105:

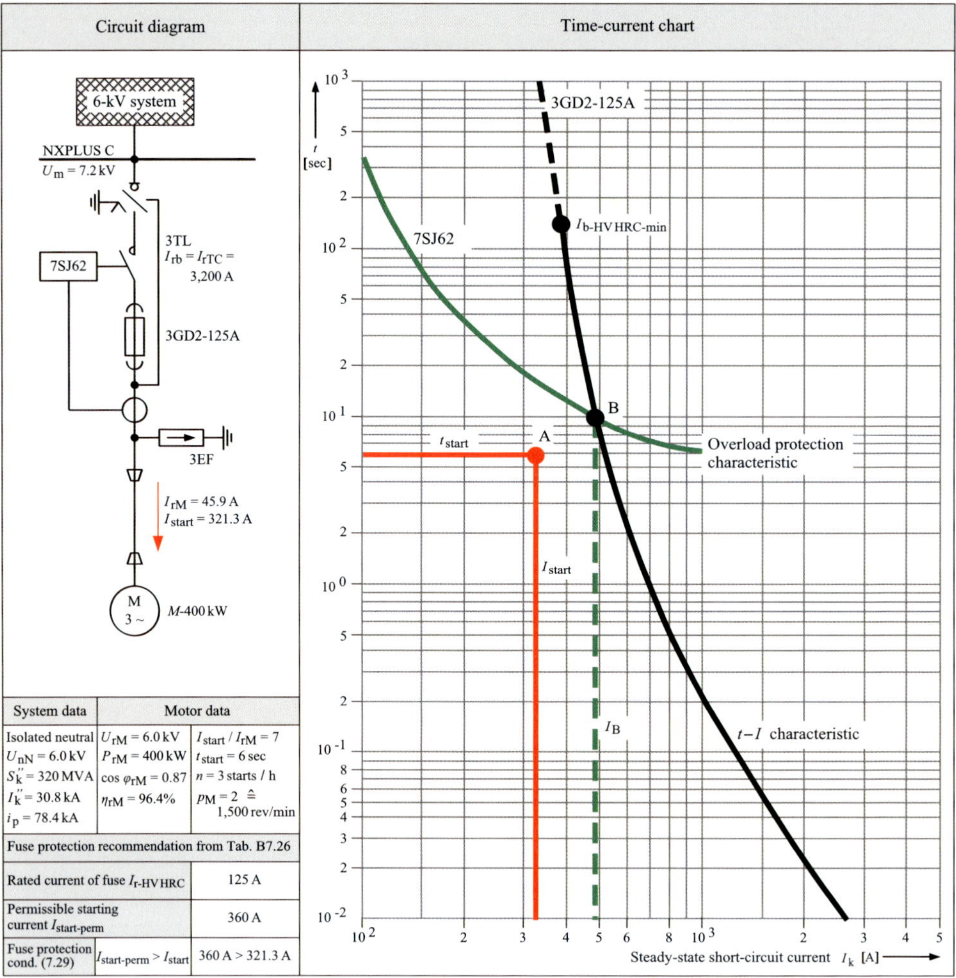

Fig. B7.28 Protection dimensioning and coordination for a 6-kV motor (Example B8)

2002-08 [7.42] (see Section 7.3.1). If Siemens 3TL vacuum contactors are used, to check the switching device condition $I_{TC} < I_{rTC}$, a rated transfer current of I_{rTC} = 3,600 A at U_m = 7.2 kV and I_{rTC} = 3,200 A at U_m = 12 kV and an opening time of T_0 = 50 msec can be used.

- The HV HRC fuse selected for protection of the HV motor limits the prospective peak short-circuit current i_p of the system to its cut-off current I_D. The maximum cut-off currents for the protection of HV motors with Siemens 3GD2 HV HRC fuses must be determined depending on the initial symmetrical short-circuit current I_k'' and the fuse rated current $I_{r\text{-HV HRC}}$ from the peak cut-off current characteristics in Fig. B7.29. The graphically determined cut-off current of the HV HRC fuse must not exceed the rated peak withstand current of the MV switchgear and the rated short-circuit making current of the vacuum contactor. If Siemens 3TL vacuum contactors are used, a maximum cut-off current of I_D = 50 kA is permissible.
- When accelerating HV motors with starting currents $I_{start} \leq$ 600 A are disconnected, high switching surges can occur. To lower the magnitude of these surges to safe values, surge limiters (e.g. Siemens type 3EF) must be used. 3EF surge limiters can

preferably be arranged parallel with the cable sealing end in the cable connection compartment.

If the described principles are followed, HV HRC fuses in combination with a vacuum contactor provide reliable and economical protection for HV motors in the power rating class $P_{rM} \leq 3$ MW at $U_m = 7.2$ kV and $P_{rM} \leq 5$ MW at $U_m = 12$ kV.

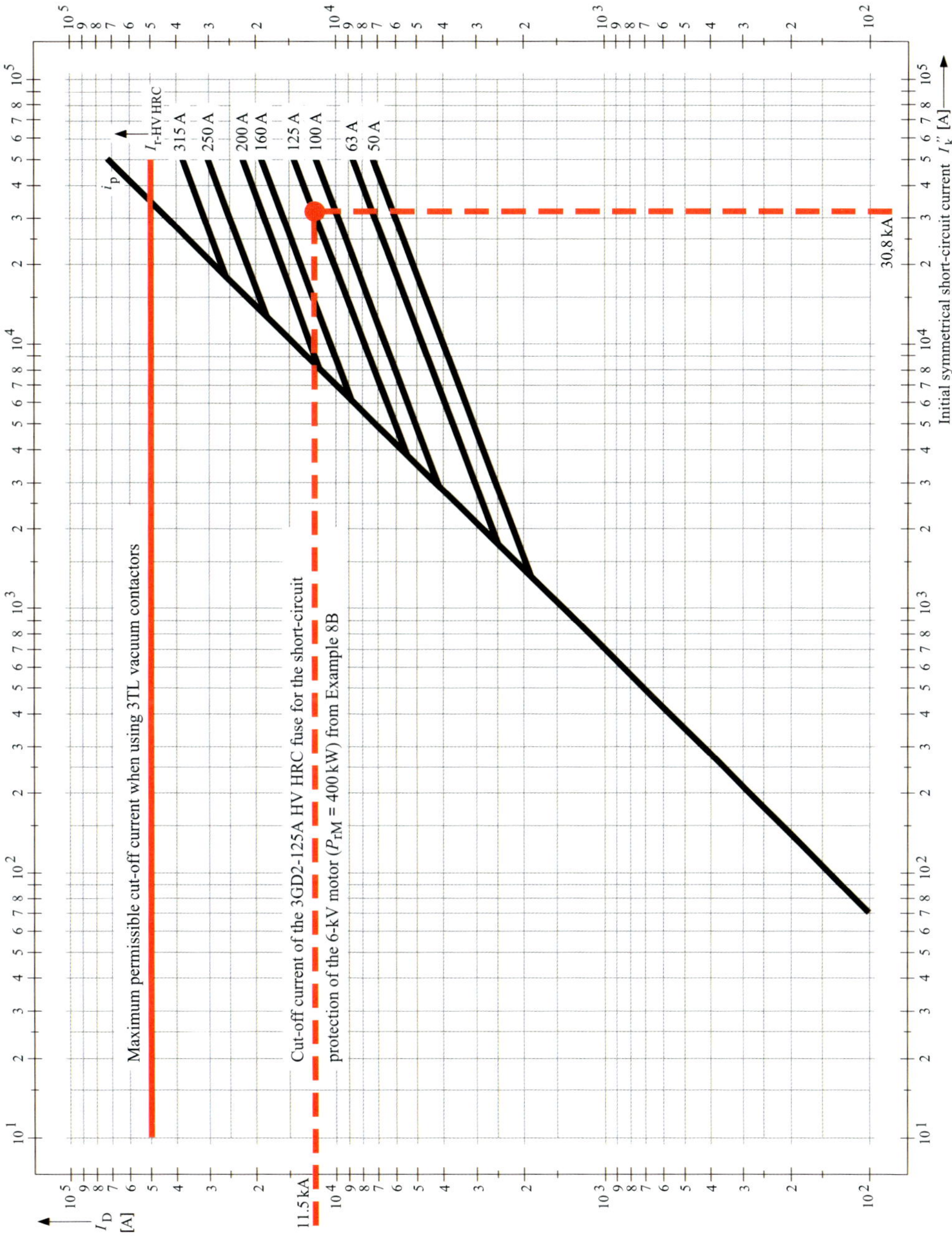

Fig. B7.29 Peak cut-off current characteristics of 3GD2 HV HRC fuses for the short-circuit protection of HV motors

7.4.2 Fuse protection of capacitors

The most important protection for MV capacitors (7.2 kV ≤ U_m ≤ 20 kV) is short-circuit protection with current-limiting HV HRC fuses. During power-up of capacitors and, in particular, when capacitors are connected in parallel, large circulating currents with short-circuit-like characteristics occur. The magnitude and duration of the current rushes that take place largely depend on the capacitor power, the natural frequency and inductance of the supplying system and the instant of closing. If closure takes place at a voltage maximum, current rushes are possible that can be 15 times the nominal current of the capacitor.

The duration of the current rushes is in the millisecond range (≤ 3 cycles). Despite the short duration, inrush currents subject HV HRC fuses to especially great stress.

Harmonic currents are a further stress factor. In addition to the 50(60)-Hz fundamental current, harmonic currents always flow through a capacitor. Assuming a regular harmonic load, the total current can be 1.3 to 1.4 times the nominal current of the capacitor.

To ensure lasting prevention of pickup and pre-damage of HV HRC fuses resulting from high harmonic and circulating currents, MV capacitors should be fuse-protected as follows:

$$2 \cdot I_{rC} \leq I_{r\text{-HV HRC}} \leq 4 \cdot I_{rC} \qquad (7.30)$$

I_{rC} nominal current of the capacitors or capacitor bank to be protected

$I_{r\text{-HV HRC}}$ fuse rated current for current-limiting capacitor protection

During capacitor switching, the components of the capacitor protection system are also subject to increased voltage stress. Owing to the switching surges that correlate with the current rushes, the rated operational voltage U_m of the HV HRC fuses and switchgear should be one voltage level higher than the normal voltage level for the equipment insulation at nominal system voltage U_{nN} (e.g. U_m = 12 kV instead of U_m = 7.2 kV at U_{nN} = 6.0 kV). The principles explained in Section 7.4.1 for coordination of the HV HRC fuses with a vacuum contactor must also be followed. It is also necessary to check whether the vacuum contactor or switching device is rated for a continuous current that is 1.43 times the rated current of the capacitor bank and suitable for switching capacitive currents. Siemens 3TL vacuum contactors are ideally suited for switching capacitive currents [7.49]. Their capacitive switching capacity is I_{rC} = 250 A in the voltage range 7.2 kV ≤ U_m ≤ 12 kV. When capacitors with 3TL vacuum contactors are switched, inrush currents up to 10 kA are reliably handled.

7.5 Protection of busbars

The largest fault currents in a power system occur on a 3-phase or line-to-earth short circuit of the busbar of the main switchgear. Because the tripping times of the DTL protection are longest for these currents for selectivity reasons, busbars are subject to especially high short-circuit stress. To limit the stress of the busbars effectively in case of a short circuit, the principle of time-overcurrent protection with reverse interlocking is convenient for simple single-busbar switchgear.

The principle of reverse interlocking is that the DTL protection of the incoming feeder(s) trips instantaneously independently of the grading time if its instantaneous tripping stage is not blocked by the overcurrent or earth-fault current starting in an outgoing feeder (Fig. B7.30).

The necessary starting current and tripping time stages and the binary interlocking input are present on the SIPROTEC time-overcurrent protection relays listed in Table B7.4,

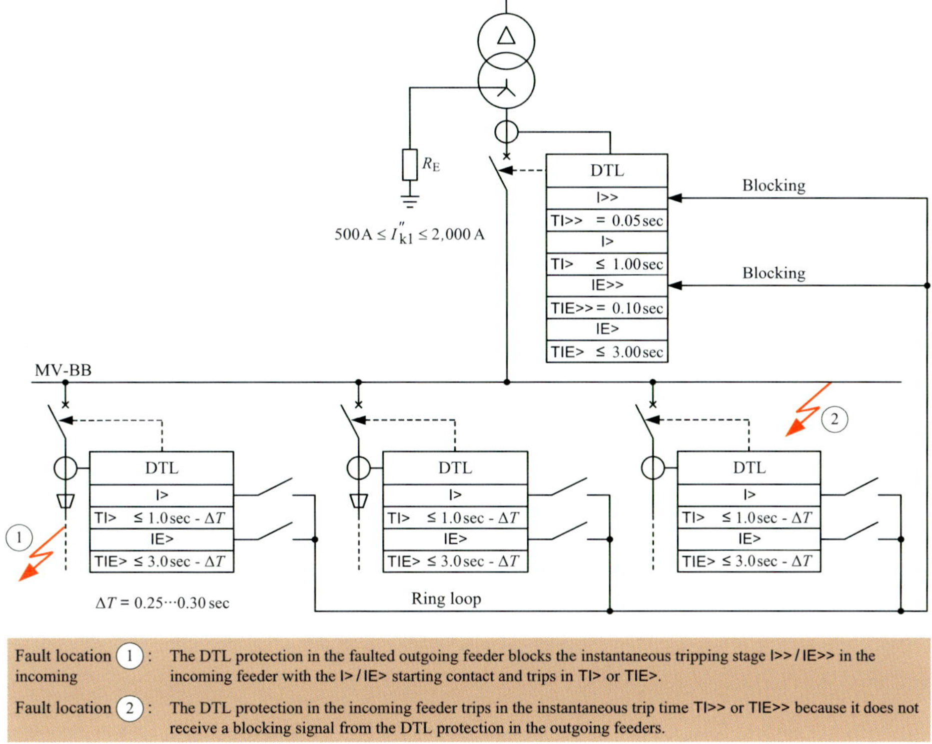

Fig. B7.30 Principle of reverse interlocking for fast short-circuit and line-to-earth short circuit protection of busbars

so that only the ring line for transmitting the blocking signal has to be installed. The principle of time-overcurrent protection with reverse interlocking therefore permits simple and low-cost busbar protection.

For especially important switchgear with double busbars, the SIPROTEC 7SS52 numerical busbar protection system is available (Fig. B7.8). This system is characterized, above all, by the fact that clearance of short circuits to limit damage to the busbar is performed extremely quickly (tripping time $t_a \leq 15$ msec). However, with today's state of the art, busbar short circuits on MV switchgear can be largely prevented with single-pole enclosure of the primary conductors (e.g. Siemens 8DA/8DB switchgear) or 1-pole solid-insulated, screened busbars (e.g. Siemens NXPLUS C switchgear). For such MV switchgear, additional busbar protection only makes sense if the tripping time is to be kept extremely short to limit the damage in the event of a line-to-earth short circuit.

7.6 Protection of lines

DTL and line differential protection devices have been shown to provide ideal general short-circuit and line-to-earth short-circuit protection for cable lines in low-impedance-earthed industrial power systems. Taking the selectivity into account, the DTL line protection must be graded such that the total clearing time in the incoming feeders is limited as far as possible to $t_{a\text{-total}} \leq 1.0$ sec. If SIPROTEC time-overcurrent protection relays are used (Table B7.4), the grading time can be $\Delta T = 250$ msec. The $I>/\vec{I}>$ overcurrent starting of the DTL line protection must be set according to Eq. (7.1). Using Eqs. (7.2) and (7.3), it is possible to set the $IE>/\overrightarrow{IE>}$ line-to-earth short-circuit starting. For the SIPROTEC numerical line differential protection, the following setting is recommended:

$$I\text{-}DIFF{>}L \geq \begin{cases} (2.0...3.0) \cdot I_C \\ 0.15 \cdot I_{perm} \end{cases} \tag{7.31}$$

$$I_C = \frac{U_{nN}}{\sqrt{3}} \cdot C'_{oper} \cdot \omega \cdot l_{cable} \tag{7.31.1}$$

$$I_{perm} = I_r \cdot \prod_i f_i \tag{7.31.2}$$

$I\text{-}DIFF{>}L$	starting current of the line differential protection
I_C	charging current of the cable
I_{perm}	permissible current-carrying capacity of the cable
I_r	rated current-carrying capacity of the cable
U_{nN}	nominal voltage of the power system
C'_{oper}	operating capacitance per unit lenght of the cable
l_{cable}	cable length
ω	angular frequency ($\omega = 2 \cdot \pi \cdot f$)
$\prod_i f_i$	product of the reduction factors for site operating conditions (i.e. differing air temperatures, grouping of cables)

The data required to calculate the charging current I_C and the permissible current-carrying capacity I_{perm} can be taken from the Siemens cable manual [7.50].

In cable networks with low-impedance neutral-point earthing, the SIPROTEC line differential protection is expected to trip instantaneously so that *TI-DIFF>L* = 0.00 sec has to be set as the tripping time.

7.6.1 Protection in the case of double-radial-line connection of system distribution substations

If system distribution substations are connected directly to the busbar of the main switchgear through parallel cables, it is not difficult to limit the total clearing time of the DTL protection in the incoming feeder to $t_{a\text{-}total} = 1.0$ sec. The precondition for this is that the distribution transformers of the system substation are protected according to Section 7.3.1 or 7.3.2.

Two practicable solutions for the general short-circuit and line-to-earth short-circuit protection in a double-radial-line connection of system substations are shown in Fig. B7.31.

In the solution for substation A, the two cable lines are protected with SIPROTEC 7SD610 numerical differential devices. The SIPROTEC 7SJ61 numerical time-overcurrent protection relay is used as the back-up protection for the line differential protection. If the line differential protection fails, the 7SJ61 DTL back-up protection relays can clear the fault alone, but not selectively. For selective clearance of faults by the back-up protection, the incoming feeder panels of the system substation A must additionally be equipped with directional time-overcurrent protection.

Directional time-overcurrent protection as the main protection is the solution for connection of the system distribution substation B to the main switchgear. Like the line differential protection, the solution with the two 7SJ62 DTL direction protection relays in substation B also ensures selective disconnection of a faulted cable line. However, a

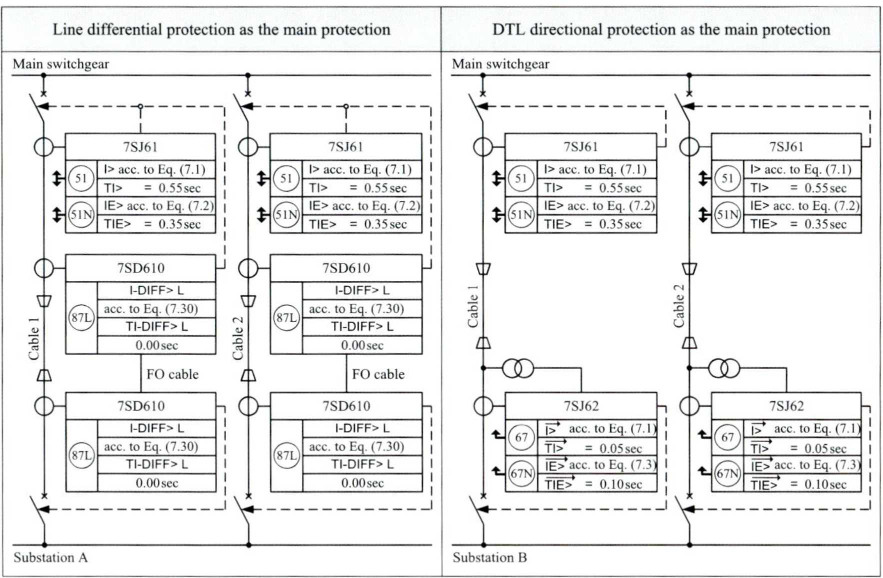

Fig. B7.31 General short-circuit and line-to-earth short-circuit protection in case of double-radial-line connection of distribution substations

short circuit is only finally cleared after *TI>* = 0.55 sec. Clearance with the 7SD610 differential protection, on the other hand, is instantaneous.

The solution with line differential protection as the main protection is therefore preferred.

7.6.2 Protection in the case of loop-in of system distribution substations

Fig. B7.32 shows the protection concept for the looping-in of two system distribution substations into a normally closed cable ring.

SIPROTEC 7SD610 differential protection for two-end lines constitutes the main protection for the two incoming feeder cables (cables 1 and 2) and the connecting cable between the substations (cable 3). The back-up protection is provided by non-directional SIPROTEC time-overcurrent protection relays (7SJ61) in the main switchgear and directional SIPROTEC time-overcurrent protection relays (7SJ62) in substations A and B. The tripping times of the DTL back-up protection devices must be time-graded in such a way that, if the line differential protection fails, all cable faults are cleared selectively.

The directional DTL back-up protection also ensures that the substation affected by a busbar fault is selectively disconnected from the power system. One low-cost alternative to line differential protection is the directional comparison protection already explained in Section 7.1. Fig. B7.33 shows the principle of operation of the directional comparison protection used in a normally closed cable ring.

According to Fig. B7.33, the following protection response occurs on a short circuit between substations A and B (cable 3):

- All DTL relays of the normally closed cable ring pick up.
- Relay 1.1 detects the short circuit in the forward direction; relay 1.2 in the reverse direction. Owing to of detection of the short circuit in the reverse direction, relay 1.2 blocks instantaneous overcurrent tripping of relay 1.1.

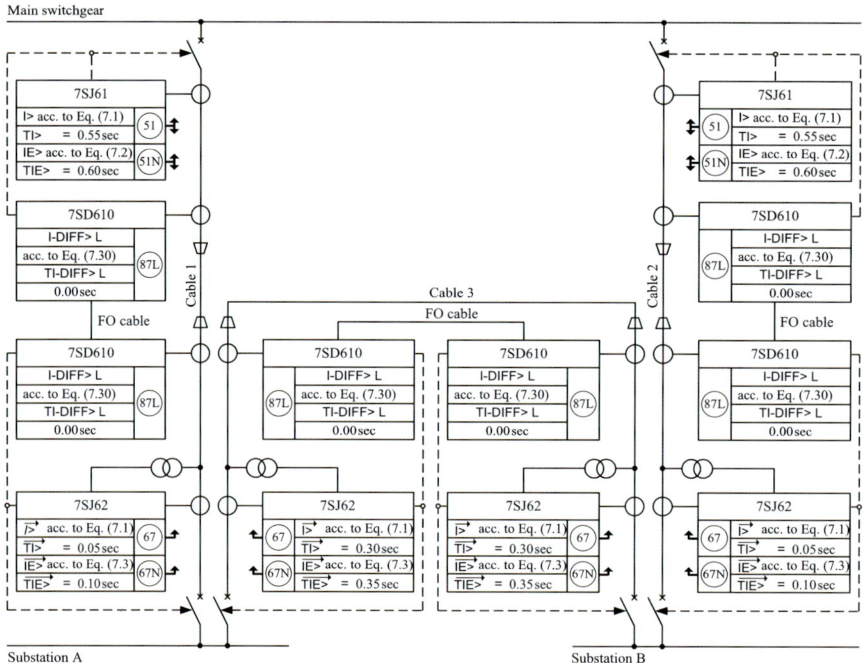

Fig. B7.32 General short-circuit and line-to-earth short-circuit protection in case of loop-in of distribution substations

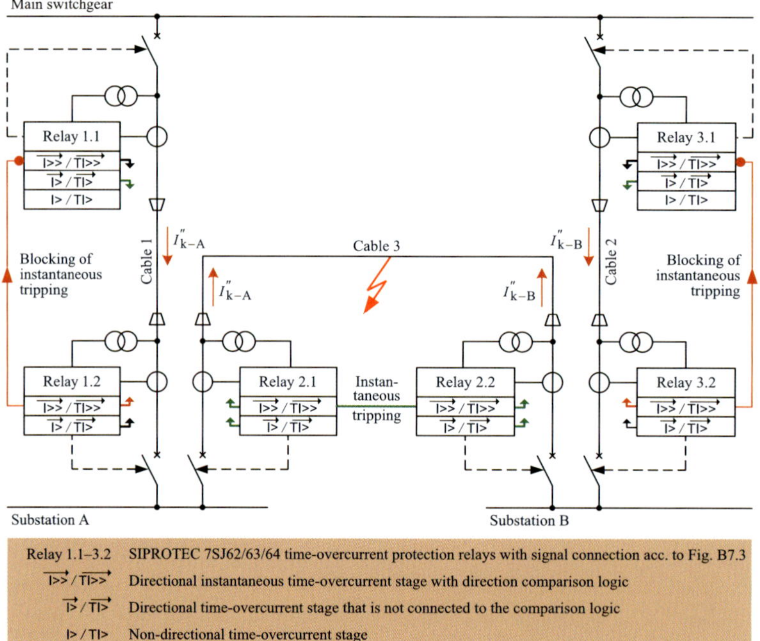

Fig. B7.33 Principle of operation of the directional comparison protection on a short circuit in a normally closed cable ring

- Relay 3.1 detects the short circuit in the forward direction; relay 3.2 in the reverse direction. Owing to detection of the short circuit in the reverse direction, relay 3.2 blocks instantaneous overcurrent tripping of relay 3.1.
- Relays 2.1 and 2.2 detect the short circuit in the forward direction. Owing to this detection of the short circuit in the forward direction both relays trip instantaneously ($\overrightarrow{TI}>>$ = 0.10 sec).

If the directional comparison logic fails, it is ensured that the short circuit is cleared by the selectively time-graded $\vec{I}>/\overrightarrow{TI}>$ settings of the DTL directional protection that is not connected with the comparison logic. $I>/TI>$ non-directional time-overcurrent tripping at the ends of the cable ring provides further back-up protection.

As an alternative to the differential protection, the directional comparison protection is a recommendable general short-circuit and line-to-earth short-circuit protection method for normally closed cable rings.

7.7 Protection concept for a fictitious 20-kV industrial power system with low-impedance neutral earthing

Taking the individual examples explained above into account in designing the power system protection, Fig. B7.34 shows the protection concept for a fictitious 20-kV industrial power system with low-impedance neutral earthing.

For connection of the industrial plant to the public 110-kV network, 110/20-kV transfer transformers are used with a delta-connected secondary winding (vector group Ynd11). For that reason, low-impedance neutral earthing is performed through two neutral earthing transformers with a connected neutral earthing resistor. The neutral earthing transformer and neutral earthing resistor are rated such that the line-to-earth short-circuit current is limited to I''_{k1} = 1,000 A. In a normally closed switching state of the bus sectionalizing circuit-breaker in the 20-kV main switchgear, only one of the two neutral earthing transformers has to be connected to the busbar.

With the protection concept shown in Fig. B7.34 by way of example, fast and selective clearing of single-phase and multi-phase faults is achieved. Compliance with the $(n-1)$ principle is achieved by isolation of the fault location using protection equipment (SR class 7, see Table A2.5).

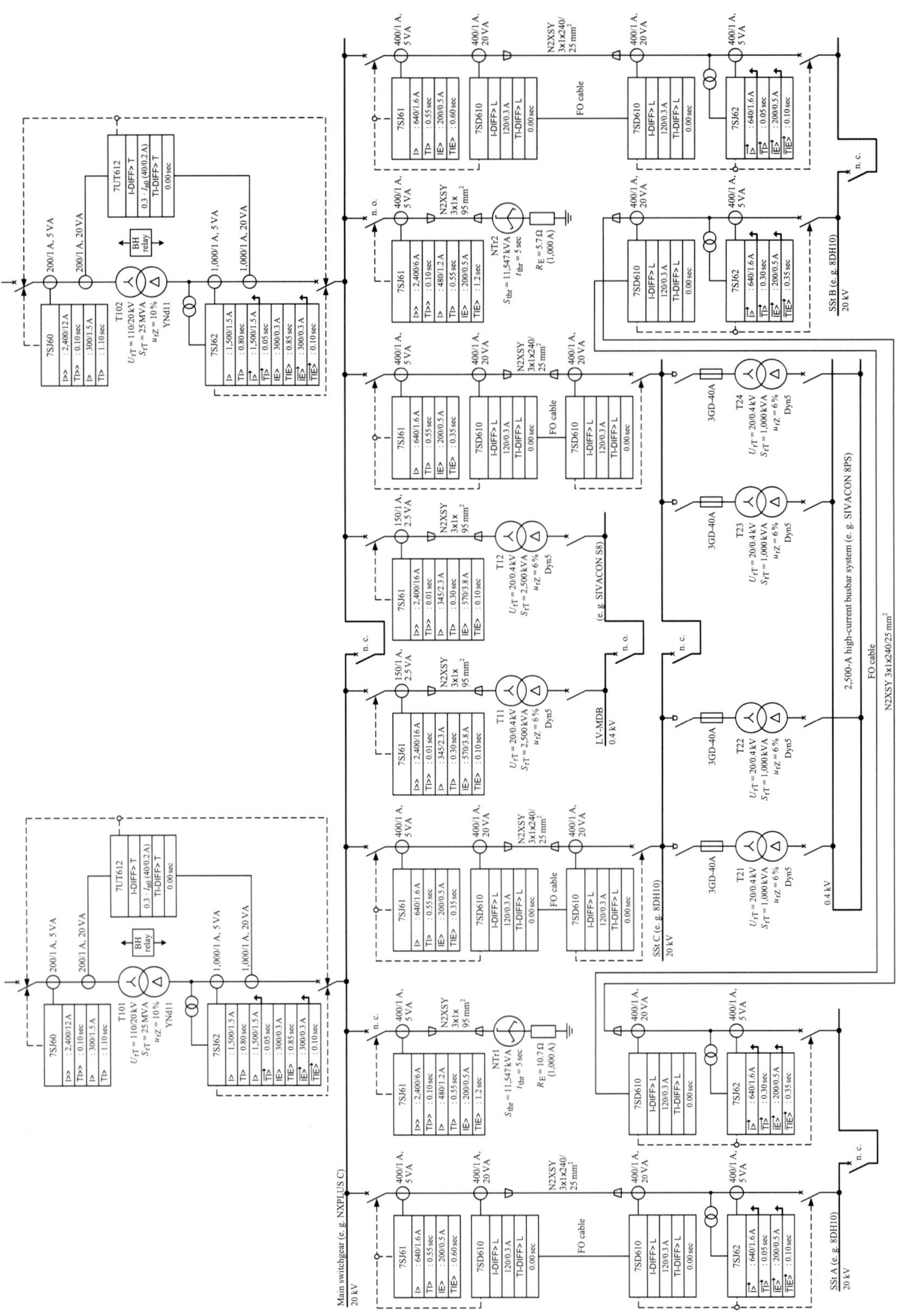

Fig. B7.34 Protection concept for a fictitious 20-kV industrial power system with low-impedance neutral earthing

C Planning recommendations for low-voltage systems

8 Choosing the LV system voltage

8.1 Categorization of the LV level as the process and load level

Of the three voltage levels used in industrial power supplies (high-voltage, medium-voltage, low-voltage), the low-voltage level (LV level) is the true process and load level. The vast majority of process-related end loads are powered from the LV level. Because of its importance for the quality of the power supply to the production processes, to integrated process automation and to computer-assisted information processing in all industries, it is also the main voltage level. This chapter discusses the system voltages available for the main voltage level according to the standards and how to select them based on technical and economic considerations.

8.2 Voltages for the process and load level

The standard DIN IEC 60038 (VDE 0175): 2002-11 [8.1] (IEC standard voltages) has been in effect in the Federal Republic of Germany since November 2002. IEC 60038: 2009-06 [8.2], which has been incorporated into the German standard VDE 0175 and was revised in June 2009, is the result of many years of effort to achieve a lasting reduction in the number of internationally applied standard voltage values for electrical power supply and consumers' installations and devices. According to this standard, the choice of voltage for LV systems with a nominal frequency of $f_N = 50$ Hz has been limited to the standard voltages 400 V/230 V 3AC and 690 V/400 V 3AC for all practical purposes (Table C8.1).

The line-to-line voltage $U_{LL} = 500$ V introduced in various industries (e.g. steel and chemicals) in about 1900 for operation of IT systems (see Section 10.2.1.1) has been removed from the standard . Because this voltage has no practical physical or mathematical relationship (division or multiplication by $\sqrt{3}$) with the system voltage 400 V, it is obsolete for use with modern equipment. To replace this impractical voltage, the standard voltages 400 V/230 V 3AC or 690 V/400 V 3AC have been mandatory for system operation since January 1, 2009.

In countries with a 60-Hz power supply (e.g. USA, Canada, Mexico, Brazil, Korea, Saudi Arabia), a larger number of standard voltages are available for operating LV networks. The voltage 208 V/120 V 3AC holds a special position among the available 60-Hz standard voltages. This voltage is used both for $f_N = 60$ Hz and for $f_N = 50$ Hz for process-related LV power supplies in the heavily Americanized IT industry (semiconductor and chip industry).

Table C8.1 Nominal voltages for alternating voltage systems in the range 100 V < U_{nN} ≤ 1,000 V according to DIN IEC 60038 (VDE 0175): 2002-11 [8.1] or IEC 60038: 2009-06 [8.2]

Three-phase four-wire or three-wire systems		Single-phase three-wire systems
Nominal system voltage U_{nN} in V		Nominal system voltage U_{nN} in V
f_N = 50 Hz	f_N = 60 Hz	f_N = 60 Hz
–	208/120	–
230	240	240/120
400/230	400/230	–
–	480/277	–
–	480	–
–	600/347	–
–	600	–
690/400	–	–
1,000	–	–
Comment on the voltage stated: e. g. 400/230 : Line-to-line / line-to-earth 480 : Line-to-line		

The circuit diagrams for generating the standard voltages defined in DIN IEC 60038 (VDE 0175): 2002-11 [8.1]/IEC 60038: 2009-06 [8.2] are shown in Figures a) to e) in Table C8.2.

For public and industrial LV distribution systems in European countries, the standard voltage 400 V/230 V 3AC is preferred [8.3]. With this voltage, above all, in the processing industry, the large number of existing loads of small and medium power (S_{load} ≤ 250 kVA) can be cost-efficiently connected to the power system. The other LV standard voltage 690 V/400 V 3AC is used, for example, in the petroleum and basic industries and in the auxiliaries systems of power stations, because their loads are on the whole larger (S_{load} > 250 kVA) and the feeder cables longer.

Individual motor drives with medium (500 kW < P_{rM} ≤ 2,000 kW) and large rated powers (P_{rM} > 2,000 kW) are supplied with power from the MV level (6(6.6) kV and 10(11) kV). In line with the state of the art, LV motor drives with rated powers up to P_{rM} ≤ 1,250 kW are also available on the market [8.4].

To supply these motors with power from the LV level, special starting devices such as star-delta starters, electronically controlled soft starters, or variable-frequency drives are used [8.5, 8.6].

The especially low-cost star-delta starters only reduce the starting torque and the starting current of the motor. Electronically controlled soft starters, on the other hand, ensure stepless and jerkless starting with a limited starting torque and starting current. This type of starting avoids sudden changes in the torque and transient current peaks. Compared with conventional starters, electronic frequency converters for drive systems permit stepless closed-loop control of the motor speed from zero to the nominal speed without falling torque. Moreover, motors controlled by variable-frequency drives can be operated above their nominal speed. Due to their advantageous control response and the possibility of extending the speed range, VFD-controlled motors are

Table C8.2 Circuit diagrams for generating the standard voltages defined in DIN IEC 60038 (VDE 0175): 2002-11 [8.1] / IEC 60038: 2009-06 [8.2]

Circuit diagram	Type of system	f_N	System voltages
a)	Three-phase four-wire system with earthed neutral (TN, TT system)	50 Hz	690 V / 400 V 400 V / 230 V (208 V / 120 V) 1)
		60 Hz	600 V / 347 V 480 V / 277 V 400 V / 230 V 208 V / 120 V
b)	Three-phase four-wire system without earthed neutral (IT system)	50 Hz	690 V / 400 V 400 V / 230 V (208 V / 120 V) 1)
		60 Hz	600 V / 347 V 480 V / 277 V 400 V / 230 V 208 V / 120 V
c)	Three-phase three-wire system without earthed neutral (IT system)	50 Hz	230 V 1,000 V
		60 Hz	240 V 480 V 600 V
d)	Three-phase three-wire system without neutral (IT system)	50 Hz	230 V 1,000 V
		60 Hz	240 V 480 V 600 V
e)	Single-phase three-wire system with earthed mid-point	60 Hz	240 V / 120 V

1) 50-Hz special voltage level for US-type load structures in the IT industry

increasingly being used for motor drives. The optimum operating voltage for motor drives depends on the following influencing factors [8.7 to 8.9]:

- number and rated power (individual power) of the motors,
- transformer limit rating for the short-circuit-proof dimensioning of the equipment,
- contribution of the motors to increasing the short-circuit current stress in the distribution system,
- voltage dip on motor starting,
- motor-induced system perturbations (voltage fluctuations, harmonics) on other loads,
- accumulated investment cost for the switchpanel, the connecting cable and the motor for alternative power supply from the MV or LV voltage level.

Taking these influencing factors into account, the standard voltages 400 V, 690 V, 6 kV and 10 kV can be used to optimize the 50-Hz operating voltage for motor loads.

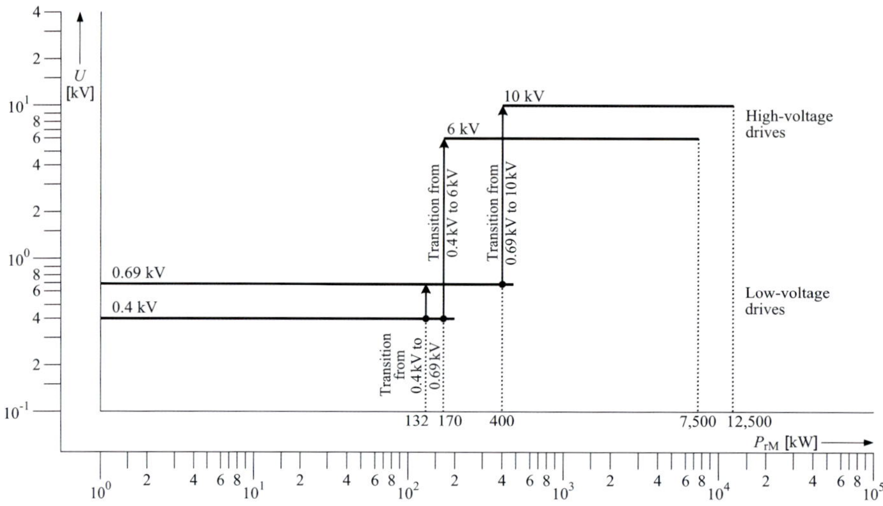

Fig. C8.3 Voltage levels for motor drives in auxiliaries systems of power stations [8.10]

Considering these standard voltages, optimization calculations were performed for the use of motor drives in auxiliaries systems of power stations in [8.10]. Fig. C8.3 shows the results of optimization calculations. According to Fig. C8.3, the following voltage-dependent power limits exist for the use of motor drives in power station auxiliaries systems:

- 132 kW for the transition from 0.4 kV to 0.69 kV,
- 170 kW for the transition from 0.4 kV to 6 kV,
- 400 kW for the transition from 0.69 kV to 10 kV.

When supplying motor drives from the 6-kV level, the permissible voltage dip during motor starting ($10\,\% < \Delta u' \leq 15\,\%$) can usually be complied with up to a power rating of $P_{rM} \leq 7.5$ MW. If the power rating of the individual drives is $P_{rM} > 7.5$ MW, the power must be supplied from the 10-kV level.

The cost-efficient power range for HV drives used in power station auxiliaries systems extends down to $P_{rM} = 170$ kW. From this power rating, supply from the LV level is preferable. In the power range of the LV level ($P_{rM} \leq 400$ kW), clear cost advantages can be ascertained for the load voltage 690 V/400 V 3AC. For this load voltage, cost advantages can be obtained largely by using smaller cable cross-sectional areas. The economic advantages continue to apply down to the lower power range ($P_{rM} < 10$ kW) but disappear as soon as the minimum cross-sectional area $A = 2.5$ mm^2 or $A = 1.5$ mm^2 is also reached for the load voltage 400 V/230 V 3AC [8.10].

In addition to the motor drives, welding machines (see Section 10.1.1.3) also have a special place in the supply of power to the LV level. By installing separate welding power systems, the technical advantages of operation with the standard voltage 690 V can be exploited to the full. Compared with 400-V welding power systems, 690-V welding power systems are characterized by smaller load currents and therefore smaller voltage drops on the incoming leads of the machine for the same impulse load. Because of the higher short-circuit power, voltage dips caused by welding are smaller for the same impulse load.

In all applications, however, the 690-V level is only ever an additional LV level alongside the 400-V level. There are always lighting systems and a large number of small loads that it is technically more convenient and cost-efficient to power with the standard voltage 400 V/230 V 3AC.

Conclusion

For the 50-Hz power supply of end loads, the LV standard voltages 400 V/230 V 3AC and 690 V/400 V 3AC are available. Because the voltage 400 V/230 V 3AC is essential for the power supply to the end loads, the only real decision facing the planner is whether to introduce a second LV level with the voltage 690 V/400 V 3AC. The 690-V level is a technically and economically advantageous LV level, especially for separately operated welding machines and motor drives in the power range $10\ \mathrm{kW} < P_{\mathrm{rM}} \leq 500\ \mathrm{kW}$. Large single motor drives ($P_{\mathrm{rM}} > 500\ \mathrm{kW}$) are preferably powered from the MV level (6 kV, 10 kV). The LV power supply in the IT industry is a special case. The voltage level 208 V/120 V 3AC may have to be introduced for special end loads in the IT industry (US-made production equipment and other devices).

9 Short-circuit power and currents in the low-voltage power system

9.1 Types and currents of faults determining the dimensioning of the system and equipment

Most of the end loads in industrial low-voltage power systems are three-phase asynchronous motors. The magnitude of the total short-circuit current in LV industrial power systems is therefore not just determined by the supplying distribution transformers but also by the three-phase asynchronous motors. When a short circuit occurs, the three-phase asynchronous motors act like generators for a few cycles and produce short-circuit current that is fed back into the industrial network. All connected running motors of the three-phase system additionally contribute to the initial symmetrical short-circuit current I_k'', peak short-circuit current i_p, symmetrical short-circuit breaking current I_b and, in asymmetrical short circuits, also to the steady-state short-circuit current I_k [9.1, 9.2]. In general, the short-circuit stress in an industrial network depends on the following characteristic quantities:

- short-circuit power S_k'' of the upstream MV network,
- rated power S_{rT} and impedance voltage u_{rZ} of the distribution transformer(s),
- vector group and method of neutral-point connection of the distribution transformer(s),
- power rating $P_{rM\text{-}\Sigma}$ of all connected running three-phase asynchronous motors,
- type of fault.

Fault types		Short-circuit current stress quantities
3-phase short circuit	a, L1; b, L2; c, L3	I_{k3}'', i_{p3}, I_{b3}, I_{k3}, I_{th3}
Line-to-line fault clear of earth	a, L1; b, L2; c, L3	I_{k2}'', i_{p2}, I_{b2}, I_{k2}, I_{th2}
Line-to-line fault with a connection to earth	a, L1; b, L2; c, L3	I_{k2E}'', i_{p2E}, I_{b2E}, I_{k2E}, I_{th2E}
Line-to-earth fault	a, L1; b, L2; c, L3	I_{k1}'', i_{p1}, I_{b1}, I_{k1}, I_{th1}

I_k'' Initial symmetrical short-circuit current
i_p Peak short-circuit current
I_b Symmetrical short-circuit breaking current
I_k Steady-state short-circuit current
I_{th} Thermal equivalent short-circuit current

Table C9.1
Fault types and short-circuit current stress quantities acc. to DIN EN 60909-0 (VDE 0102): 2002-07 [9.3] or IEC 60909-0: 2001-07 [9.4]

Table C9.2 Comparison of the stress quantities for the short-circuit-proof dimensioning of LV power systems with supplying Dyn5 transformers and motors contributing short-circuit current in case of a fault

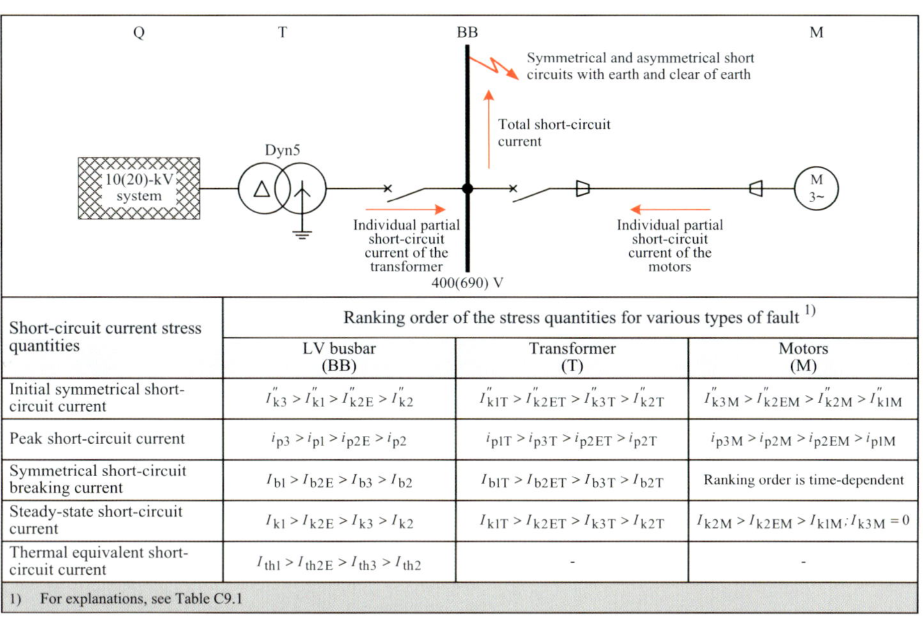

Short-circuit current stress quantities	Ranking order of the stress quantities for various types of fault [1]		
	LV busbar (BB)	Transformer (T)	Motors (M)
Initial symmetrical short-circuit current	$I''_{k3} > I''_{k1} > I''_{k2E} > I''_{k2}$	$I''_{k1T} > I''_{k2ET} > I''_{k3T} > I''_{k2T}$	$I''_{k3M} > I''_{k2EM} > I''_{k2M} > I''_{k1M}$
Peak short-circuit current	$i_{p3} > i_{p1} > i_{p2E} > i_{p2}$	$i_{p1T} > i_{p3T} > i_{p2ET} > i_{p2T}$	$i_{p3M} > i_{p2M} > i_{p2EM} > i_{p1M}$
Symmetrical short-circuit breaking current	$I_{b1} > I_{b2E} > I_{b3} > I_{b2}$	$I_{b1T} > I_{b2ET} > I_{b3T} > I_{b2T}$	Ranking order is time-dependent
Steady-state short-circuit current	$I_{k1} > I_{k2E} > I_{k3} > I_{k2}$	$I_{k1T} > I_{k2ET} > I_{k3T} > I_{k2T}$	$I_{k2M} > I_{k2EM} > I_{k1M}; I_{k3M} = 0$
Thermal equivalent short-circuit current	$I_{th1} > I_{th2E} > I_{th3} > I_{th2}$	-	-

1) For explanations, see Table C9.1

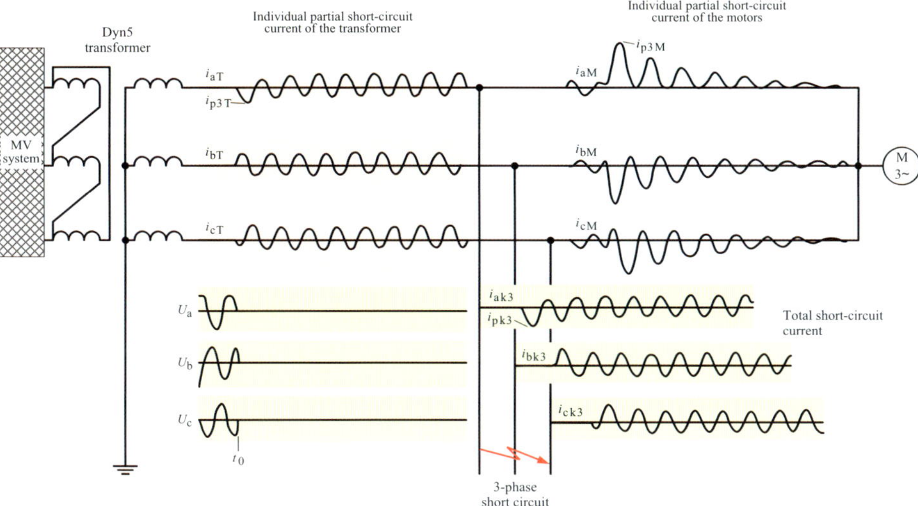

Fig. C9.3 Time characteristic of the currents and voltages on a three-phase short circuit in the central incoming supply of an LV system acc. to [9.2]

Today, industrial LV power systems are preferably operated as a TN system with a directly earthed transformer neutral point (see Section 10.2.1.3). In such power systems, symmetrical and asymmetrical short circuits to earth and clear of earth can occur.

The short-circuit current stress quantities that occur depending on the type of fault according to DIN EN 60909-0 (VDE 0102): 2002-07 [9.3] or IEC 60909-0: 2001-07 [9.4] are listed in Table C9.1. With the maximum values occurring during a short circuit, the stress quantities listed in Table C9.1 form the basis for short-circuit-proof dimensioning of the LV power system.

Table C9.2 compares the stress quantities for various types of fault for the purpose of short-circuit-proof dimensioning of LV power systems used to supply electricity to three-phase asynchronous motors and fed through transformers with the preferred vector group Dyn5. These stress quantities are arranged according to their magnitude. The type of short circuit that is followed by the highest dynamic or thermal short-circuit current stress is the determining factor for power system and equipment dimensioning.

The comparison in Table C9.2 shows that a three-phase short circuit on the LV busbar causes the largest initial symmetrical short-circuit power ($I''_{k\text{-max}} = I''_{k3}$) and the largest peak short-circuit current ($i_{p\text{-max}} = i_{p3}$).

The line-to-earth short circuit results in the largest symmetrical short-circuit breaking current ($I_{b\text{-max}} = I_{b1}$), the largest steady-state short-circuit current ($I_{k\text{-max}} = I_{k1}$) and the largest thermal equivalent short-circuit current ($I_{th\text{-max}} = I_{th1}$). Other short-circuit current conditions prevail on the secondary side of the supplying Dyn5 transformer. Here, the line-to-earth short circuit causes the largest dynamic and thermal stress values over the whole time period. The fault current stress dominated by the line-to-earth short circuit in the incoming feeder can be traced back to the impedance ratio $Z_{0T}/Z_{1T} \approx 0.95$ for transformers of the vector group Dyn5.

Whereas the maximum individual partial short-circuit currents flow in the transformer incoming feeder in a line-to-earth short circuit, the maximum short-circuit stress due to three-phase asynchronous motors contributing short-circuit current to the system occurs in a three-phase short circuit. As the curve for the individual partial short-circuit current of the motors involved in the short circuit of the LV system illustrated in Fig. C9.3 shows, the peak short-circuit current i_{p3M} is especially pronounced. Because of its prevalence, it is the short-circuit current quantity that determines the additional short-circuit stress in systems with motors contributing short-circuit current to the system in the event of a fault. When dimensioning systems and equipment for the supply of electrical power to three-phase asynchronous motors, special attention must therefore be paid to meeting the short-circuit current conditions $i_p \leq I_{cm}$ (Eq. 2.7) and $i_p \leq I_{pk}$ (Eq. 2.8).

Table C9.4 lists the short-circuit types and currents used as the basis for dimensioning LV systems with supplying transformers with the less common vector groups Yyn and Dzn.

In the case of short circuits in feeders of the LV power system, the short-circuit currents are always reduced by the supplementary impedances of the cables, which are always present. The current-reducing effect of the supplementary impedance, above all, depends on the cable cross-sectional area and the cable length (distance from the fault location).

Table C9.5 shows an example of dependency of the short-circuit current on the cable cross-sectional area and the cable length. The example shows the calculation performed in [9.2] of the peak short-circuit current for three-phase and line-to-earth short circuits in two different cable feeders of a 660-V industrial power system. As the generalizable calculation results in Table C9.5 show, the short-circuit currents are more strongly reduced in line-to-earth short circuits in cable feeders of LV industrial power systems than in three-phase short circuits.

The large reduction in line-to-earth short circuits (in the calculated example to 10.6 % if $A = 50\ \text{mm}^2$ and $l = 100$ m) is due to the prevalence of the cable zero-sequence impedance Z_{0C} that is especially high for small cable cross-sectional areas. The prevalent effect of the cable impedances in the case of far-from-infeed line-to-line and line-to-earth short circuits was also demonstrated in [9.5] for public LV networks.

Table C9.4 Types of fault and fault currents that are decisive for short-circuit-proof dimensioning of LV power systems with supplying Yyn/Dzn transformers

Vector group of the supplying transformer	Data decisive for system dimensioning: Fault type [1]	Short-circuit current stress quantities [1]
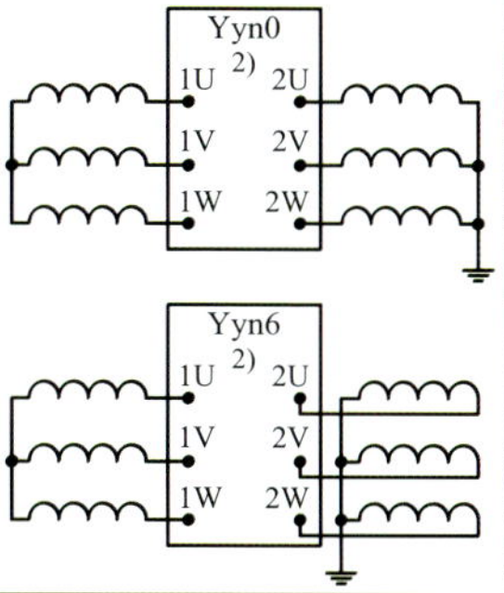 Yyn0 [2] (1U 2U, 1V 2V, 1W 2W) Yyn6 [2] (1U 2U, 1V 2V, 1W 2W)	3-phase short circuit	$I''_{k3} > I''_{k2}\,,\ I''_{k2E}\,,\ I''_{k1}$ $i_{p3} > i_{p2}\,,\ i_{p2E}\,,\ i_{p1}$ $I_{b3} > I_{b2}\,,\ I_{b2E}\,,\ I_{b1}$ $I_{k3} > I_{k2}\,,\ I_{k2E}\,,\ I_{k1}$ $I_{th3} > I_{th2}\,,\ I_{th2E}\,,\ I_{th1}$
	Because the ratio of the zero-sequence to the positive-sequence impedance is in the range $Z_{0T}/Z_{1T} = 7 \cdots 100$ [3], the fault currents clear of earth are many times greater than the line-to-earth fault currents $(3 \cdot I''_{k1} \leq I''_{k3} \leq 34 \cdot I''_{k1})$.	
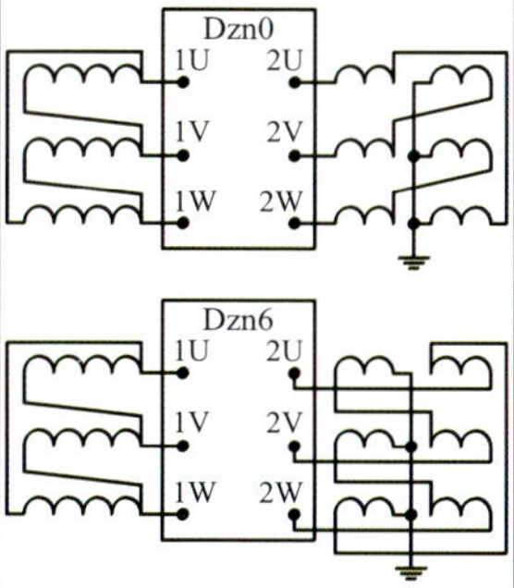 Dzn0 (1U 2U, 1V 2V, 1W 2W) Dzn6 (1U 2U, 1V 2V, 1W 2W)	Line-to-earth fault / Line-to-line fault with a connection to earth	$I''_{k1} > I''_{k2E} > I''_{k3} > I''_{k2}$ $i_{p1} > i_{p2E} > i_{p3} > i_{p2}$ $I_{b1} > I_{b2E} > I_{b3} > I_{b2}$ $I_{k1} > I_{k2E} > I_{k3} > I_{k2}$ $I_{th1} > I_{th2E} > I_{th3} > I_{th2}$
	Because the ratio of the zero-sequence to the positive-sequence impedance is in the range $Z_{0T}/Z_{1T} = 0.10 \cdots 0.28$, the line-to-earth fault currents far exceed the fault currents clear of earth $(1.3 \cdot I''_{k3} \leq I''_{k1} < 1.5 \cdot I''_{k3})$. For short-circuit-proof network dimensioning, it is usually sufficient to calculate the line-to-earth short-circuit currents.	

1) For explanations, see Table C9.1
2) Transformers with vector group Yyn are not suitable for protection against indirect contact (protection by automatic disconnection in the TN system)
3) Primary-side transformer neutral point not earthed

Table C9.5 Example comparison of the peak short-circuit currents i_{p3} and i_{p1} in two cable feeders of a 660-V industrial power system [9.2]

Length (distance from fault location)	Cable type of the LV feeder (U_m = 1.2 kV)							
	NYY 4 x 50 mm²				NYY 4 x 185 mm²			
	3-phase short circuit		Line-to-earth short circuit		3-phase short circuit		Line-to-earth short circuit	
ℓ [m]	i_{p3} [kA]	i'_{p3} [%]	i_{p1} [kA]	i'_{p1} [%]	i_{p3} [kA]	i'_{p3} [%]	i_{p1} [kA]	i'_{p1} [%]
0	65.6	100	63.4	100	65.6	100	63.4	100
10	46.1	70.3	35.4	55.8	57.4	87.5	50.2	79.2
25	31.7	48.3	20.9	33.0	48.3	73.6	38.4	60.6
50	20.6	31.4	12.3	19.4	38.3	58.4	27.4	43.2
75	15.3	23.3	8.7	13.7	31.6	48.2	21.3	33.6
100	12.1	18.4	6.7	10.6	27.0	41.2	17.4	27.4

Conclusion

The short-circuit current quantities occurring during a short circuit according to DIN EN 60909-0 (VDE 0102): 2002-07 [9.3] or IEC 60909-0: 2001-07 [9.4] with their maximum values form the basis for the short-circuit withstand capability of the equipment. In LV systems used to supply three-phase asynchronous motors with electrical power, transformers and motors contribute to the short-circuit stress on the equipment. If transformers of the preferred standard vector group Dyn5 are used and motors contribute to the short-circuit current during a fault, three-phase and line-to-earth short circuits cause the greatest dynamic and thermal stress. The three-phase short circuit causes the largest initial symmetrical short-circuit currents ($I''_{k\text{-max}} = I''_{k3}$) and peak short-circuit currents ($i_{p\text{-max}} = i_{p3}$). During a line-to-earth short circuit, on the other hand, the largest symmetrical short-circuit breaking currents ($I_{b\text{-max}} = I_{b1}$), steady-state short-circuit currents ($I_{k\text{-max}} = I_{k1}$) and thermal equivalent short-circuit currents ($I_{th\text{-max}} = I_{th1}$) occur. This summary of the maximum short-circuit stress applies to 400-V and 690-V industrial power systems.

9.2 Use of equipment reserves to handle short-circuit currents

One important criterion for cost-efficent use of equipment is the greatest possible utilization of the short-circuit current-carrying capacity. In industrial LV power systems, Dyn5 dry-type transformers are mostly used that are installed containerized in the load centres of the production and function areas. Installation in containers has the advantage that, for example, GEAFOL cast-resin transformers (Section 11.1) in AF operation (air forced cooling) can be placed under permanent load with fans switched on at 140 % of their rated current I_{rT}. For AF cooling, the rated current I_n of the secondary-side transformer circuit-breaker must be defined according to the size of the AF load current $I_{T\text{-AF}} = 1.4 \cdot I_{rT}$. With reference to the AF load current $I_{T\text{-AF}}$, the line-to-earth and three-phase short-circuit currents of the transformer have the following maximum values:

- $I''_{k1(3)T} \leq 20 \cdot I_{T\text{-AF}}$ and $i_{p1(3)T} \leq 43 \cdot I_{T\text{-AF}}$
  for an impedance voltage at rated current of $u_{rZ} = 4\ \%$,
- $I''_{k1(3)T} \leq 13 \cdot I_{T\text{-AF}}$ and $i_{p1(3)T} \leq 31 \cdot I_{T\text{-AF}}$
  for an impedance voltage at rated current of $u_{rZ} = 6\ \%$.

For controlling the short-circuit currents $I''_{k1(3)T}$ and $i_{p1(3)T}$, the rated ultimate short-circuit breaking capacity I_{cu} ($I_{cu} > I''_{k1(3)T}$) and rated short-circuit making capacity I_{cm} ($I_{cm} > i_{p1(3)T}$) of the low-voltage circuit-breaker selected for AF operation of the transformer are decisive.

Using the example of the Siemens circuit-breaker 3WL1 [9.6] from the range of LV switching devices available today, Table C9.6 provides information about the rated-current-related short-circuit breaking (I_{cu}/I_n) and making capacity (I_{cm}/I_n). According to the table, the switching capacity of the SENTRON 3WL1 circuit-breaker is a multiple of its rated current I_n (for size I at $U_e = 415$ V, for example, $I_{cu} = (34...105) \cdot I_n$ and $I_{cm} = (75...230) \cdot I_n$). With regard to a high voltage stability, the breaking capacity of the circuit-breakers is never fully utilized with a single transformer in radial operation.

The short-circuit power at the power system nodes is a measure of the voltage stability. Full use of the short-circuit power that the switching devices can handle is essential, particularly in non-steady system operation (starting and restarting of motors, impulse loads due to welding machines and presses). The SENTRON 3WL1 circuit-breakers are divided into switching capacity classes based on the special requirements for operation of LV industrial power systems (Table C9.6). The division into switching capacity classes N (ECO switching capacity), S (standard switching capacity), H (high

Table C9.6 Rated-current-related short-circuit breaking (I_{cu}/I_n) and making capacity (I_{cm}/I_n) of the SENTRON 3WL1 circuit-breaker

Circuit-breaker		SENTRON 3WL1							
Standardized circuit-breaker rated currents		630 A–800 A–1,000 A–1,250 A–1,600 A–2,000 A–2,500 A–3,200 A–4,000 A–5,000 A–6,300 A							
Switch size		I		II			III		
Type		3WL11		3WL12			3WL13		
Size-specific rated current I_n in A		630 – 1,600		800 – 4,000			4,000 – 6,300		
Switching capacity class		N	S	N	S	H	H	C 3-pole	C 4-pole
Rated ultimate short-circuit breaking capacity I_{cu} [kA]	$U_e \leq 415$ V~	55	66	66	80	100	100	150	130
	$U_e \leq 500$ V~	55	66	66	80	100	100	150	130
	$U_e \leq 690$ V~	42	50	50	75	85	85	150	130
	$U_e \leq 1{,}000$ V~	-	-	-	-	45	50	70	70
Rated short-circuit making capacity I_{cm} [kA]	$U_e \leq 415$ V~	121	145	145	176	220	220	330	286
	$U_e \leq 500$ V~	121	145	145	176	220	220	330	286
	$U_e \leq 690$ V~	88	105	105	165	187	187	330	286
	$U_e \leq 1{,}000$ V~	-	-	-	-	95	105	154	154
Short-circuit breaking capacity as a ratio to the rated current I_{cu}/I_n	$U_e \leq 415$ V~	34···87	41···105	16···82	20···100	25···125	16···25	24···37	21···32
	$U_e \leq 500$ V~	34···87	41···105	16···82	20···100	25···125	16···25	24···37	21···32
	$U_e \leq 690$ V~	26···66	31···79	12···62	19···94	21···106	13···21	24···37	21···32
	$U_e \leq 1{,}000$ V~	-	-	-	-	11···56	8···12	11···17	11···17
Short-circuit making capacity as a ratio to the rated current I_{cm}/I_n	$U_e \leq 415$ V~	75···192	90···230	36···181	44···220	55···275	35···55	52···82	45···71
	$U_e \leq 500$ V~	75···192	90···230	36···181	44···220	55···275	35···55	52···82	45···71
	$U_e \leq 690$ V~	55···140	65···166	26···131	41···206	46···233	29···46	52···82	45···71
	$U_e \leq 1{,}000$ V~	-	-	-	-	23···118	16···26	24···38	24···38

U_e AC rated operational voltage of the circuit-breaker (AC = alternating current)
N Circuit-breaker with ECO switching capacity (ECO = economic)
S Circuit-breaker with standard switching capacity
H Circuit-breaker with high switching capacity
C Circuit-breaker with very high switching capacity

switching capacity) and C (very high switching capacity) allows selection of circuit-breakers with the optimum ratio of short-circuit making/breaking capacity and rated current for dimensioning the power system, that is, $(I_{cu}/I_n)_{opt}$ and $(I_{cm}/I_n)_{opt}$.

The SENTRON 3WL1 circuit-breaker can also be modified with rating plugs on the over-current release to convert it to a switching device with a smaller rated current. For example, it is possible to convert a 800-A circuit-breaker of size II and switching capacity H into a 630-A circuit-breaker with a high switching capacity (I_{cu} = 100 kA and I_{cm} = 220 kA).

The reduction of the rated currents of circuit-breakers with a high and very high switching capacity can be extremely useful. This is especially the case if outgoing feeder circuit-breakers with small rated currents and high breaking capacity have to be used in industrial LV power systems. A very high system short-circuit power (e.g. for parallel operation of multiple distribution transformers) and the functional setting of the circuit-breaker overcurrent release system determine the use of such modified outgoing feeder circuit-breakers.

The settings on the overcurrent release (overload protection (L), time-delay overcurrent protection (S), instantaneous overcurrent protection (I)) always refer to the rated current I_n of the switching device. For that reason, circuit-breakers with large rated currents and a large breaking capacity can only be used as outgoing feeder circuit-breakers under certain conditions.

Due to a functional setting of the circuit-breaker overcurrent release system, it will nearly always be necessary to reduce the rated currents. With the available rated current modules, the rated current of SENTRON 3WL1 circuit-breakers of sizes I and II can be reduced to $I_n = 250$ A. Using the 3WL1 circuit-breaker of switch size III, reduction of the rated current to $I_n = 1{,}250$ A is possible. For use as an outgoing feeder circuit-breaker, the SENTRON 3VL compact circuit-breaker with a rated current of $16 \text{ A} \leq I_n \leq 160 \text{ A}$ and a rated ultimate short-circuit breaking capacity of $40 \text{ kA} \leq I_{cu} \leq 100 \text{ kA}$ at $U_e = 415$ V AC is also available [9.6].

Today's LV HRC fuses, which have a rated short-circuit breaking capacity of $I_{cu} = 120$ kA, are a low-cost alternative to outgoing feeder circuit-breakers [9.7, 9.8].

The number ratio of the short-circuit strength values to the rated current is large not only for circuit-breakers and LV HRC fuses but also for all other power system components (e.g. cable, distribution boards, busbar systems). It is therefore economically imperative to make full use of the available short-circuit current-carrying capacity of the equipment used in dimensioning industrial LV power systems [9.9]. Making full use of the short-circuit currents $I''_{k1(3)}$ that the equipment can handle also makes it easier to comply with the "protection by automatic disconnection of supply" measure for protection against indirect contact according to DIN VDE 0100-410 (VDE 0100-410): 2007-06 [9.10] or IEC 60364-4-41: 2005-12 [9.11]. This is especially important for outgoing feeders with large LV HRC fuses ($800 \text{ A} \leq I_n \leq 1{,}250$ A).

Conclusion

In the dimensioning of industrial LV power systems, it is important for technical and economic reasons to ensure that the short-circuit current-carrying capacity of all the equipment used (circuit-breakers, LV HRC fuses, cables, distribution boards, busbar systems) is fully utilized. Superfluous equipment reserves to handle short-circuit current stress quantities $I''_{k1(3)}$, $i_{p1(3)}$ and $I_{th1(3)}$ should be avoided.

10 Designing a low-voltage power system to meet requirements

10.1 Analysis of the load structure

Analysis of the load structure is an important precondition for designing and dimensioning industrial power systems. It includes determining the functional characteristics of the electrical load-consuming equipment items and the process-specific requirement profile of the loads. Owing to the variety of load-consuming equipment items and the many forms of energy consumption, it is convenient to form groups of loads. Load groups can be formed according to certain classification principles. Such principles include:

- groups of loads that cause system perturbations (e.g. welding machines, three-phase asynchronous induction motors, VFD-controlled motors, converter-fed drives),
- groups of loads with clearly defined supply reliability requirements (see Table A2.5),
- groups of loads with standardized compatibility levels for the voltage quality (see Table A2.12),
- groups of loads that form a technological or process-related unit in terms of the sequence of production steps in space or time (e.g. car manufacture with the subsystems press shop, body shop, paint shop and final assembly).

10.1.1 Characteristic load groups in the metal-processing industry

For the power supply from the LV level (100 V < U_{nN} ≤ 1,000 V), six significant load groups can be classified in the metal-processing industry. These load groups, shown in Fig. C10.1, differ by their:

- maximum power demand of the load-consuming equipment items,
- determining type of load (continuous load, impulse load, intermittent load),
- system perturbations caused (flicker, harmonics, voltage unbalance, voltage fluctuations),
- required supply reliability and/or voltage quality,
- electromagnetic compatibility (EMC) of the load-consuming equipment items,
- level of production automation.

These differences require a system design and dimensioning that are tailored to the specific load group. The following sections provide specific planning recommendations for this.

10.1.1.1 Toolmaking and mechanical workshops

In toolmaking and mechanical workshops, a large number of motor loads with different power ratings P_{rM} are evenly distributed over the production floor area. Compared with the power ratings of industrial distribution transformers (250 kVA ≤ S_{rT} ≤ 2,500 kVA), the power ratings of motors are relatively small (P_{rM} ≤ 30 kW). Dimensioning of the LV power system is therefore determined not by the influence of individual loads but by the maximum load from all loads in the interconnected power systems. The maximum

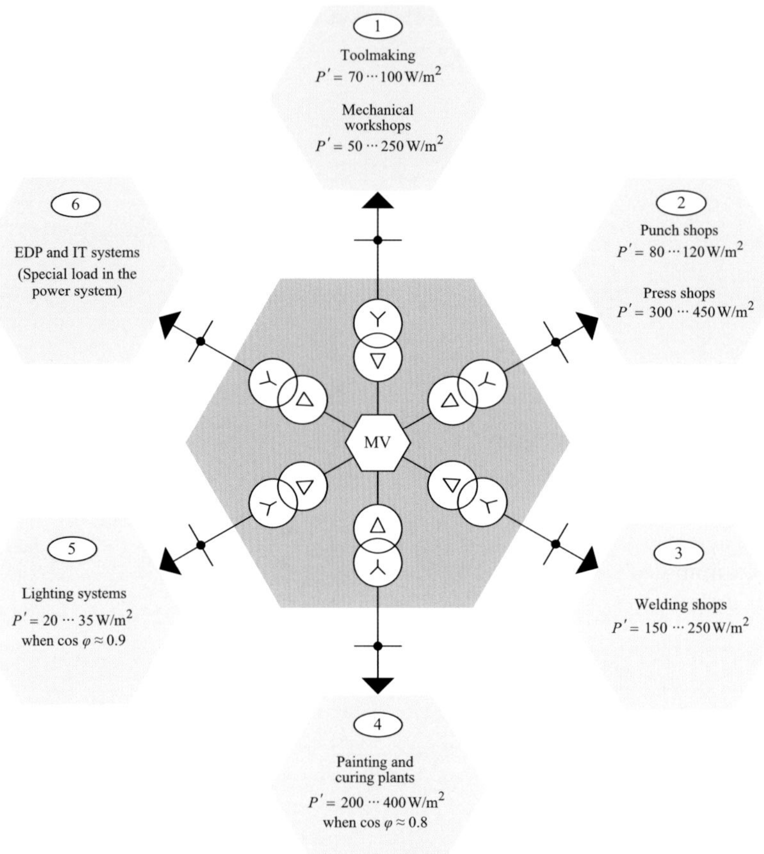

Fig. C10.1 Classifiable load groups in the metal-processing industry acc. to [10.1]

load of a load group mainly comprising motor drives including the workshop lighting can be calculated as follows:

$$S_{\text{max}} = \frac{1}{\cos\bar{\varphi}}(P_{\text{max-M}} + P_{\text{light}}) \tag{10.1}$$

$$P_{\text{max-M}} = g \cdot \sum_{\text{i}} P_{\text{max-M}_{\text{i}}} = g \cdot \sum_{\text{i}} \frac{P_{\text{rM}_{\text{i}}} \cdot a_{\text{i}}}{\eta_{\text{i}}} \tag{10.1.1}$$

S_{max}	maximum load of the LV system in kVA
$P_{\text{max-M}}$	maximum active power demand of the motors in kW
P_{light}	active power demand of the workshop lighting in kW
$P_{\text{rM}_{\text{i}}}$	power rating of the motor or the motor group i in kW
$\cos\bar{\varphi}$	average power factor of the load-consuming equipment items to be powered
g	coincidence factor in the power supply to the load group
η_{i}	efficiency of the motor or the motor group i
a_{i}	capacity utilization factor of the motor or the motor group i

As Eq. (10.1.1) shows, the efficiency η_i, capacity utilization factor a_i and coincidence factor g have a substantial influence on calculation of the expected active power demand of the motors. The coincidence factor g is a measure of the proportion that the motors of the load group contribute to the maximum load given that they do not all run concurrently. The capacity utilization factor a_i takes account of the idle times and the mechanical load within one duty cycle, which is usually below the motor power rating P_{rMi}. In toolmaking and for mechanical workshops, the product $g \cdot a_i$ yields a value that experience has shown to be between 0.25 and 0.45 [10.1].

Using the following example calculation, it is simple to ascertain the system dimensioning in toolmaking and for mechanical workshops based on the maximum load.

Example C1

A load group mainly comprising motor drives that forms a technological unit for the sequence of production steps over space and time in toolmaking must be supplied with electrical power.

By calculating the maximum load S_{max}, it is possible to determine the transformer power rating that is required for the system supply. For calculation of S_{max}, a list of load-consuming equipment items was provided (Table C10.2, columns 1 to 7) that contains all motor drives including the active and reactive power demand of the lighting system.

Table C10.2 Sizes of load-consuming equipment items for calculating the maximum load S_{max}

Consumer and process-specific parameters							Calculation values	
1	2	3	4	5	6	7	8	9
i	Quantity n [1]	Motor	P_{rM_i} [kW]	$\cos\varphi_i$ [1]	η_i [%]	$g \cdot a_i$ [1]	$P'_{max\text{-}M_i}$ [1)] [kW]	$Q'_{max\text{-}M_i}$ [2)] [kvar]
1	10	Leadscrew and feed shaft lathe	11.0	0.88	89.5	0.40	49.2	26.6
2	8	Plate edging machine	7.5	0.89	88.0	0.25	17.0	8.7
3	24	Drilling machine	5.5	0.89	86.5	0.30	45.8	23.5
4	16	Slotting machine	5.5	0.89	86.5	0.25	25.4	13.0
5	12	Circular saw and jigsaw	4.0	0.86	86.0	0.35	19.5	11.6
6	8	Column-type grinder	5.5	0.89	86.5	0.35	17.8	9.1
7	16	Universal milling machine	5.5	0.89	86.5	0.30	31.0	15.9
8	8	Plate shears	3.0	0.85	84.0	0.35	10.0	6.2
9	4	Hacksaw machine	1.5	0.85	80.0	0.40	3.0	1.9
10	4	Crane motor	2.2	0.85	82.0	0.30	3.2	2.0
11	24	Hot-air heater	0.55	0.82	70.0	0.45	8.5	5.9
12	4	Low-pressure compressor	30.0	0.89	92.3	0.30	39.0	20.0
Maximum active power demand of the motors $P_{max\text{-}M} = \sum_i P'_{max\text{-}M_i}$							266.4	
Active power demand of the lighting system P_{light} if $\cos\varphi = 0.9$							60	
Maximum reactive power demand of the motors $Q_{max\text{-}M} = \sum_i Q'_{max\text{-}M_i}$								144.4
Reactive power demand of the lighting system Q_{light} if $\sin\varphi = 0.436$								29.1

1) $P'_{max\text{-}M_i} = n \cdot P_{rM_i} \cdot g \cdot a_i / \eta_i$

2) $Q'_{max\text{-}M_i} = P'_{max\text{-}M_i} \cdot \dfrac{\sin\varphi_i}{\cos\varphi_i}$

By expanding the list of load-consuming equipment items with columns 8 and 9, it is possible to ascertain the active and reactive power demand of all 12 motor groups $P_{max\text{-}M}$ and $Q_{max\text{-}M}$. Separation into active and reactive power is necessary to calculate the average power factor $\cos\bar{\varphi}$ of the load-consuming equipment items to be powered. For calculation of $\cos\bar{\varphi}$ the following applies:

$$\bar{\varphi} = \arctan\left(\frac{Q_{max\text{-}M} + Q_{light}}{P_{max\text{-}M} + P_{light}}\right) \quad (10.1.2)$$

$$\bar{\varphi} = \arctan\left[\frac{(144.4 + 29.1)\ \text{kvar}}{(266.4 + 60.0)\ \text{kW}}\right] = \arctan 0.532 = \underline{\underline{28.0°}}$$

From $\bar{\varphi} = 28.0°$ it follows that the average power factor is $\cos\bar{\varphi} \approx 0.88$. According to Eq. (10.1), the following maximum load occurs with this power factor:

$$S_{max} = \frac{1}{\cos\bar{\varphi}}(P_{max\text{-}M} + P_{light}) = \frac{1}{0.88} \cdot (266.4 + 60.0)\ \text{kVA} = \underline{\underline{370.9\ \text{kVA}}}$$

For the maximum load of $S_{max} = 370.9$ kVA, a transformer with the power rating $S_{rT} = 400$ kVA must be chosen for the system supply.

The necessary transformer power rating for a system supply with $(n-1)$ redundancy can be calculated based on the load capacity condition (10.2). The following applies:

$$S_{perm}(n-1) \leq \frac{(n-1) \cdot k_{AF} \sum_{i=1}^{n} S_{rT_i}}{n} \quad (10.2)$$

$S_{perm}(n-1)$ permissible maximum load, applying the $(n-1)$ principle
S_{rT} transformer power rating
n number of transformers installed
k_{AF} overload factor ($k_{AF} = 1.4$ for GEAFOL cast-resin transformers in a housing with forced-air circulation)

If $S_{perm}(n-1) = S_{max} = 370.9$ kVA, $n = 2$ and $k_{AF} = 1.4$, rearranging Eq. (10.2) will yield the following transformer power rating with $(n-1)$ redundancy:

$$\sum_{i=1}^{2} S_{rT_i} \geq \frac{n \cdot S_{max}}{k_{AF} \cdot (n-1)} = \frac{2 \cdot 370.9\ \text{kVA}}{1.4 \cdot (2-1)} = \underline{\underline{530\ \text{kVA}}}$$

$$S_{rT} \geq \frac{1}{n} \cdot \sum_{i=1}^{n} S_{rT_i} = \frac{1}{2} \cdot \sum_{i=1}^{2} S_{rT_i} \geq \frac{1}{2} \cdot 530\ \text{kVA} = \underline{\underline{265\ \text{kVA}}}$$

$$\Rightarrow \underline{\underline{S_{rT} = 315\ \text{kVA}}}$$

To handle the $(n-1)$ fault in the power supply to the load-consuming equipment items from Table C10.2, two 315-kVA GEAFOL cast-resin transformers with AF cooling must be installed rather than one 400-kVA GEAFOL cast-resin transformer.

The power rating of the distribution transformers to be installed can be reduced by means of reactive-power compensation (Chapter 12). For example, the load-consuming equipment items of Example C1, after improving the power factor from $\cos\bar{\varphi}_1 = 0.88$ to $\cos\bar{\varphi}_2 = 0.95$, could also be powered through two AF-cooled 250-kVA GEAFOL cast-resin transformers with $(n-1)$ redundancy.

10.1.1.2 Punch and press shops

In punch and press shops too, a large number of motor drives are required for working sheet and solid metal and for punching and cutting. However, the individual power rating of specific drives is relatively large ($P_{rM} \geq 100$ kW) compared with the total power demand, which subjects the LV system to frequent impulse loads.

In addition to dimensioning the supplying transformers based on the expected total power demand, in this case the starting of large motors and the fluctuating load during punching and working the sheet-metal are the main factors influencing the required dimensioning of the system supply. For that reason, it is necessary to check whether the load changes caused by the motor loads (motor starting, load cycles during punching, working and nibbling) do not result in impermissible voltage fluctuations and flicker in the power system.

Voltage fluctuations that cause flicker are generally only relevant to the common operation of lighting and power circuits. Table C10.3 lists the compatibility levels for voltage changes and flicker intensity to be observed during combined operation of light and power circuits. To avoid straining the human eye unduly with fluctuations of luminous flux, the following must apply:

$$P_{st} \leq P_{st\text{-}perm} \tag{10.3}$$

and

$$P_{lt} \leq P_{lt\text{-}perm} \tag{10.4}$$

$P_{st\text{-}perm}$ permissible short-term flicker intensity (see Table C10.3)
$P_{lt\text{-}perm}$ permissible long-term flicker intensity (see Table C10.3)
P_{st} short-term flicker intensity of the load group (see Eq. 10.5)
P_{lt} long-term flicker intensity of the load group (see Eq. 10.6)

The short-time flicker intensity of a load group that subjects the system to an impulsive and/or pulsating load can be determined as follows [10.6]:

$$P_{st} = \frac{\Delta u'}{\Delta u'_{ref}(r)} \cdot P_{st\text{-}ref} \tag{10.5}$$

$\Delta u'$ relative voltage change on impulsive or fluctuating load (see Eq. 10.7)
$\Delta u'_{ref}(r)$ relative voltage change depending on the repeat rate r according to the flicker reference curve $P_{st\text{-}ref} = 1$ (see Fig. C10.4)
$P_{st\text{-}ref}$ reference value for flicker perception ($P_{st\text{-}ref} = 1$)

The long-term flicker intensity P_{lt} is calculated from a sequence of $n = 12$ P_{st} values over a 2-hour interval according to the following equation [10.6]:

$$P_{lt} = \sqrt[3]{\sum_{n=1}^{12} \frac{P_{st\,n}^3}{12}} \tag{10.6}$$

Table C10.3 Compatibility level for voltage changes and flicker intensity in industrial LV power systems

Supply class of the LV system	Compatibility level for voltage changes $\Delta u'_{perm}$ [%]	Compatibility level for the flicker	
		Permissible short-term flicker intensity $P_{st\text{-}perm}$ [1]	Permissible long-term flicker intensity $P_{lt\text{-}perm}$ [1]
1	For in-plant points of coupling of class 1, protection by means of DVR or DDUPS is a necessary precondition. Voltage dips and fluctuations must therefore be ruled out.		
2	10 [1)]	1.0 [1) 2)]	0.8 [1) 2)]
3	15 [1)]	--	--

1) Limit values acc. to DIN EN 61000-2-4 (VDE 0839-2-4): 2003-05 [10.2] or IEC 61000-2-4: 2002-06 [10.3]

2) Voltage fluctuations resulting in flicker are generally only important for lighting systems. They should only be connected to supply systems of class 2. The compatibility levels for the flicker in public LV systems according to DIN EN 61000-2-2 (VDE 0839-2-2): 2003-02 [10.4] or IEC 61000-2-2: 2002-03 [10.5] apply.

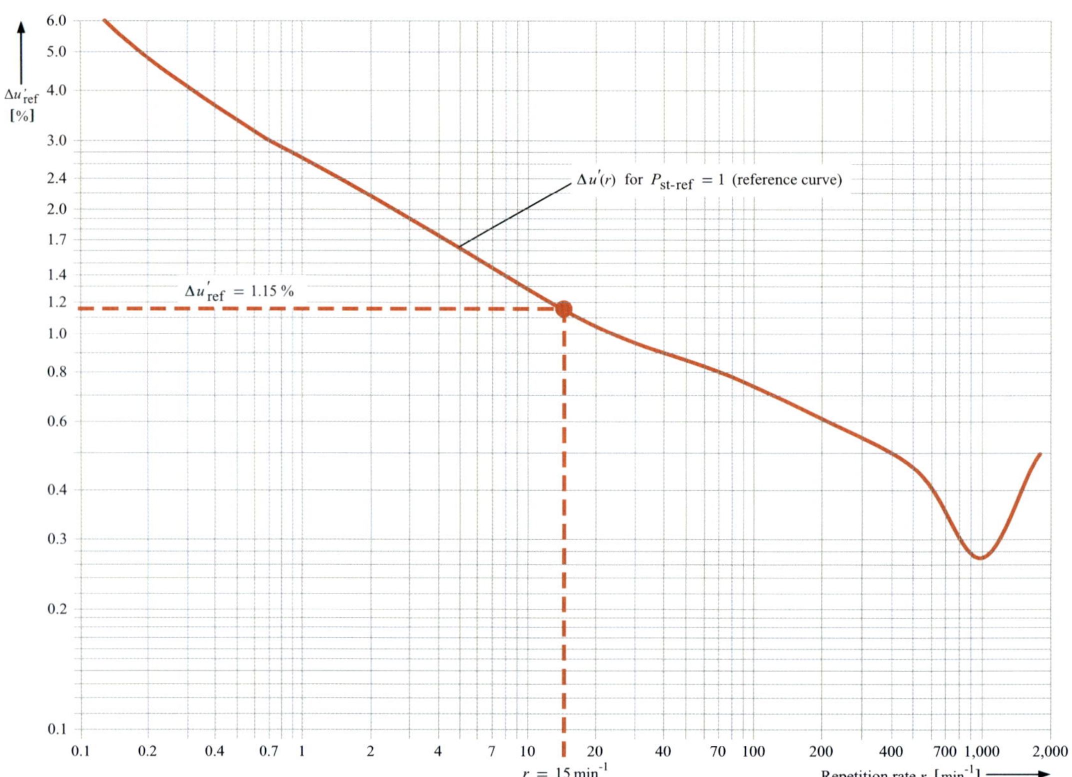

Fig. C10.4 Reference curve for regular, rectangular voltage changes depending on the repeat rate $r\ \Delta u'_{ref}(r)$

The long-term flicker intensity value calculated according to Eq. (10.6) (P_{lt}) is the flicker value that is decisive for evaluating the voltage quality. The voltage quality under impulsive or fluctuating loads primarily depends on the short-circuit power at the system nodes (points of connection (PCs) and points of common coupling (PCCs)). This dependency is expressed in Eq. (10.7).

$$\Delta u'_{PC/PCC} = k_{\Delta S} \cdot \frac{\Delta S}{S''_{k\text{-}PC/PCC}} \cdot 100\,\% \qquad (10.7)$$

$\Delta u'_{PC/PCC}$ voltage change at the decisive point of connection or point of common coupling

$k_{\Delta S}$ factor for symmetrical (Eq. 10.10) and asymmetrical (Table C10.16) voltage changes

ΔS load change (symmetrical or asymmetrical)

$S''_{k\text{-}PC/PCC}$ short-circuit power at the decisive point of connection or point of common coupling ($S''_{k\text{-}PC/PCC}$ is preferably to be calculated according to Eq. (4.1) as per DIN EN 60909-0 (VDE 0102): 2002-07 [10.7] or IEC 60909-0: 2001-07 [10.8]. For a comparison of methods for calculation of $S''_{k\text{-}PC/PCC}$ see Table B4.2)

Motor drives in punch shops and press shops cause symmetrical voltage changes. Symmetrical voltage changes are caused by load changes that are symmetrical, that is, identical in all three line conductors. Calculation of symmetrical load changes is possible with the Equations (10.8) and (10.9) from Table C10.5.

Symmetrical load change	Cause	Calculation
impulsive	Motor starting	$\Delta S = \frac{I_{start}}{I_{rM}} \cdot \frac{P_{rM}}{\cos\varphi_{rM} \cdot \eta_{rM}}$ (10.8)
fluctuating	Alternating load operation	$\Delta S = \sqrt{3} \cdot \Delta I \cdot U_{nN}$ (10.9)

P_{rM} Rated power of the motor

I_{start}/I_{rM} Ratio of the starting current to the rated current of the motor

$\cos\varphi_{rM}$ Nominal power factor of the motor

η_{rM} Nominal efficiency of the motor

(For recommended values, see Table C10.6)

ΔI Current fluctuation during alternating load operation

U_{nN} Nominal voltage of the system

Table C10.5 Calculation of symmetrical load changes

The factor $k_{\Delta S}$ arising on symmetrical load changes can be calculated as follows:

$$k_{\Delta S} = \frac{\frac{R}{X} \cdot \cos\varphi_{\Delta S} + \sin\varphi_{\Delta S}}{\sqrt{1 + \left(\frac{R}{X}\right)^2}} \qquad (10.10)$$

R/X ratio of resistance to reactance at the determining nodal point in the system (point of connection or point of common coupling)

$\cos\varphi_{\Delta S}$ active power factor occurring at the instant of the load change ΔS (for guidance values, see Table C10.6)

$\sin\varphi_{\Delta S}$ reactive power factor occurring at the instant of the load change ΔS ($\sin\varphi_{\Delta S} = \sqrt{1 - \cos^2\varphi_{\Delta S}}$)

Table C10.6 Electrical motor parameters for calculating symmetrical load and voltage changes (guidance values)

Parameter	Motor	P_{rM} in kW	Recommended values for the parameters [1)]
$\frac{I_{start}}{I_{rM}}$	LV: 400(690) V	0.55 ··· 5.5 ··· 315	4.0 ··· 6.5 ··· 7.0 [2)]
	MV: 6(10) kV	400 ··· 2,500	5.5 ··· 7 [2)]
η_{rM}	LV: 400(690) V	0.55 ··· 5.5 ··· 315	0.70 ··· 0.85 ··· 0.97
	MV: 6(10) kV	400 ··· 2,500	0.96 ··· 0.98
$\cos\varphi_{rM}$	LV: 400(690) V	0.55 ··· 315	0.85 ··· 0.91
	MV: 6(10) kV	400 ··· 2,500	0.90 ··· 0.92
$\cos\varphi_{\Delta S}$ [3)]	LV: 400(690) V	0.55 ··· 30 ··· 75 ··· 315	0.60 ··· 0.55 ··· 0.45 ··· 0.25
	MV: 6(10) kV	400 ··· 2,500	0.15 ··· 0.10
M_{start}/M_n [4)]	LV: 400(690) V	0.55 ··· 315	2.5 ··· 1.6
	MV: 6(10) kV	400 ··· 2,500	0.9

1) Compiled from manufacturers' documentation

2) Applies to n_{syn} = 3,000 min^{-1}; as correction factors, 0.85 (n_{syn} = 1,500 min^{-1}) and 0.70 (n_{syn} = 750 min^{-1}) can be used

3) Corresponds to the power factor in the locked-rotor condition (short-circuit power factor); this can be approximately calculated as follows: $\cos\varphi_{\Delta S} \approx (M_{start}/M_n) : (I_{start}/I_{rM})$

4) Ratio of the starting torque to the nominal torque of the motor

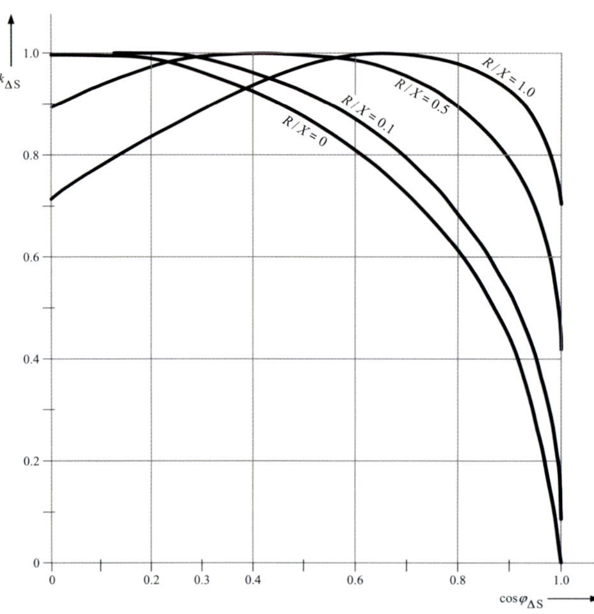

Fig. C10.7 Factor $k_{\Delta S}$ for calculation of symmetrical voltage changes

Moreover, the factor $k_{\Delta S}$ can be read off from Fig. C10.7. It has a value no higher than $k_{\Delta S}$ = 1.0. This makes it possible to use the ratio $\Delta S/S_k''$ for relevant evaluations of the voltage quality.

The following application example shows how a detailed evaluation of the voltage quality on symmetrical load changes with impulsive and fluctuating characteristics can be performed.

Example C2

The voltage quality of the power supply shown in Fig. C10.8 with three supplying 1,250-kVA transformers is to be evaluated for the press shop of a car factory.

Compliance with the compatibility levels of supply class 2 for voltage fluctuations and flicker intensity according to DIN EN 61000-2-4 (VDE 0839-2-4): 2003-05 [10.2] or IEC 61000-2-4: 2002-06 [10.3] is to be verified to ascertain whether common operation of power and lighting circuits is possible (for the compatibility levels to be complied with, see Table C10.3).

The press drives are only switched on once and then remain in operation all day. During switch-on, the drives are mutually interlocked. The greatest voltage dip $\Delta u'$ is

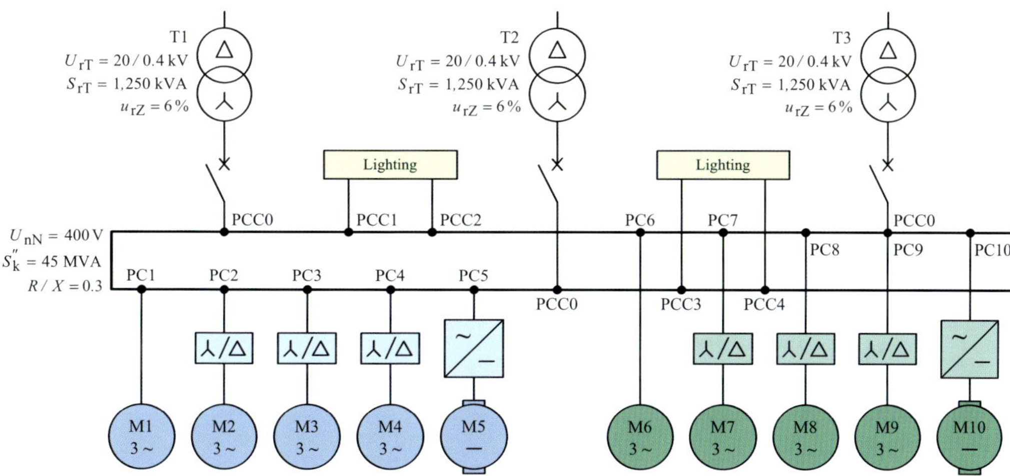

Process data	Press line 1			Press line 2		
	Presses 1–4		Special press	Presses 5–8		Tryout press
	Average cycle time for body component production: T = 4 sec (repetition rate r = 15 min^{-1})					
	Current fluctuations due to alternating load operation of the press drives: $\Delta I = I_{max} - I_{min}$ = 2,800 A − 2,100 A = 700 A ($\cos\varphi_{\Delta S}$ = 0.85)					
P_{rM} in kW	160	200	250	160	200	132
I_{start} / I_{rM}	7	2.3 when ⅄/Δ starting	≤ 1.0	7	2.3 when ⅄/Δ starting	≤ 1.0
η_{rM} in %	95	96	98	95	96	98
$\cos\varphi_{rM}$	0.90	0.91	≤ 1.0	0.90	0.91	≤ 1.0
$\cos\varphi_{\Delta S}$	0.40	0.60	-	0.40	0.60	-

PC Point of connection of the disturbing load
PCC Point of common coupling of the influenced load or system supply

Fig. C10.8 Process, power system and load-consuming equipment data of the power supply for the press shop of a car factory (Example C2)

caused by the impulse load ΔS on direct-on-line starting of motor M1 or M6. According to Eq. (10.8), the following start-up load impulse can be expected:

$$\Delta S = \frac{I_{start}}{I_{rM}} \cdot \frac{P_{rM}}{\cos\varphi_{rM} \cdot \eta_{rM}} = \frac{7 \cdot 160 \cdot 10^3\ \text{VA}}{0.90 \cdot 0.95} = \underline{\underline{1{,}310\ \text{kVA}}}$$

The factor $k_{\Delta S}$ correlating with this impulse load can be calculated with Eq. (10.10).

$$k_{\Delta S} = \frac{\frac{R}{X} \cdot \cos\varphi_{\Delta S} + \sin\varphi_{\Delta S}}{\sqrt{1 + \left(\frac{R}{X}\right)^2}} = \frac{0.3 \cdot 0.4 + 0.9165}{\sqrt{1 + 0.3^2}} = \underline{\underline{0.99}}$$

To calculate the voltage dip during direct-on-line starting of the motor, the impulse load ΔS = 1,310 kVA and the factor $k_{\Delta S}$ = 0.99 must be inserted in Eq. (10.7).

$$\Delta u' = k_{\Delta S} \cdot \frac{\Delta S}{S_k''} \cdot 100\,\% = 0.99 \cdot \frac{1{,}310 \cdot 10^3\ \text{VA}}{45 \cdot 10^6\ \text{VA}} \cdot 100\,\% = \underline{\underline{2.88\,\%}}$$

For a short-circuit power of S_k'' = 45 MVA at the nodal points (PCs and PCCs) a DOL-starting-induced voltage dip of $\Delta u'$ = 2.88 % occurs. This impulsive voltage change is smaller than the voltage change to be complied with according to Table C10.3 (2.88 % < 10 %). The compatibility level of the supply class 2 for voltage changes would even be complied with if press drives M1 and M6 were switched on simultaneously (5.76 % < 10 %). In addition to the DOL-starting-induced voltage dips, voltage fluctuations in the press shop are also caused by the load cycles of the press drives. Load cycle operation in the press shop in this example taken from the automotive industry is characterized by

- a repeat rate of r = 15 min^{-1} (15 changes in load per minute),
- effective current fluctuations of ΔI = 700 A and
- a power factor of $\cos\varphi_{\Delta S}$ = 0.85 occurring during load changes.

Applying equations (10.7), (10.9) and (10.10), for ΔI = 700 A, $\cos\varphi_{\Delta S}$ = 0.85, R/X = 0.3 and S_k'' = 45 MVA, the following symmetrical voltage fluctuations in the system result:

$$\Delta u' = k_{\Delta S} \cdot \frac{\Delta S}{S_k''} \cdot 100\,\% = 0{,}75 \cdot \frac{485 \cdot 10^3\ \text{VA}}{45 \cdot 10^6\ \text{VA}} \cdot 100\,\% = \underline{\underline{0.81\,\%}}$$

From the reference curve for $P_{st\text{-}ref}$ = 1 in Fig. C10.4, for r = 15 min^{-1}, a maximum permissible voltage change of $\Delta u'_{ref}(r)$ = 1.15 % can be read. The short-term flicker intensity for the power system of the press shop can be calculated by inserting the numerical values $\Delta u'$ = 0.81 % and $\Delta u'_{ref}(r)$ = 1.15 % in Eq. (10.5).

$$P_{st} = \frac{\Delta u'}{\Delta u'_{ref}(r)} \cdot P_{st\ ref} = \frac{0.81\,\%}{1.15\,\%} \cdot 1.0 = \underline{\underline{0.70}}$$

According to the condition $P_{st} < P_{st\text{-}perm}$ (Eq. 10.3), the short-term flicker emission with P_{st} = 0.70 is in the permissible range (0.7 < 1.0). Due to the continuous sequence of load changes, the short-term flicker can be assumed to be constant. This results in the following long-term flicker intensity (Eq. 10.6):

$$P_{lt} = \sqrt[3]{\sum_{n=1}^{12} \frac{P_{st\,n}^3}{12}} = \sqrt[3]{\frac{12 \cdot 0.70^3}{12}} = \underline{\underline{0.70}}$$

The long-term flicker emission $P_{lt} = 0.7$ also falls within the permissible limit of $P_{lt\text{-}perm} = 0.8$ (see Table C10.3 and Eq. 10.4). Because the short-term and long-term emission are in the permissible range, the power and lighting circuits of the press shop can be operated on a common system.

In addition to the compatibility levels for voltage changes and flicker intensity, the compatibility levels for harmonics must also be complied with in press shops (Table A2.12). Today, converter-fed three-phase drives are increasingly being used for the production process in press shops. Because of their use as “electronic gearing“ for adjusting and controlling the speed, converter-fed drives generate harmonic oscillations that adversely affect the sinusoidal curve of the system voltage. In press shops with a high proportion of converter-fed three-phase drives (converter proportion of the total load > 15 %), the reactive-power compensation should be implemented with reactor-connected capacitors (see Section 12.3). The optimum detuning factor of the reactive-power compensation system is based on the harmonics that mainly occur with the hth order (e.g. $h = 5, 7, 11, 13, 17, 19, 23$ and 25 for three-phase bridge circuits). In practice, reactor-connected capacitors are preferred with a detuning factor between $p = 5\,\%$ and $p = 7\,\%$.

10.1.1.3 Welding shops

Electrical welding machines do not comprise a continuous load because of their intermittent operation, that is, weld-pause-weld. They mainly subject the supply system to an impulse load.

In the case of impulsive loads, the thermal stress of equipment with a large time constant (transformers, cables, busbar trunking systems) can be determined by calculating a thermal equivalent current. This thermal equivalent current to be calculated must correspond in its heat effect to the current pulses of the welding cycle shown in Fig. C10.9.

Based on the load duty cycle depicted in Fig. C10.9, the thermal equivalent current can be calculated as a root mean square value. The following applies:

$$I_{th} = I_{wsm}\sqrt{\frac{t_W}{T}} = I_{wsm}\sqrt{ED} \qquad (10.11)$$

I_{wsm}	RMS welding current of a single machine
t_W	welding time ($t_W = 5...20$ cycles for spot and projection welding machines)
T	cycle time ($T \le 10$ sec)
ED	duty ratio ($ED = \frac{t_W}{T} \cdot 100$)

In practice, the duty ratios for single machines are $ED = 1...10\,\%$ [10.9]. If multiple resistance welding machines are connected to the power system, the instantaneous value of the resulting total current and, by extension, the equivalent current that determines thermal stress will vary constantly depending on the machine operation scheduling. As regards the thermal stress of equipment in welding power systems, the two extreme cases A and B exist (Fig. C10.10).

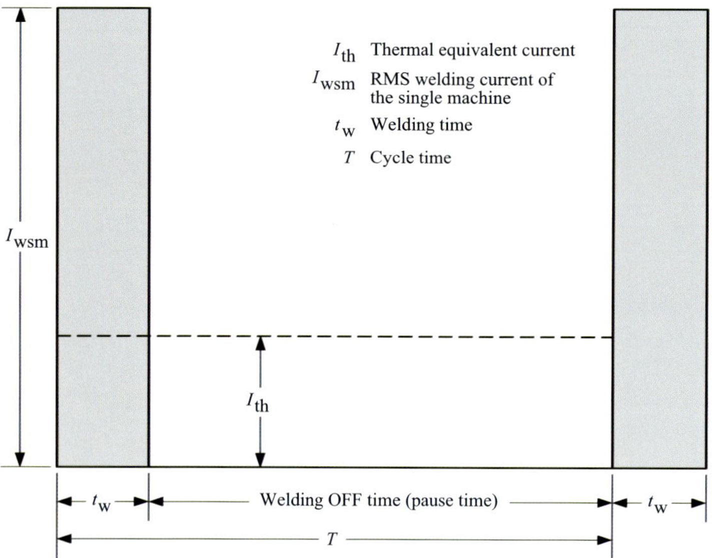

Fig. C10.9 Load duty cycle of a single welding machine

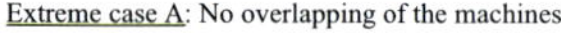

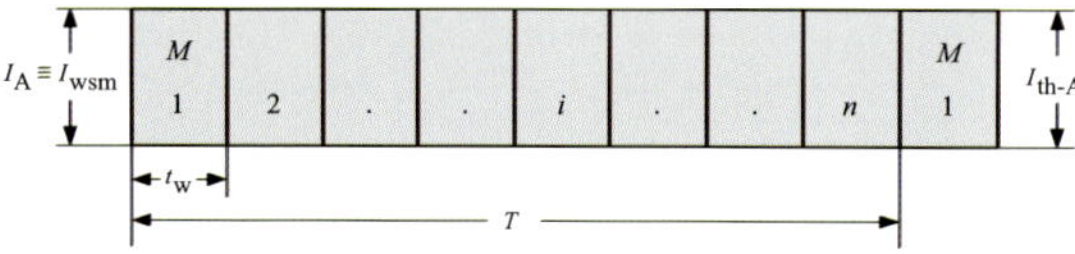

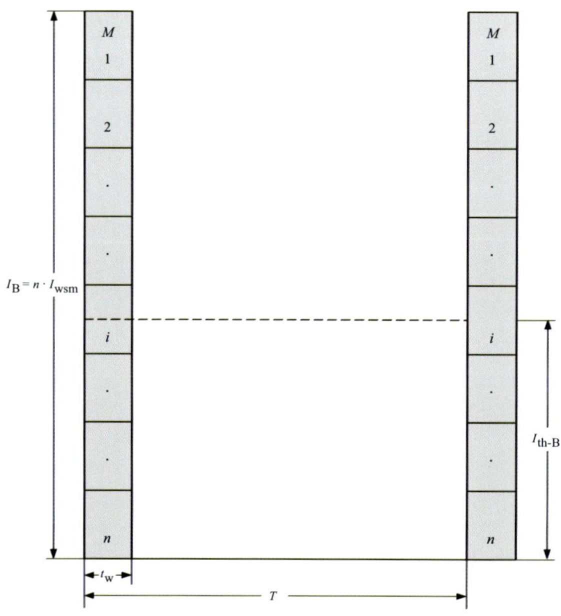

Fig. C10.10 Extreme cases during operation of n welding machines connected to two line conductors of a three-phase power system

In the extreme case A, the n single machines of a welding group are mutally interlocked in such a way that only one machine at a time is ever welding. The thermal equivalent current is then as follows:

$$I_{\text{th-A}} = I_{\text{A}} \cdot \sqrt{ED_{\text{A}}} = I_{\text{wsm}} \sqrt{n \cdot ED_{\text{sm}}} \tag{10.12}$$

$I_{\text{th-A}}$	thermal equivalent current in the extreme case A (no overlapping of machines)
I_{A}	RMS welding current for the thermal stress in extreme case A
ED_{A}	duty ratio of the machines in extreme case A
ED_{sm}	duty ratio of a single machine
n	number of single machines in a welding group

If all the machines are welding simultaneously as in extreme case B, Eq. (10.13) must be used to calculate the thermal equivalent current.

$$I_{\text{th-B}} = I_{\text{B}} \cdot \sqrt{ED_{\text{B}}} = n \cdot I_{\text{wsm}} \sqrt{ED_{\text{sm}}} \tag{10.13}$$

$I_{\text{th-B}}$	thermal equivalent current in extreme case B (all machines welding simultaneously)
I_{B}	RMS welding current for the thermal stress in extreme case B
ED_{B}	duty ratio of the machines in extreme case B

In practice, the thermal stress of the equipment will always be between the two extreme values, that is, between the value for welding operation not overlapping in time and the value for welding operation fully overlapping in time (Eq. 10.14).

$$I_{\text{wsm}} \sqrt{n \cdot ED_{\text{sm}}} \leq I_{\text{th}} \leq n \cdot I_{\text{wsm}} \sqrt{ED_{\text{sm}}} \tag{10.14}$$

I_{wsm}	RMS welding current of a single machine

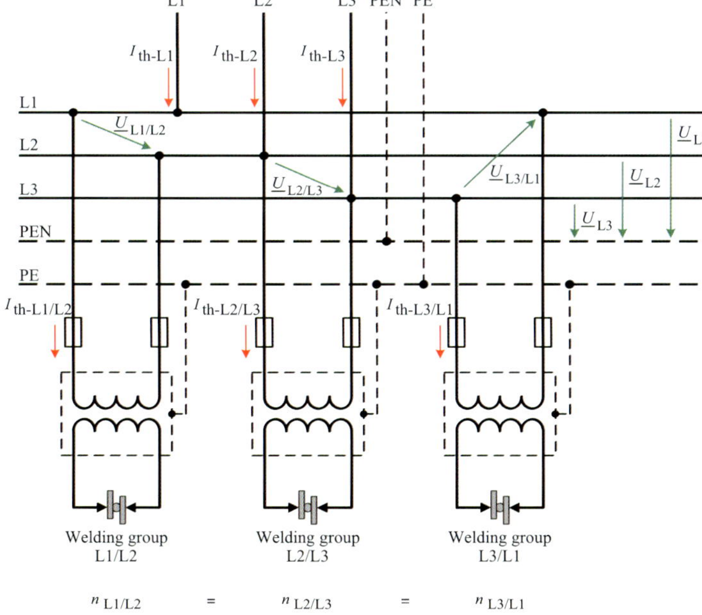

Fig. C10.11 Symmetrical distribution of the welding machines over the line conductors of a 400-V three-phase system

Welding power systems usually have to be dimensioned for mass operation of welding machines. To achieve as even a system load as possible, the welding machines intended for mass deployment should be distributed as symmetrically as possible over the line conductors of the three-phase system (Fig. C10.11).

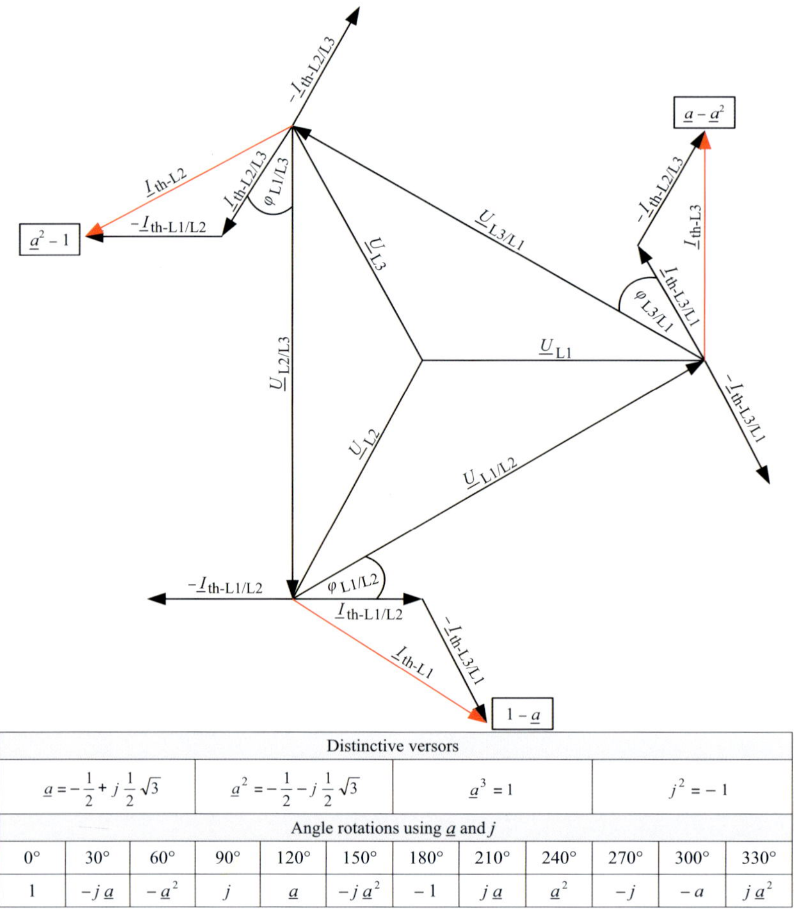

Distinctive versors			
$\underline{a}=-\frac{1}{2}+j\frac{1}{2}\sqrt{3}$	$\underline{a}^2=-\frac{1}{2}-j\frac{1}{2}\sqrt{3}$	$\underline{a}^3=1$	$j^2=-1$

Angle rotations using $\underline{a}$ and j											
0°	30°	60°	90°	120°	150°	180°	210°	240°	270°	300°	330°
1	$-j\underline{a}$	$-\underline{a}^2$	j	$\underline{a}$	$-j\underline{a}^2$	-1	$j\underline{a}$	$\underline{a}^2$	$-j$	$-a$	$j\underline{a}^2$

Fig. C10.12 Vector diagram for the thermal equivalent current calculation

The thermal equivalent current calculation for distribution of the welding machines over the line conductors of the 400-V three-phase system in groups can be performed based on the vector diagram shown in Fig. C10.12. The thermal equivalent currents of the three welding groups L1/L2, L2/L3 and L3/L1 can be calculated for each of the line conductors L1, L2 and L3 of the 400-V three-phase system as follows:

$$\underline{I}_{\text{th-L1}} = \underline{I}_{\text{th-L1/L2}} - \underline{I}_{\text{th-L3/L1}} = 1 \cdot I_{\text{th-L1/L2}} - \underline{a} \cdot I_{\text{th-L3/L1}}$$

$$I_{\text{th-L1}} = \sqrt{I^2_{\text{th-L1/L2}} + I_{\text{th-L1/L2}} \cdot I_{\text{th-L3/L1}} + I^2_{\text{th-L3/L1}}} \qquad (10.15)$$

$$I_{\text{th-L1}} = I_{\text{th-L1/L2}} \cdot \sqrt{3} \qquad \text{in the case of absolute symmetry}$$

$$\underline{I}_{\text{th-L2}} = \underline{I}_{\text{th-L2/L3}} - \underline{I}_{\text{th-L1/L2}} = \underline{a}^2 \cdot I_{\text{th-L2/L3}} - 1 \cdot I_{\text{th-L1/L2}}$$

$$I_{\text{th-L2}} = \sqrt{I^2_{\text{th-L2/L3}} + I_{\text{th-L2/L3}} \cdot I_{\text{th-L1/L2}} + I^2_{\text{th-L1/L2}}} \tag{10.16}$$

$I_{\text{th-L2}} = I_{\text{th-L2/L3}} \cdot \sqrt{3}$ in the case of absolute symmetry

$$\underline{I}_{\text{th-L3}} = \underline{I}_{\text{th-L3/L1}} - \underline{I}_{\text{th-L2/L3}} = \underline{a} \cdot I_{\text{th-L3/L1}} - \underline{a}^2 \cdot I_{\text{th-L2/L3}}$$

$$I_{\text{th-L3}} = \sqrt{I^2_{\text{th-L3/L1}} + I_{\text{th-L3/L1}} \cdot I_{\text{th-L2/L3}} + I^2_{\text{th-L2/L3}}} \tag{10.17}$$

$I_{\text{th-L3}} = I_{\text{th-L3/L1}} \cdot \sqrt{3}$ in the case of absolute symmetry

The system of equations (10.15) to (10.17) is the basis for selecting and dimensioning the equipment according to the thermal load caused by the welding machines. The LV HRC fuses for impulse load due to resistance welding machines can be selected based on the diagram in Fig. C10.13.

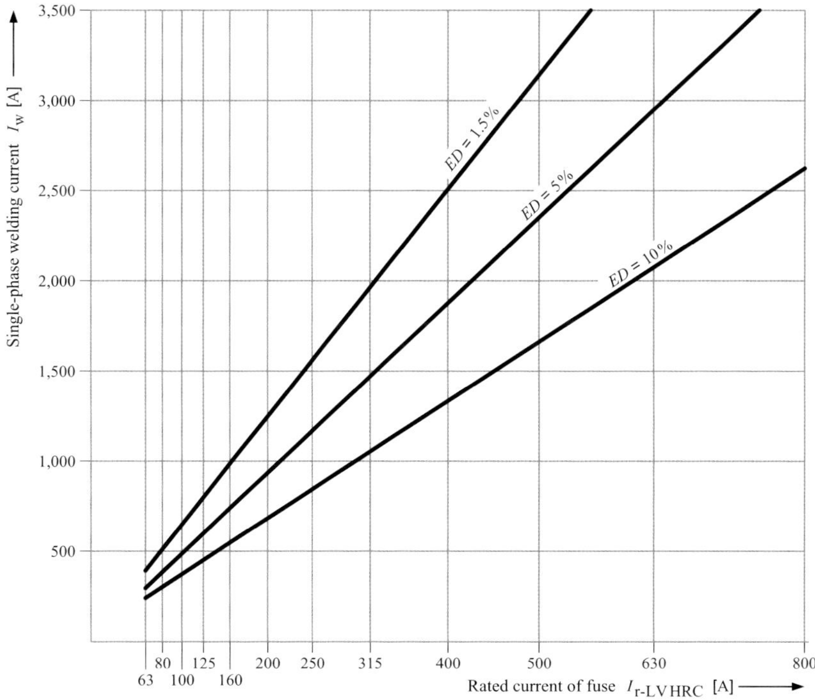

Fig. C10.13 Selection of the LV HRC fuses for impulse load due to resistance welding machines (fuse protection recommendation)

The impulse load of the resistance welding machines has a particularly strong influence on the definition of the power rating and the number of distribution transformers that must be installed in the welding shop. The single-phase welding machines connected in groups to two different line conductors of the three-phase system

(Fig. C10.11) are usually operated in parallel without interlocking. Without a power limiting control (PLC), therefore, overlapping of the individual welding currents and excessive resulting total currents can occur. The higher resulting total welding currents also give rise to an increase in the voltage dips $\Delta u'$ occurring in the supply system. Essentially the voltage dips $\Delta u'$ caused by randomly overlapping welding pulses must not adversely affect the quality of the product (e.g. vehicle bodywork).

To avoid incorrect welding, a certain minimum voltage must always be applied to the terminals of the resistance welding machines. To ensure perfect functioning of the power control and voltage control of resistance welding machines, a minimum voltage of $0.85 \cdot U_{nN}$ is generally necessary. With this minimum voltage, which is typical of the device, and a limit value for the steady-state load voltage of $0.95 \cdot U_{nN}$, the voltage dip $\Delta u'_{perm} = 10\,\%$ is permissible [10.9 to 10.11]. Accordingly, the following probabilistic voltage stability criterion must be fulfilled for the quality welding of individual sheet-metal parts to form sheet-metal structures (e.g. vehicle bodywork):

$$\overline{\Delta u'} \leq \Delta u'_{perm} \tag{10.18}$$

$\overline{\Delta u'}$ statistically secured mean value of the voltage dips in non-interlocked simultaneous operation of the welding machines

$\Delta u'_{perm}$ permissible voltage dip at the terminals of the welding machines ($\Delta u'_{perm} = 10\,\%$)

During the welding process, which is stochastic, cases of overlapping are also possible in which the resulting total welding current causes voltage dips $\Delta u' > 10\,\%$. The permissible voltage dip $\Delta u'_{perm} = 10\,\%$, however, may only be exceeded if the number of voltage-induced welding errors is smaller than the number of permissible welding errors that may occur for other reasons. Generally, an overlap voltage dip larger than 10 % is permitted for 1 ‰ of spot welds [10.9].

A dimensioning criterion that takes this stochastic problem into account is the permissible failure rate for voltage-induced welding errors. For the greatest reliability and quality of the welding processes, the following applies:

$$\lambda(\Delta u' > 10\,\%) < \lambda_{perm} \tag{10.19}$$

$\lambda(\Delta u' > 10\,\%)$ failure rate for welding errors due to random voltage dips $\Delta u' > 10\,\%$

λ_{perm} permissible failure rate for voltage-induced welding errors ($\lambda_{perm} = 1$ ‰)

The dimensioning criteria (10.18) and (10.19) must be verified in line with the rules of probability calculation. For mass operation of welding machines, this is achieved using the Bernoulli formula (Eq. 10.20). To apply this formula, it is first necessary to express the various types of welding equipment as a single equivalent welding machine with

- identical RMS peak welding current I_W,
- identical duty ratio *ED* and
- identical power factor $\cos\varphi$.

Example C3

Final result of a welding machine equivalent calculation (Fig. C10.14) for power system dimensioning according to the probabilistic voltage stability criterion (Eq. 10.18) and the λ dimensioning criterion (Eq. 10.19)

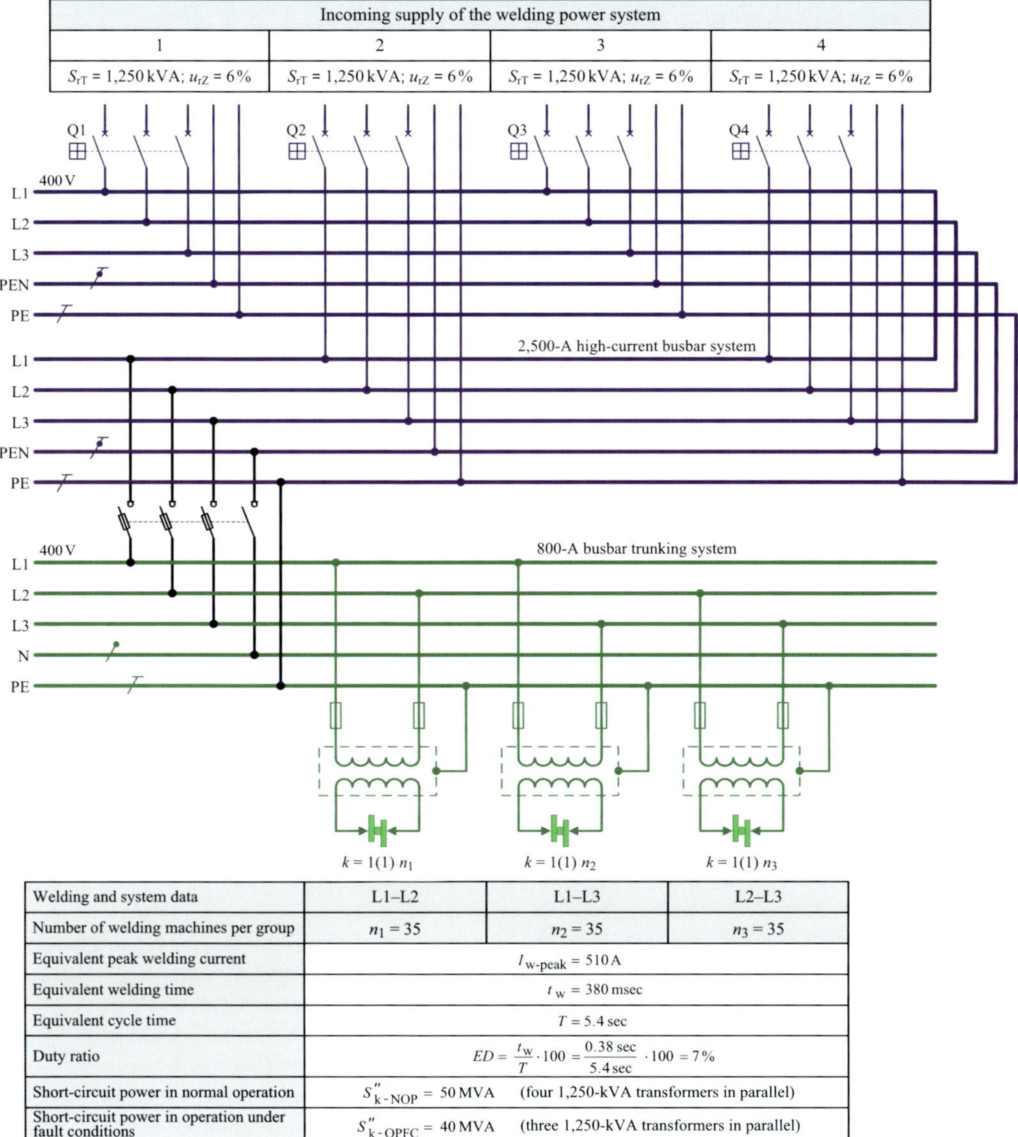

Welding and system data	L1–L2	L1–L3	L2–L3
Number of welding machines per group	$n_1 = 35$	$n_2 = 35$	$n_3 = 35$
Equivalent peak welding current	$I_{\text{w-peak}} = 510\,\text{A}$		
Equivalent welding time	$t_{\text{w}} = 380\,\text{msec}$		
Equivalent cycle time	$T = 5.4\,\text{sec}$		
Duty ratio	$ED = \frac{t_{\text{w}}}{T} \cdot 100 = \frac{0.38\,\text{sec}}{5.4\,\text{sec}} \cdot 100 = 7\%$		
Short-circuit power in normal operation	$S''_{\text{k-NOP}} = 50\,\text{MVA}$ (four 1,250-kVA transformers in parallel)		
Short-circuit power in operation under fault conditions	$S''_{\text{k-OPFC}} = 40\,\text{MVA}$ (three 1,250-kVA transformers in parallel)		

Fig. C10.14 Example 400-V welding system with equivalent welding machines (Example C3)

The probability that of the $n = 35$ equivalent welding machines in Fig. C10.14, k machines are welding simultaneously can be calculated as follows according to the Bernoulli formula:

$$p_k(n) = \binom{n}{k} \cdot ED^k \cdot (1 - ED)^{n-k} \quad (10.20)$$

$$p_k(n = 35) = \binom{35}{k} \cdot 0.07^k \cdot 0.93^{35-k}$$

The binomial distribution calculated using Eq. (10.20) for the simultaneous operation of $n = 35$ welding machines with a duty ratio of $ED = 7\,\%$ is shown in Fig. C10.15. For the expected value of the binomially distributed total welding current, the following applies:

$$I_{W\mu} = n \cdot I_W \cdot ED \quad (10.21)$$

$$I_{W\mu} = 35 \cdot 510\text{ A} \cdot 0.07 = \underline{\underline{1{,}249.5\text{ A}}}$$

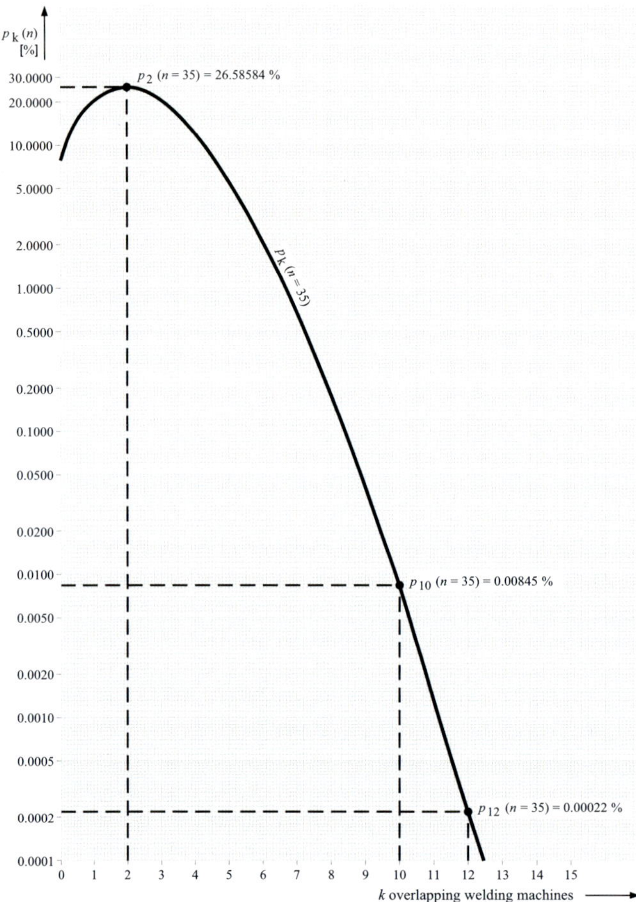

Fig. C10.15 Binomially distributed probabilities $p_k(n)$ of simultaneous welding by k welding machines (Example C3: $n = 35$, $ED = 7\,\%$, $k = 0(1)n$)

Using the variance $I_{S\sigma}^2$ it is possible to ascertain the mean deviation from the peak welding current of a machine group.

$$I_{W\sigma}^2 = n \cdot I_W^2 \cdot (ED - ED^2) \tag{10.22}$$

$$I_{W\sigma}^2 = 35 \cdot 510^2\ \text{A}^2 \cdot (0.07 - 0.07^2) = \underline{\underline{592{,}638\ \text{A}^2}}$$

The square root of the variance is called the standard deviation. Geometric addition of the standard deviation and expected value yields the peak welding current as a root mean square.

$$I_{W\text{-RMS}} = \sqrt{I_{W\mu}^2 + I_{W\sigma}^2} \tag{10.23}$$

$$I_{W\text{-RMS}} = \sqrt{1{,}249.5^2\ \text{A}^2 + 592{,}638\ \text{A}^2} = \underline{\underline{1{,}468\ \text{A}}}$$

The root mean square of the total welding current multiplied by the line-to-line voltage (U_{LL} = 400 V) yields the root mean square of the asymmetrical impulse load.

$$\Delta S_{\text{ASYM-RMS}} = I_{W\text{-RMS}} \cdot U_{\text{LL}} \tag{10.24}$$

$$\Delta S_{\text{ASYM-RMS}} = 1{,}468\ \text{A} \cdot 400\ \text{V} = 0.59\ \text{MVA}$$

Based on the root mean square of the asymmetric impulse load $\Delta S_{\text{ASYM-RMS}}$, it is possible to calculate the voltage dip $\overline{\Delta u'}$ occurring during welding as a statistical mean. For the calculation, the following simplified equation applies:

$$\overline{\Delta u'} = k_{\Delta S} \cdot \frac{\Delta S_{\text{ASYM-RMS}}}{S_k''} \cdot 100\ \% \tag{10.25}$$

Normal operation: $\overline{\Delta u'}_{\text{NOP}} = 2.1 \cdot \frac{0.59\ \text{MVA}}{50\ \text{MVA}} \cdot 100\ \% = \underline{\underline{2.5\ \%}}$

Operation under fault conditions: $\overline{\Delta u'}_{\text{OPFC}} = 2.1 \cdot \frac{0.59\ \text{MVA}}{40\ \text{MVA}} \cdot 100\ \% = \underline{\underline{3.1\ \%}}$

The factor $k_{\Delta S}$ for calculation of asymmetrical voltage changes [10.12] can be taken from Table C10.16.

Table C10.16 Orientation values for the factor $k_{\Delta S}$ for calculation of asymmetrical voltage changes [10.12]

Connection of the load in the LV system between	Node in the system at which $\Delta u'$ is to be calculated in the	Vector group of the supplying feeder distribution transformer	$k_{\Delta S}$ for voltage changes $\Delta u'$ between	
			line conductor and earth	two line conductors
line and neutral conductor	LV system near-to-transformer	Yz5	2.2	1.8
		Dy5, Dy11	2.8	1.8
	LV system far-from-transformer	Yz5, Dy5, Dy11	5.5	1.8
two line conductors	LV system	Yz5, Yy0, Dy5, Dy11	1.8	2.1
	MV system	Yz5, Dy5, Dy11	2.1	1.8
		Yy0	1.8	2.1

For the welding power system shown in Fig. C10.14 ($U_{nN} = 400$ V, $S_k'' = 40...50$ MVA, $n = 35$ welding machines per group, $I_W = 510$ A, $ED = 7$ %), statistically secured voltage dips of $\overline{\Delta u'_{NOP}} = 2.5$ % in normal operation (four 1,250-kVA transformers in parallel) and of $\overline{\Delta u'_{OPFC}} = 3.1$ in operation under fault conditions (three 1,250-kVA transformers in parallel) result. The probabilistic voltage stability criterion (Eq. 10.18) during welding in this example of a welding system is therefore reliably complied with (2.5(3.1) % < 10 %).

To verify compliance with the permissible failure rate λ_{perm} for voltage-induced welding errors, the following probabilistic approach can be taken:

The single probability calculated according to the Bernoulli formula $p_k(n)$ means that within a period of $1/p_k(n)$ time intervals ΔT, presumably an overlap of k out of n welding machines can be expected. It is therefore also possible to calculate with what probability more than k^* welding machines will weld simultaneously. For calculation of this probability, the following applies:

$$p_{k^*}(k>k^*) = 1 - p_{k^*}(k<k^*) \tag{10.26}$$

$$p_{k^*}(k<k^*) = p_0(n) + p_1(n) + p_2(n) + ... + p_{k^*}(n) \tag{10.27}$$

For the example of a welding process with $n = 35$ machines and a duty ratio of $ED = 7$ %, the probabilities $p_{k^*}(k>k^*)$ and voltage dips $\Delta u'_{k^*}(k>k^*)$ are shown graphically in Fig. C10.17. As the probabilistic curve in the figure shows, overlaps of more than 11 welding machines in normal operation result in voltage dips $\Delta u' > 10$ %. In operation under fault conditions (one supplying 1,250-kVA transformer failed), the permissible voltage dip is already exceeded when more than 9 welding machines overlap. The probability that more than 9 welding machines overlap is $p_9(k>9) = 0.01015$ %. With a very low probability of $p_{11}(k>11) = 0.00025$ %, more than 11 welding machines will simultaneously overlap. Based on the probabilities $p_{k^*}(k>k^*)$, the time $\Delta T(k>k^*)$ between two overlaps of more than k^* welding machines can be calculated:

$$\Delta T(k>k^*) = \frac{1}{p_{k^*}(k>k^*)} \cdot \frac{T - t_W}{3.600} \tag{10.28}$$

$p_{k^*}(k>k^*)$ probability that more than k^* machines will weld simultaneously
T cycle time in sec
t_W welding time in sec

In evaluating the welding quality, only the overlap time intervals for voltage dips > 10 % are relevant. For simultaneous welding with the machines of the 400-V example system, this overlap time during normal operation is ΔT_{NOP} ($\Delta u' > 10$ %) = 559.8 h and in operation under fault conditions ΔT_{OPFC} ($\Delta u' > 10$ %) = 13.8 h (Fig. C10.18). With the overlap time intervals ΔT ($\Delta u' > 10$ %) for voltage dips $\Delta u' > 10$ %, the failure rate $\lambda(\Delta u' > 10$ %) for voltage-induced welding errors can be calculated. The following applies [10.13]:

$$\lambda(\Delta u' > 10\,\%) = \frac{k(\Delta u' > 10\,\%) \cdot T}{n \cdot \Delta T \cdot (\Delta u' > 10\,\%)} \cdot \frac{1}{3.6} \tag{10.29}$$

Normal operation: $\lambda_{NOP}(\Delta u' > 10\,\%) = \dfrac{12 \cdot 5.4\ \text{sec}}{35 \cdot 559.8\ \text{h}} \cdot \dfrac{1}{3.6} = \underline{\underline{0.0009\ ‰}}$

Operation under fault conditions: $\lambda_{OPFC}(\Delta u' > 10\,\%) = \dfrac{10 \cdot 5.4\ \text{sec}}{35 \cdot 13.8\ \text{h}} \cdot \dfrac{1}{3.6} = \underline{\underline{0.03\ ‰}}$

In normal operation, the 400-V example welding system (Fig. C10.14) exhibits a failure rate for voltage-induced welding errors of $\lambda_{NOP}(\Delta u' > 10\,\%) = 0.0009\,‰$. In operation under fault conditions, this failure rate is increased to $\lambda_{OPFC}(\Delta u' > 10\,\%) = 0.03\,‰$.

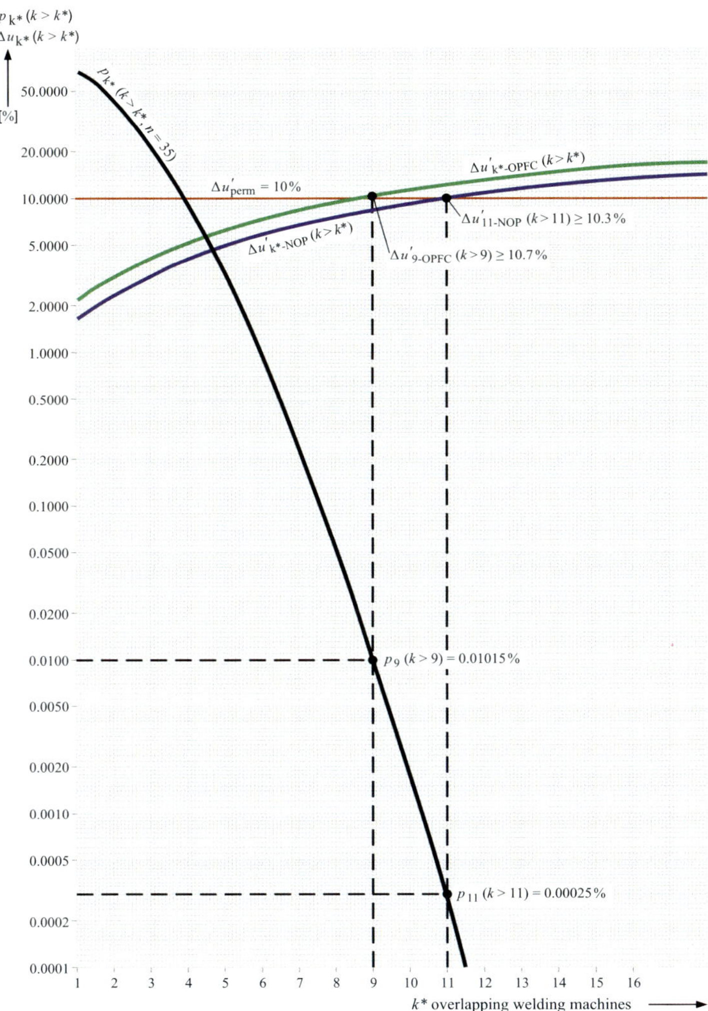

Fig. C10.17 Probabilities $p_{k^*}(k>k^*)$ and voltage dips $\Delta u'_{k^*}(k>k^*)$ for simultaneous welding by more than k^* welding machines (Example C3: $n = 35$, $ED = 7\,\%$, $k^* = 1(1)n-1$)

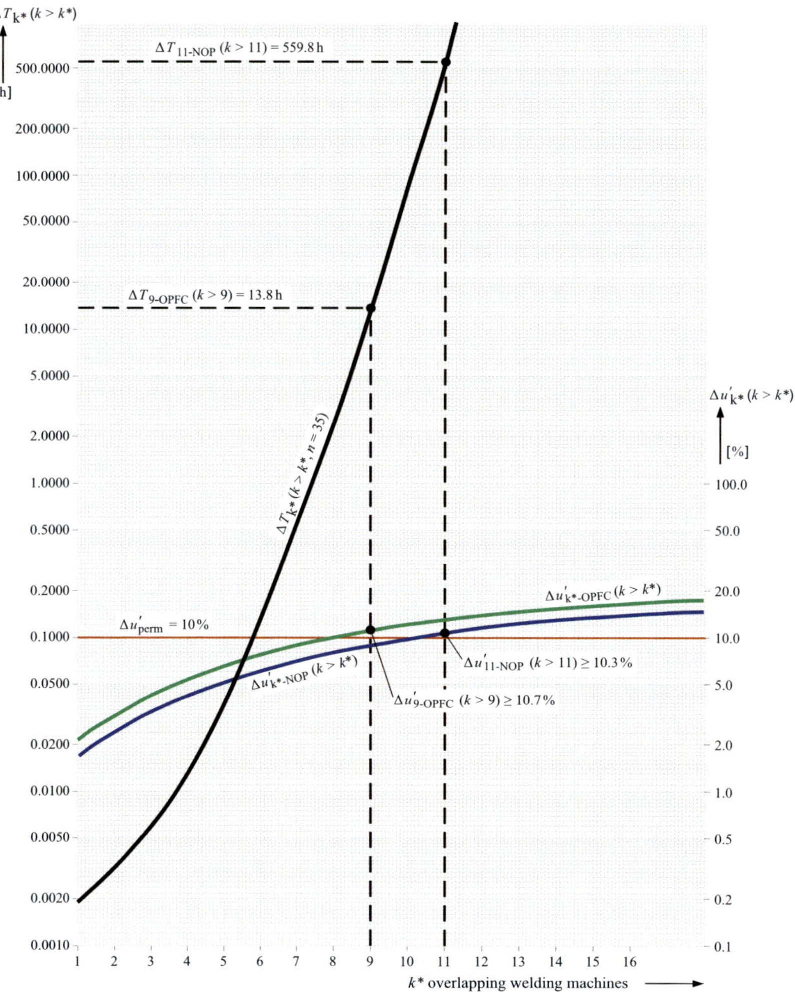

Fig. C10.18 Time intervals $\Delta T_{k^*}(k>k^*)$ and voltage dips $\Delta u'_{k^*}(k>k^*)$ between two overlaps of more than k^* welding machines (Example C3: $n = 35$, $ED = 7\,\%$, $k^* = 1(1)n-1$)

Both in normal operation and in operation under fault conditions, the λ dimensioning criterion (Eq. 10.19) is reliably complied with (0.0009 (0.03) ‰ < 1 ‰).

Failure rates for voltage-induced welding errors < 1 ‰ are an unmistakeable indication of the high welding quality expected by all car manufacturers [10.13, 10.14]. High welding quality always necessitates high voltage stability. Experience has shown that only the required voltage stability determines the power rating of the distribution transformers to be used for supplying welding machines. The thermal equivalent current is almost always a minor consideration.

If the power rating of the distribution transformers is defined according to the necessary voltage stability, it is possible to dispense with dynamic compensation equipment and power limiting controls (PLCs). For static reactive-power compensation, capacitor units without reactors can be used in most welding systems. In 400-V welding systems, the use of capacitors with a rated voltage of $480\text{ V} \le U_{rC} \le 525\text{ V}$ is recommended.

10.1.1.4 Painting and curing plants

The typical operating characteristics of painting and curing plants are a large load density ($P' \leq 400$ W/m^2) and a high duty ratio of the loads (50 % ≤ *ED* < 100 %).

The loads that are decisive for system dimensioning in curing plants are the resistance furnaces with rated powers $P_r \leq 100$ kW. Moreover, high-frequency ($f \leq 450$ kHz) and medium-frequency curing plants ($f \leq 25$ kHz) with rated powers $P_r \leq 200$ kW and phase-angle control or multi-cycle control are used. Connection of resistance furnaces, as well as of high-frequency and medium-frequency curing plants, to the 400-V system does not usually cause any major problems. However, phase-controlled curing plants do absorb non-sinusoidal currents and, in the case of a purely resistive load, also phase control reactive power. These disadvantages can be avoided by the use of curing plants with multi-cycle control. Fig. C10.19 illustrates the principle of phase-angle control and multi-cycle control.

The power demand of painting plants mainly depends on the fan power to be installed (power of the fans for heating and exhaust air removal). To relieve the LV system of the paint shop of harmonics, the power rectifiers for eletrophoretic or cathodic painting preferably draw power directly from the MV network through separate transformers (e.g. three-winding transformers) (Fig. C10.20).

The harmonics caused by rectifiers are generally immaterial for the MV system because the rectifier power rating ($S_{rPR} \leq$ 1,000 kVA) is small relative to the short-circuit power applied on the MV side ($S_k'' = 350{,}000...750{,}000$ kVA).

Strict compliance with supply reliability requirements of the painting process is also important for adhering to the reliable compatibility level for harmonics according to DIN EN 61000-2-4 (VDE 0839-2-4): 2003-05 [10.2] or IEC 61000-2-4: 2002-06 [10.3]. This includes the uninterrupted handling of an (n – 1) fault by isolation of the fault location by means of protection equipment (SR class 7 of Table A2.5). For uninterrupted handling of a single fault, the AF-cooled GEAFOL cast-resin transformers of the power supply concept shown in Fig. C10.20 provide an instantaneous reserve or "hot standby" redundancy. Moreover, for failure-critical loads in the paint shop (e.g. safety-related fan and sprinkler systems), a standby generating system is provided. The standby generating systems used in paint shops are usually diesel generator sets with automatic starting. To back up the time that the diesel generator sets require for automatic starting (5 sec < $t_{start} \leq$ 15 sec), modern standby generating systems are equipped with an additional battery-based UPS for uninterruptible power supply.

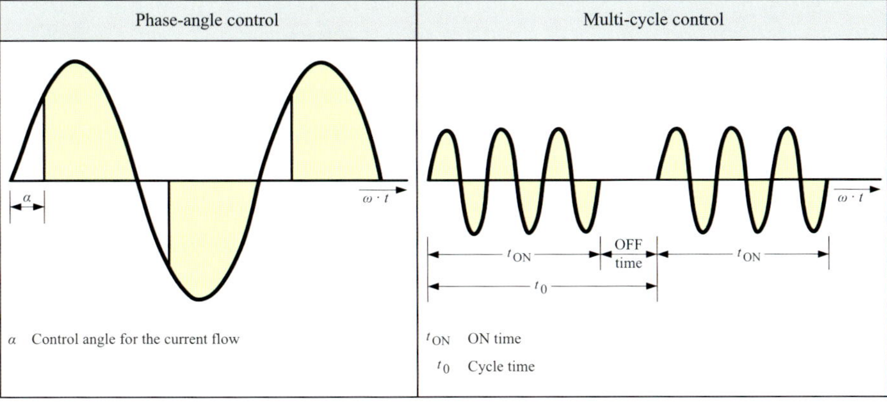

Fig. C10.19 Principle of phase-angle control and multi-cycle control

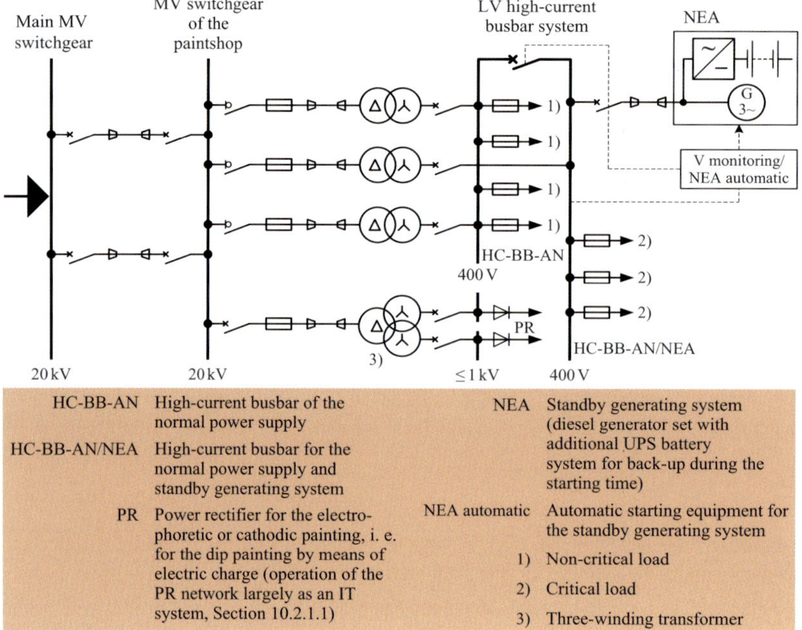

Fig. C10.20 Example of a power supply concept for a paint shop

10.1.1.5 Lighting systems

The lighting systems for industrial buildings must be planned according to the lighting engineering stipulations of the standard DIN EN 12464-1 (DIN 12464-1): 2003-03 [10.15] or CIE S 008/1: 2001-01 [10.16] and the electrical stipulations of the standard DIN VDE 0100-0100 (VDE 0100-0100): 2009-06 [10.17] or IEC 60364-1: 2005-11 [10.18]. The requirements for the electrical lights in planning lighting systems are contained in the standard DIN EN 60598-1 (VDE 0711-1): 2009-09 [10.19] or IEC 60598-1: 2008-04 [10. 20].

The important aspect for system dimensioning is that lighting installations are single-phase continuous loads evenly distributed over the production area. Depending on the size of the production area or zone to be illuminated and the mean illuminance prescribed for industrial workplaces by DIN VDE 12464-1 (VDE 12464-1): 2003-03 [10.15] $\bar{E}_m$ (e.g. $\bar{E}_m = 200\ldots750$ lx for workplace lighting in the metal-processing industry), the connected powers of the subdistribution boards for supply of the 230-V lighting circuits is in the range 50 kW $\leq P_{pr} \leq$ 80 kW.

Even though the consumption of the lighting circuits is relatively small compared with the power circuits, power costs can be saved with modern lighting installations. However, utilization of the potential for energy saving in the lighting circuits does load the LV system with additional harmonic currents. The causes of these harmonics include:

- electronic ballast (EB) for energy-saving lamps (e.g. compact fluorescent lamps, Fig. C10.21),
- dimmable fluorescent lamps for adapting the illuminance to the incident daylight detected using light sensors,
- converters for low-volt halogen lamps.

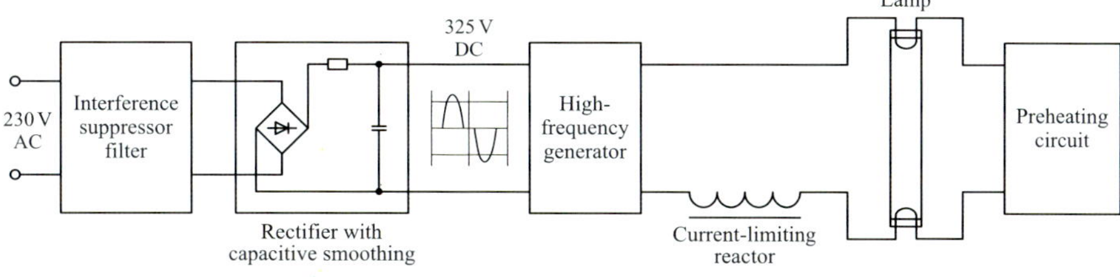

Example for the Fourier analysis of the lamp current [10.21]					
Frequency f in Hz	50	150	250	350	450
Current harmonic I_h in mA	66	50	25	10	9
Ratio of the current harmonic to the fundamental I_h/I_1 in %	100	76	38	15	14

Fig. C10.21 Schematic diagram of electronic ballast for energy-saving lamp with a Fourier analysis of its load current

As the result of the Fourier analysis listed in the table in Fig. C10.21 shows, the spectrum of current harmonics caused by the operation of energy-efficient lighting installations exhibits a high proportion of 3rd-order current harmonics (150-Hz current). The system perturbation that is characteristic of the 3rd-order current harmonic takes the form that the 150-Hz line conductor currents are summated in the neutral conductor (N conductor) even with perfectly symmetrical distribution of the single-phase loads over the line conductors of the three-phase system (Fig. C10.22). A high proportion of current harmonics whose harmonic order h is divisible by 3 ($h = 3, 9, 15, 21, \ldots, 3 + (n-1) \cdot 6$)) can cause overloading of the neutral conductor and a high risk of fire in the case of excessive heating.

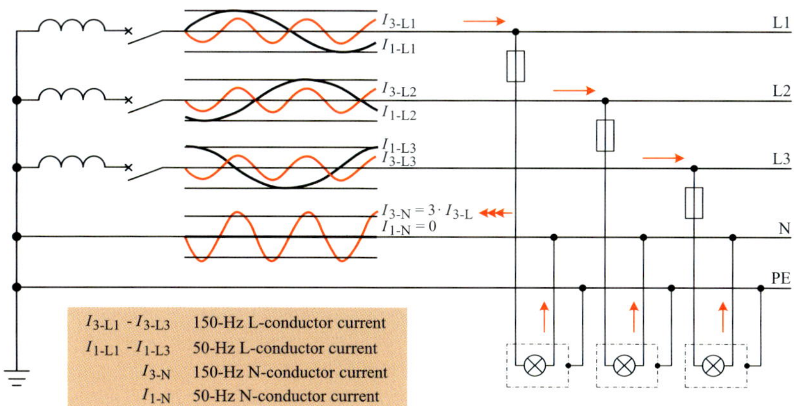

Fig. C10.22 Neutral conductor loading on symmetrical connection of energy-saving lamps

The measures and precautions by which overloading of the neutral conductor can be avoided are defined in the standards

- DIN VDE 0100-430 (VDE 0100-430): 1991-11 [10.22] or IEC 60364-4-43: 2008-08 [10.23] and
- DIN VDE 0100-520 (VDE 0100-520): 2003-06 [10.24] or IEC 60364-5-52: 2009-10 [10.25]

Dimensioning of the neutral conductor cross-sectional area according to the 150-Hz current is a technically and economically advantageous precaution against overloading the neutral conductor [10.26]. It is now state of the art for the busbar trunking systems (Section 11.2) of modern building installations to be dimensioned according to the neutral conductor load exerted by 150-Hz currents. For this dimensioning, the three load cases explained in Table C10.23 are of interest [10.27].

Table C10.23 Selection and dimensioning of busbar trunking systems according to the 150-Hz current

Case	Harmonic load due to 150-Hz current I_3	Dominant-loaded conductor of the three-phase system	Dimensioning rules for busbar trunking systems acc. to [10.27]
1	$I_3 \leq 0.15 \cdot I_{load}$	Line conductor	• In principle, neutral conductor can be considered not under a load • Rated current I_r is based on the load current I_{load} in the line conductors ($I_r > I_{load}$) • $A_L = A_N$, i. e. all conductors have the same cross-sectional area (neutral conductors with a smaller cross-section than the line conductors ($A_N < A_L$) are outmoded)
2	$0.15 \cdot I_{load} < I_3 \leq 0.33 \cdot I_{load}$	Line conductor	• Neutral conductor must be considered to be under a significant load • Rated current I_r results from the quotient $I_{load} / 0.84$ [1) ($I_r > I_{load} / 0.84$) • $A_L = A_N$, i. e. rating of the busbar trunking system with equal cross-sections acc. to the Joule heat of the 3 line conductors and the neutral conductor
3	$I_3 > 0.33 \cdot I_{load}$	Neutral conductor	• Neutral conductor is considered to be under a disproportionately large load • Rated current I_r is based on the neutral-conductor current $I_{load\text{-}N}$ ($I_r > I_{load\text{-}N}$) • $A_N = A_L$, i. e. rating of the busbar trunking system with equal cross-sections acc. to the neutral-conductor current load • Setting of the overload protection to trip the circuit-breaker to the small L-conductor current

I_3 3rd-order harmonic current (150-Hz current)
I_{load} Load current of the three-phase system
I_r Rated current or rated current step of the busbar trunking system
A_L Line conductor cross-section
A_N Neutral conductor cross-section
1) The divisor 0.84 is the result of measurements and temperature-rise tests conducted in the laboratory of Schneider Electric [10.28]. The divisor 0.84 ensures rating according to the maximum value principle. This takes care of the worst case of harmonic current loading.

The neutral conductor load exerted by 150-Hz currents has to be considered not only in selecting busbar trunking systems but also in dimensioning cable systems for power networks with energy-efficient lighting equipment. Consideration of 150-Hz neutral conductor currents in the dimensioning of cable systems is explained in Section 11.3.1.

For joint operation of lighting and power circuits, it is decisive whether the compatibility level for voltage changes and flicker intensity defined in the standards can be complied with (Table C10.3). In powerful LV systems ($3.750\ \text{kVA} \le \sum_i S_{rT_i} \le 4.000\ \text{kVA}$), loads in power circuits do not usually disturb operation of lighting installations (example C2, p. 181 et seq.). Separate power systems may have to be provided for the lighting installations only in the case of extreme load cycles of press drives (large load changes with high repeat frequencies ($r > 60\ \text{min}^{-1}$)) and diverse mass operation of welding machines and robots.

10.1.1.6 EDP and IT systems

The EDP and IT systems used in the metal-processing industry in the automation of the production processes place especially high demands on the quality of supply. In addition to the high supply reliability (SR class 9 in Table A2.5), above all the high quality of the power supply voltage (system class 1 in Table A2.11) must be ensured. Table C10.24 shows an overview of the relevant quality disturbances of the power supply voltage and their adverse effect on EDP and IT systems. The disturbances listed in this table cannot be absolutely ruled out, either in the MV nor in the LV network of the normal power supply. Only modern UPS systems can provide optimum protection against all types of network perturbation.

However, only the on-line method with double conversion (AC/DC–DC/AC conversion) provides a truly uninterruptible power supply without any disturbance in the voltage quality. The other principles, such as off-line UPS or line-interactive UPS, cause switching gaps of several milliseconds [10.29]. Fig. C10.25 shows an example of an on-line functional diagram of a static battery UPS system.

The main components of the on-line UPS shown in Fig. C10.25 are a power rectifier, DC link for energy storage (battery system and electronic battery disconnector) and a power inverter. The power rectifier, which converts the alternating voltage of the normal power supply to direct voltage, supplies the power inverter with power in normal operation. For this purpose, the battery of the DC link is maintained in the fully charged state and recharged if necessary. The power inverter generates a precisely sinusoidal power supply voltage with a constant amplitude and constant power frequency from the internal direct voltage of the UPS. During short or long interruptions of the normal power supply, the power inverter is automatically supplied with power from the battery. Continuation of supply by the battery is performed without interruption. Return to normal operation on recovery of the system voltage within the battery operation mode is also performed without interruption. On an overload or converter fault, the electronic bypass switch ensures automatic load transfer to the standby system (second power system). In this way, the EDP/IT loads can continue to be supplied without interruption even on an overload at the output of the power inverter or in the event of a fault within the UPS [10.30].

The battery operation mode for backing up long interruptions of the normal power supply is limited by the back-up time of the static UPS system. To extend the back-up time for powering EPD/IT loads, static UPS systems can be combined with standby generating systems (diesel generator sets). With a combination of static UPS system and standby generating system, definition of the battery capacity is based on the starting

Table C10.24 Quality disturbances of the power supply voltage and their effects on EDP/IT systems

Type of disturbance	Oscillogram of the disturbance	Effects
Voltage dip	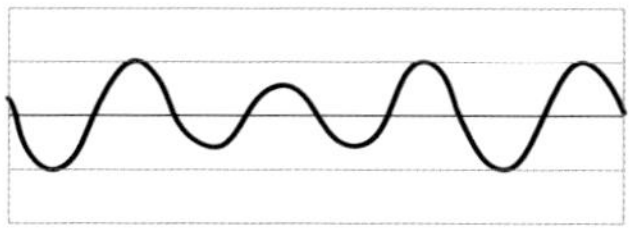	• Difficult data errors • Possible system crash
Transient overvoltages and voltage peaks	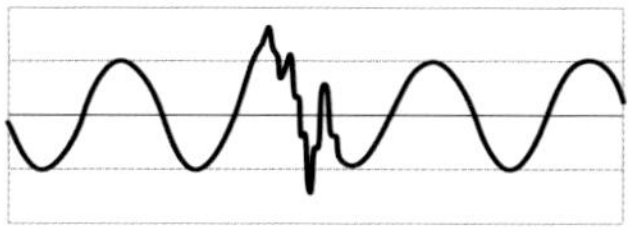	• Damage to hardware • Difficult data errors
High-frequency periodic and random deviations (harmonics)	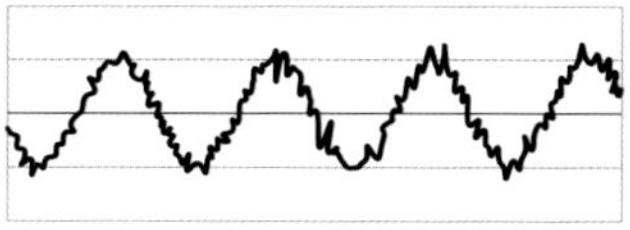	• Difficult data errors and losses • Inconsistent databases
Voltage fluctuations	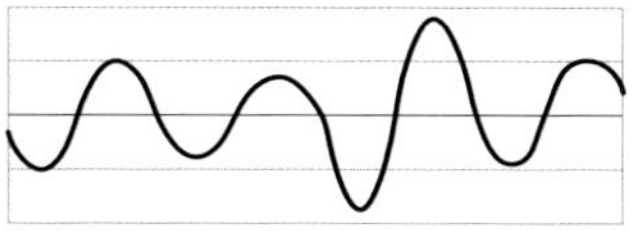	• Possible system crash • Difficult data errors
Frequency fluctuations	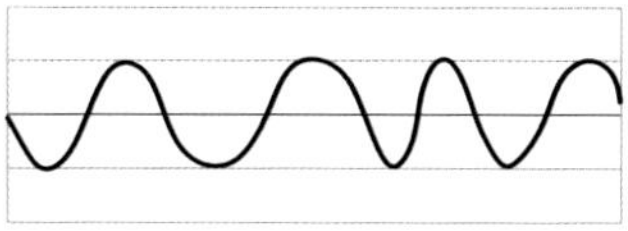	• Difficult data errors and losses • Possible system crash
Power-frequency overvoltages	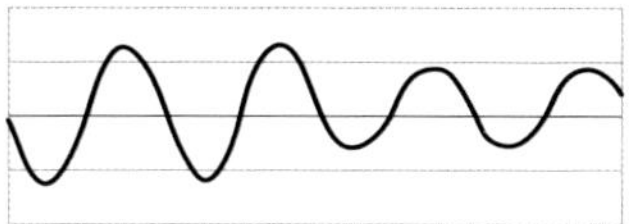	• Damage to hardware • Difficult data errors
Voltage interruptions	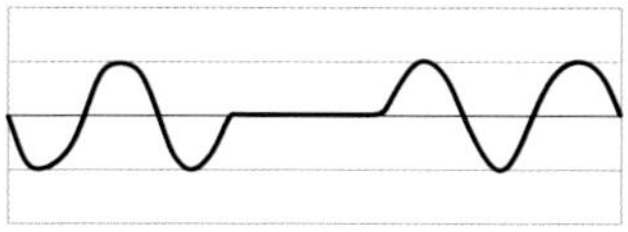	• Difficult data losses and errors • Possible system crash
Voltage breakdowns	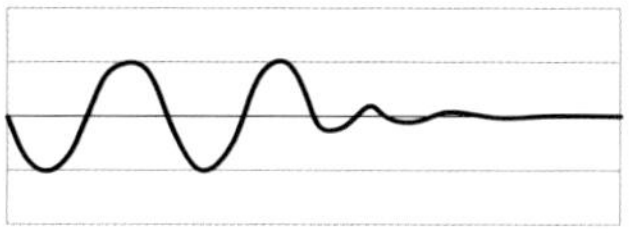	• Interruptions of program execution • Loss of data • So-called head crash possible

time that is reliably achieved with the emergency diesel generating set used. To achieve the full performance of diesel generator sets with rated powers $S_{rG} \leq 2$ MVA, starting times of $t_{start} = 6...8$ sec are typical [10.31].

Dynamic diesel UPS systems are a convenient alternative to a combination of a static UPS system and a standby generating system [10.32 to 10.34]. The mode of operation of a dynamic diesel UPS system is depicted schematically in Fig. C10.26.

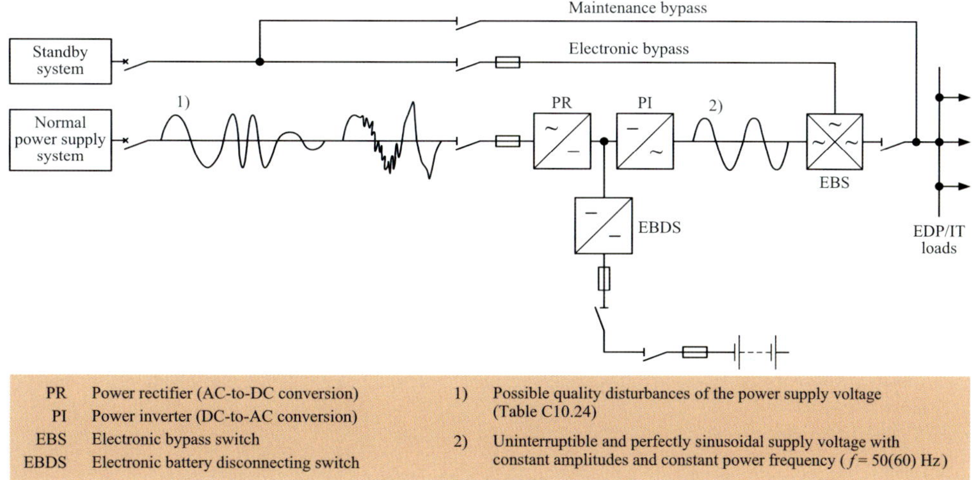

Fig. C10.25 On-line functional diagram of the Masterguard S static battery UPS system [10.30]

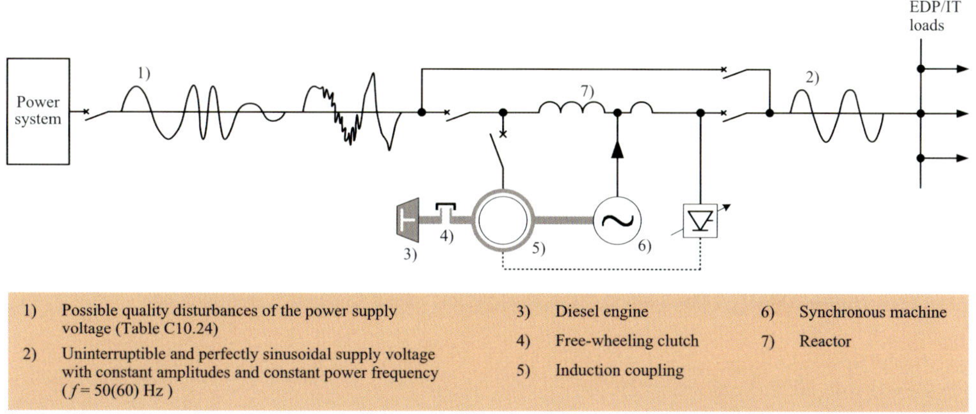

Fig. C10.26 Schematic mode of operation of the dynamic diesel UPS system from HITEC [10.34]

As Fig. C10.26 shows, the double conversion of the power supply voltage is not required in dynamic diesel UPS systems. In this case, a combination of synchronous generator (6) and reactor (7) serve as the active filter for a supply voltage that meets the quality requirements. The most important technical and ecological advantage of dynamic diesel UPS systems is the absence of a battery because it uses a kinetic energy store to back up the necessary starting time of the diesel generator sets on a power outage.

Both for static battery UPS and for dynamic diesel UPS systems, the LV network of the normal power supply is the primary power source. It is especially important for this network to comply with the EMC requirements when it is used to power EDP and IT equipment. According to [10.35], 80 % of all EMC disturbances on electrical and IT equipment are due to incorrect implementation of the LV network. The main prob-

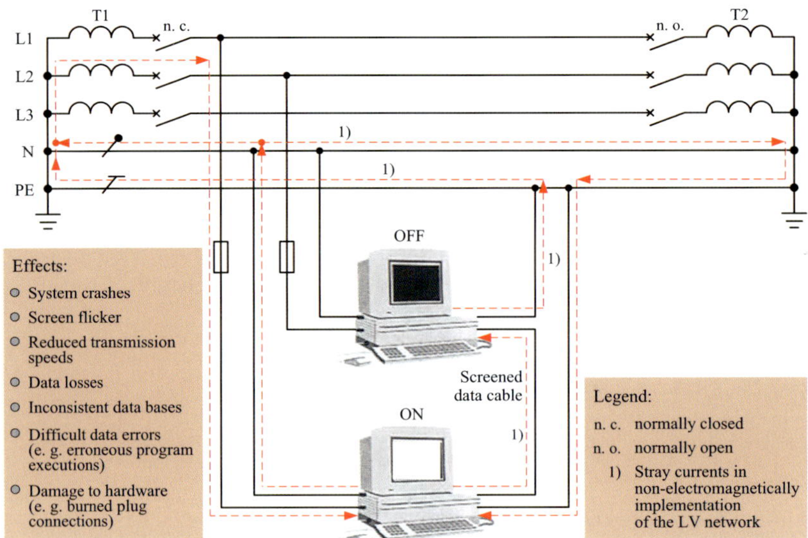

Fig. C10.27 Conducted EMC disturbances due to stray currents and their effects on EDP/IT systems

lem arising from incorrect implementation of the LV network takes the form of stray currents, that is, currents flowing along protective conductors (PE conductors) and shields of the data and information cables (Fig. C10.27).

Stray currents can only be avoided by building LV networks that meet the EMC requirements [10.36 to 10.38]. The conducted EMC depends on the earthing conditions prevailing in the LV network. The type of system earthing, that is, the type of earthing connections chosen, defines the earthing conditions that prevail in the LV network.

10.2 Choosing the type of LV system earthing

The earthing system types for implementation of LV networks to supply production processes with power differ in the

- type and number of conductors (live conductors L1-L3 and N, protective conductor PE, combined protective and neutral conductor PEN) and the
- type of connection to earth (IT, TT and TN earthing arrangements).

The type of system earthing must be chosen carefully because the earthing conditions of the LV network determine the cost of the protection measures and the level of electromagnetic compatibility (EMC). The planning details for a judicious choice of earthing system type are expained below.

10.2.1 System types possible according to the standards

The types of system earthing that can be considered in the planning of LV networks are defined in DIN VDE 0100-100 (VDE 0100-100): 2009-06 [10.17] or IEC 60364-1: 2005-11 [10.18]. Depending on the type of earth connection of the power source in the network and the type of earth connection of the conductive parts of the consumer‘s electrical installation, a distinction is made between IT, TT and TN systems (Table C10.28). The

Table C10.28 Types of LV system earthing according to DIN VDE 0100-100 (VDE 0100-100): 2009-06 [10.17] or IEC 60364-1: 2005-11 [10.18]

Type of system	TN system			TT system	IT system	
Characteristics	• One point of the system is directly earthed at the source. • The exposed conductive parts of the installation being connected to that point by protective conductors.			• One point of the system is directly earthed. • The exposed conductive parts of the installation are connected to earth electrodes.	• All live parts are isolated from earth or one point is connected to earth through a high impedance. • The exposed conductive parts of the installation are connected to earth electrodes.	
Variant(s)	TN-S system	TN-C system	TN-C-S system			
Characteristics	• Neutral conductor (N) and protective conductor (PE) are routed separately throughout the power system.	• Neutral conductor and protective conductor functions are combined in a single conductor, the PEN conductor, throughout the power system.	• Neutral conductor and protective conductor function are combined in a single conductor, the PEN conductor, in a part of the power system.	• Exposed conductive parts are connected directly to earth • Multiple exposed conductive parts are connected to earth through a common protective conductor.	• Exposed conductive parts of the individual electrical equipment items are connected directly to earth.	• Multiple exposed conductive parts are connected to earth through a common protective conductor.

Notes

- The first letter identifies the relationship of the power system to earth, i. e.
  - T direct connection of the point to earth,
  - I all live parts are isolated from earth or one point is connected to earth through a high impedance.
- The second letter identifies the relationship of the exposed conductive parts of the installation to earth, i. e.
  - T direct electrical connection of the exposed conductive parts to earth, independently of the earthing of any point of the power system,
  - N direct electrical connection of the exposed conductive parts with the earthed point of the power system.
- The other letters identify the arrangement of the neutral and protective conductor, i. e.
  - S protective function provided by a conductor separate from the neutral conductor (N,) or from the earthed line conductor (PE,),
  - C neutral and protective function combined in a single conductor (PEN conductor,).
- L1, L2, L3 denote the line conductors of a system type.

protection to be applied against elecric shock in these system types is defined by the standard DIN VDE 0100-410 (VDE 0100-410): 2007-06 [10.39] or IEC 60364-4-41: 2005-12 [10.40]. Basically, every protection measure must consist of

- a suitable combination of two independent protection precautions, that is, one basic protection precaution (protection against direct contact) and one fault protection precaution (protection against indirect contact), or of one
- enhanced protection precaution that provides both basic protection and fault protection.

The following measures of protection are permitted:

- protection by automatic disconnection of supply,
- protection by double or reinforced insulation,
- protection by electrical separation for supply to one item of current-using equipment,
- protection by means of extra-low voltage (SELV and PELV).

The most frequently used protection measure in electrical installations is automatic disconnection of supply [10.39 to 10.41]. With protection by automatic disconnection of supply, the disconnecting times defined in the standards must be complied with in the final circuits with a nominal current $I_n \leq 32$ A. These disconnecting times are listed in Table C10.29.

In distribution circuits and circuits not classified according to Table C10.29, the following disconnecting times are permitted depending on the system type:

- $t_a = 5$ sec in a TN system,
- $t_a = 1$ sec in a TT system.

If all extraneous conductive parts of an installation are connected to the protective equipotential bonding through the main earthing terminal, the longer disconnecting times of the TN system can also be used in a TT system [10.39, 10.40].

To verify the effectiveness of protection measures against personal injury in case of indirect contact, the operating current relative to the required disconnecting time of the overcurrent protective devices or the maximum permissible fault loop impedance of the LV network implemented as a TN or TT system must be known. Tables C10.30 and C10.31 contain a representative selection of these fault protection sizes. Miniature circuit-breakers (MCBs) with characteristics B and C according to DIN EN 60898-1 (VDE 0641-11): 2006-03 [10.42] or IEC 60898-1: 2003-07 [10.43] and LV HRC fuses of the utilization category gG according to DIN EN 60269-1 (VDE 0636-1): 2010-03 [10.44] or IEC 60269-1: 2009-07 [10.45] were selected as examples of overcurrent protective devices. For MCBs with characteristics B and C, the different disconnecting times are always achieved by means of instantaneous tripping according to the device standard [10.46].

A residual current-operated device (RCD) with a rated differential current $I_{\Delta n} \leq 30$ mA must be installed in addition to the basic protection (protection against direct contact) and fault protection (protection against indirect contact) for

- socket-outlets with a rated current of $I_r \leq 20$ A that is intended for use by ordinary persons and for general use, or
- final circuits for portable equipment used outdoors with a rated current of $I_r \leq 32$ A

[10.39, 10.40]. As regards additional protective measures in TN and TT systems, possible differences in incorporation into national standards must be considered.

Dimensioning of the protective conductors included in the circuits must be performed by applying the measure "protection by automatic disconnection of supply" according to DIN VDE 0100-540 (VDE 0100-540): 2007-06 [10.47] or IEC 60364-5-54: 2002-06

Table C10.29 Maximum disconnecting times t_a in final circuits with a nominal current of $I_n \leq 32$ A [10.39, 10.40]

Type of system earthing	Nominal alternating voltage line to earth			
	$50\,V < U_0 \leq 120\,V$	$120\,V < U_0 \leq 230\,V$	$230\,V < U_0 \leq 400\,V$	$U_0 > 400\,V$
TN	0.8 sec	0.4 sec	0.2 sec	0.1 sec
TT	0.3 sec	0.2 sec	0.07 sec	0.04 sec

Table C10.30 Operating currents I_a and maximum permissible loop impedances Z_s in TN or TT systems with overcurrent protective devices $I_n \leq 63$ A

$U_0 = 230\,V\,AC$			Nominal current of the overcurrent protective device I_n in A										
			2	4	6	10	16	20	25	32	40	50	63
Miniature circuit-breaker (MCB) B characteristic	I_a in A for a value of t_a	5 sec / 1 sec / 0.4 sec / 0.2 sec	10	20	30	50	80	100	125	160	200	250	315
	Z_s in Ω for a value of t_a	5 sec / 1 sec / 0.4 sec / 0.2 sec	23	11.5	7.67	4.6	2.88	2.3	1.84	1.44	1.15	0.92	0.73
Miniature circuit-breaker (MCB) C characteristic	I_a in A for a value of t_a	5 sec / 1 sec / 0.4 sec / 0.2 sec	20	40	60	100	160	200	250	320	400	500	630
	Z_s in Ω for a value of t_a	5 sec / 1 sec / 0.4 sec / 0.2 sec	11.5	5.75	3.83	2.3	1.44	1.15	0.92	0.72	0.58	0.46	0.37
LV HRC fuse utilization category gG	I_a in A for a value of t_a	5 sec	9.2	18.5	28	46.5	65	85	110	150	190	250	320
		1 sec	13	26	38	65	90	120	145	220	260	380	440
		0.4 sec	16	32	47	82	107	145	180	265	310	460	550
		0.2 sec	19	38	56	97	130	170	220	310	380	540	650
	Z_s in Ω for a value of t_a	5 sec	25	12.43	8.21	4.95	3.54	2.71	2.09	1.53	1.21	0.92	0.72
		1 sec	17.69	8.85	6.05	3.54	2.56	1.92	1.59	1.05	0.88	0.61	0.52
		0.4 sec	14.38	7.19	4.89	2.80	2.15	1.59	1.28	0.87	0.74	0.50	0.42
		0.2 sec	12.11	6.05	4.11	2.37	1.77	1.35	1.05	0.74	0.61	0.43	0.35

Table C10.31 Operating currents I_a and maximum permissible loop impedances Z_s in TN or TT systems with overcurrent protective devices $80\,A \leq I_n \leq 1{,}250$ A

$U_0 = 230\,V\,AC$			Nominal current of the overcurrent protective device I_n in A												
			80	100	125	160	200	250	315	400	500	630	800	1,000	1,250
LV HRC fuse utilization category gG	I_a in A for a value of t_a	5 sec	425	580	715	950	1,250	1,650	2,200	2,840	3,800	5,100	7,000	9,500	13,000
		1 sec	595	812	1,001	1,330	1,750	2,310	3,080	3,976	5,320	7,140	9,800	13,300	18,200
		0.4 sec	723	986	1,216	1,615	2,125	2,805	3,740	4,828	6,460	8,670	11,900	16,150	22,100
		0.2 sec	850	1,160	1,430	1,900	2,500	3,300	4,400	5,680	7,600	10,200	14,000	19,000	26,000
	Z_s in Ω for a value of t_a	5 sec	0.541	0.397	0.322	0.242	0.184	0.139	0.105	0.081	0.061	0.045	0.033	0.024	0.018
		1 sec	0.387	0.283	0.230	0.173	0.131	0.100	0.075	0.058	0.043	0.032	0.023	0.017	0.013
		0.4 sec	0.318	0.233	0.189	0.142	0.108	0.082	0.061	0.048	0.036	0.027	0.019	0.014	0.010
		0.2 sec	0.271	0.198	0.161	0.121	0.092	0.070	0.052	0.040	0.030	0.023	0.016	0.012	0.009

[10.48]. The cross-sectional area of the protective conductor can either be selected according to Table C10.32 or calculated for the specific application.

Table C10.32 Minimum cross-sectional areas of protective conductors, depending on the cross-section of the line conductor

Cross-sectional area of the line conductor A_L	Minimum cross-sectional area of the corresponding protective conductor A_{PE}	
	Protective conductor consists of the same material as the line conductor	Protective conductor does not consist of the same material as the line conductor
$A_L \le 16\,mm^2$	A_L	$\frac{k_1}{k_2} \cdot A_L$
$16\,mm^2 < A_L \le 35\,mm^2$	$16\,mm^2$ 1)	$\frac{k_1}{k_2} \cdot 16\,mm^2$
$A_L > 35\,mm^2$	$\frac{1}{2} \cdot A_L$ 1)	$\frac{1}{2} \cdot \frac{k_1}{k_2} \cdot A_L$

k_1 Material-related value of the line conductor
k_2 Material-related value of the protective conductor
The k values of the material are given in the tables in Annex A of DIN VDE 0100-540 (VDE 0100-540): 2007-06 [10.47] or IEC 60364-5-54: 2002-06 [10.48].

1) Cross-section reduction of a PEN conductor is only permitted in compliance with the rules for sizing of the neutral conductor according to DIN VDE 0100-520 (VDE 0100-520): 2003-06 [10.24] or IEC 60364-5-52: 2009-10 [10.25]. In modern installation systems with a high proportion of non-linear loads, reduction of the PEN conductor cross-section is not recommended because of the harmonic load caused by 150-Hz currents. For mechanical design reasons, PEN conductors shall have a cross-sectional area of $A_{PEN} \ge 10\,mm^2$ Cu or $A_{PEN} \ge 16\,mm^2$ Al. Unprotected installation of Al conductors is not permitted.

To design a single-fed LV network as a TN or TT system, it is usually enough to choose the conductor cross-sectional area according to Table C10.32. The conductor cross-sectional areas necessary to design an EMC-compliant TN system with multiple incoming supply (Section 10.2.2), on the other hand, should be calculated. The following formula must be used for the calculation:

$$A_{PE} \ge \frac{\sqrt{I_F^2 \cdot t_F}}{k} \quad (10.30)$$

$$0.1\ sec \le t_F \le 5\ sec \quad (10.30.1)$$

A_{PE} required conductor cross-sectional area in mm^2

I_F RMS value of the fault current for a fault of negligible impedance, which can flow through the protective device, in A

t_F operating/tripping time of the protective device (including trip tolerance) for automatic disconnection of power supply in seconds

k k value, which depends on the material of the protective conductor, the insulation and other parts and the initial and final temperature of the conductor, in $A \cdot \sqrt{sec}/mm^2$ (the k values of the materials are given in the tables in Annex A of DIN VDE 0100-540 (VDE 0100-540): 2007-06 [10.47] or IEC 60364-5-54: 2002-06 [10.48].)

In practice, the following terms are still used in connection with protection measures against personal injury on indirect contact:

- classic multiple-earthing protection (measure of protection in the TN-C system by overcurrent protective devices),
- modern multiple-earthing protection (measure of protection in the TN-S system by overcurrent protective devices),
- RCD protection (measure of protection using residual current protective devices),
- protective conductor system (measure of protection that is today termed an IT system with insulation monitoring).

The type of system earthing that can be selected when planning LV networks must permit not only reliable protection against electric shock but also effective protection against EMC disturbances.

Planning system recommendations to avoid EMC disturbances due to galvanic coupling are listed in Table C10.33. Details of further LV networks that meet the EMC requirements are explained in Section 10.2.2.

Table C10.33 System recommendations to avoid EMC disturbances according to [10.35] and DIN EN 50310 (VDE 0800-2-310): 2006-10 [10.49]

<table>
<tr><th colspan="3">Type of system earthing that can be selected for</th><th rowspan="2">Notes on electromagnetic compatibility (EMC)</th></tr>
<tr><th>outdoor distribution system</th><th colspan="2">indoor consumer's installation</th></tr>
<tr><td>TN-S</td><td colspan="2">TN-S</td><td>best solution</td></tr>
<tr><td>TN-C</td><td colspan="2">TN-S</td><td>recommended</td></tr>
<tr><td>TN-C</td><td colspan="2">TN-C</td><td>not recommended</td></tr>
<tr><td>TN-C</td><td colspan="2">TN-C-S</td><td>not recommended</td></tr>
<tr><td>TN-C</td><td>TN-C
in the basement as far as the main earthing terminal</td><td>TN-S
between and within storeys of the building</td><td>recommended</td></tr>
<tr><td>TT</td><td colspan="2">TT</td><td>• suitable for EDP systems inside a building
• not suitable for connecting cables between buildings with EDP systems</td></tr>
<tr><td>TT</td><td colspan="2">an isolating transformer must be installed to configure a TN-S system</td><td>good for EMC</td></tr>
<tr><td>IT</td><td colspan="2">IT</td><td>• less frequently used system
• suitable for EDP systems inside a building
• not suitable for connection cables between buildings with EDP systems</td></tr>
<tr><td>IT</td><td colspan="2">an isolating transformer must be installed to configure a TN-S system</td><td>good for EMC</td></tr>
</table>

10.2.1.1 IT system

IT systems [10.17, 10.18, 10.50 to 10.55] are used in particular when consumers' installations are subject to stringent requirements for the availability of electrical power. For example, DIN VDE 0100-710 (VDE 0100-710): 2002-11 [10.56] or IEC 60364-7-710: 2002-11 [10.57] require IT systems for power supplies in buildings with rooms used for medical purposes (e.g. hospitals and large practices for human healthcare and dentistry). Also, in certain industries such as chemicals, petroleum and steel, LV networks ($400\ V \leq U_{nN} \leq 1{,}000\ V$) are preferably implemented as an IT system.

In IT systems, no hazardous fault current can flow through a person touching an earth-faulted item of equipment when a single fault (short circuit to exposed conductive part or earth fault) occurs. For that reason, no automatic disconnection is required on the first fault of a line conductor to an exposed conductive part or earth. Because disconnection of the first short circuit to an exposed conductive part or earth fault is not necessary, operation can continue without a supply interruption.

However, the IT system only provides the advantage of a largely uninterruptible power supply if occurrence of a second fault is avoided by speedy elimination of the first fault [10.50]. Occurrence of a second fault (double earth fault) must result in immediate disconnection of the power supply. To

- signal the first fault (earth fault) and
- disconnect the second fault (double earth fault)

IT systems must be designed as follows:

- power source insulated against earth (e.g. transformer or generator),
- exposed conductive parts of the electrical installation earthed individually or in groups (for the second fault, the disconnection conditions of the TT system (Section 10.2.1.2) apply),
- exposed conductive parts interconnected by protective conductors and earthed collectively (for the second fault, the disconnection conditions of the TN system (Section 10.2.1.3) apply),
- insulation monitoring device for signalling the first fault,
- overcurrent protective device to disconnect the second fault,
- supplementary protective equipotential bonding (local equipotential bonding) on non-compliance with the disconnecting times applicable to TT or TN systems.

Fig. C10.34 shows the basic design of IT systems for protection against electric shock on indirect contact. To enable the IT systems shown in Fig. C10.34 to continue operation without hazard after occurrence of the first short circuit to the exposed conductive part or earth fault, the following condition must be fulfilled [10.39, 10.40]:

$$R_A \cdot I_d \leq 50\ V \qquad (10.31)$$

R_A sum of resistances of the protection earth electrode and of the protective conductor for the exposed conductive parts in Ω (total resistance of the earth electrode for exposed conductive parts)

I_d fault current on the first fault with negligible impedance between a line conductor and an exposed conductive part in A (the value I_d considers the leakage currents and the total earthing impedance of the electrical installation)

In the case of protection by signalling, insulation monitoring devices (IMDs) with characteristic values according to DIN EN 61557-2 (VDE 0413-2): 2008-02 [10.58] or IEC 61557-2: 2007-01 [10.59] must be used.

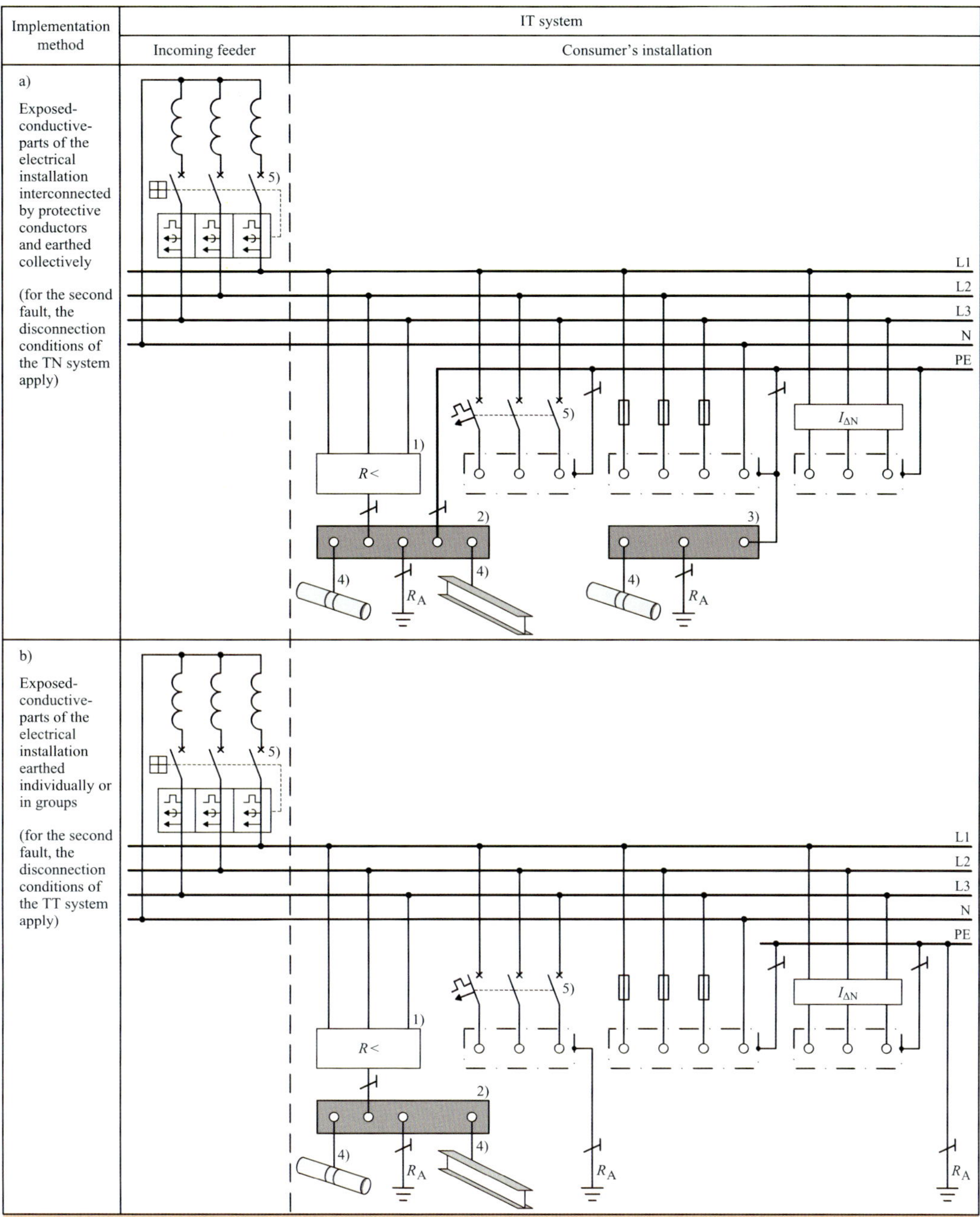

1) Insulation monitoring device (preferred connection behind the incoming feeder circuit-breaker)
2) Protective equipotential bonding through the main earthing terminal (main equipotential bonding)
3) Supplementary equipotential bonding (local equipotential bonding); supplementary equipotential bonding is necessary whenever the fault cannot be cleared in the required time (for faults in final circuits according to Table C10.29)
4) Protective bonding conductor (protective conductor provided for protective equipotential bonding)
5) Circuit-breaker with IT-compatible breaking capacity

Fig. C10.34 Methods of implementation of the IT system (depending on the type of earthing of the exposed conductive part of the electrical installation)

The value to be set for the response of the insulation monitoring device (on a decrease in the insulation resistance) is usually 100 Ω/V, that is, 23 kΩ in 230/400-V power systems.

A residual current monitor (RCM) according to DIN EN 62020 (VDE 0663): 2005-11 [10.60] or IEC 62020: 2003-11 [10.61] or an insulation fault locator according to DIN EN 61557-9 (VDE 0413-9): 2009-11 [10.62] or IEC 61557-9: 2009-01 [10.63] may also be provided to signal the occurrence of a first fault between a live part and exposed conductive parts or earth, unless a protective device is installed to disconnect the power supply on the first fault [10.39, 10.40]. In [10.50], the restriction defined in the standards on use of RCMs to signal a first fault is not considered relevant to safety because, even if a single fault is disconnected by the protective device, it can be convenient to signal the beginning of a fault as indicated by the impedance. As the fault indication, the insulation monitoring device (use of which is mandatory in Germany) and the residual current monitor must produce an audible and/or visual signal that must remain until the fault is eliminated.

The time required to eliminate the first fault is not standardized. However, it is recommended that the first fault be eliminated as quickly as is practically possible. This recommendation is intended to ensure largely uninterruptible operation because propagation to a second fault (double earth fault) must result in automatic disconnection of supply. For disconnection of the power supply in the event of a second fault, the following protective devices can be used:

- miniature circuit-breakers with characteristics B and C according to DIN EN 60898-1 (VDE 0641-11): 2006-03 [10.42] or IEC 60898-1: 2003-07 [10.43],
- LV HRC fuses of utilization category gG according to DIN EN 60269-1 (VDE 0636-1): 2010-03 [10.44] or IEC 60269-1: 2009-07 [10.45],
- circuit-breakers according to DIN EN 60947-2 (VDE 0660-101): 2010-04 [10. 64] or IEC 60947-2: 2009-05 [10.65],
- residual current-operated protective devices (RCDs) according to DIN EN 61543 (VDE 0664-30): 2006-06 [10.66] or IEC 61543: 1995-04 [10.67].

RCDs must be provided individually for each load. Such RCDs include:

- residual current-operated circuit-breakers without integral overcurrent protection (RCCBs) according to DIN EN 61008-1 (VDE 0664-10): 2010-01 [10.68] or IEC 61008-1: 2010-02 [10.69],
- residual current-operated circuit-breakers with integral overcurrent protection (RCBOs) according to DIN EN 61009-1 (VDE 0664-20): 2010-01 [10.70] or IEC 61009-1: 2010-02 [10.71].

If overcurrent protective devices are used (LV HRC fuses, MCBs, circuit-breakers) to disconnect the power supply in case of a second fault, the disconnection conditions will depend on how the IT system is implemented (Fig. C10.34). Depending on the method of implementation of the IT system, the following disconnection conditions are relevant:

- *Implementation method a:*

  Exposed conductive parts of the electrical installation interconnected by protective conductors and earthed collectively (Fig. C10.34a).

  For the second fault, the disconnection conditions of the TN system apply. To effect automatic disconnection of supply, the following conditions must be met:

  a1) AC systems without a neutral conductor

$$Z_S \leq \frac{U}{2 \cdot I_a} \tag{10.32}$$

a2) AC systems with a distributed neutral conductor

$$Z_S^* \leq \frac{U_0}{2 \cdot I_a} \quad (10.33)$$

Z_S	impedance of the fault loop, comprising the line conductor and the protective conductor of the circuit
Z_S^*	impedance of the fault loop, comprising the neutral conductor and the protective conductor of the circuit
U	nominal alternating voltage between lines (line-to-line voltage)
U_0	nominal alternating voltage between line and neutral (line-to-earth voltage)
I_a	current that causes the protective device to operate within the disconnecting time required for TN systems

- *Implementation method b:*

  Exposed conductive parts of the equipment earthed individually or in groups (Fig. C10.34b).

  With this method of implementation, the disconnection conditions of the TT system apply to the second fault. For disconnection of the power supply within the disconnecting time required for TT systems, the following condition must be met:

$$R_A \leq \frac{50\ \text{V}}{I_a} \quad (10.34)$$

R_A	sum of the resistances of the protection earth electrode and of the protective conductor for the exposed conductive parts
I_a	current that causes the protective device to operate within the disconnecting time required for TN systems

According to the stipulations for protection by automatic disconnection as per DIN VDE 0100-410 (VDE 0100-410): 2007-06 [10.39] or IEC 60364-4-41: 2005-12 [10.40], a protective conductor must be included in every circuit, unless "protection by double or reinforced insulation" can be used as the only protective measure [10.72]. Inclusion of an additional protective conductor makes implementation method a) of the IT system ideal for automatic disconnection of supply in case of a second fault.

For automatic disconnection of supply in the event of a second fault, the breaking capacity of the switching devices used must be IT-compatible. The breaking capacity for use of circuit-breakers in IT systems is verified accordingly in Annex H of DIN EN 60947-2 (VDE 0660-101): 2010-04 [10.64] or IEC 60947-2: 2009-05 [10.65]. However, passing a test defined in Annex H of the aforesaid standard does not provide sufficient information about whether the circuit-breaker is really suitable for a specific installation location in the IT system. Its suitability will ultimately depend on the short-circuit current conditions at the installation location.

In a worst-case scenario of double earth faults, that is, first earth fault before the circuit-breaker pole of a line conductor and second earth fault after the circuit-breaker pole of another line conductor, the line conductor pole with the downstream earth fault must break the fault current $I''_{k2E} \leq I''_{k2} = \sqrt{3}/2 \cdot I''_{k3}$ of the line-to-line short circuit with a connection to earth.

Unlike the three-pole and line-to-line short circuit clear of earth in which voltage $U = U_{nN}/\sqrt{3}$ or $U = U_{nN}/2$ is applied to the circuit-breaker pole, the voltage stress of the circuit-breaker pole in double faults increases to $U = U_{nN}$. So the breaking capacity of the circuit-breaker must be checked against the fault current of the line-to-line short circuit with a connection to earth at the installation location when $U_{nN} \cdot \sqrt{3}$, that is, 690 V in a 400-V IT system. Formally, the following must apply:

$$I_{cu}\big|_{\sqrt{3} \cdot U_{nN}} \geq I''_{k2E}\big|_{U_{nN}} \tag{10.35}$$

I_{cu} rated ultimate short-circuit breaking capacity of the circuit-breaker at $\sqrt{3} \cdot U_{nN}$ (e.g. I_{cu} at 690 V in a 400-V system)

I''_{k2E} fault current of the line-to-line short circuit with a connection to earth at the installation location of the circuit-breaker at the nominal system voltage (e.g. I''_{k2E} at $U_{nN} = 400$ V)

Selection and dimensioning of switching devices based only on formal compliance with the (10.35) condition can result in impractical use of circuit-breakers in IT systems. It should be noted that the breaking capacity of the circuit-breakers for disconnecting double faults according to Annex H of DIN EN 60947-2 (VDE 0660-101): 2010-04 [10.64] or IEC 60947-2: 2009-05 [10.65] only has to be tested up to a maximum fault current of $I_{F\text{-}IT} = 50$ kA. It is therefore assumed that the fault currents to be disconnected during double faults in IT systems are usually smaller than 50 kA. As practical experience shows, an earth fault is usually resistive. Moreover, double faults in IT systems can almost be ruled out because the operator of such a system is notified of occurrence of the first earth fault and is prompted to eliminate the fault as fast as practically possible after the earth-fault indication.

10.2.1.2 TT system

For applications in industrial power supplies, the TT system [10.17, 10.18, 10.50, 10.73 to 10.76] is of only minor importance. In the public power supplies of certain countries, on the other hand, TT and TN systems still coexist. In the German-speaking countries, that is, Germany, Austria and Switzerland (D-A-CH area), TN systems are primarily used in public power supplies (especially in Switzerland and Germany). Austria was the last country in the D-A-CH area to decide to make the transition from TT to TN systems [10.75]. The TT system is still used relatively frequently for LV-side power supplies in southern Europe (e.g. Spain, Italy and Turkey) and on the Arabian peninsula (e.g. Qatar, Kuwait and Saudi Arabia).

If an LV network earthed at the transformer neutral point is implemented as a TT system, the earthing conditions in the consumer‘s installation may differ as follows:

- all exposed conductive parts of the installation interconnected by protective conductors and earthed collectively (Fig. C10.35a),
- all exposed conductive parts of the installation earthed individually or in groups (Fig. C10.35b).

Due to of the earthing conditions in the TT system, the fault current only flows back to the power source through earth in the event of an earth fault in the consumer‘s installation (Fig. C10.36). Because of this, the fault current can be limited to the extent that the normal overcurrent protective devices (fuses, MCBs, circuit-breakers) are not able to disconnect earth faults in the time prescribed by the standard.

For automatic disconnection of supply in the TT system, therefore, residual current-operated protective devices (RCDs) are usually used. RCDs are characterized by the fact that they can also reliably disconnect especially small fault currents. The chosen rated residual current $I_{\Delta N}$ of the RCD used is decisive for reliable fault clearance.

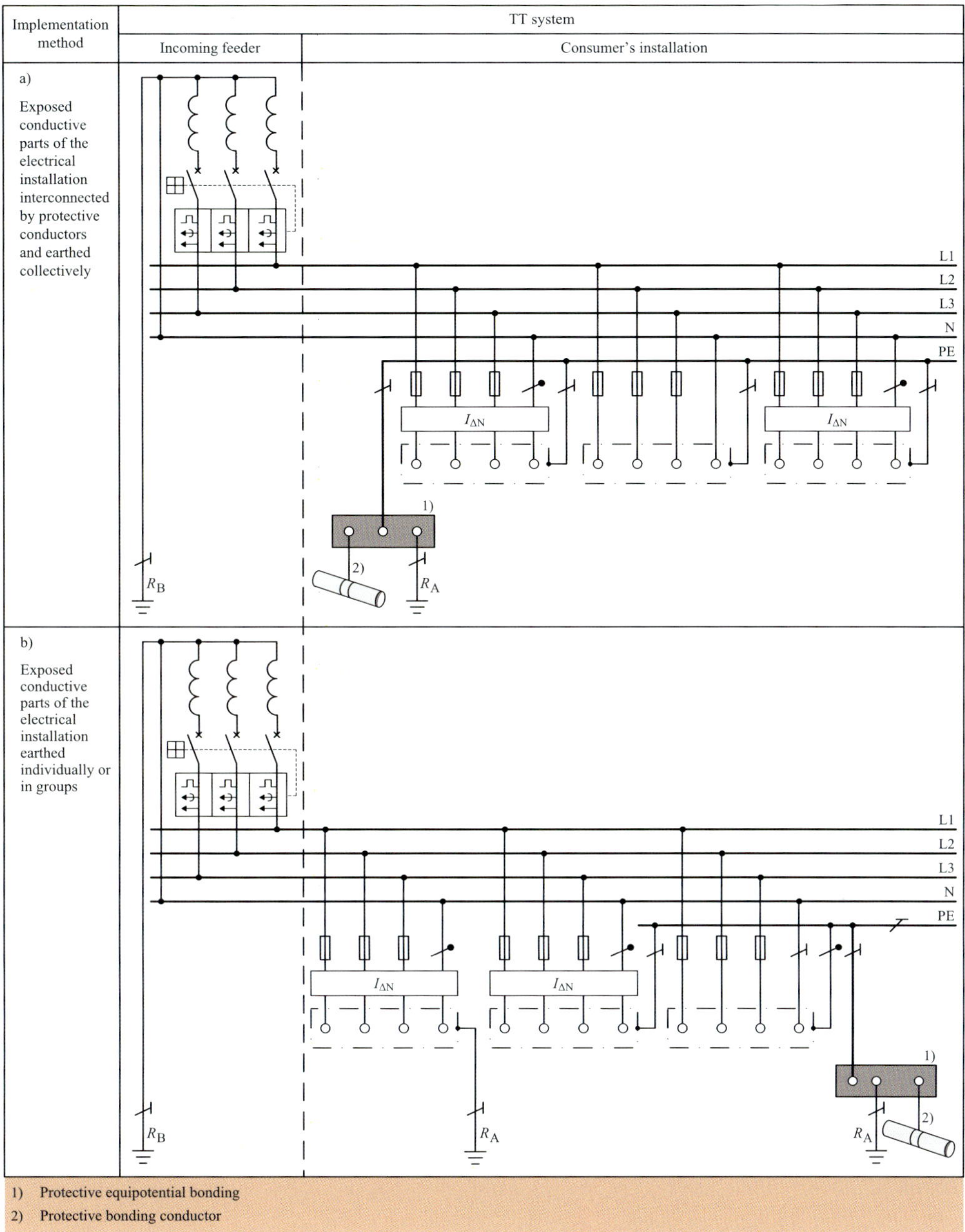

Fig. C10.35 Methods of implementation of the TT system (depending on the type of earthing of the exposed conductive parts of the electrical installation)

The rated residual current $I_{\Delta N}$ of an RCD for use in a TT system must be chosen so that, observing the valid disconnecting time (for disconnecting times t_a for final circuits, see Table C10.29), the following condition can be met:

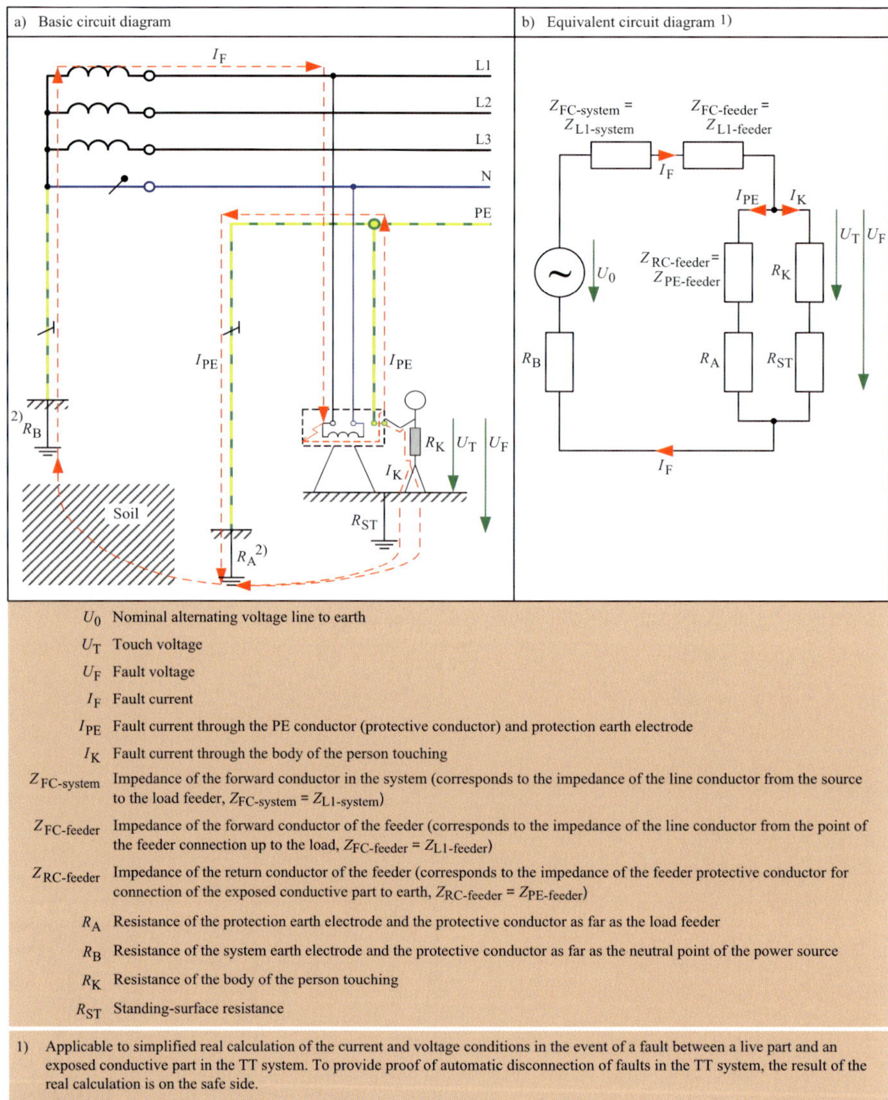

Fig. C10.36 Short circuit to exposed conductive part in the final circuit of a TT system (Example C4)

$$R_A \leq \frac{50\ \text{V}}{I_{\Delta N}} \qquad (10.36)$$

R_A sum of the resistances of the earth protection electrode and of the protective conductor for the exposed conductive parts in Ω

$I_{\Delta N}$ rated residual current of the RCD in A

If condition (10.36) has been met, at a line-to-earth voltage of $U_0 = 230$ V in the event of a fault, a fault current of (230 V/50 V) · $I_{\Delta N} = 4.6 \cdot I_{\Delta N}$ flows at which compliance with the disconnecting time according to Table C10.29 can be ensured [10.39, 10.40].

If an overcurrent protective device is used for protection by automatic disconnection of supply, the following condition must be met:

$$Z_{s\text{-TT}} \leq \frac{U_0}{I_a} \tag{10.37}$$

$Z_{s\text{-TT}}$ impedance of the fault loop in the TT system, consisting of
- the power source,
- the line conductor up to the point of the fault,
- the protective conductor of the exposed conductive parts,
- the earthing conductor (connection of the protective conductor to the earth electrode of the installation),
- the earth electrode of the installation (protection earth electrode),
- the earth electrode of the power source (system earth electrode).

U_0 nominal line-to-earth alternating voltage

I_a current that causes automatic operation of the overcurrent protective device within the required disconnecting time

Example C4

Short circuit to exposed conductive part in the final circuit of a TT system

The restricted suitability of overcurrent protective devices for protection against indirect contact in the TT system can be verified using an example calculation based on Fig. C10.36. The calculation Example C4 is based on the following system parameters:

$U_0 = 230\ \text{V}$

$Z_{\text{FC-system}} = Z_{\text{L1-system}} = 0.05\ \Omega$

$Z_{\text{FC-feeder}} = Z_{\text{L1-feeder}} = 0.50\ \Omega$

$Z_{\text{RC-feeder}} = Z_{\text{PE-feeder}} = 0.50\ \Omega$

$R_K = 750\ \Omega$

$R_A = 4.00\ \Omega$

$R_B = 2.00\ \Omega$

According to the equivalent circuit in Fig. C10.36, the impedance of the fault loop can be calculated as follows for the example of a final circuit in the TT system:

$$Z_{s\text{-TT}} = Z_{\text{L1-system}} + Z_{\text{L1-feeder}} + \frac{(Z_{\text{PE-feeder}} + R_A) \cdot (R_K + R_{ST})}{(Z_{\text{PE-feeder}} + R_A) + (R_K + R_{ST})} + R_B \tag{10.37.1}$$

Because of the relation between resistances $R_K >> R_{\text{PE-feeder}}$, R_A, R_{ST}, calculation of the loop impedance can be simplified as follows:

$$Z_{s\text{-TT}} \approx Z_{\text{L1-system}} + Z_{\text{L1-feeder}} + Z_{\text{PE-feeder}} + R_A + R_B \tag{10.37.2}$$

$$Z_{s\text{-TT}} \approx (0.05 + 0.50 + 0.50 + 4.00 + 2.00)\ \Omega = 7.05\ \Omega$$

The touch voltage U_T and the fault current I_F can be calculated using the loop impedance $Z_{s\text{-}TT}$. The following applies:

$$U_T = \frac{Z_{PE\text{-feeder}} + R_A}{Z_{s\text{-}TT}} \cdot U_0 \tag{10.37.3}$$

$$U_T = \frac{(0.50 + 4.00)\ \Omega}{7.05\ \Omega} \cdot 230\ \text{V} = \underline{\underline{146.8\ \text{V}}}$$

$$I_F = \frac{U_0}{Z_{s\text{-}TT}} \tag{10.37.4}$$

$$I_F = \frac{230\ \text{V}}{7.05\ \Omega} = \underline{\underline{32.624\ \text{A}}}$$

By applying the current divider rule and Kirchhoff's first law (Kirchhoff's current law), it is possible to calculate the fault-induced partial currents I_{PE} and I_K for the short circuit to exposed conductive part shown in Fig. C10.36.

$$I_{PE} = \frac{R_K}{Z_{PE\text{-feeder}} + R_A + R_K} \cdot I_F \tag{10.37.5}$$

$$I_{PE} = \frac{750\ \Omega}{(0.5 + 4.0 + 750.0)\ \Omega} \cdot 32.624\ \text{A} = \underline{\underline{32.429\ \text{A}}}$$

$$I_K = I_F - I_{PE} \tag{10.37.6}$$

$$I_K = 32.624\ \text{A} - 32.429\ \text{A} = \underline{\underline{0.195\ \text{A}}}$$

As the calculation Example C4 shows, the permissible touch voltage $U_{Tp} = 50$ V for short circuits to exposed conductive parts in the TT system is clearly exceeded. In the ratio of $R_A/R_B = 9$, the touch voltage U_T in the TT system is in fact twice as high as in the TN system. TT systems with a resistance ratio of $R_A/R_B = 9$ are primarily found in public power supplies [10.74].

If overcurrent protective devices are used for protection against electric shock, it proves especially problematic that occurrence of a high touch voltage U_T is associated with a strong limitation of the fault current I_F. The magnitude of the fault current during an earth fault in the TT system is largely determined by the resistance of the protection earth electrode R_A and of the system earth electrode R_B. With a series resistance of $R_A + R_B = 6\ \Omega$, a fault current of $I_F = 32.624$ A results for the final circuit of the example of a TT system (Eq. 10.37.4). To disconnect this fault current in the time $t_a = 0.2$ sec defined in the standard, a 6-A MCB with characteristic B or an LV HRC fuse of utilization category gG with a nominal current of $I_n = 2$ A is required (Table C10.30).

Because of the load currents flowing in consumers' installations, the use of overcurrent protective devices with very small nominal currents ($I_n \leq 6$ A in Example C4) is only possible in exceptional cases. For that reason, protection against electric shock on indirect contact in the TT system is usually ensured by residual current-operated protective devices (RCDs). The overcurrent protective devices used in the TT systems (MCBs, LV HRC fuses, circuit-breakers) are primarily for overload and short-circuit protection.

10.2.1.3 TN system

The TN system [10.17, 10.18, 10.50, 10.77 to 10.82] is the type of system earthing that is preferred in design of LV networks for both public and industrial power supplies.

Fig. C10.37 shows the methods of implementation of the TN system (TN-C, TN-C-S and TN-S). In all three TN system configurations, the exposed conductive parts of the electrical installation are connected to the neutral point of the power source through protective conductors. The fault current on a short circuit to exposed conductive part in the TN system therefore flows along the PEN/PE conductor back to the power source (Fig. C10.38). Owing to the live and PEN/PE conductor connections, the fault loop impedance is low so that a short circuit to exposed conductive part in the TN system produces a short-circuit-like fault current.

Because the short circuit to exposed conductive part in the TN system is a short-circuit-like fault, it can also be disconnected using protective devices with relatively high operating currents, such as LV HRC fuses (Table C10.31). Moreover, it is advantageous that, in many cases, the overcurrent protective devices which are installed in the LV network anyway for overcurrent and short-circuit protection, such as MCBs, fuses or circuit-breakers, are also suitable for ensuring protection against electric shock by automatic disconnection of supply. In addition to overcurrent protective devices, residual current-operated protective devices (RCDs) can also be used for automatic disconnection of supply.

To ensure that the potentials of the protective conductors deviate as little as possible from the earth potential in case of a fault, the protective conductor must be earthed at as many positions as possible.

Multiple earthing of the protective conductor gives rise to a large, low-impedance equipotential area that prevents the occurrence of high touch voltages U_T. In combination with residual current-operated protective devices (RCDs), the TN system is therefore an optimum solution for protection against indirect contact. The design of LV networks as a TN system is subject to the following requirements:

a) Protective devices and conductor cross-sectional areas

must be chosen such that, in 400/230-V power systems, automatic disconnection of supply on a short circuit to exposed conductive parts is performed with negligible impedance within $t_a = 0.4$ sec (final circuits with a nominal current of $I_n \leq 32$ A) or $t_a = 5$ sec (final circuits with a nominal current of $I_n > 32$ A and distribution circuits). This requires that the following disconnection condition be met:

$$Z_{s\text{-TN}} \leq \frac{U_0}{I_a} \tag{10.38}$$

$Z_{s\text{-TN}}$ impedance of the fault loop in the TN system, consisting of
- the power source,
- the line conductor up to the point of the fault,
- the protective conductor between the point of the fault and the power source.

U_0 nominal line-to-earth alternating voltage

I_a current that causes automatic operation of the disconnecting protective device (overcurrent protective device or residual current-operated protective device (RCD)) within the required time

If a residual current-operated protective device (RCD) is used, the disconnection condition (10.38) is especially easy to meet. In case of a short circuit to exposed conductive part in the TN system, the fault current I_F is many times greater than the rated residual current $I_{\Delta N}$ of the RCD used ($I_F >> I_{\Delta N}$).

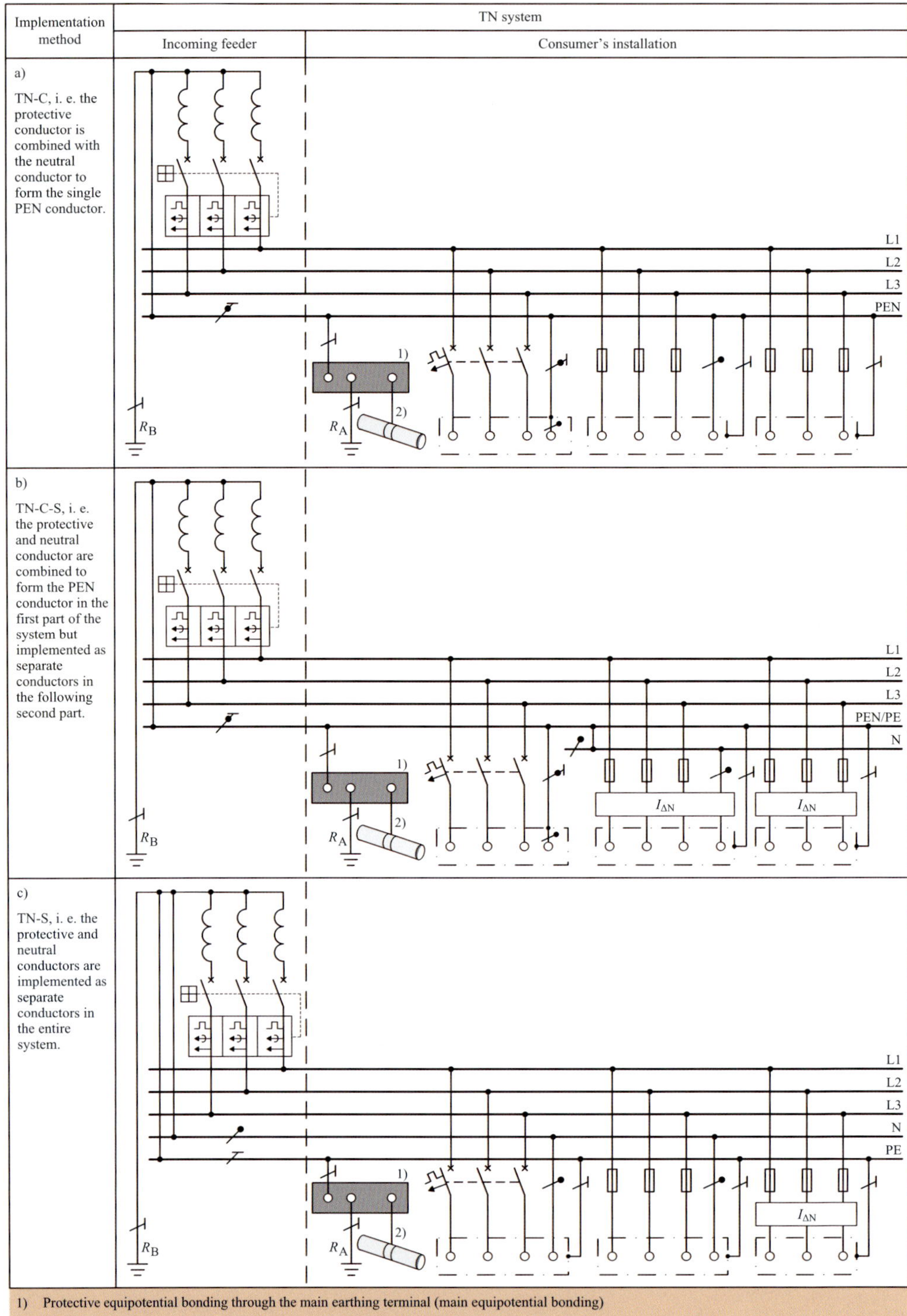

Fig. C10.37 Methods of implementation of the TN system (depending on the type of earth connection)

The disconnecting time according to Table C10.29 is always ensured. The rated residual current $I_{\Delta N}$ of each residual current-operated protective device must be used as the fault current I_F for automatic disconnection of supply (e.g. $I_{\Delta N}$ = 30 mA for RCD type S).

If an overcurrent protective device is used, the fault current I_F must be calculated using the loop impedance $Z_{s\text{-}TN}$. By means of the fault current I_F, the necessary operating or tripping current I_a of the overcurrent protective device can be determined. To clear a short circuit to exposed conductive part in a distribution or final circuit in the disconnecting time prescribed by the standard (Tables C10.30 and C10.31), the fault current I_F must be larger than the required operating (tripping) current I_a ($I_F > I_a$).

Example C5

Short circuit to exposed conductive part in the final circuit of a TN system

Fig. C10.38 shows an example of a basic and equivalent circuit diagram for a short circuit to exposed conductive part occurring in the final circuit of a TN system. This short circuit to exposed conductive part is to be disconnected by an overcurrent protective device based on the following system parameters:

$U_0 = 230\ \text{V}$

$Z_{FC\text{-}system} = Z_{L1\text{-}system} = 0.05\ \Omega$

$Z_{FC\text{-}feeder} = Z_{L1\text{-}feeder} = 0.50\ \Omega$

$Z_{RC\text{-}feeder} = Z_{PE\text{-}system} = 0.10\ \Omega$

$Z_{RC\text{-}feeder} = Z_{PE\text{-}feeder} = 0.50\ \Omega$

$R_K = 750\ \Omega$

$R_B = 2.00\ \Omega$

According to the equivalent circuit diagram (Fig. C10.38b), the impedance of the fault loop for the example of a final circuit in the TN system must be calculated as follows:

$$Z_{s\text{-}TN} = Z_{L1\text{-}system} + Z_{L1\text{-}feeder} + \frac{(Z_{PE\text{-}feeder} + Z_{PE\text{-}system}) \cdot (R_K + R_{ST} + R_B)}{(Z_{PE\text{-}feeder} + Z_{PE\text{-}system}) + (R_K + R_{ST} + R_B)} \qquad (10.38.1)$$

Because $R_K >> R_{PE\text{-}system}$, $R_{PE\text{-}feeder}$, R_B, R_{ST}, calculation of the loop impedance can be simplified. For simplified calculation, the following applies:

$$Z_{s\text{-}TN} \approx Z_{L1\text{-}system} + Z_{L1\text{-}feeder} + Z_{PE\text{-}feeder} + Z_{PE\text{-}system} \qquad (10.38.2)$$

$$Z_{s\text{-}TN} \approx (0.05 + 0.50 + 0.50 + 0.10)\ \Omega = 1.15\ \Omega$$

Once the loop impedance of the final circuit $Z_{s\text{-}TN}$ is known, the touch voltage U_T and fault current I_F can be calculated. The following applies:

$$U_T = \frac{Z_{PE\text{-}feeder} + Z_{PE\text{-}sysrem}}{Z_{s\text{-}TN}} \cdot U_0 \qquad (10.38.3)$$

$$U_T = \frac{(0.50 + 0.10)\ \Omega}{1.15\ \Omega} \cdot 230\ \text{V} = \underline{\underline{120\ \text{V}}}$$

$$I_F = \frac{U_0}{Z_{s\text{-}TN}} \qquad (10.38.4)$$

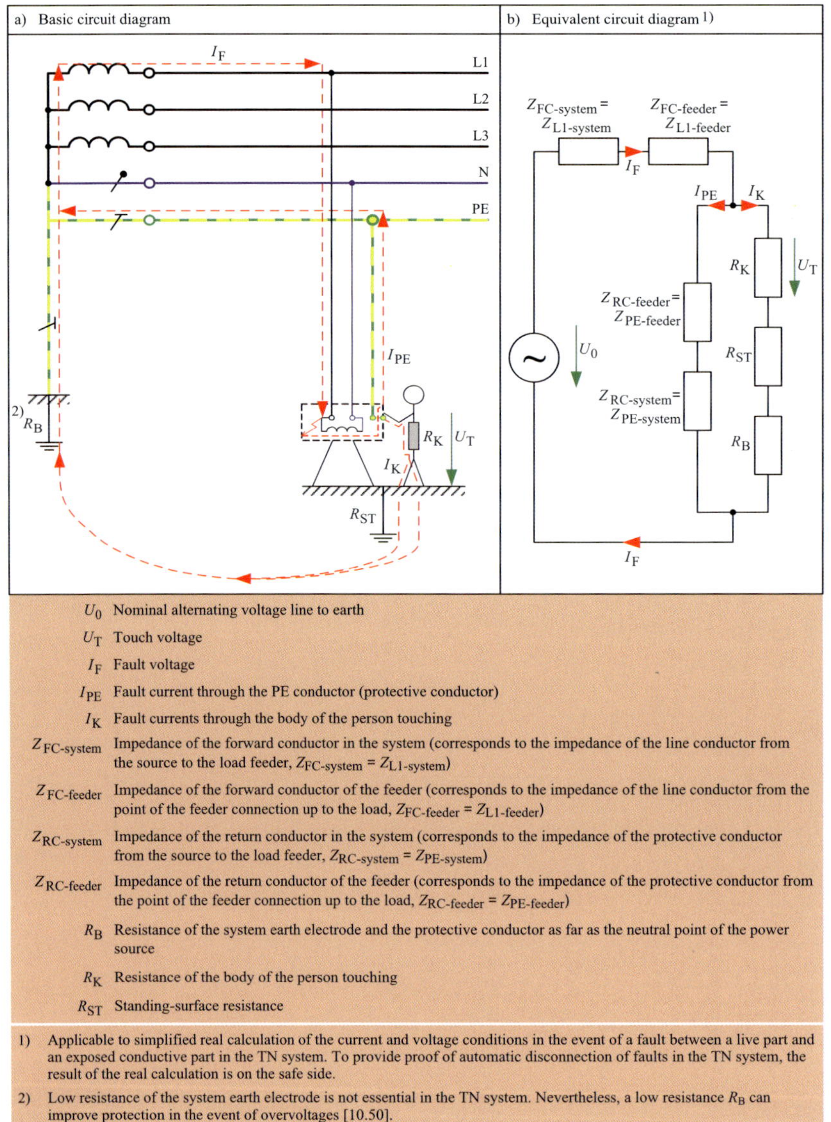

Fig. C10.38 Short circuit to exposed conductive part in the final circuit of a TN system (Example C5)

$$I_F = \frac{230\ \text{V}}{1.15\ \Omega} = \underline{\underline{200\ \text{A}}}$$

Using the current divider rule and Kirchhoff's current law, the following fault-induced partial currents I_{PE} and I_K can be calculated (Fig. C10.38):

$$I_{PE} = \frac{R_K}{Z_{PE\text{-feeder}} + Z_{PE\text{-system}} + R_K} \cdot I_F \tag{10.38.5}$$

$$I_{PE} = \frac{750\ \Omega}{(0.5 + 0.1 + 750.0)\ \Omega} \cdot 200\ \text{A} = \underline{\underline{199.840\ \text{A}}}$$

$$I_K = I_F - I_{PE} \quad (10.38.6)$$

$$I_K = 200.000\ \text{A} - 199.840\ \text{A} = \underline{\underline{0.160\ \text{A}}}$$

The fault current $I_F = 200$ A of the TN final circuit (Example C5) is much higher than the fault current $I_F = 32.6$ A of the equivalent TT final circuit (Example C4). As the example calculation shows, the TN system always exhibits a lower impedance than the TT system in short circuits to exposed conductive part. For that reason, the fault current I_F flowing during a short circuit to exposed conductive part in the TN system is usually enough to operate the overcurrent protective devices within the prescribed disconnecting time.

According to Table C10.29, a disconnecting time of $t_a = 0.4$ sec applies to the example of a TN final circuit. Within this time, a fault current of $I_F = 200$ A can reliably be disconnected by a 32-A MCB with characteristic B ($I_a = 160$ A) or a 25-A LV HRC fuse of utilization category gG ($I_a = 180$ A) (Table C10.30). Reliable disconnection of short circuits to exposed conductive parts by the overcurrent protective devices, which in any case have been installed in the circuits of LV networks, is an important advantage of TN systems over TT systems. Because of this advantage, it is often unnecessary to use additional residual current-operated protective devices (RCDs) for protection against electric shock. The operating currents I_a of overcurrent protective devices required to meet the disconnection condition (10.38) for protection against electric shock are listed in Table C10.30 (protective devices with nominal currents $I_n \leq 63$ A) and Table C10.31 (protection devices with nominal currents $80\ \text{A} \leq I_n \leq 1{,}250$ A).

b) Each exposed conductive part of an electrical installation item

must be connected to a protective conductor that can be connected with the earthed neutral point directly, or through PEN conductors with the earthed power source. The protective conductor cross-sectional area A_{PE} must be either selected according to Table C10.32 or calculated according to Eq. (10.30).

c) PEN conductors

must only be used in fixed electrical installations and they must have a minimum cross-sectional area of $A_{PEN} \geq 10\ \text{mm}^2$ Cu or $A_{PEN} \geq 16\ \text{mm}^2$ Al for reasons of mechanical strength.

d) The splitting of a PEN conductor

into protective conductor and neutral conductor must be performed at the point in the installation at which the cross-sectional areas of the line conductors fall below $A_{PEN} \geq 10\ \text{mm}^2$ Cu or $A_{PEN} \geq 16\ \text{mm}^2$ Al or from which a TN-S system is required.

e) In the PEN conductor

there must be no switching or disconnecting devices, because such equipment would impair the effectiveness of the protective measure “protection by automatic disconnection of supply“. For example, a switching device contact that did not close properly in the PEN conductor would prevent the automatic disconnection of a short circuit to exposed conductive part in the TN-C system.

f) In the TN-C system

no residual current-operated protective devices (RCDs) must be used. For protection against electric shock by automatic disconnection, only overcurrent protective devices must be used (Fig. C10.37a).

g) For use of a residual current-operated protective device (RCD)

in a TN-C-S system, the common PEN conductor must be split into a neutral and protective conductor before the RCD. No PEN conductor must be used on the supply side of the RCD (Fig. C10.37b).

h) The earth contact resistance R_B

should be as low as possible to limit, in case of an earth fault of one line conductor, the voltage rise against earth of all other conductors, in particular of the protective or PEN conductor. The following condition must be met:

$$\frac{R_B}{R_E} \leq \frac{50\text{ V}}{U_0 - 50\text{ V}} \tag{10.39}$$

R_B resistance of the system earth electrode and of the protective conductor as far as the neutral point of the power source

R_E minimum contact resistance with earth of extraneous conductive parts not connected to a protective conductor or an equipotential bonding conductor, through which a fault between line conductor and earth may occur

U_0 nominal alternating voltage between line conductor and earth

Usually, the occurrence of a hazardous touch voltage is prevented during an earth fault lasting a long time by the protective equipotential bonding system (protective equipotential bonding through main earthing terminal (main equipotential bonding) and supplementary protective equipotential bonding (local equipotential bonding)) of the consumer‘s installation. Whether an extraneous conductive part needs to be incorporated into the protective equipotential bonding system by connection to the supplying PEN conductor/protective conductor depends on the condition (10.39), the so-called “voltage balance“. This condition can be rearranged so that the resistance R_E of the voltage balance can be determined. With reference to the normal standard voltage 400 V/230 V 3AC, the following applies:

$$R_E \geq \frac{U_0 - 50\text{ V}}{50\text{ V}} \cdot R_B = 3.6 \cdot R_B \tag{10.39.1}$$

R_B resistance of the system earth electrode and of the protective conductor as far as the neutral point of the power source

R_E minimum contact resistance with earth of extraneous conductive parts not connected to a protective conductor or an equipotential bonding conductor, through which a fault between line conductor and earth may occur

U_0 nominal alternating voltage between line conductor and earth

In the case of a usual resistance of the system earth electrode of $R_B = 2\ \Omega$, $R_E \geq 7.2\ \Omega$ would have to apply in order to be able to dispense with integration of the extraneous conductive part into the protective equipotential bonding system. It can be proven by calculation that if $R_E \geq 7.2\ \Omega$, no impermissible touch voltage U_T will occur in the event of an earth fault. The earth fault of the line conductor L1 in an TN-C-S system shown by way of example in Fig. C10.39 is used to provide proof by calculation.

Example C6

Fulfilment of the voltage balance during an earth fault of the line conductor L1 with an extraneous conductive part in the TN-C-S system (Fig. C10.39)

To simplify, it is assumed that the conducted impedances (line conductors from the power source to the fault location) are negligible compared with R_B and R_E.

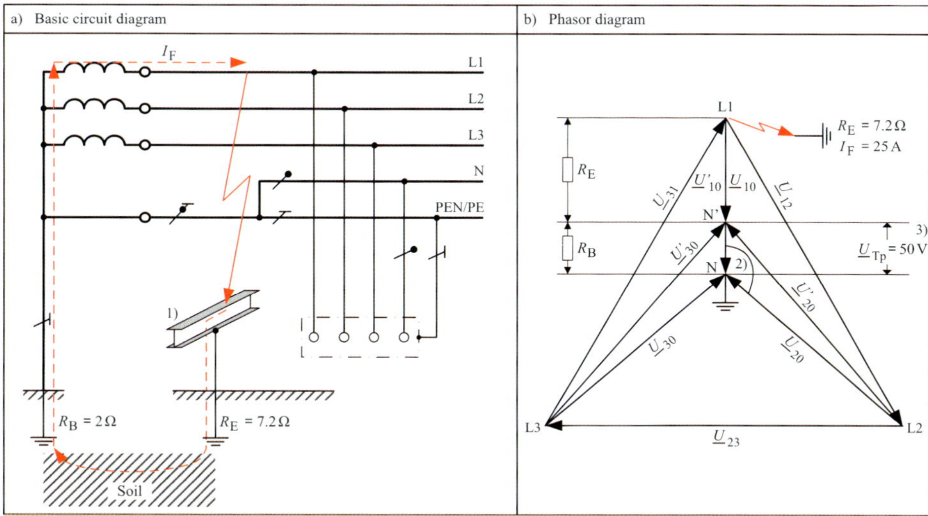

1) Extraneous conductive part that is in contact with earth and is not connected to the existing PEN/PE conductor of the system

2) $\measuredangle(\underline{U}_{Tp}, \underline{U}_{20}) = 120°$

3) Touch voltage permitted in the standards (According to DIN VDE 0100-410 (VDE 0100-410): 2007-06 [10.39] / IEC 60364-4-41: 2005-12 [10.40], the continuously permitted touch voltage is U_{Tp} = 50 V in the case of alternating voltage and U_{Tp} = 120 V in the case of direct voltage.)

Fig. C10.39 Voltage limitation according to the voltage balance during an earth fault of the line conductor L1 with an extraneous conductive part in the TN-C-S system (Example C6)

The following fault current arises in this simplification:

$$I_F = \frac{U_0}{R_B + R_E} \tag{10.39.2}$$

$$I_F = \frac{230\ \text{V}}{(2 + 7.2)\ \Omega} = \underline{\underline{25\ \text{A}}}$$

The voltage $\underline{U}'_{10}$ of the line conductor L1 affected by the fault is reduced by the magnitude of the touch voltage $\underline{U}_T$. For $\underline{U}_T = \underline{U}_{Tp}$ the following applies:

$$|\underline{U}'_{10}| = |\underline{U}_{10}| - |\underline{U}_T| \tag{10.39.3}$$

$$|\underline{U}'_{10}| = 230\ \text{V} - 50\ \text{V} = \underline{\underline{180\ \text{V}}}$$

The voltages $\underline{U}'_{20}$ and $\underline{U}'_{30}$ of the faultless line conductors L1/L2 are increased as follows:

$$|\underline{U}'_{20}| = |\underline{U}'_{30}| = \sqrt{|\underline{U}_{20}|^2 + |\underline{U}_T|^2 - 2|\underline{U}_{20}| \cdot |\underline{U}_T| \cdot \cos\angle(\underline{U}_T, \underline{U}_{20})} \tag{10.39.4}$$

$$|\underline{U}'_{20}| = |\underline{U}'_{30}| = \sqrt{(230^2 + 50^2 - 2 \cdot 230 \cdot 50 \cdot \cos 120°)\ \text{V}^2} = \underline{\underline{258.7\ \text{V}}}$$

The line-to-line voltages $\underline{U}_{12}$, $\underline{U}_{23}$ and $\underline{U}_{31}$ of the TN-C-S system affected by the earth fault, on the other hand, do not change. Their magnitude 400 V remains the same in the event of an earth fault.

Faults between energized conductors and extraneous conductive parts with high resistance ($R_E \geq 3.6\ R_B$) or low-resistance earth connections ($R_E < 3.6 \cdot R_B$) can be considered

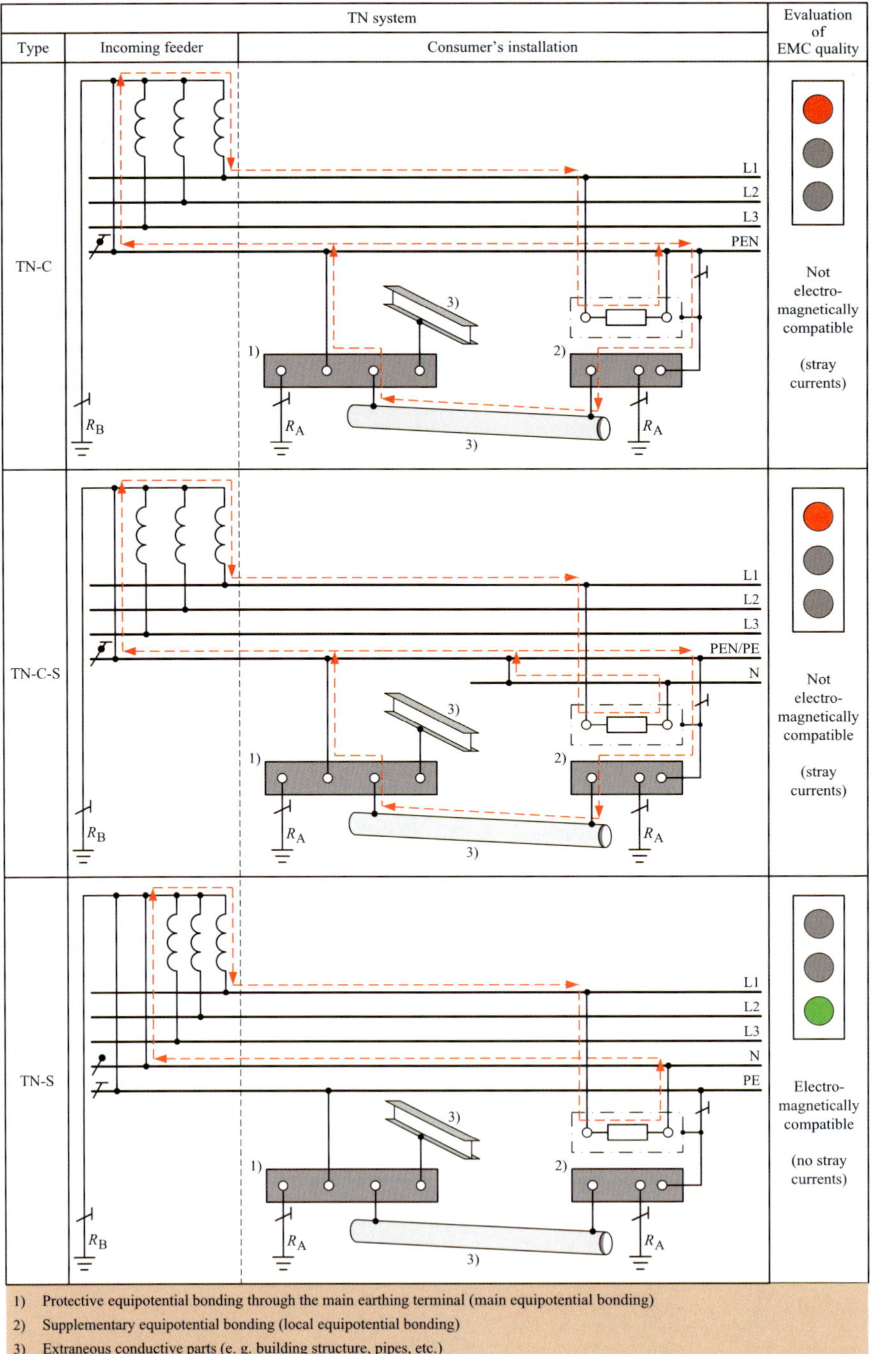

1) Protective equipotential bonding through the main earthing terminal (main equipotential bonding)
2) Supplementary equipotential bonding (local equipotential bonding)
3) Extraneous conductive parts (e. g. building structure, pipes, etc.)

Fig. C10.40 Inherent EMC quality of single-fed TN systems (TN single source systems) an extremely improbable event [10.50]. Nevertheless, to fulfil the voltage balance (Eq. 10.39), all extraneous conductive parts with a low-resistance earth connection $R_E < 3.6 \cdot R_B$ must be incorporated into the protective equipotential bonding system of the consumer's installation.

i) Designing LV networks as TN systems

must be performed in such a way that personal safety and high electromagnetic compatibility (EMC) are reconciled. Today's production and process engineering can only be managed with the highest quality and reliability when all types of EMC disturbances are ruled out. EMC disturbances are caused by:

- galvanic coupling (conducted interference),
- inductive coupling (magnetic interference),
- capacitive coupling (electrostatic interference),
- radiation coupling (radio and/or low-frequency interference).

The characteristic EMC interference in LV networks takes the form of stray currents. Stray currents are components of the load current that do not flow through the electrical power system (L1, L2, L3, N) during operation but flow without control through conductive parts of the installation and building back to the power source [10.36].

They impair the reliability of electrical and electronic equipment by:

- corrosion and pitting,
- emission of electromagnetic fields,
- immission of low-frequency fields,
- contact erosion and erosion of screens of signal cables.

One important cause of stray currents are the multiply earthed PEN conductors that are usual in consumers' installations. Because the PEN conductor combines the functions of protective and neutral conductor, it also carries the load current. Because of the protective conductors and protective equipotential bonding conductors that are connected in parallel to and picked off the PEN conductors, part of the load current automatically flows back to the power source as stray current. As has been shown, the three implementation methods of TN system (TN-C, TN-C-S, TN-S) are only equivalent in terms of personal protection, not in terms of their inherent EMC properties.

As shown in Fig. C10.40, only the TN-S system provides sufficient protection from EMC disturbances caused by galvanic or inductive coupling. Use of the TN-S system avoids stray currents because the load current only flows back to the power source through the separate neutral conductor (N conductor).

Conclusion

To reconcile reliable personal protection with high electromagnetic compatibility, LV networks must be designed as TN-S systems. The TN-S system should always be preferred in planning of LV networks.

10.2.2 EMC-compliant TN systems with multiple incoming supply

It is not possible to design LV networks as a pure TN-S system if they are multiply fed because, in a TN-S system, the neutral and protective conductors must be implemented as separate conductors from the power source. However, splitting into a neutral conductor and a protective conductor from the power sources of the multiple incoming supply would not result in a TN-S system but only in an impermissible parallel connection of the N and PE conductors (cf. Fig. C10.27). The impermissible parallel connection of the N and PE conductors can only be avoided in case of a multiple incoming supply by designing centrally earthed TN systems. Such systems are technically termed

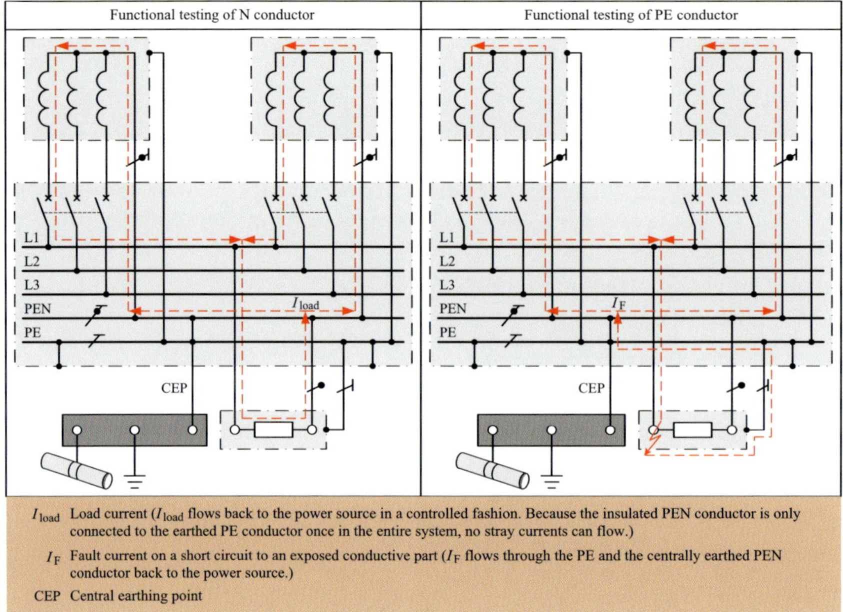

Fig. C10.41 Presence of the PEN conductor in an EMC-compliant TN system with multiple incoming supply

TN-S system with insulated PEN conductor [10.82]. This term is not physically accurate because TN-S systems are 5-wire systems and only comprise the 3 line conductors L1-L3, the neutral conductor N and the protective conductor PE. In the centrally earthed TN system, the conductor from the power source to the central earthing point (CEP) is not an N conductor but a PEN (Fig. C10.41).

Unlike the classic TN-C/TN-C-S system, the PEN conductor of the EMC-compliant TN system with multiple incoming supply is installed insulated against earth and is only connected to the earthed PE conductor at one central point. TN systems with isolated and centrally earthed PEN conductor that meet the EMC requirements are termed TN-EMC systems.

10.2.2.1 TN-EMC system with centralized multiple incoming supply

In the case of a centralized multiple incoming supply, the distribution transformers (Section 11.1) operated in parallel are installed centrally in an enclosed electrical operating room. This type of multiple source system is only recommended for supply to small-area consumer installations. Today, centralized TN-EMC multiple source systems are principally used to supply semiconductor production, data processing and computer centres with electrical power meeting the EMC requirements. Fig. C10.42 shows the basic structure of a TN-EMC system with centralized multiple incoming supply.

Special attention must be paid to fulfilling the following requirements when designing TN-EMC systems with centralized multiple incoming supply:

- Earthing of all transformer neutral points must be effected through a single central earthing point (CEP) in the low-voltage main distribution board (LV-MDB).
- The PEN conductors emanating from the transformer neutral points must be installed insulated against earth over their entire length. They may only be connected to the earthed protective conductor at the CEP.

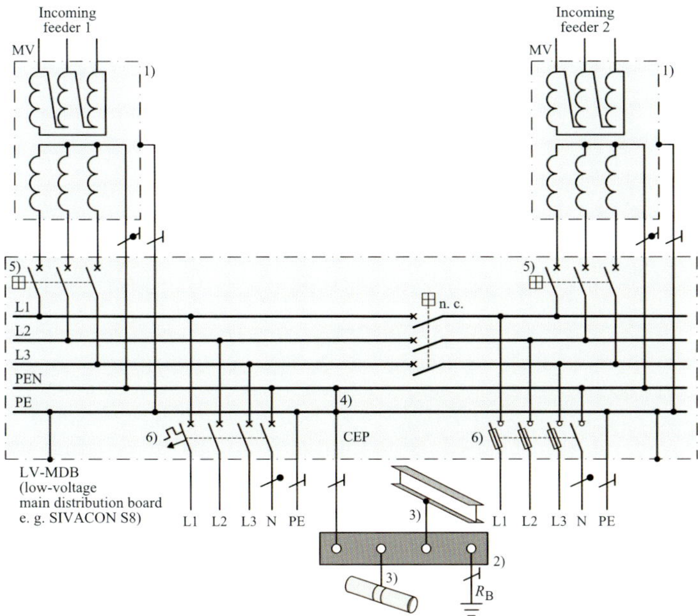

1) GEAFOL cast-resin transformers installed centrally in an enclosed electrical operating room
2) Protective equipotential bonding of the centralized multiple incoming supply through main earthing terminal (main equipotential bonding)
3) Protective bonding conductor (protective conductor provided for protective equipotential bonding)
4) Jumper between the insulated PEN conductor and earthed protective conductor (This jumper is a close central earthing point (CEP) for both transformers.)
5) Only 3-pole switching devices in the incoming feeder because the PEN conductor must not be switched or isolated
6) 4-pole switching devices in the outgoing feeders should be recommended especially for redundant supply to subdistribution boards. Apart from that, 3-pole switching devices can also be used.

Fig. C10.42 TN-EMC system with centralized multiple incoming supply (centralized TN-EMC multiple source system)

- In the incoming-feeder panels of the LV-MDB, only 3-pole switching devices may be used because the PEN conductor must not be switched. The outgoing-feeder panels of the the LV-MDB can be equipped with 4-pole switching devices (L1, L2, L3, N). To ensure that the phase currents (50-Hz currents) add up to zero in the neutral conductor, if possible, the use of 4-pole switching devices is recommended, especially for a redundant supply to subdistribution boards.
- The colour coding of the PEN conductor must comply with the standards. According to IEC 60364-5-51: 2005-04 [10. 83], PEN conductors must be marked green/yellow throughout their entire length with, in addition, light blue markings at the terminations or light blue throughout their length with, in addition, green/yellow markings at the terminations. DIN VDE 0100-510 (VDE 0100-510): 2007-06 [10. 84] stipulates that the IEC alternative with continuous light blue colouring and additional green/yellow marking at the terminations is forbidden in Germany.
- The transformer box must be connected to the PE conductor (colour-coded green/yellow) of the LV-MDB. The cross-sectional area of the PE conductor connection between the transformer box and the LV-MDB must be rated for the maximum short-circuit current that can occur.
- At the PEN conductor of the LV-MDB, only outgoing N conductors may be connected, <u>not</u> PE conductors.

- The jumper between the insulated PEN conductor and the earthed PE conductor in the LV-MDB may only be inserted and removed again using a special tool. The warning "Removing the jumper terminates the protection measure" should also be applied to the central earthing point (CEP).

Centralized TN-EMC multiple source systems that meet the requirements stated above and the conditions for protection against electric shock by automatic disconnection according to DIN VDE 0100-410 (VDE 0100-410): 2007-06 [10.39] or IEC 60364-4-41: 2005-12 [10.40] provide the same protection against damage and personal protection as the classic TN-S system.

10.2.2.2 TN-EMC system with decentralized multiple incoming supply

In case of a decentralized multiple incoming supply (for a comparison of decentralized and centralized multiple incoming supplies, see Section 5.2), the distribution transformers are installed containerized in the load centres of large-area factory and production halls. LV-side interconnection of distributed dry-type transformers installed in the containers is today preferably performed using high-current busbar systems (see Section 10.3.1.5).

Due to the high level of automation in modern industrial plants, processes running in large-area production halls respond very sensitively to EMC disturbances. To rule out such disturbances, decentralized TN-EMC multiple source systems must be designed. Decentralized TN-EMC multiple source systems are chiefly found in modern car factories [10.14]. Fig. C10.43 shows the structure of a TN-EMC system specially designed for the automotive industry.

The structure of the TN-EMC system depicted in Fig. C10.43 is basically subject to the same requirements as the centralized TN-EMC multiple source system. Displacing the central earthing point (CEP) from the low-voltage main distribution board (LV-MDB) into one of the distributed transformer load centre substations (load centre substation 1 in the example) however results in higher loop impedances for automatic disconnection of supply in case of a short circuit to an exposed conductive part. When verifying the disconnection condition (10.38), it must be noted that the fault current does not directly flow back to the neutral point of all supplying transformers. The fault current from transformers with a neutral point located remotely from the CEP always has to take the indirect route via the PEN-PE jumper. For these transformers, the fault loop consists of the

- live conductor from the power source to the fault location,
- PE conductor from the fault location to the PEN-PE jumper and the
- PEN conductor from the PEN-PE jumper to the power source.

With the hall dimensions that are usual in the automotive industry today, the return conductor of a fault loop can be as long as $l_{RC} \leq 450$ m [10.85]. Because the impedance of the return conductors very heavily damps the fault current of the supplying transformers, decentralized TN-EMC multiple source systems must be planned especially carefully. The system planner must exercise great care, above all, in ensuring personal protection. With the following additional measures, it is possible to raise protection against damage to property and personal injury to a sustainably safe level:

- Permanent monitoring of the PEN-PE jumper using an instrument transformer and a residual current monitor (RCM) according to DIN EN 62020 (VDE 0663): 2005-11 [10.60] or IEC 62020: 2003-11 [10.61]. The PE system should additionally be monitored. RCMs from W. Bender GmbH & Co. KG [10.86] can be used for monitoring.

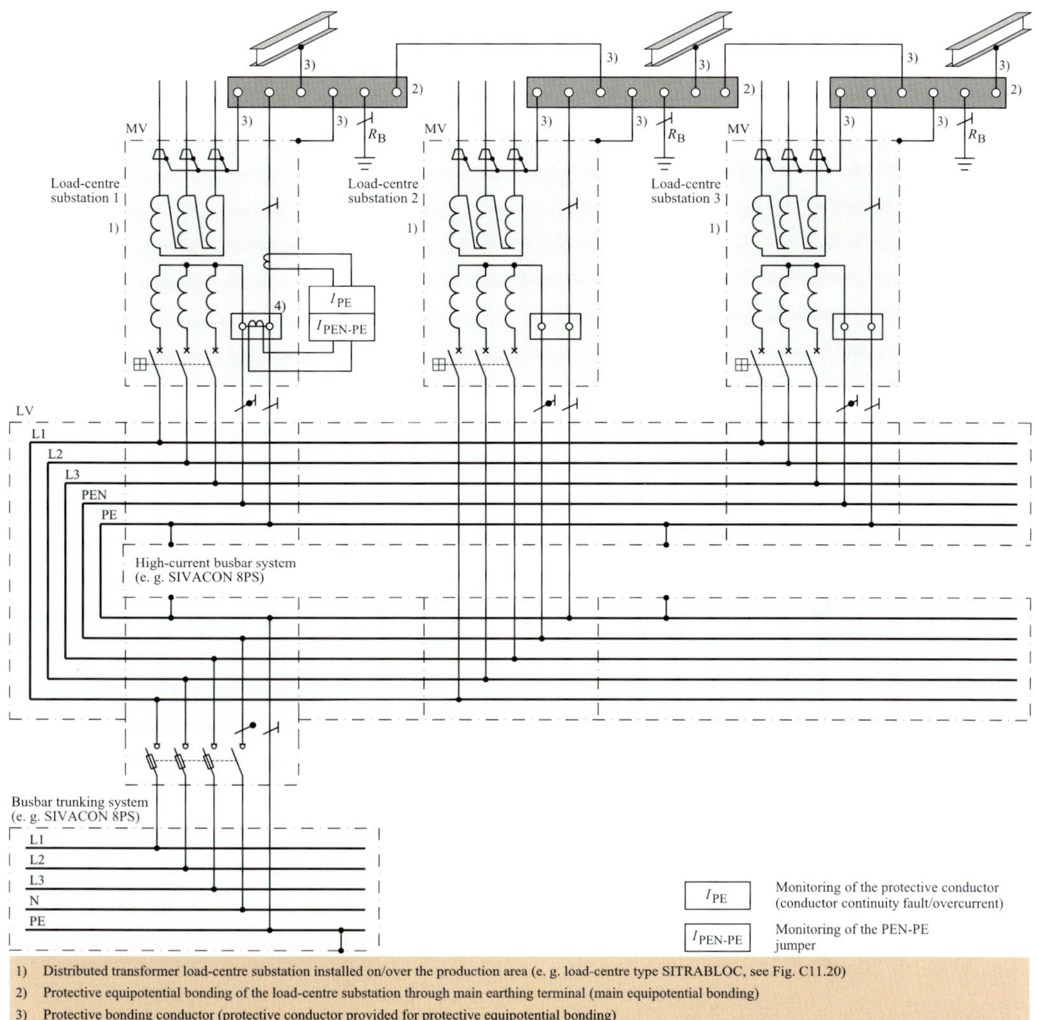

Fig. C10.43 TN-EMC system with decentralized multiple incoming supply (decentralized TN-EMC multiple source system)

- The measured jumper current I_{PEN-PE} is a tripping criterion for general disconnection of the multiple incoming supply. General disconnection must be performed on a line-to-earth fault on the high-current busbar system or on failure of an overcurrent protective device in the feeder circuits. An LV circuit-breaker intertripping circuit must be provided for general disconnection of supply.
- To create an adequate protective equipotential bonding system, the PE conductor must be connected to earthed parts of the housing and structure as many times as possible. The main earthing terminals of the distributed load-centre substations must be interconnected by protective bonding conductors.

- Reduced cross-sectional areas for PEN and N conductors must be avoided. PEN and N conductors must be rated for the expected load due to current harmonics, whose harmonic order *h* is divisible by three (see Tables C10.23 and C11.30).
- Compliance with the disconnection condition (10.38) must be verified not only by calculation but also by measurement. For this purpose, the loop impedance for each transformer of the decentralized multiple incoming supply must be measured. Measurement must be performed after installation.

The latest edition of DIN VDE 0100-410 (VDE 0100-410): 2007-06 [10.39] or IEC 60364-4-41: 2005-12 [10.40] no longer requires the neutral point of the power source to be earthed closely or in the immediate vicinity of the power source. Decentralized TN-EMC multiple source systems therefore no longer violate the standards. If protection of property and persons are reconciled, TN-EMC systems are the only recommended alternative to the classic decentralized TN-C and TN-C-S multiple source systems still widely used in industry.

10.3 Definition of the network configuration

Defining the network configuration is an important part of the planning decision. The network configurations that a planner can consider for process-related power supply and distribution are explained and evaluated below.

10.3.1 Network configurations for power supply and distribution

In industrial plants with large-area production halls and high load densities, cost efficiency requires that energy transmission to the load centres be performed wherever possible with medium voltage rather than with low voltage. To illustrate this, Fig. C10.44 provides a simple example of supply of a load centre with three different supply system variants. As the costs $totex_i$ show, energy transmission at medium voltage is already more cost-efficient at a power demand of $S_{max} > 50$ kVA.

The MV system for the process-related power supply should preferably be implemented as a load-centre system. Compared with a ring system, a load-centre system is the more cost-efficient and more reliable solution for process-related energy supply (see Section 5.3).

The network configurations that can be considered for process-related energy distribution on the LV side are:

- radial networks,
- networks with switchover reserve capacity,
- radial networks joined in an interconnected system and
- meshed networks.

In industrial plants, all networks are implemented as cable or busbar trunking systems.

10.3.1.1 Simple radial network

The simple radial network (Fig. C10.45) is characterized by its clear structure, low investment costs and simple protection coordination at the current-time level. Its drawbacks are the low supply reliability and poor voltage stability.

One important precondition for use of a simple radial network is acceptance of supply interruptions that can only be eliminated by repairing or replacing the damaged equipment. Radial networks are therefore preferably used where clear system management is considered important and supply reliability, voltage stability and adaptability of the network are lesser considerations.

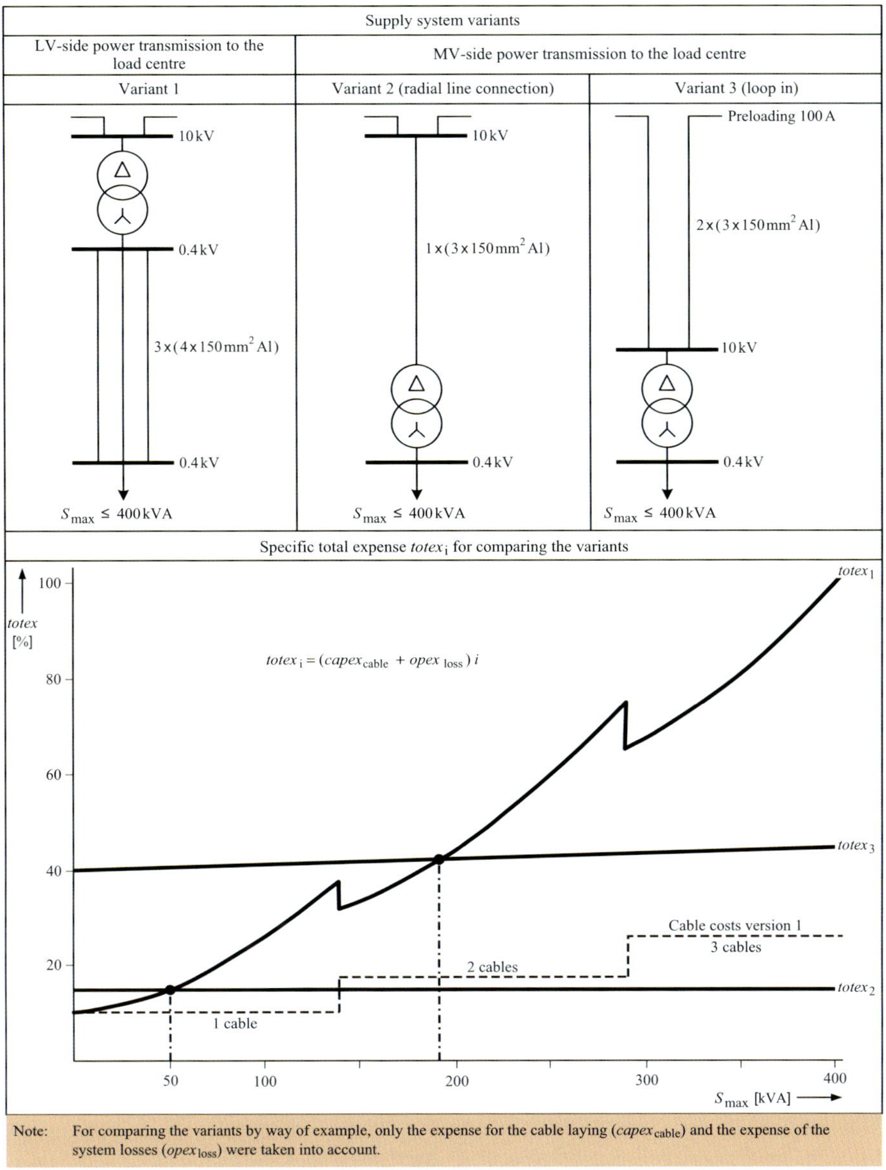

Fig. C10.44 Example comparison of variants for cost-efficient energy transmission to the load centres of a large-area factory hall

10.3.1.2 Radial network with switchover reserve capacity

Radial networks with switchover reserve are characterized by normally open (n.o.) cable connections between the low-voltage main distribution boards (LV-MDBs) of individual load centres. Depending on the dimensioning of the supplying transformers and the ratings of the interconnecting cable that can be energized in case of a power failure, a distinction is made between radial networks with partial-load reserve capacity and those with full-load reserve capacity (Fig. C10.46). In radial networks with full-load reserve capacity, the (n–1) failure is handled by switchover ("cold" standby redundan-

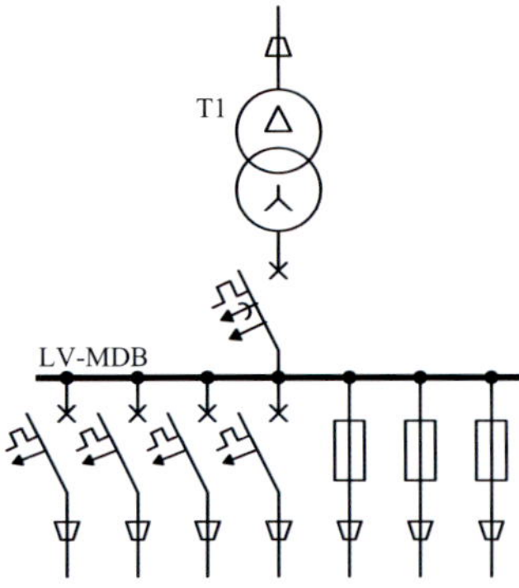

Fig. C10.45
Simple radial network

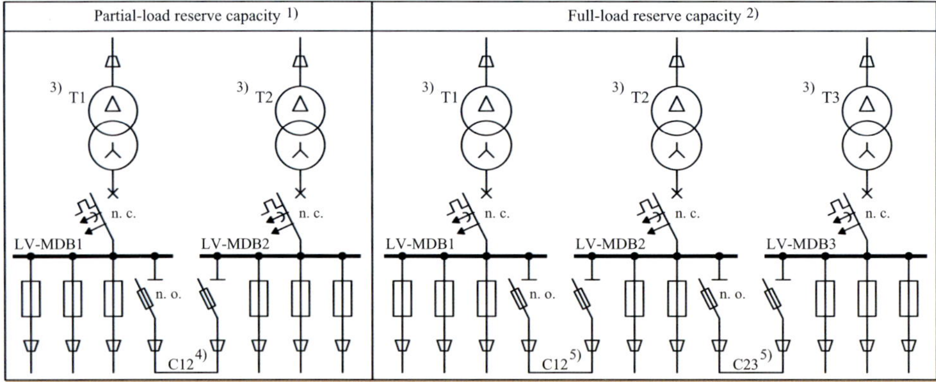

1) Continued operation of selected loads after failure of a transformer
2) Continued operation of all loads after failure of a transformer
3) Selection and dimensioning of transformers according to Section 11.1
4) Cable must be rated for power of the loads still to be supplied
5) Cable must be rated for a transmission power of $S_{cable} = S(n-1) = a \cdot S_{rT}$
S_{rT} Rated power of a transformer
a Permissible capacity utilization to handle the $(n-1)$ fault event ($a = n-1/n$ for dry-type transformers with natural air circulation (AN cooling) and $a = 1.4 \cdot (n-1)/n$ for dry-type transformers with forced-air circulation (AF cooling))

Fig. C10.46 Example radial network with switchover reserve capacity

cy). This network configuration therefore generally complies with the minimum requirements for reliability of the power supply in industrial plants.

To handle the $(n-1)$ failure, the radial network with the full-load reserve capacity shown in Fig. C10.46 must be dimensioned such that two faultless transformers can always take over the load of a failed transformer. To handle the $(n-1)$ failure in the incoming supply of the LV system with $n = 3$ AN-cooled dry-type transformers, their capacity utilization is limited to $a \leq 0.67$ in normal operation. If $n = 3$ supplying dry-type transformers with AF-cooling ($S_{AF} = 1.4 \cdot S_{rT}$) are used, the permissible capacity utilization is increased to $a \leq 0.93$. According to the requirements for handling the $(n-1)$ failure, the interconnecting cables C_{12} and C_{23} must be able to transmit the failure power $S(n-1) = a \cdot S_{rT}$.

The double radial network with bus sectionalizers in the LV main distribution boards and LV sub-distribution boards (Fig. C10.47), which is the solution of choice in the chemical and petroleum industries, offers full-load power reserve as well. While the

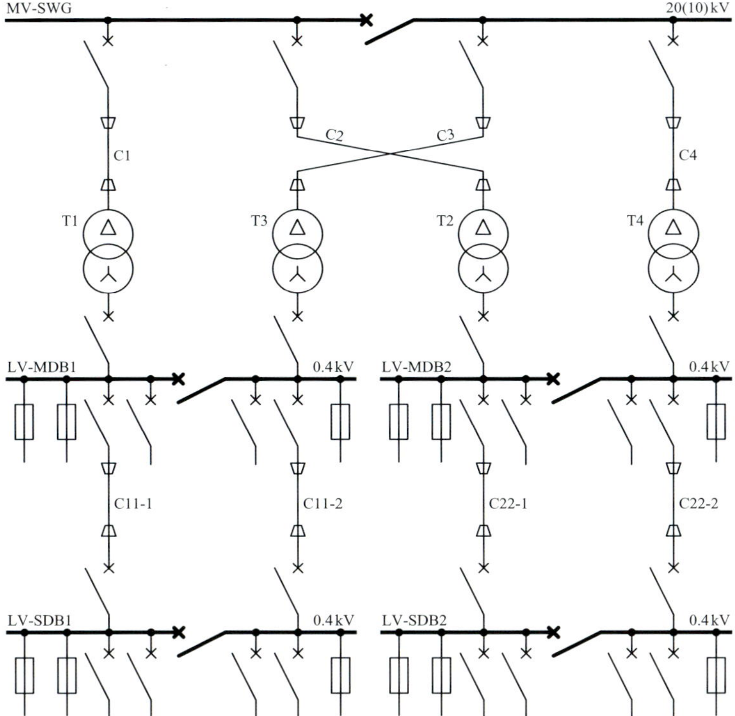

Fig. C10.47 Example double radial network

bus sectionalizer is closed, the double radial network provides all loads with hot standby redundancy, and while the bus sectionalizer is open, with cold standby redundancy just like the simple radial network with switchover reserve capacity. To provide hot or cold standby redundancy power for all loads however requires a capacity utilization during normal operation of the supplying dry-type transformers of $a \leq 0.5$ for AN cooling and $a \leq 0.7$ for existing AF cooling.

10.3.1.3 Radial networks in an interconnected cable system

The radial network in an interconnected cable system (Fig. C10.48) with regard to its configuration and layout is similar to the radial network with full-load reserve capacity.

Unlike the radial network with switchover reserve capacity, with this network configuration the cable connections between the low-voltage main distribution boards (LV-MDBs) are closed during normal operation. Operation as normally closed (n.c.) ring-main offers special advantages:

- higher supply reliability, that is, instantaneous reserve or hot standby redundancy on failure of a transformer or LV-MDB connecting cable,
- better voltage stability because of higher system short-circuit power,
- lower system losses due to the more uniform utilization of the transformers,
- easier compliance with the disconnection condition for protection against indirect contact in the TN system (see Section 10.2.1.3).

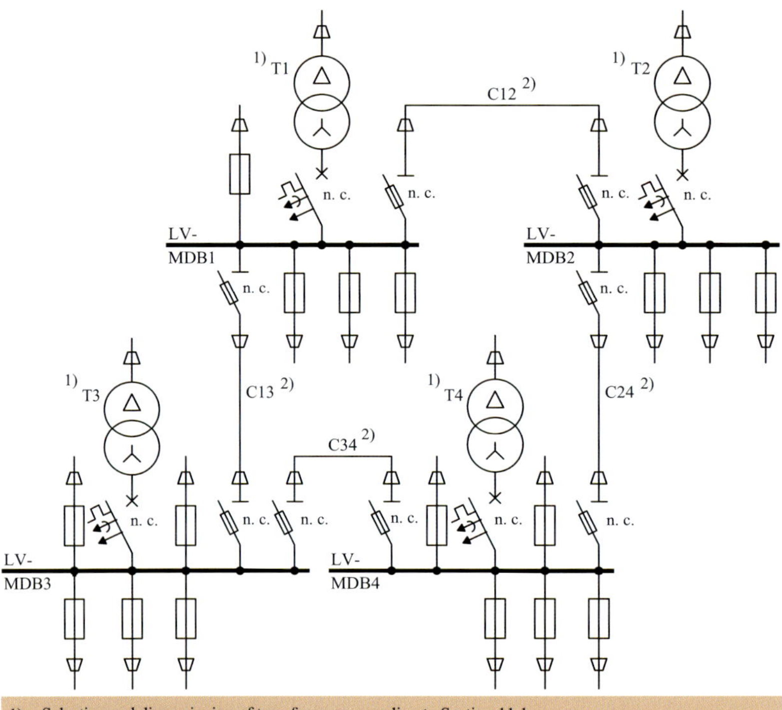

1) Selection and dimensioning of transformers according to Section 11.1

2) Cables must be rated for a transmission power of $S_{cable} = S\,(n-1) = a \cdot S_{rT}$

S_{rT} Rated power of a transformer

a Permissible capacity utilization to handle the $(n-1)$ fault event ($a = n-1/n$ for dry-type transformers with natural air circulation (AN cooling) and $a = 1.4 \cdot (n-1)/n$ for dry-type transformers with forced-air circulation (AF cooling))

Note: For further implementation details see Fig. C11.19

Fig. C10.48 Radial networks in an interconnected cable system

10.3.1.4 Multi-end-fed meshed network

The meshed network is the system configuration with the highest supply reliability. The variant of a multi-end-fed meshed network shown in Fig. C10.49 offers a reliability level that permits an instantaneous reserve up to the individual loads of the production process.

Because of its high supply reliability, use of a meshed network was particulary strongly recommended in the 1960s and 1970s [10.87 to 10.90]. Now, however, the once highly favoured meshed network has fallen into almost complete disuse. This is largely due to the high investment costs of the meshed system cables and their high fire load, and bad experience with operation (e.g. with replacing meshed network fuses or restarting after a mains failure). Modified technologies and manufacturing techniques (e.g. production lines with loads installed in rows instead of workshop-like production with loads covering the entire area) have also contributed to turning away from the meshed network.

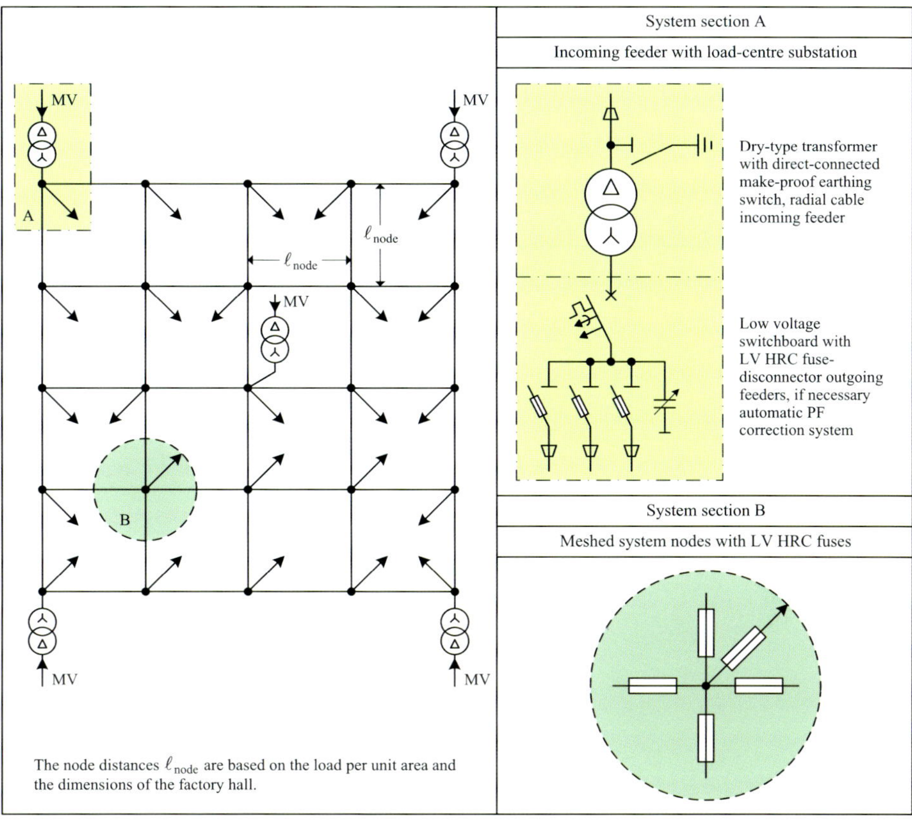

Fig. C10.49 Multi-end-fed meshed network

10.3.1.5 Radial networks interconnected through busbar trunking systems

Radial networks interconnected through busbar trunking systems are not a network configuration in their own right. In principle, this is a modified interconnected ring-main system in which the cable network with main distribution boards and subdistribution boards usually utilized to supply loads is replaced by busbar trunking systems (e.g. Siemens SIVACON 8PS). Depending on the current-carrying capacity and the operation purpose, a distinction is made between high-current busbar systems ($I_r \geq 1{,}000$ A) and busbar trunking systems ($I_r \leq 800(1{,}250)$ A). By combining the two busbar systems, it is possible to construct a supply system that is physically tailored to the factory hall and can be adapted to changing production processes. Fig. C10.50 illustrates the structure of such a supply system.

The busbar trunking systems of a supply network according to Fig. C10.50 are exclusively used for process-related connection of LV loads. The use of busbar trunking systems is above all advantageous if machines or industrial robots arranged in rows (e.g. welding or painting robots) have to be connected. This type of machine configuration means that the electrical loads can be connected relatively simply with plug-in connections to the busbar trunking system routed alongside the line of machines. The energy transmission to the busbar trunking system is effected through a high-current busbar system that is supplied from distributed dry-type transformers. Point loads can also be connected to the high-current busbar system used preferentially for energy transmission. The top-off units of modern high-current busbar systems (e.g. Siemens

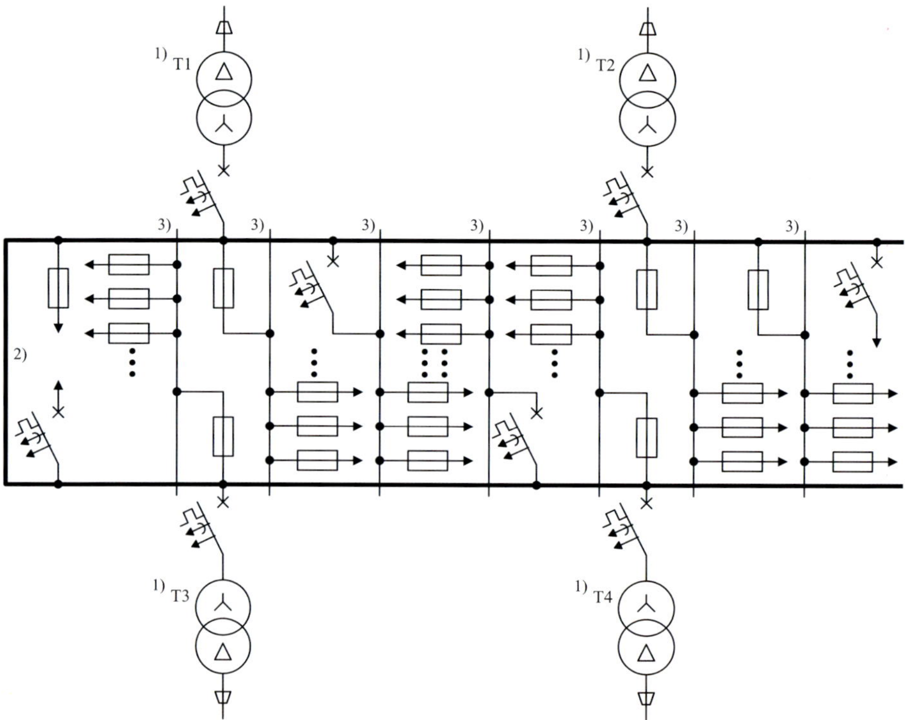

1) Selection and dimensioning of transformers according to Section 11.1
2) High-current busbar system (1,000 A ≤ I_r ≤ 6,300 A)
3) Busbar trunking system (I_r ≤ 800 (1,250) A)
Note: For further implementation details see Fig. C11.21

Fig. C10.50 Radial networks interconnected through busbar trunking systems

SIVACON 8PS) have rated currents from $I_r = 35$ A up to $I_r = 1{,}250$ A (Section 11.2.4, Table C11.18).

Radial networks interconnected through busbar trunking systems offer advantages to the planner, above all, where the basic system for supplying the production hall with power has to be planned at an early stage although the size and location of the loads is still not certain (e.g. missing or incomplete machine installation drawings). Another reason for choosing this network configuration might also be because a low fire load is required or that frequent changes to the production processes necessitate fast and simple changing of the load connections.

10.3.2 Selecting the economically and technically most favourable network configuration

Taking the supply reliability requirements of a specific production process into account (Section 2.3.2), the economically and technically most advantageous network configuration is selected based on the optimality criterion (see Eqs. 2.17 and 2.18). As an additional decision-making aid, Table C10.51 contains multiple-objective-oriented evaluation of the network configurations that can be considered for process-related energy distribution on the LV side. Table C10.51 shows that radial networks in an interconnected cable system and radial networks interconnected through busbar trunking

Table C10.51 Multiple-objective-oriented evaluation of LV-side network configurations for a process-related power supply

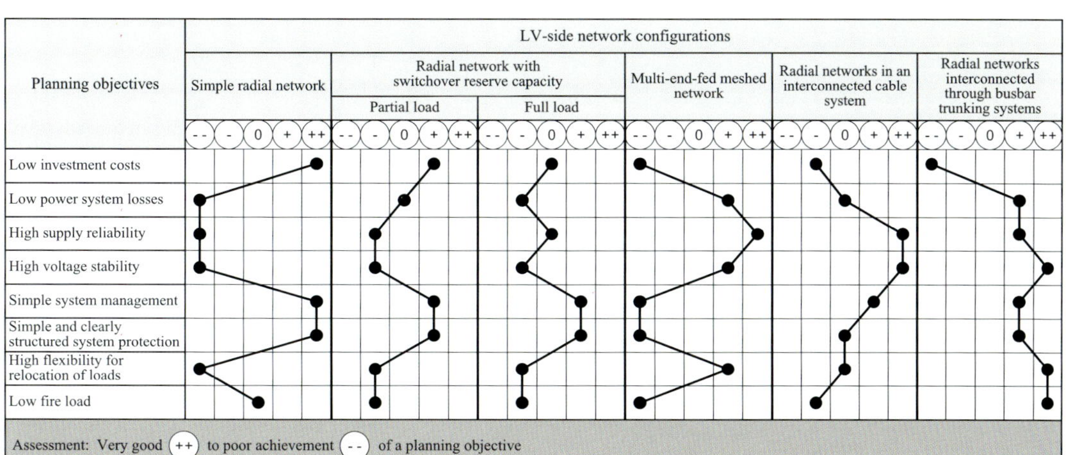

Planning objectives	Simple radial network	Radial network with switchover reserve capacity – Partial load	Radial network with switchover reserve capacity – Full load	Multi-end-fed meshed network	Radial networks in an interconnected cable system	Radial networks interconnected through busbar trunking systems
Low investment costs	++	+	0	- -	-	- -
Low power system losses	- -	0	-	+	0	+
High supply reliability	- -	-	0	++	++	+
High voltage stability	- -	-	-	+	++	++
Simple system management	++	+	+	- -	+	+
Simple and clearly structured system protection	++	+	+	- -	0	+
High flexibility for relocation of loads	- -	-	-	+	0	++
Low fire load	0	-	-	- -	-	++

Assessment: Very good (++) to poor achievement (- -) of a planning objective

systems are the configurations that best meet the criteria for a power supply that complies with the quality requirements.

The radial networks interconnected through busbar trunking systems have proven a very advantageous configuration for the process-related power supply to production processes in the automotive industry despite the high investment costs [10.14].

11 Selecting and dimensioning the electrical equipment

11.1 Distribution transformers

Distribution transformers link the MV to the LV system and step down the electrical power to be distributed to the intended low-voltage level (e.g. 400 V/230 V 3AC). They are available as three-phase oil-immersed or dry-type transformers in the standardized power range 50 kVA $\leq S_{rT} \leq$ 2,500 kVA.

To supply power to the load centres and main loads in industrial plants, dry-type transformers according to DIN EN 60076-11 (VDE 0532-76-11): 2005-04 [11.1] or IEC 60076-11: 2004-05 [11.2] are preferred. In industrial power systems, GEAFOL cast-resin transformers have been successfully used as dry-type transformers for many years [11.3]. Their use has the following advantages:

- low fire load due to design with little insulating material (less than 10 % of the weight is accounted for by the insulants),
- no special fire protection measures required (cast-resin moulding material GEAFOL is fire-retardant and self-extinguishing once the energy supply has been cut off),
- no risks that would make a fire more serious (e.g. toxicity risk due to release of poisonous gases in case of a fire),
- measures to protect the ground water (e.g. oil collecting throughs or traps) are not required,
- continuous overload capacity up to 140(150) % of the rated power due to built-on, temperature-dependently controlled radial-flow fans,
- utilization of the continuous overload capacity as "hot standby" redundancy to increase the supply reliability,
- no loss of service life when continuous overload capacity is used,
- no danger of impermissible switching overvoltages due to resonance excitation of the windings on switch-on and switch-off with a vacuum switch (use of multi-layer windings with a very small range for the resonance frequency).

To select cast-resin transformers, the following electrical quantities must be determined:

a) rated voltage U_{rT} (primary and secondary side),

b) impedance voltage at rated current u_{rZ},

c) vector group,

d) rated power S_{rT}.

The appropriate values of these electrical quantities a) to d) depends on the use to which the transformer will be put.

a) Rated voltage U_{rT}

The required rated voltage U_{rT} on the primary and secondary side depends on the choice of voltage (Chapter 3 and 8). Table C11.1 lists preferred values for the rated voltage of distribution transformers. It is important to note that the rated voltage of the transformer on the secondary side has values that are 5 % higher than the nominal system voltage of the LV level. This largely compensates for the internal voltage drops of the transformer when a load is applied. Distribution transformers can also be adapted to the prevailing system conditions using taps.

Table C11.1 Preferred rated voltages of distribution transformers

Rated voltage U_{rT} of the	
primary side in kV	secondary side in V
10 (11)	420 / 240
	725 / 420
20 (22)	420 / 240
	725 / 420

Taps are additional primary-side winding terminations used to change the transformation ratio k_{Tr}. The tapping range is the range between the MV nominal system voltage U_{nN} and the highest and lowest adjustable voltage of a winding. The tapping range is stated as a positive or negative percentage of the nominal system voltage U_{nN}. The transformation ratio k_{Tr} can be changed to prevent excessively low or excessively high voltages reaching the loads on the LV level due to the voltage conditions prevailing in the LV network (Table C11.2).

Table C11.2 Matching the transformation ratio k_{Tr} to the voltage conditions in a 20/0.4-kV network

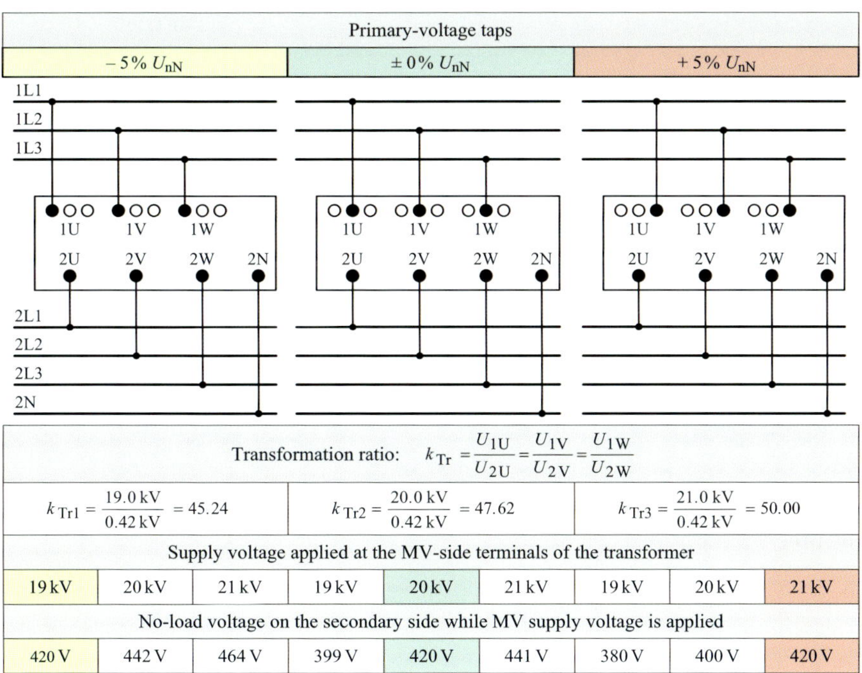

Transformation ratio: $k_{Tr} = \frac{U_{1U}}{U_{2U}} = \frac{U_{1V}}{U_{2V}} = \frac{U_{1W}}{U_{2W}}$								
$k_{Tr1} = \frac{19.0\,kV}{0.42\,kV} = 45.24$			$k_{Tr2} = \frac{20.0\,kV}{0.42\,kV} = 47.62$			$k_{Tr3} = \frac{21.0\,kV}{0.42\,kV} = 50.00$		
Supply voltage applied at the MV-side terminals of the transformer								
19 kV	20 kV	21 kV	19 kV	20 kV	21 kV	19 kV	20 kV	21 kV
No-load voltage on the secondary side while MV supply voltage is applied								
420 V	442 V	464 V	399 V	420 V	441 V	380 V	400 V	420 V

b) Impedance voltage at rated current u_{rZ}

The impedance voltage at rated current is the voltage that is applied to the primary winding of the transformer when the rated current I_{rT} is flowing in the short-circuited secondary winding. It is expressed as a percentage of the primary-side rated voltage.

In calculation of the short-circuit stress of electrical equipment, the impedance voltage at rated current u_{rZ} must not be confused with the impedance voltage u_Z because the impedance voltage is a calculation value for operation deviating from the rated current values. It can be expressed as a decimal multiple of the impedance voltage at rated current u_{rZ}. The following applies:

$$u_Z = u_{rZ} \cdot \frac{I_{load}}{I_{rT}} \tag{11.1}$$

I_{load} load current
I_{rT} rated current of the transformer

The impedance voltage at rated current u_{rZ} must always be used as the standardized calculation value for the short-circuit stress of the electrical equipment. It is also a measure of how “voltage stiff“ the distribution transformer is. If the impedance voltage at rated current is small, the transformer is “voltage stiff“; if it is large, the transformer is not “voltage stiff“. GEAFOL cast-resin transformers are manufactured with an impedance voltage at rated current of $u_{rZ} = 4\,\%$ and/or $u_{rZ} = 6\,\%$ as standard (Table C11.3).

Table C11.3 Standard impedance voltages at rated current in the power range of GEAFOL cast-resin transformers that is relevant to industrial applications

S_{rT} in kVA	250		315		400		500		630		800		1,000		1,250	1,600	2,000	2,500
u_{rz} in %	4	6	4	6	4	6	4	6	4	6	4	6	4	6	6	6	6	6

In the automotive industry, non-standard GEAFOL cast-resin transformers with a reduced impedance voltage at rated current $u_{rZ} = 2.8\,\%$ were previously used. They served only to power welding systems in car factories but their use has now ceased for economic reasons. The necessary supply quality is today preferably assured by parallel operation of groups of distribution transformers with an identical standard impedance voltage at rated current u_{rZ} [11.4, 11.5].

Parallel operation of distribution transformers with an impedance voltage at rated current of $u_{rZ} = 4\,\%$ is largely found in networks with loads that cause unwanted power system perturbations. These include, for example, asynchronous motors with an individual power rating that is large relative to the total power demand (Section 10.1.1.2) and welding machines (Section 10.1.1.3). If the transformer power rating is large enough (e.g. $S_{rT} \geq 1{,}250$ kVA), it is also possible to achieve the necessary voltage stability using distribution transformers with an impedance voltage at rated current of $u_{rZ} = 6\,\%$.

c) Vector group

The vector group indicates how the phases of the two windings of a transformer are connected and the phase position of their respective voltage vectors. It consists of letters and a phase angle number.

The upper-case letter of the vector group denotes the type of connection of the primary winding; the lower-case letter, that of the secondary winding. “D“ or “d“ means delta connection; “Y“ or “y“ means star connection; and “Z“ or “z“ means zigzag connection. An additional letter “N“ or “n“ indicates that the neutral point of a Y(y) winding or Z(z) winding is brought out and therefore accessible.

The phase angle number of the vector group denotes the integer multiple of 30° by which the secondary voltage lags behind the primary voltage of each phase in the anti-clockwise direction. In the vector group YynO, the primary and secondary voltages are in phase in each winding phase. In the vector groups Dyn5, Ynd5, and Yzn5, the secondary voltage lags 150° behind the primary voltage. Table C11.4 provides a list of the most common vector groups used in three-phase transformers.

Table C11.4 Preferred vector groups when using three-phase transformers

Designation		Vector diagram		Circuit diagram		Continuous neutral loading capacity
Vector group	Phase angle number	primary-side	secondary-side	primary-side	secondary-side	
Yyn0 [1]	0	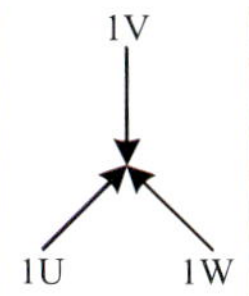 1V 1U 1W	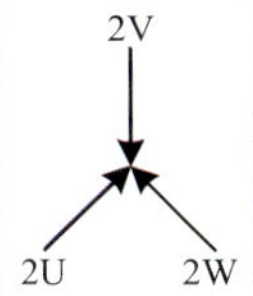 2V 2U 2W	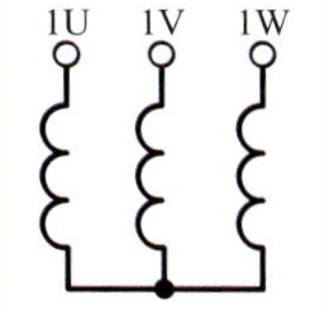 1U 1V 1W	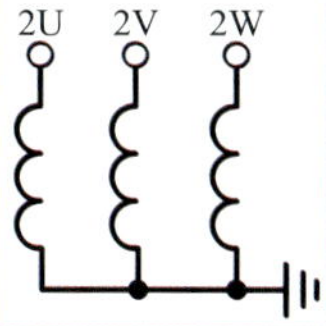 2U 2V 2W	10% of the rated current ($0.10 \cdot I_{rT}$)
Dyn5 [2]	5	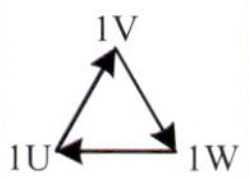 1V 1U 1W	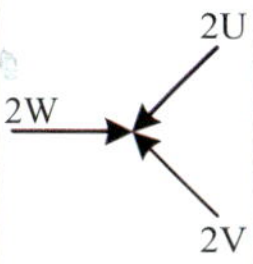 2U 2W 2V	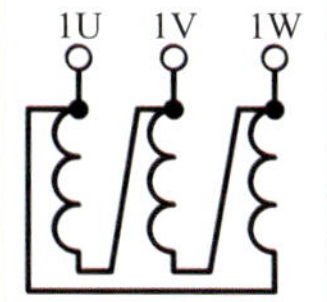 1U 1V 1W	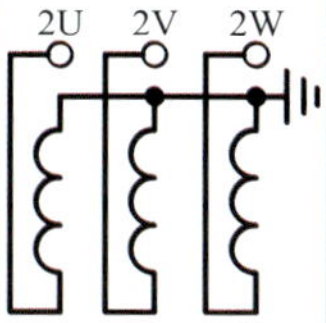 2U 2V 2W	Rated current ($1.0 \cdot I_{rT}$)
YNd5 [3]	5	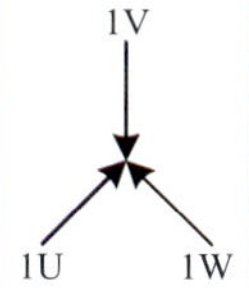 1V 1U 1W	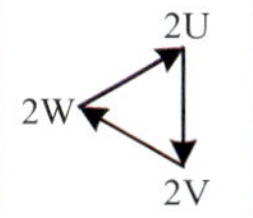 2U 2W 2V	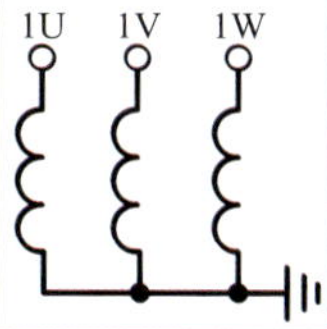 1U 1V 1W	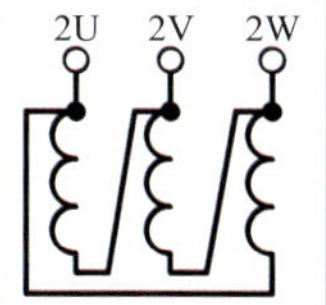 2U 2V 2W	Rated current ($1.0 \cdot I_{rT}$)
Yzn5 [4]	5	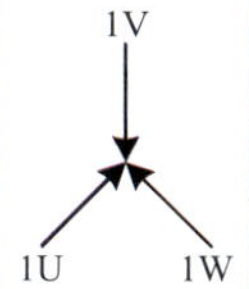 1V 1U 1W	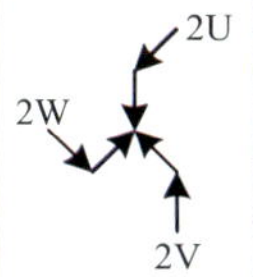 2U 2W 2V	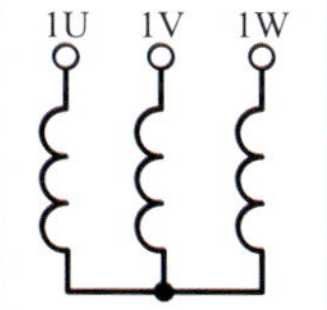 1U 1V 1W	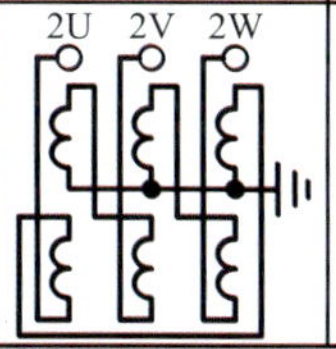 2U 2V 2W	Rated current ($1.0 \cdot I_{rT}$)

1) Not suitable for LV systems in which the protection against indirect contact is ensured by overcurrent protective devices and/or that contain a high proportion of single-phase loads (Because of the star-connected windings on the primary side, the current linkage is no longer balanced on the secondary side in the case of a heavily unbalanced load. The phase voltages on the secondary side can therefore vary greatly [11.6].)

2) Preferred vector group for distribution transformers with rated powers $S_{rT} \geq 250\,\text{kVA}$ in industrial power systems

3) Common vector group for generator transformers in power stations or transfer power transformers in 110-kV/MV substations

4) Preferred vector group of distribution transformers with rated powers $S_{rT} < 250\,\text{kVA}$ because the zigzag connection is more favourably able to handle an unbalanced load. Unbalanced loads are especially pronounced in small three-phase systems with single-phase loads [11.7].

Both economic and technical considerations determine the choice of vector group. For reasons concerning the insulation, the star connection is preferred at high nominal system voltages because the insulation of a star-connected winding only has to be dimensioned for $1/\sqrt{3}$ times the line-to-line voltage. For high load currents, on the other hand, a delta connection is more favourable. The delta winding is characterized by the fact that its winding phases are only subjected to $1/\sqrt{3}$ times the phase current. This means that smaller cross-sectional areas can be used for the winding wires than in a star connection, which saves costs for materials. For these reasons, the vector group YNd5 is used for generator transformers.

Unlike generator transformers, distribution transformers supplying a low-voltage system have to be star-connected on the secondary side. A brought-out neutral on the low-voltage side is essential so that the neutral conductor can be connected to provide the voltage for single-phase loads, which usually require 230 V.

When power is supplied to single-phase loads, however, unbalanced loads must be expected. For this reason, the necessary delta connection can only be implemented on the primary side of distribution transformers. The preferred vector group for distribution transformers is therefore Dyn5. For use in industrial power systems, distribution transformers with vector group Dyn11 can also be considered. [11.8] states that the mixed use of vector groups Dyn11 and Dyn5 when interconnecting distribution transformers relieves the power system of harmonics of the orders h = 5 and h = 7. In fact, in the interconnection of Dyn11 and Dyn5 transformers as shown in Fig. C11.5, the 5th-order and 7th-order harmonics on the secondary side are offset by 180°. However, the angular offset of the 5th-order and 7th-order harmonics on the secondary side shown in Fig. C11.6 no longer applies on the primary side of the interconnected transformers. With a secondary-side angular position of 30° the 5th-order harmonic on the primary side of the Dyn5 transformers results in an angular position of $30° + 5 \cdot 30° = 180°$. On the primary side of the Dyn11 transformers, the angular position $-150° + 11 \cdot 30° = 180°$ results for the 5th-order harmonic. The result for the harmonic load on the primary side of the interconnected transformers is thus a uniform total angle of 180°. Instead of being subtracted in the MV network, the 5th-order harmonics generated in LV network 1 and LV network 2 must therefore be added. The same applies to the 7th-order harmonics generated on the LV side. Mixed use of vector groups Dyn11 and Dyn5 when interconnecting distribution transformers, which is sometimes recommended, is therefore not a suitable method for relieving the MV network of harmonics generated on the LV side.

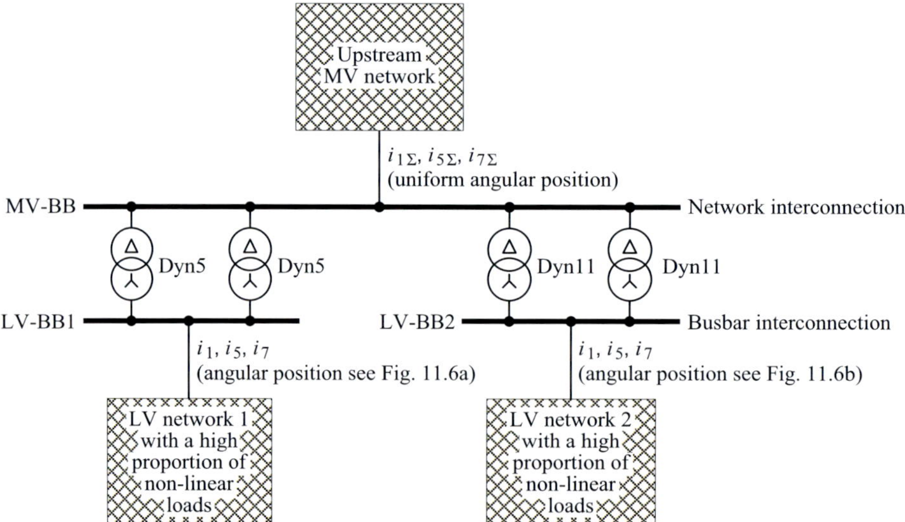

Fig. C11.5 Interconnection of distribution transformers with vector groups Dyn5 and Dyn11

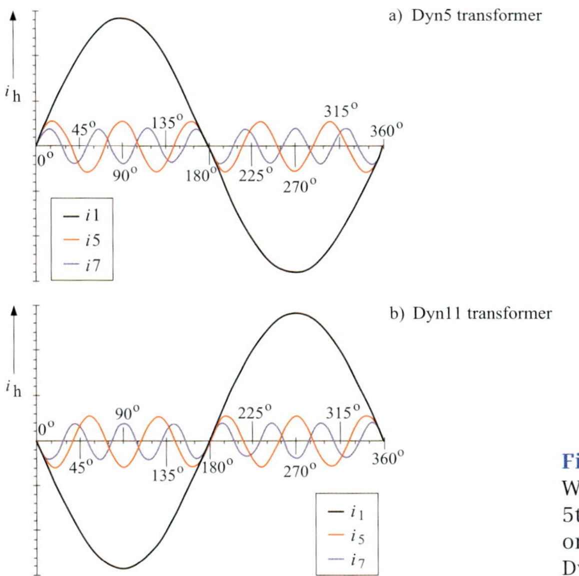

Fig. C11.6
Waveform of the fundamental, 5th-order and 7th-order harmonics on the secondary side of Dyn5 and Dyn11 transformers [11.8]

d) Rated power S_{rT}

The following influencing factors determine the choice of rated power S_{rT}:

- expense for cables and switchboards to distribute the power to the loads,
- transformer loads optimized for losses in parallel operation (busbar interconnection of the transformers),
- maximum power demand of the loads that form a technological and process-related unit (e.g. production or function area),
- maximum impulse load caused by individual consumers (e.g. large asynchronous motors) or consumers operated in groups (e.g. welding machines),
- necessary power reserve to adhere to the $(n-1)$ principle in case of a transformer fault,
- maximum possible short-circuit capacity of the LV operational equipment.

In accordance with the importance of each of these influencing factors, optimization calculations to determine the most economical rated transformer power were performed in [11.9, 11.10] for the radial networks in an interconnected cable system (Section 10.3.1.3) widely used in the metal-processing industry. The basis for these calculations is an area to be supplied with power within a factory with an average load per unit area of $\overline{P}' = 300...350\ \text{VA/m}^2$, a nominal system voltage of $U_{nN} = 400$ V and a permissible short-circuit load of $I_k \leq 100$ kA. For a supplied area with these system parameters, the rated power $S_{rT} = 800$ kVA proved the most cost-efficient solution. For smaller loads per unit area ($P' < 300\ \text{VA/m}^2$) or higher nominal system voltages (e.g. $U_{nN} = 690$ V), smaller rated powers S_{rT} and for larger loads per unit area ($P' > 350\ \text{VA/m}^2$) or lower nominal system voltages (e.g. $U_{nN} = 208$ V), larger rated powers S_{rT} are more cost-efficient.

For a power supply adjusted to the load centres, dry-type transformers from the power range 500 kVA $\leq S_{rT} \leq$ 1,250 kVA should preferably be chosen because, with large transformer units, the power reserve required to handle the $(n-1)$ fault case increases.

Table C11.7 Power calculation for parallel operation of GEAFOL cast-resin transformers with restrictive consideration of the short-circuit capacity

Number of parallel supplying transformers	Rated transformer power	Available individual powers		Available total power during parallel operation in an interconnected system		Transformer utilization in	
		AN mode	AF mode	normal operation	operation under fault conditions	normal operation	operation under fault conditions
		• Radial-flow fans are switched off	• Radial-flow fans are switched on	• n transformers are in operation • Radial-flow fans are switched off	• n–1 transformers are in operation • Radial-flow fans are switched on	• n transformers are in operation • Radial-flow fans are switched off	• n–1 transformers are in operation • Radial-flow fans are switched on
n 1) [1]	S_{rT_i} [kVA]	$S_{AN_i} = S_{rT_i}$ [kVA]	$S_{AF_i} = 1.4 \cdot S_{rT_i}$ [kVA]	$S_{perm\text{-}NOP} = \sum_{i=1}^{n} S_{AN_i}$ [kVA]	$S_{perm\text{-}OPFC} = \sum_{i=1}^{n-1} S_{AF_i}$ [kVA]	a_{NOP} 2) [%]	a_{OPFC} 3) [%]
8	500	500	700	4,000	4,900	100	114
6	630	630	882	3,780	4,410	100	120
5	800	800	1,120	4,000	4,480	100	125
4	1,000	1,000	1,400	4,000	4,200	100	133
3	1,250	1,250	1,750	3,750	3,500	93	140
2	1,600	1,600	2,240	3,200	2,240	70	140
2	2,000	2,000	2,800	4,000	2,800	70	140

1) In case of busbar interconnection of multiple transformers on the LV side, the maximum permissible short-circuit current capacity of $I_{k\text{-perm}}$ = 100 kA is generally reached at $\sum_i S_{rT_i}$ = 4,000 kVA installed transformer power output.

2) During normal operation the radial-flow fans must remain switched off for noise abatement reasons. If $S_{perm\text{-}NOP} < S_{perm\text{-}OPFC}$, the utilization must be limited to a_{NOP} = 100 %.

3) Since the radial-flow fans must remain switched off during normal operation, the full overload capability of the GEAFOL cast-resin transformers cannot be utilized when $S_{perm\text{-}NOP} < S_{perm\text{-}OPFC}$. For the utilization factor during operation under fault conditions when $S_{perm\text{-}NOP} < S_{perm\text{-}OPFC}$, the following applies: $a_{OPFC} = (n / n{-}1) \cdot 100$.

As Table C11.7 shows, adherence to the (n–1) principle results in poorer capacity utilization of the distribution transformers in normal operation as the rated power (S_{rT} > 1,250 kVA) increases. For that reason, larger transformer units (1,600 kVA ≤ S_{rT} ≤ 2,500 kVA) should only be used to supply power to large single loads and impulse loads and where the load density is particularly high. For the maximum number of distribution transformers that can be connected in parallel with the same rated power S_{rT}, the physical limit of the short-circuit capacity for the equipment used in the LV system is decisive. In an LV system without motors that contribute short-circuit current in the case of a fault, the physical limit of the short-circuit load capacity is generally reached with an installed transformer power of $\sum_i S_{rT_i} = 4$ MVA.

Table C11.7 is based on this value to provide a power calculation to handle the (n–1) fault case in parallel operation of GEAFOL cast-resin transformers. Using the power calculation provided in Table C11.7, industrial LV power systems can be dimensioned according to the (n–1) criterion. The electrical parameters of GEAFOL cast-resin transformers required for system dimensioning are provided in Tables C11.8a and C11.8b.

11.2 Low-voltage switchboards and distribution board systems

Low-voltage switchboards are system components used for the functional connection of electrical equipment (generators, transformers, cables) and loads (e.g. motors, solenoid valves, heating apparatus, lighting, air conditioning, electronic data processing). In modern industrial plants with ever more complex processes, low-voltage switchboards and distribution board systems can only fully perform their function in the provision and distribution of electrical power if it is possible to include them in the process control and monitoring.

Table C11.8a Electrical parameters of GEAFOL 4GB cast-resin transformers with the vector group Dyn5 (S_{rT} = 250...630 kVA)

Rated power	Siemens type designation	Rated voltage	Insulation level	Impedance voltage at rated current at 120°C	No-load current	No-load losses	Load losses (at 75°C and U_s=400 V)	Load losses (at 120°C and U_s=400 V)	Ratio of max. inrush current to rated current	Time to half value
S_{rT} [kVA]		U_{rT1}/U_{rT2} [kV]		u_{rZ} [%]	i_0 [%]	P_0 [W]	P_{k75} [W]	P_{k120} [W]	I_{E1}/I_{rT} [1]	$K_{T0.5}$ [Cycles]
250	4GB5444-3CA	10/0.42	AC28-LI75	4	0.90	820	2,800	3,200	19.8	7
	4GB5444-3GA				0.30	600	2,800	3,200	15.2	8
	4GB5444-3DA			6	0.78	700	2,900	3,300	13.3	9.5
	4GB5444-3HA				0.28	560	2,900	3,300	11.6	10.5
	4GB5464-3CA	20/0.42	AC50-LI95	4	0.90	880	2,800	3,200	20	7
	4GB5464-3GA				0.30	800	2,900	3,300	15.4	7.5
	4GB5464-3DA			6	0.91	880	3,000	3,400	15.2	8
	4GB5464-3HA				0.38	650	3,000	3,400	12.5	8.5
	4GB5467-3DA		AC50-LI125		0.83	880	3,300	3,800	14.4	8
315	4GB5544-3CA	10/0.42	AC28-LI75	4	0.81	980	3,000	3,500	19.2	8
	4GB5544-3GA				0.29	730	3,000	3,500	15.6	8.5
	4GB5544-3DA			6	0.96	850	3,400	3,900	13.7	10
	4GB5544-3HA				0.38	670	3,200	3,700	11.5	11
	4GB5564-3CA	20/0.42	AC50-LI95	4	1.28	1,250	3,000	3,500	20.5	8
	4GB5564-3GA				0.42	930	3,000	3,500	17.6	9
	4GB5564-3DA			6	0.72	1,000	3,300	3,800	14.5	10.5
	4GB5564-3HA				0.38	780	3,300	3,800	13	11
	4GB5567-3DA		AC50-LI125		0.78	1,000	3,600	4,200	14.7	9.5
400	4GB5644-3CA	10/0.42	AC28-LI75	4	0.85	1,150	3,800	4,400	18.5	8.5
	4GB5644-3GA				0.33	880	3,800	4,400	15.3	9
	4GB5644-3DA			6	0.72	1,000	4,300	4,900	13.7	10.5
	4GB5644-3HA				0.27	800	4,300	4,900	11.9	11.5
	4GB5664-3CA	20/0.42	AC50-LI95	4	0.98	1,270	3,300	3,800	18.7	9
	4GB5664-3GA				0.37	1,100	3,300	3,800	16.5	10
	4GB5664-3DA			6	1.03	1,200	3,700	4,300	14.7	10
	4GB5664-3HA				0.39	940	3,700	4,300	12.6	12
	4GB5667-3DA		AC50-LI125		0.77	1,200	4,100	4,700	14.6	9
500	4GB5744-3CA	10/0.42	AC28-LI75	4	1.05	1,300	5,200	6,000	17.6	9
	4GB5744-3GA				0.37	1,000	5,200	6,000	15.0	10
	4GB5744-3DA			6	1.05	1,200	5,700	6,600	13.8	10
	4GB5744-3HA				0.37	950	5,700	6,600	12	11
	4GB5764-3CA	20/0.42	AC50-LI95	4	1.76	1,650	4,400	5,000	20.2	9
	4GB5764-3GA				0.60	1,300	4,300	5,000	17.9	10
	4GB5764-3DA			6	1.34	1,400	5,200	6,000	14.6	13
	4GB5764-3HA				0.48	1,100	5,200	6,000	13	14
	4GB5767-3DA		AC50-LI125		0.96	1,400	5,200	6,000	14.5	10
630	4GB5844-3CA	10/0.42	AC28-LI75	4	0.91	1,500	6,100	7,000	17.2	10
	4GB5844-3GA				0.31	1,150	6,100	7,000	14.5	11
	4GB5844-3DA			6	0.73	1,370	6,300	7,200	13.3	13
	4GB5844-3HA				0.30	1,100	6,300	7,200	11.5	14
	4GB5864-3CA	20/0.42	AC50-LI95	4	1.71	2,000	5,700	6,600	19.1	10
	4GB5864-3GA				0.55	1,500	5,700	6,600	16.9	10
	4GB5864-3DA			6	1.26	1,650	6,300	7,200	14.2	12
	4GB5864-3HA				0.43	1,250	6,300	7,200	12.3	14
	4GB5867-3DA		AC50-LI125		0.77	1,650	6,300	7,200	13.7	12

AC Abbreviation for the rated power-frequency withstand voltage stated in kV

LI Abbreviation for the rated lightning impulse withstand voltage stated in kV

Notes: The figures of Table C11.8a can also be used as approximate values for the rated voltages 11 kV/0.42 kV and 22 kV/0.42 kV. For calculation of the line-to-earth short-circuit current the ratios X_{0T} / X_{1T} = 0.95 and R_{0T} / R_{1T} = 1 are valid.

Similarly, as in the medium-voltage field, a change has taken place in the low-voltage field from the individually, on-site-assembled cubicles to the type-tested, factory-assembled switchboard. The type testing is performed according to DIN EN 61439-1 (VDE 0660-600-1): 2010-06 [11.11] or IEC 61439-1: 2009-01 [11.12] and DIN EN 61439-2

Table C11.8b Electrical parameters of GEAFOL 4GB cast-resin transformers with the vector group Dyn5 (S_{rT} = 800...2500 kVA)

Rated power S_{rT} [kVA]	Siemens type designation	Rated voltage U_{rT1}/U_{rT2} [kV]	Insulation level	Impedance voltage at rated current at 120°C u_{rZ} [%]	No-load current i_0 [%]	No-load losses P_0 [W]	Load losses (at 75°C and U_s=400V) P_{k75} [W]	Load losses (at 120°C and U_s=400V) P_{k120} [W]	Ratio of max. inrush current to rated current I_{E1}/I_{rT} [1]	Time to half value $K_{T0.5}$ [Cycles]
800	4GB5944-3CA	10/0.42	AC28-LI75	4	0.87	1,800	7,000	8,000	17	12
	4GB5944-3GA				0.31	1,400	7,000	8,000	14	13
	4GB5944-3DA			6	0.85	1,700	7,200	8,300	13.3	15
	4GB5944-3HA				0.29	1,300	7,200	8,300	11.2	17
	4GB5964-3CA	20/0.42	AC50-LI95	4	1.52	2,300	7,400	8,500	18.4	10
	4GB5964-3GA				0.54	1,750	7,400	8,500	16.3	11
	4GB5964-3DA			6	1.24	1,950	7,200	8,300	13.5	15
	4GB5964-3HA				0.43	1,500	7,200	8,300	12	16
	4GB5967-3DA		AC50-LI125		0.82	1,950	7,200	8,300	13.7	14
1,000	4GB6044-3CA	10/0.42	AC28-LI75	4	0.84	2,100	7,900	9,000	15.8	13
	4GB6044-3GA				0.28	1,600	7,900	9,000	13.3	15
	4GB6044-3DA			6	0.77	2,000	8,300	9,500	13	17
	4GB6044-3HA				0.24	1,500	8,300	9,500	10.8	19
	4GB6064-3CA	20/0.42	AC50-LI95	4	1.40	2,600	8,700	10,000	17.3	11
	4GB6064-3GA				0.48	2,000	8,700	10,000	15.3	12
	4GB6064-3DA			6	1.19	2,300	8,200	9,400	13.3	16
	4GB6064-3HA				0.47	1,800	8,200	9,400	12.1	17
	4GB6067-3DA		AC50-LI125		0.98	2,300	8,200	9,400	13.3	15
1,250	4GB6144-3DA	10/0.42	AC28-LI75	6	0.84	2,400	9,600	11,000	12.5	18
	4GB6144-3HA				0.26	1,800	9,600	11,000	10.6	20
	4GB6164-3DA	20/0.42	AC50-LI95	6	1.18	2,700	10,500	12,000	12.4	19
	4GB6164-3HA				0.43	2,100	10,500	12,000	11	20
	4GB6167-3DA		AC50-LI125		0.88	2,700	9,300	10,600	13.2	18
1,600	4GB6244-3DA	10/0.42	AC28-LI75	6	0.63	2,800	11,000	12,600	12.3	22
	4GB6244-3HA				0.19	2,100	11,000	12,600	10.2	25
	4GB6264-3DA	20/0.42	AC50-LI95	6	0.67	3,100	11,200	12,800	12.5	19
	4GB6264-3HA				0.26	2,400	11,200	12,800	10.8	21
	4GB6267-3DA		AC50-LI125		0.73	3,100	11,200	12,800	12.4	19
2,000	4GB6344-3DA	10/0.42	AC28-LI75	6	0.75	3,500	14,200	16,200	12.3	22
	4GB6344-3HA				0.23	2,600	14,200	16,200	10.2	24
	4GB6364-3DA	20/0.42	AC50-LI95	6	1.01	4,000	13,600	15,500	12.3	24
	4GB6364-3HA				0.31	2,900	13,600	15,500	10.4	26
	4GB6367-3DA		AC50-LI125		0.83	4,000	13,600	15,500	13.2	21
2,500	4GB6444-3DA	10/0.42	AC28-LI75	6	0.79	4,300	16,700	19,000	11.6	26
	4GB6444-3HA				0.20	3,000	16,700	19,000	9.5	30
	4GB6464-3DA	20/0.42	AC50-LI95	6	1.01	5,000	16,300	18,500	12.9	26
	4GB6464-3HA				0.29	3,600	16,300	18,500	10.9	28
	4GB6467-3DA		AC50-LI125		0.99	5,000	16,300	18,500	12.5	26

AC Abbreviation for the rated power-frequency withstand voltage stated in kV

LI Abbreviation for the rated lightning impulse withstand voltage stated in kV

Notes: The figures of Table C11.8b can also be used as approximate values for the rated voltages 11 kV/0.42 kV and 22 kV/0.42 kV. For calculation of the line-to-earth short-circuit current the ratios X_{0T} / X_{1T} = 0.95 and R_{0T} / R_{1T} = 1 are valid.

(VDE 0660-600-2): 2010-06 [11.13] or IEC 61439-2: 2009-01 [11.14]. The rated operational voltage is max. 1,000 V AC or 1,500 V DC. To test the behaviour of the switchboard in case of an internal arc, a further test according to DIN EN 60439-1 Bbl. 2 (VDE 0660-500 Bbl. 2): 2009-05 [11.15] or IEC/TR 61641: 2008-01 [11.16] can optionally be conducted. The type-tested switchboards and distribution board systems are selected according to the following criteria [11.17]:

- Rated currents:
  - rated current I_r of the busbar,
  - rated current I_r of the incoming feeders,

  - rated current I_r of the outgoing feeders,
  - rated short-time withstand current I_{cw} of the busbar,
  - rated peak withstand current I_{pk} of the busbar.
- Protection and type of installation:
  - Degree of protection according to DIN EN 60529 (VDE 0470-1): 2000-09 [11.18] or IEC 60529: 2001-02 [11.19],
  - protection against electric shock (safety class) according to DIN VDE 0100-410 (VDE 0100-410): 2007-06 [11.20] or IEC 60364-4-41: 2005-12 [11.21],
  - material of the enclosure,
  - mounting type (against a wall or free-standing),
  - number of operating fronts,
- Type of device installation:
  - fixed-mounting, withdrawable units, plug-in system,
  - snap-on mounting on top-hat rail.
- Application:
  - main switchboard or main distribution board,
  - subdistribution board,
  - linear distribution,
  - motor control board,
  - distribution cabinet,
  - industry-type distribution board,
  - light and power distribution board,
  - reactive-power compensation,
  - open-loop control.

Using Table C11.9, it is possible to select low-voltage switchboards and distribution boards from the Siemens production program for infrastructure and industrial applications based on these criteria.

Depending on the type of power distribution, point (radial) and linear distribution systems can be selected (Fig. C11.10). Point (radial) distribution systems are switchboards and distribution boards that supply the electrical power to the remote loads radially via cables from distribution boards functioning as a hub. The required switching, protective and measuring devices are grouped together centrally in the switchboard or distribution board. In case of linear distribution systems, sometimes also known as busbar trunking systems, the power is taken along enclosed busbars right up to the immediate vicinity of the loads. The loads located along the length of the busbar run are connected to the busbar via relatively short stub-feeder cables and through tap-off units with LV HRC fuses [11.22].

A main switchboard is powered directly from a transformer or generator. Because the various types of distribution boards do not differ greatly, the distribution boards downstream from the main switchboard or main distribution board, such as motor control boards, distribution cabinets and distribution boards for controls, lighting, heating, air-conditioning etc. are considered to be subdistribution boards.

Because of the system configurations (radial networks in an interconnected cable system (Section 10.3.1.3) and radial networks interconnected through busbar trunking systems (Section 10.3.1.5)) used in industry today, utilization of low-voltage switchboards and distribution boards is concentrated on the SIVACON S8 switchboard and the SIVACON 8PS busbar trunking system.

Table C11.9 Selection of low-voltage switchboards and distribution board systems for industrial applications (Siemens type program)

Selection criteria	Switchboards and distribution board systems from the Siemens type program			
	Switchboards	Distribution board systems		
	Point (radial) distribution			Linear distribution
1 Rated currents				
1.1 Rated current I_r of the busbar	≤ 7,010 A	≤ 630 A	≤ 1,000 A	≤ 6,300 A
1.2 Rated current I_r of the incoming feeders	≤ 6,300 A	≤ 630 A	≤ 1,000 A	≤ 6,300 A
1.3 Rated current I_r of the outgoing feeders	≤ 6,300 A	≤ 630 A	≤ 800 A	≤ 1,250 A
1.4 Rated short-time withstand current I_{cw} (t_{cw} = 1 sec)	≤ 150 kA	≤ 20 kA	≤ 40 kA	≤ 150 kA
1.5 Rated peak withstand current I_{pk} of the busbar	≤ 330 kA	≤ 61.3 kA	≤ 80 kA	≤ 286 kA
2 Protection and type of installation				
2.1 Degree of protection 1)	Up to IP 54	Up to IP 55	IP 65	Up to IP 68
2.2 Protection against electric shock 2)	Safety class 1	Safety class 1 Safety class 2	Safety class 2	Safety class 1
2.3 Enclosure material	Metal	Metal	Moulded plastic	Metal Aluminium Moulded plastic
2.4 Mounting type (indoors)	Against a wall, free-standing, back-to-back, double-front	Against a wall or free-standing	Against a wall	Suspended from ceiling, wall-mounted, sub-floor-mounted
2.5 Number of operating fronts	1 or 2	1	1	1 or 2
3 Type of device installation				
3.1 Fixed-mounting	X	X	X	--
3.2 Withdrawable unit	X	--	--	--
3.3 Plug-in system	X	--	--	X
3.4 Snap-on mounting	--	X	X	--
4 Application				
4.1 Main switchboard	X	--	--	--
4.2 Main distribution board	X	X	--	X
4.3 Subdistribution board	X	X	--	X
4.4 Busbar trunking system	--	--	--	X
4.5 Motor control board	X	--	--	X
4.6 Distribution cabinet	--	X	--	X
4.7 Industry-type distribution board	X	--	X	X
4.8 Light and power distribution boards	--	X	X	X
4.9 Reactive-power compensation	X	--	--	--
4.10 Open-loop control	X	X	X	--
Siemens system type	SIVACON S8	ALPHA 630	ALPHA 8HP	SIVACON 8PS

1) The first characteristic numeral after the code letters IP (International Protection) indicates the degree of protection against access to hazardous parts and against the ingress of solid foreign objects. The second characteristic numeral identifies the degree of protection by enclosures with respect to harmful effects due to the ingress of water.

2) Safety class 1 = Protective earth connection, Safety class 2 = Protective insulation

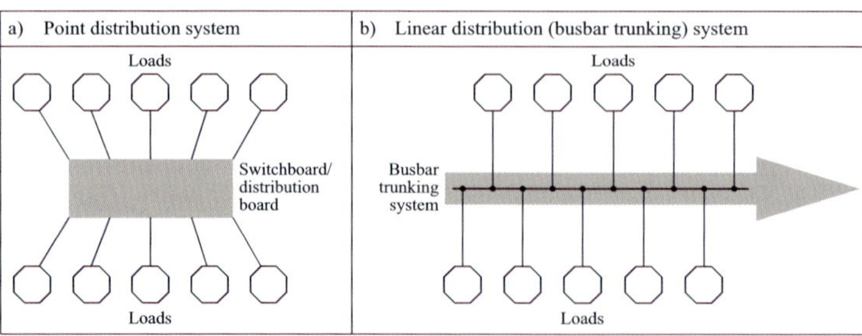

Fig. C11.10 Schematic diagram of a point (radial) and linear distribution system [11.22]

Details of the various Siemens switchboards and distribution boards are explained in the following sections.

11.2.1 SIVACON S8 switchboard

The SIVACON S8 low-voltage switchboard is a cost-efficient, suitable and type-tested low-voltage switchgear and controlgear assembly (TTA). It is characterized by a high level of safety for persons and equipment when used to build LV power systems. The following safety features are decisive [11.23, 11.24]:

- certificate of safety for each specifically developed switchboard,
- section-to-section safety due to partitions between sections,
- safety in the test and disconnected position (switchboard degree of protection is retained up to IP54). This permits enhanced protection of operating personnel and prevents harmful deposits in the switchboard),
- standardized operation of the withdrawable units with integrated maloperation protection prevents incorrect operation and shortens or facilitates training,
- resistance to internal arcing according to DIN EN 60439-1 Bbl. 2 (VDE 0660-500 Bbl. 2): 2009-05 [11.15] or IEC/TR 61641: 2008-01 [11.16] by testing the response to internal faults (up to $U_e \leq 440$ V, $I_{cw} \leq 65$ kA, $t_{cw} \leq 300$ msec) and implementation of a graduated concept with additive components (arc protection barriers, insulated busbars) for active and passive limitation of the fault.

In industrial power systems, the type-tested switchgear assembly (TTA) of Siemens type SIVACON S8 can be used for all power levels up to 6,300 A as

- main switchboard or main distribution board,
- motor control centre (MCC) and
- subdistribution board.

The SIVACON S8 is very versatile due to a combination of different mounting designs in one section and variable ways of subdividing the switchboard internally.

The great scope of variation permits flexible adaptation to the prevailing conditions of use (e.g. necessary degree of protection) and customer requirements (e.g. cost of preventative and corrective maintenance, cost of live working). Tables C11.11 and C11.12 provide a summary of use of the internal separations and mounting designs of the SIVACON S8 switchboard that can be selected.

According to Table C11.12, the mounting designs described below can be used for the switchboard arrangement [11.23]:

Circuit-breaker design

The incoming feeders, outgoing feeders and bus couplers of the circuit-breaker design are equipped with SENTRON® 3WL air circuit-breakers of withdrawable or fixed-mounted design up to 6,300 A, or alternatively with SENTRON® 3VL moulded-case circuit-breakers up to 1,600 A.

Universal mounting design

For many applications, space-optimized installation of the switchboards is required. This makes it necessary to combine different mounting designs in one section. By combining cable outgoing feeders using the fixed-mounted design (SENTRON 3VL circuit-breaker up to 630 A, SIRIUS 3RV circuit-breaker up to 100 A, SENTRON 3K switch-disconnector with 3N fuse-links up to 630 A, SENTRON 3NP switch-disconnector with fuse-links 3N up to 630 A) and cable outgoing feeders using the 3NJ6 in-line de-

Table C11.11 Forms of internal separation when using the SIVACON S8 switchboard

Form	Schematic diagram	Explanations	
1	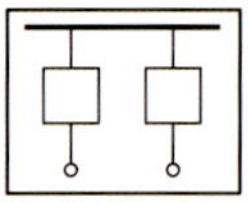	No internal separation, form 1plus with shock protection cover of the main busbar	
2a *)	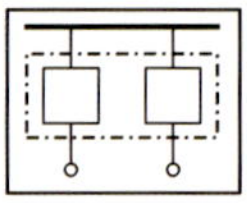	Separation between busbars and functional units	No separation between connections and busbars
2b	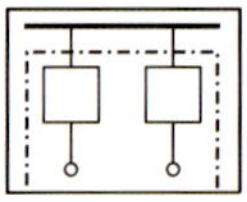		Separation between connections and busbars
3a	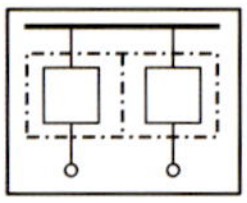	Separation between busbars and functional units + Separation between functional units + Separation between connections and functional units	No separation between connections and busbars
3b	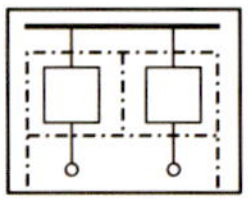		Separation between connections and busbars
4a	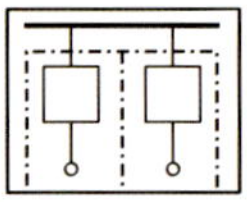		Connections in the same separation that is used for the connected functional unit
4b	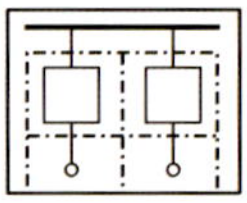		Connections not in the same separation that is used for the connected functional unit
Legend	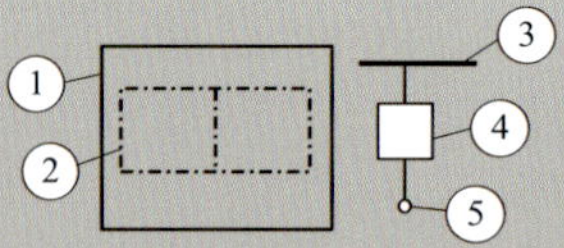		(1) Enclosure (2) Internal separation (3) Busbars (4) Functional unit(s) (5) Connection point(s) for external conductors
*) Not used for SIVACON S8 switchboard			

sign, plug-in type (SENTRON 3NJ6 switch-disconnector with 3N fuse-links up to 630 A), a high level of safety and variability is achieved.

3NJ in-line design

The sections for switchboards using the in-line design with incoming-feeder-side plug-in contact up to 630 A provide an economical alternative to the plug-in design, and their modular equipment form enables quick and easy conversion or replacement under operational conditions.

Table C11.12 Relevant mounting design for the SIVACON S8 LV switchboard

Type of cubicle	Circuit-breaker design	Universal mounting design	3NJ6 in-line design	Fixed-mounted design	3NJ4 in-line design	Reactive-power compensation
Mounting design	Fixed-mounted design, with-drawable design	Fixed-mounted design, plug-in design, with-drawable design	Plug-in design	Fixed-mounted with compartment doors	Fixed-mounted design	Fixed-mounted design
Function	Incoming feeders, outgoing feeders, bus couplers	Cable outgoing feeders, motor outgoing feeders	Cable outgoing feeders	Cable outgoing feeders	Cable outgoing feeders	Central VAR compensation
Rated current or power	≤ 6,300 A	≤ 630 A ≤ 250 kW	≤ 630 A	≤ 630 A	≤ 630 A	≤ 600 kvar
Connection	front side and rear side	front side and rear side	front side	front side	front side	front side
Cubicle width in mm	400/600/800/ 1,000/1,400	600/1,000/1,200	1,000/1,200	1,000/1,200	600/800	800
Internal separation [1)]	1; 2b; 3a; 4b	2b; 3b; 4a; 4b	1; 3b; 4b	1; 2b; 3b; 4a; 4b	1; 2b	1; 2b
Busbars	rear/top	rear/top	rear/top	rear/top	rear	rear/top/without

1) For forms of internal separation, see Table C11.11

Fixed-mounted design with compartment doors

In a certain range of applications, it is not necessary to replace components under operational conditions or short downtimes are acceptable. In this case, the sections for the cable outgoing feeders are implemented by the fixed-mounted design with compartment doors as additional mechanical protection against electric shock.

3NJ4 in-line design, fixed mounted

The sections for the cable outgoing feeders in the fixed-mounted design are equipped with switchable in-line fuse switch-disconnectors. With their compact and modular design, they provide optimum installation conditions with respect to the packing density that can be achieved.

Reactive-power compensation

Reactive-power compensation (Chapter 12) can be performed through either pure capacitor units or reactor-capacitor units with defining factors of $p = 5.67\,\%$, $p = 7\,\%$ and $p = 14\,\%$. In one cubicle capacitor, powers up to

- $Q_c = 400$ kvar using capacitor units or reactor-capacitor units with a switch-disconnector,
- $Q_c = 500$ kvar using reactor-capacitor units without a switch-disconnector,
- $Q_c = 600$ kvar using capacitor units without a switch-disconnector

can be installed. The switch-disconnector is used as the disconnecting point between the main busbar and the vertical cubicle distribution rail.

If motor drives are present, LV switchboards are used not only as power distribution boards for the normal power supply but also as motor control centres (MCC). A motor control centre is an LV switchboard with cubicles in fixed-mounted or withdrawable design that are equipped for connection of motor feeders with a door-locking main switch and a motor-starter combination.

SIVACON S8 motor control centre

The main circuit-breaker in each motor feeder unit has motor-current switching capacity (6 to 8 I_{rM}) and ensures that the compartment door in front of the withdrawable unit can only be open e.g. during the motor run-up period (possibly with locked rotor, or with welded contactor and associated current) once it has been switched off and the main power to the unit has been safely disconnected [11.22]. Because of the high switching frequency of motor feeders, power contactors of type 3RT or 3TF are used for operational switching on and off of the motors. These contactors are implemented as

- contactors for normal and DOL starting,
- contactor combinations for reversing circuits for changing the direction,
- contactor combinations for star-delta starting circuits.

The overload and short-circuit protection of the motor feeders can be of the either fused or fuseless design.

Fuseless design

a) Withdrawable MCC unit with a circuit-breaker for motor protection:
   - SIRIUS 3RV circuit-breaker up to 100 A or
   - 3VL circuit-breaker up to 630 A.

   The circuit-breaker of the MCC withdrawable unit provides both short-circuit and overload protection.

b) Withdrawable MCC unit with a circuit-breaker and overload relay:
   - SIRIUS 3V circuit-breaker up to 100 A or
   - 3VL circuit-breaker up to 630 A,
   - 3RU thermal overload relay up to 80 A or
   - 3RB electronic overload relay up to 630 A or
   - SIMOCODE pro 3UF7 motor management and control device up to 630 A.

The circuit-breaker of the MCC withdrawable unit provides the short-circuit protection. For the overload protection, both the overload relay in its thermal or electronic versions and the SIMOCODE control device can be used.

Fused design

c) Withdrawable MCC unit with a fuse-switch-disconnector and overload relay:
   - SENTRON 3K switch-disconnector with 3N fuse-links up to 630 A,
   - 3RU thermal overload relay up to 80 A or
   - 3RB electronic overload relay up to 630 A or
   - SIMOCODE pro 3UF7 motor management and control device up to 630 A.

The fuses provide the short-circuit protection. For the overload protection, both the overload relay in its thermal or electronic versions and the SIMOCODE control device can be used.

SIMOCODE control devices provide not only the classic overload protection but also numerous additional protection and control functions. With the additional protection functions it is possible to implement electronic full motor protection. For implementation of a multifunctional full motor protection, the following functions are available [11.25]:

- inverse-time delayed overload protection with adjustable tripping characteristics,
- phase failure/load unbalance protection,

- stall protection,
- thermistor motor protection,
- internal and external earth fault monitoring,
- monitoring of adjustable limit values for the motor current,
- voltage monitoring,
- monitoring of the active power,
- monitoring of the power factor (motor idling/load shedding),
- phase sequence detection,
- temperature monitoring,
- monitoring of process quantities (e.g. liquid filling level, fouling factor of filters),
- monitoring of operating hours, downtimes and number of starts,
- motor control for direct-on-line, reversing and star-delta starters, direction reversal, shut-off valve operation, valve operation, operation for soft starters, etc.

With this range of functions, an intelligent connection between the higher-level automation equipment and the motor feeder can be established if SIMOCODE pro control devices are installed in the SIVACON S8 motor control centre. For that reason, the process industries (e.g. chemical, petrochemical, steel, cement, glass, paper and pharmaceutical) are an area where SIMOCODE pro systems for motors can be most beneficially deployed.

To protect motors with motor-starter combinations, the requirements according to DIN EN 60947-4-1 (VDE 0660-102): 2006-04 [11. 26] or IEC 60947-4-1: 2009-09 [11.27] must be met.

Standard-compliant overload protection

Different tripping classes are defined for the overload protection of motors. The tripping classes describe the time intervals within which the protective device (overload release of the circuit-breaker or overload relay) must trip from the cold condition on a symmetrical three-phase load with 7.2 times the current setting. These tripping times are

- CLASS 5 between 0.5 sec and 5 sec,
- CLASS 10A between 2 sec and 10 sec,
- CLASS 10 between 4 sec and 10 sec,
- CLASS 20 between 6 sec and 20 sec,
- CLASS 30 between 9 sec and 30 sec,
- CLASS 40 between 30 sec and 40 sec.

For standard applications (normal starting), devices with the tripping class 5 or 10 or 10 A are used. The classes 5, 10, and 10 A are frequently also referred to as normal starting classes. For the overload protection of motors that start heavily, that is, motors with high starting currents or long run-up times (e.g. compressor motors), overload protection devices of class 20 or higher must be used. The tripping classes in the range CLASS 20 to CLASS 40 are called heavy starting classes. By the splitting into normal starting and heavy starting classes, the tripping time can be sufficiently precisely adapted to the load torque of the motor so that the motor is better utilized.

Standard-compliant short-circuit protection

Starters and/or contactors in motor feeder units do not usually have to break short-circuit currents. Circuit-breakers (fuseless design) or switch-fuse units (fused design) are always used for the short-circuit protection.

The permissible damage to fuseless or fused motor-starter combinations, resulting from a short circuit, is defined by the following two types of coordination [11.26, 11.27]:

- Coordination type "1"

  No danger may be caused to persons on neighbouring parts of the installation.
  The destruction of the contactor and the overload relay is permissible. If necessary, the contactor and/or the overload relay must be repaired or replaced. The equipment need not be suitable for further service after the short-circuit condition.

- Coordination type "2"

  No danger may be caused to persons on neighbouring parts of the installation.
  The overload relay may not suffer any damage. Welding of the contacts in the contactor is only permissible if such welding can be broken easily and the contacts can be separated. A replacement of the contactor must not be necessary. The combination must be suitable for further service after the short-circuit condition.

The protective device used for short-circuit protection of the motor feeder unit (overcurrent release of the circuit-breaker or fuses) must always trip in case of a fault. If fuses are used as the protective device, they must be replaced after a short circuit. This fuse replacement still complies with the standard even if the motor-starter combination meets the more stringent requirements of coordination type "2".

Motor starting methods

To limit system perturbations due to impulse currents, star-delta-connected contactor combinations are often used to start three-phase motors. If a star-delta connection is used, a motor normally operated in a delta connection is temporarily operated in a star connection. In a star connection, the values of the starting current and starting torque are reduced to a third of the values that occur during direct on-line starting in a delta connection. Because the starting torque is reduced to a third, the star-delta connection requires a starting operation with a load torque that remains low (e.g. running-up of machine tools without a load). Moreover, this connection can only be used for three-phase asynchronous motors whose rated motor voltage in delta connection matches the nominal system voltage (e.g. $U_\Delta = U_{nN} = 400$ V or $U_\Delta = U_{nN} = 690$ V) and whose wind-

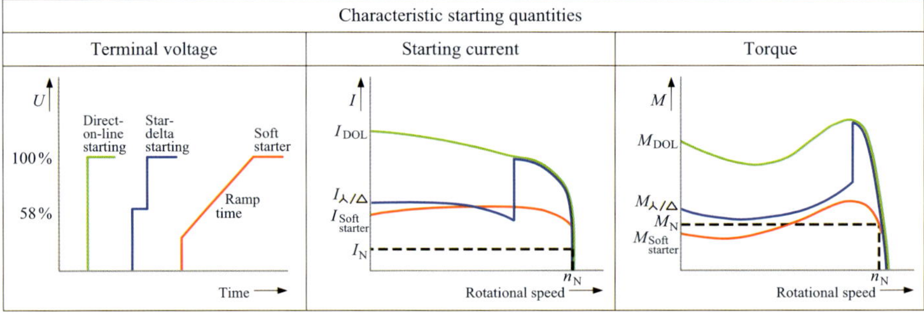

Fig. C11.13 Qualitative comparison of the starting characteristics for different starting methods

ing ends are separately routed to the terminal board [11.22]. The overload protection of the motor must be active both in the star connection and in the delta connection. The overload relay of the star-delta starting connection must be chosen for 58 % ($\hat{=} 1/\sqrt{3}$) of the rated motor current.

Electronically controlled soft motor starters (e.g. the SIRIUS soft starter [11.28]) are an alternative to star-delta starting. Electronic soft starters ensure controlled motor starting without transient current peaks and sudden changes in torque because they control the motor voltage according to the phase angle control principle. By continuously increasing the motor voltage, soft starters relieve the network, motor and drive systems of excessive loads during starting. Soft starters can usually be easily combined with existing contactors, circuit-breakers and overload relays in motor feeder units. They are available up to a nominal system voltage of U_{nN} = 690 V.

The starting behaviour of motors during direct-on-line starting, star-delta starting and electronic soft starting is illustrated in Fig. C11.13, which shows the voltage, current and torque.

Each motor-starter combination of the SIVACON S8 motor control centre, which can be implemented as a fixed-mounted or withdrawable unit, is a separate module. Separate MCC modules are available as withdrawable units in the following designs and sizes:

- ¼-size miniature withdrawable unit with a height of 150 mm or 200 mm,
- ½-size miniature withdrawable unit with a height of 150 mm or 200 mm,
- normal-size withdrawable unit with a height of 100 mm to 700 mm.

¼ size means that 4 miniature withdrawable units fit into one tier of the MCC cubicle. ½ size means that only 2 miniature units fit into one withdrawable MCC cubicle row. The normal-size withdrawable unit fills one whole row of the MCC cubicle. Based on a cubicle equipment height of 1,800 mm, 48 ¼-size miniature withdrawable units or 24 ½-size miniature withdrawable units can be mounted in one MCC cubicle.

Conclusion

Because of the many variants available, its rated current ($I_r \leq$ 6,300 A), its short-circuit current ratings ($I_{cw} \leq$ 150 kA for t_{cw} = 1 sec, $I_{pk} \leq$ 330 kA) and the rated short-circuit making/breaking capacity of the incoming feeder, outgoing feeder and coupler circuit-breakers ($I_{cu} \leq$ 150 kA, $I_{cm} \leq$ 330 kA), the SIVACON S8 low-voltage switchboard is the ideal point (radial) energy distribution system for building centralized or decentralized multiply-fed industrial networks.

11.2.2 ALPHA 630 floor-mounted distribution board

The ALPHA 630 floor-mounted distribution board can be used as a main distribution board or subdistribution board not only in administrative and commercial buildings but also in industrial buildings. It can also be used as a control cabinet with a cubicle-

Table C11.14 Rated voltage and currents of the ALPHA 630 floor-mounted distribution board

Electrical ratings	Type of installation distribution system	
	ALPHA 630 DIN	ALPHA 630 Universal
Rated operational voltage U_e	690 V AC	690 V AC
Rated current I_r	630 A	630 A
Rated short-time withstand current I_{cw} (t_{cw} = 1 sec)	20 kA	25 kA
Rated peak withstand current I_{pk}	≤ 61.3 kA	53 kA

height mounting plate. The distribution boards and components are assembled according to the modular principle [11.29].

The largest fused switching devices that can be mounted are LV HRC fuse-switch-disconnectors of size NH3, 630 A. Installation kits for moulded-case circuit-breakers of the type series 3VL, 63 A to 630 A are available for fuseless feeders. The most important electrical ratings of the ALPHA 630 floor-mounted distribution board are listed in Table C11.14.

According to the short-circuit current ratings given in Table C11.14, the short-circuit power at the installation location of a 400-V network must not be greater than 17 MVA. This corresponds to an installed rated transformer power of approximately 1,000 kVA. In industrial networks with a high load density, there are therefore strict limits to the use of the ALPHA 630 floor-mounted distribution board.

11.2.3 ALPHA 8HP moulded-plastic distribution board

The ALPHA 8HP moulded-plastic distribution board [11.29, 11.30] is ideal for use as a main distribution board or subdistribution board in humid, dusty and corrosive environments. For that reason, the steel, chemical and mining industries are among those that make the most use of the ALPHA 8HP moulded-plastic distribution boards.

Table C11.15 Rated voltage and currents of the ALPHA 8HP moulded-plastic distribution board

Rated operational voltage U_e		690 V AC			
Rated current of the outgoing feeders $I_{r\text{-outgoing}}$		≤ 800 A			
Rated current of the busbars $I_{r\text{-BB}}$		250 A	400 A	630 A	1,000 A
Rated current of the incoming feeder $I_{r\text{-incoming}}$	single-end infeed	250 A	400 A	630 A	1,000 A
	centralized infeed	400 A	800 A	1,000 A	1,800 A
Short-circuit strength of busbar system	Rated short-time withstand current I_{cw} (t_{cw} = 1 sec)	10 kA	20 kA	30 kA	40 kA
	Rated peak withstand current I_{pk}	40 kA	70 kA	70 kA	80 kA

With the 8HP, which is a type-tested modular system, it is possible to build customized floor-mounted distribution boards with degree of protection IP65. All parts of the enclosure and operating mechanisms of the 8HP system are designed to comply with the conditions of the protective measure “Total Insulation“ according to DIN VDE 0100-410 (VDE 0100-410): 2007-06 [11.20] or IEC 60364-4-41: 2005-12 [11. 21] while in the operationally enclosed assembly. Table C11.15 provides a list of the most important electrical ratings.

Based on the ratings from Table C11.15, ALPHA 8HP moulded-plastic distribution boards can be used in 400-V networks with a short-circuit power of $S_k'' = 25$ MVA. Without motors that contribute partial short-circuit currents to the total short-circuit current in the event of a fault, this short-circuit power corresponds to a rated transformer power installed in the system of $\sum_i S_{rT_i} = 1.500$ kVA (e.g. 3 × 500 kVA). The maximum rated power of a transformer feeding into the ALPHA 8HP distribution board in the centre is limited to $S_{rT} = 1{,}250$ kVA.

11.2.4 SIVACON 8PS busbar trunking system

The SIVACON 8PS busbar trunking system [11.31 to 11.34] is a linear distribution system for the transmission and distribution of electrical power in the current range 25 A ≤ I_r ≤ 6,300 A. Compared with a classic cable installation, the SIVACON 8PS busbar trunking system provides numerous advantages in terms of power system and installation engineering (Table C11.16).

Because of its advantages in terms of power system and installation engineering, the 8PS busbar trunking system is an excellent solution for versatile power supply to loads in industrially and commercially used buildings. The 8PS busbar trunking system comprises six subsystems structured for specific applications and operating conditions. The following subsystems are available [11.31]:

- CD-L system (25 A to 40 A)

  for supplying power to lighting systems and low-rating loads in shopping centres, logistics warehouses and buildings of every kind.

Table C11.16 Comparison of the 8PS busbar trunking system with a classic cable installation [11.33, 11.34]

Comparison feature	Busbar trunking system	Cable installation
Network topology	Line topology with load feeders connected in series via tap-off units	Significant cable cluster at the infeed point due to star topology of the supply to the loads
Safety	Safe operation due to type-tested low-voltage switchgear and controlgear assemblies acc. to EN/IEC 61439-1 (2) [11.11–11.14]	Safety of operation depends on the quality of implementation
Flexibility for changes, expansions, or relocation of main loads	Very high flexibility due to variable tap-off units that can be changed, added to or replaced depending on requirements; live working also possible	Mostly new installation or considerable effort required due to new splices, terminals, joints, parallel cables, etc.; installation work only possible in the de-energized state
Fire load	Very low fire load	For PVC cables the fire load is up to 10 times greater and for PE cables up to 30 times greater than for busbar trunking systems
Electromagnetic interference	Low interference due to sheet-steel enclosure	Relatively high for standard cables; for single-core cables, it largely depends on the type of bundling
Current-carrying capacity	High ampacity due to type-tested low-voltage switchgear and controlgear assemblies acc. to EN/IEC 61439-1 (2) [11.11–11.14]	Limit values must be calculated depending on the type of laying, grouping, bundling and operating conditions
Free of halogens/PVC	Trunking units are always halogen-free and PVC-free	Standard cables are not free of halogens and PVC; halogen-free cables are very expensive
Space requirement	Low space requirement due to compact design, junction units for horizontal and vertical installation and ampacity	Large space requirement due to bending radii, type of laying, grouping and current-carrying capacity
Weight	Only ½ to ⅓ of the weight of a comparable cable	Up to 3 times the weight of a comparable busbar trunking system
Installation work	Simple installation with few tools; short installation times	High installation effort with many tools, long installation times

Table C11.17 Technical data of the subsystems of the 8PS busbar trunking system

Technical data	8PS subsystems					
	CD-L	BD01	BD2	LD	LX	LR
Rated operational voltage U_e	400 V AC	400 V AC	690 V AC	1,000 V AC	690 V AC	1,000 V AC
Degree of protection	IP55	IP54; IP55	IP52; IP54; IP55	IP34; IP54	IP54; IP55	IP68
Rated current I_r	25 A and 40 A	40 A up to 160 A	160 A up to 1,250 A	1,100 A up to 5,000 A	800 A up to 6,300 A	630 A up to 6,300 A
Rated short-time withstand current I_{cw} (t_{cw} = 1 sec)	≤ 6.16 kA	≤ 2.5 kA	≤ 34 kA	≤ 116 kA	≤ 150 kA	≤ 100 kA
Rated peak withstand current I_{pk}	≤ 10.6 kA	≤ 15.3 kA	≤ 90 kA	≤ 286 kA	≤ 255 kA	≤ 220 kA
Conductor configurations for the transmission and distribution of electrical power	L1, N, PE L1, L2, N, PE L1, L2, L3, N, PE	L1, L2, L3, N, PE	L1, L2, L3, N, ½ PE L1, L2, L3, N, PE	L1, L2, L3, N, PE L1, L2, L3, ½ N, ½ PE L1, L2, L3, N, ½ PE L1, L2, L3, PEN L1, L2, L3, ½ PEN	L1, L2, L3, PE L1, L2, L3, PEN L1, L2, L3, N, PE L1, L2, L3, 2 N, PE L1, L2, L3, N, CE PE 1) L1, L2, L3, 2 N, CE PE 1) L1, L2, L3, N, 2 PE (Co) L1, L2, L3, 2 N, 2 PE (Co)	L1, L2, L3, PEN L1, L2, L3, N, PE
Ambient temperature ϑ	– 5 °C min and + 40 °C max					
Tap-off points	Every 0.5 m, 1 m and 1.5 m on one or both sides	Alternatively 0.5 m or 1 m on one side	Alternatively offset every 0.25 m or 0.5 m on both sides	Every 1 m on one side	Every 0.5 m on both sides	Every 1 m on one side
Tap-off units can be modified while energized	≤ 16 A	≤ 63 A	≤ 630 A	≤ 1,250 A	≤ 630 A	-
Conductor material	Insulated Cu conductor	Insulated Cu or Al conductor	Al or Cu conductor	Al or Cu busbar (conductor surfaces galvanized), epoxy-resin coating of the conductors	Insulated Al or Cu busbar	Cu busbar
Enclosure material	Sheet steel galvanized and unpainted/painted	Sheet steel galvanized and painted	Sheet steel galvanized and painted	Sheet steel painted	Aluminium painted	Epoxy resin
Fire load	≤ 0.75 kWh/m	≤ 0.76 kWh/m	≤ 2.0 kWh/m	≤ 8.83 kWh/m	≤ 16.6 kWh/m	≤ 77.3 kWh/m
Communication skills	yes	yes	yes	yes	yes	-

1) Clean earth, additionally separately laid PE conductor

- BD01 system (40 A to 160 A)

  for supplying power to power tools in workshops and to lighting systems.
- BD2 system (160 A to 1,250 A)

  for the transmission and distribution of power in office and industrial buildings.
- LD system (1,100 A to 5,000 A, ventilated)

  for the transmission and distribution of power in exhibition halls and workshops in the automotive and heavy industry.
- LX system (800 A to 6,300 A, sandwich)

  for the transmission and distribution of power with a high load per unit area, that is, in computer centres, data centres and chip and semiconductor factories.
- LR system (630 A to 6,300 A, encapsulated)

  for the transmission of large quantities of energy in hostile environmental conditions, for tunnel power supplies or the networking of buildings outdoors and for the transmission of power in the chemical industry.

Table C11.17 lists the most important technical data of the CD-L, BD01, BD2, LD, LX and LR subsystems of the 8PS busbar trunking system. Table C11.18 provides information on the options for equipping the 8PS tap-off units.

Table C11.18 Equipment of the 8PS tap-off units

8PS subsystem	Moulded-case circuit-breaker MCCB	Miniature circuit-breaker MCB	Residual current-operated protective device RCD	LV HRC fuses	Neozed fuses	Diazed fuses	Cylindrical fuses	3K switch-disconnector	3NP switch-disconnector	CEE socket-outlet	Schuko (German type) socket-outlet
CD-L	---	---	---	---	---	---	≤ 16 A	---	---	---	---
BD01	---	≤ 63 A	≤ 63 A	---	≤ 63 A D02	≤ 63 A DIII	≤ 32 A	---	---	≤ 32 A	≤ 16 A
BD2	≤ 630 A [1)]	≤ 125 A	≤ 125 A	≤ 630 A NH3	≤ 63 A D02	≤ 63 A DIII	≤ 125 A	≤ 630 A NH3	≤ 630 A NH3	≤ 63 A	≤ 16 A
LD	≤ 1,250 A [2)]	---	---	≤ 630 A NH3	---	---	---	---	≤ 630 A NH3	---	---
LX	≤ 1,250 A [2)]	---	---	≤ 630 A NH3	---	---	---	≤ 630 A NH3	---	---	---
LR	≤ 630 A	---	---	≤ 630 A NH3	---	---	---	≤ 630 A NH3	≤ 630 A NH3	---	---

1) If I_r = 630 A, the maximum load current is 530 A

2) If $I_r \geq$ 800 A, the maximum load current is limited to $0.9 \cdot I_r$

The tap-off units of the CD-L, BD01, BD2, LD and LX systems can be complemented with the communicative ancillary equipment units for

- lighting control (CD-L, BD01, BD2),
- switching, signalling, remote operation and remote monitoring (BD2, LD, LX) and
- consumption measurement and recording (BD2, LD, LX).

The combination of tap-off units, communicative ancillary equipment units and interoperable bus systems (e.g. PROFIBUS-DP, AS interface and KNX/EIB) permits the intelligent networking of power distribution, consumption and automatic process control. The SIVACON 8PS busbar trunking system therefore meets all the requirements of modern, automated production sites for a suitable low-voltage distribution system.

Conclusion

Only intelligent power distribution concepts with communicative busbar trunking systems meet the stringent requirements for flexibility of the installation and system automation in industrial power supplies. With its communication capability and many variants, the SIVACON 8PS busbar trunking system is an ideal linear energy distribution system for the cost-efficient building of decentralized multiply-fed industrial networks.

11.2.5 Transformer load-centre substation with SIVACON S8/8PS

The transformer load-centre substation tested by PEHLA according to DIN EN 62271-202 (VDE 0671-202): 2007-08 [11.35] or IEC 62271-202: 2006-06 [11.36] is a safe and cost-efficient module for decentralized power supplies. For the construction of radial networks in an interconnected cable system (Section 10.3.1.3), load-centre substations equipped

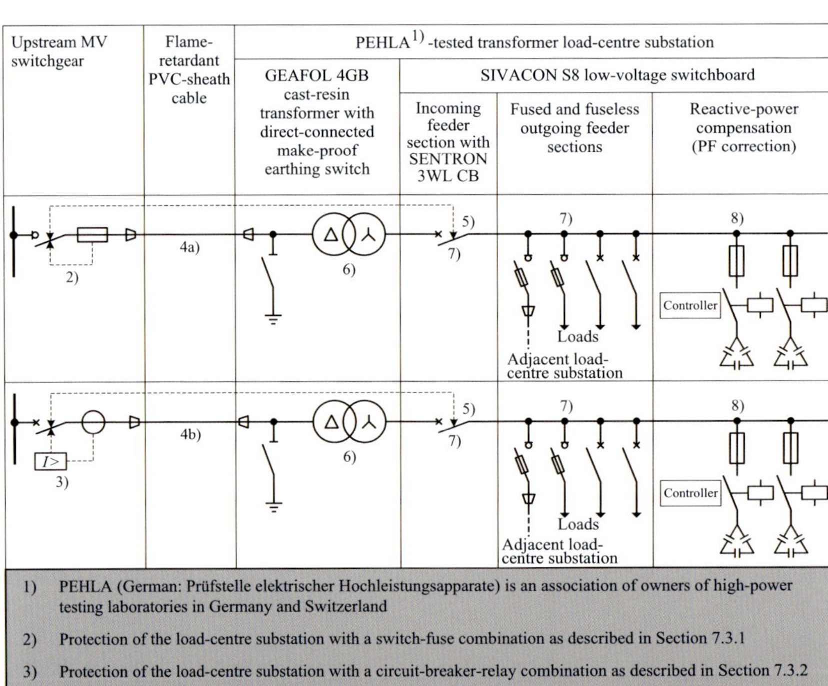

1) PEHLA (German: Prüfstelle elektrischer Hochleistungsapparate) is an association of owners of high-power testing laboratories in Germany and Switzerland

2) Protection of the load-centre substation with a switch-fuse combination as described in Section 7.3.1

3) Protection of the load-centre substation with a circuit-breaker-relay combination as described in Section 7.3.2

4a) Rating of the cable cross-sectional area according to the AF-load current of the GEAFOL cast-resin transformer. All distribution transformers with rated powers S_{rT} < 2,500 kVA should preferably be connected to the MV substation through a cable of type N2XSY 3x1x35/16 mm^2.

4b) Rating of the cable cross-sectional area according to the short-circuit power of the MV network. In 20-kV networks with a short-circuit power of $S_k'' \leq$ 500 MVA, the distribution transformers can be connected through a cable of type N2XSY 3x1x95/16 mm^2 (see Section 4.2.2, example B2).

5) Transfer tripping (breaker intertripping) for safe disconnection of a faulty load-centre substation from the network.

6) Short-circuit withstand capability according to Section 4.2.3, selection and dimensioning according to Section 11.1.

7) Selection and dimensioning complying with the load current condition (Eq. 2.4) and the short-circuit current conditions (Eqs. 2.6, 2.7 and 2.8).

8) With an average expected power factor of $\cos\varphi_1 = 0.7$, a correction to $\cos\varphi_2 = 0.90 - 0.95$ requires a capacitor rating of about 40 % of the transformer rating ($Q_c = 0.4 \cdot S_{rT}$).

Fig. C11.19 Implementation of radial networks in an interconnected cable system with a GEAFOL cast-resin transformer and the communicative SIVACON S8 low-voltage switchboard have proven successful. Fig. C11.19 shows some practically tested concepts for implementing radial networks in an interconnected cable system with load-centre substations and an integrated SIVACON S8 low-voltage switchboard.

In line with the current state of the art in industrial power supplies, radial networks interconnected through busbar trunking systems (Section 10.3.1.5) are today the preferred solution for decentralized power distribution rather than radial networks in an interconnected cable system. The SITRABLOC load-centre substation [11.38] (Fig. C11.20) was developed as a compact module for constructing radial networks interconnected through busbar trunking systems. It comprises:

- a transformator enclosure with fans for AN ($I_{T\text{-}AN} = 1.0 \cdot I_{rT}$) and AF operation ($I_{T\text{-}AF} \leq 1.4 \cdot I_{rT}$),
- a GEAFOL cast-resin transformer of type 4GB with make-proof earthing switch
- a SENTRON circuit-breaker of type 3WL,
- automatic reactive-power compensation system,
- a control and measurement unit and an interface for remote data transmission,
- universal connection options for a high-current busbar system (e.g. SIVACON 8PS).

Due to the PEHLA testing [11.37] of the MV and LV module, the SITRABLOC substation provides a high level of safety for persons and equipment. The following test results are certified:

- MV module (10 kV and 20 kV): 25 kA/1 sec,
- LV module (400 V): 120 kA/300 msec.

The rated current I_r of the SENTRON 3WL circuit-breaker for supplying the high-current busbar system must be defined according to the AF-load current of the GEAFOL cast-resin transformer ($I_{T\text{-}AF} \leq 1.4 \cdot I_{rT}$). For that reason, the 3WL incoming feeder circuit-breaker is usually chosen based not on the necessary rated short-circuit breaking/making capacity I_{cu} or I_{cm} but on the required current rating I_r.

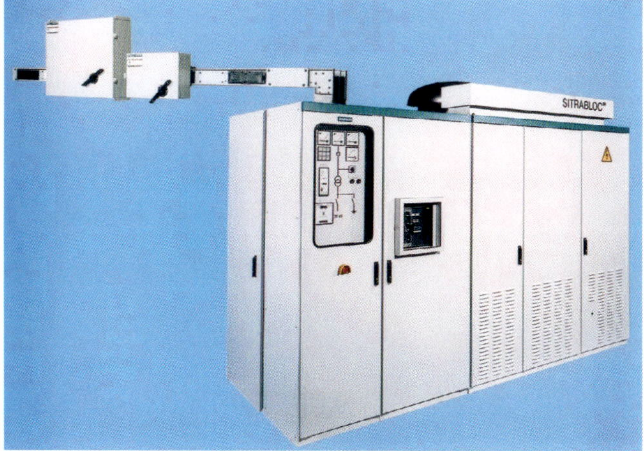

Fig. C11.20 Front view of the SITRABLOC load-centre substation with high-current busbar connection

The short-circuit withstand capability in decentralized multiply-fed networks constructed with SITRABLOC substations interconnected through the SIVACON 8PS busbar trunking system is limited to 120 kA. Fig. C11.21 shows how such a network configuration is implemented.

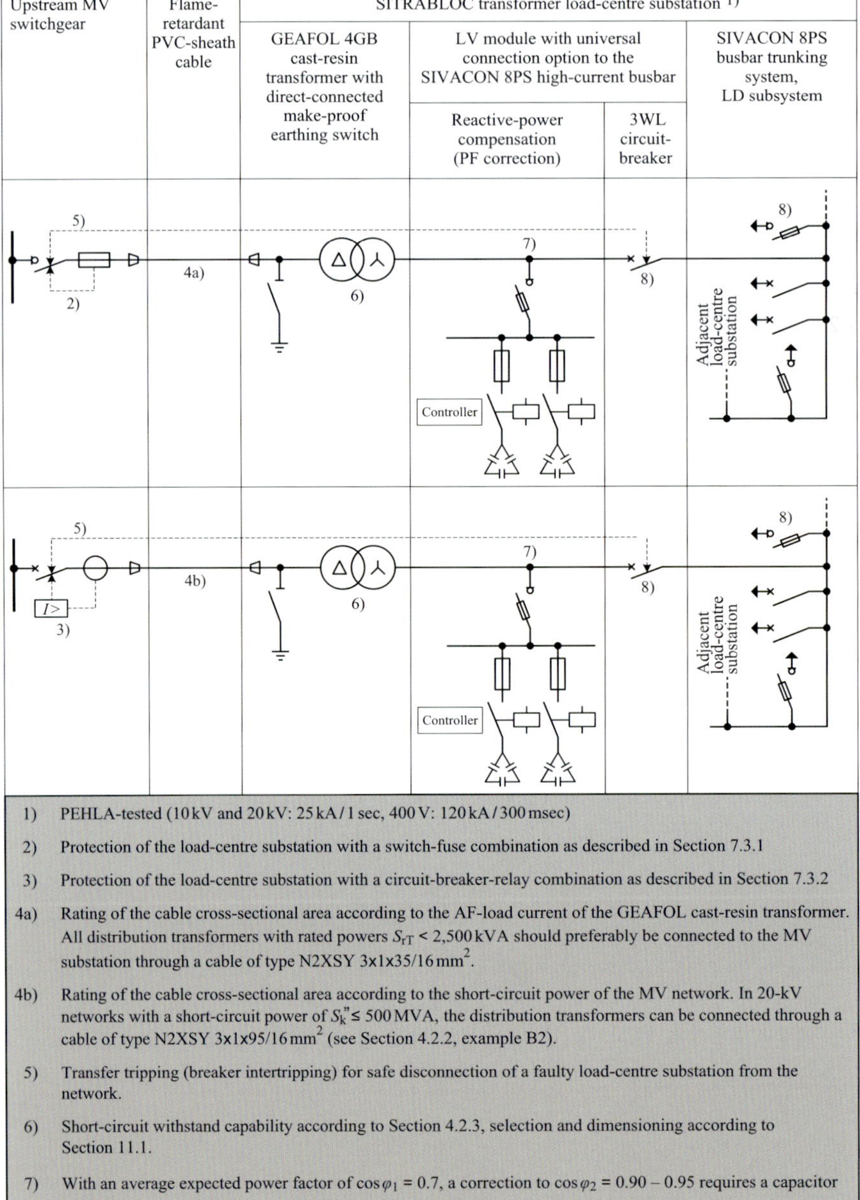

1) PEHLA-tested (10 kV and 20 kV: 25 kA / 1 sec, 400 V: 120 kA / 300 msec)

2) Protection of the load-centre substation with a switch-fuse combination as described in Section 7.3.1

3) Protection of the load-centre substation with a circuit-breaker-relay combination as described in Section 7.3.2

4a) Rating of the cable cross-sectional area according to the AF-load current of the GEAFOL cast-resin transformer. All distribution transformers with rated powers S_{rT} < 2,500 kVA should preferably be connected to the MV substation through a cable of type N2XSY 3x1x35/16 mm^2.

4b) Rating of the cable cross-sectional area according to the short-circuit power of the MV network. In 20-kV networks with a short-circuit power of $S_k'' \leq$ 500 MVA, the distribution transformers can be connected through a cable of type N2XSY 3x1x95/16 mm^2 (see Section 4.2.2, example B2).

5) Transfer tripping (breaker intertripping) for safe disconnection of a faulty load-centre substation from the network.

6) Short-circuit withstand capability according to Section 4.2.3, selection and dimensioning according to Section 11.1.

7) With an average expected power factor of $\cos \varphi_1 = 0.7$, a correction to $\cos \varphi_2 = 0.90 - 0.95$ requires a capacitor rating of about 40 % of the transformer rating ($Q_c = 0.4 \cdot S_{rT}$).

8) Selection and dimensioning complying with the load current condition (Eq. 2.4) and the short-circuit current conditions (Eqs. 2.6, 2.7 and 2.8).

Fig. C11.21 Implementation of radial networks interconnected through busbar trunking systems

Conclusion

PEHLA-tested transformer load-centre substations are a safe and cost-efficient system module for the decentralized power suppy in industrial plants. They are essential for implementing radial networks in an interconnected cable system and radial networks interconnected through busbar trunking systems.

11.3 Cables

For dimensioning cables in high-current electrical power installations up to 1,000 V, the following criteria must be checked and fulfilled:

a) *Permissible current-carrying capacity* in different installation and ambient conditions for fixed wiring in and around buildings according to DIN VDE 0298-4 (VDE 0298-4): 2003-08 [11.39] or IEC 60364-5-52: 2009-10 [11.40] and generally for power distribution systems according to DIN VDE 0276-1000 (VDE 0276-1000): 1995-06 [11.41],

b) *Protection against overload* according to DIN VDE 0100-430 (VDE 0100-430): 1991-11 [11.42] or IEC 60364-4-43: 2008-08 [11.43],

c) *Protection against short circuit* according to DIN VDE 0100-430 (VDE 0100-430): 1991-11 [11.42] / IEC 60364-4-43: 2008-08 [11.43] and DIN VDE 0298-4 (VDE 0298-4): 2003-08 [11.39] / IEC 60364-5-52: 2009-10 [11.40],

d) *Protection against electric shock* according to DIN VDE 0100-410 (VDE 0100-410): 2007-06 [11.20] or IEC 60364-4-41: 2005-12 [11.21],

e) *Compliance with the permissible voltage drop* according to DIN VDE 0100-520 (VDE 0100-520): 2003-06 [11.44] / IEC 60364-5-52: 2009-10 [11.40] and DIN IEC 60038 (VDE 0175): 2002-11 [11.46] / IEC 60038: 2009-06 [11.47].

The dimensioning of the necessary conductor cross-sectional area for shorter cables ($l \leq 100$ m) is primarily influenced by criteria a), b) and c). For longer cables, rating criteria d) and e) are decisive.

11.3.1 Permissible current-carrying capacity

The current-carrying capacity of a cable for fixed wiring in and around buildings must be determined according to DIN VDE 0298-4 (VDE 0298-4): 2003-08 [11.39] or IEC 60364-5-52: 2009-10 [11.40] depending on the operating conditions, that is

- type of operation (continuous, short-time, intermittent),
- installation conditions (installation in ground or air, grouping) and
- ambient temperature (air temperatures deviating from 30 °C).

Standardized reference operating conditions that are listed in Table C11.22 apply to the rated value I_r of the current-carrying capacity.

For operating conditions that differ from the reference conditions in Table C11.22, the permissible current-carrying capacity must be calculated as follows [11.48]:

$$I_{perm} = I_r \cdot \prod_i f_i \tag{11.2}$$

I_{perm} permissible current-carrying capacity of the cable

I_r rated value of the current-carrying capacity of the cable for the reference method of installation A1, A2, B1, B2, C, D, E, F or G (Tables C11.23 to C11.26)

f_i rating factors for site operating conditions according to DIN VDE 0298-4 (VDE 0298-4): 2003-08 [11.39] or IEC 60364-5-52: 2009-10 [11.40]

Table C11.22 Reference operating conditions for the rated value I_r of the current-carrying capacity of LV cables and lines

Type of operation		
Continuous operation, i. e. operation with constant current in thermal equilibrium (load factor m = 1)		
Cable installation conditions		
Reference method of installation		Categorization
Installation in thermally insulated walls	A1	Single-core cables in a conduit in a thermally insulated wall
	A2	Multi-core cable or multi-core sheathed cable in a conduit in a thermally insulated wall
Installation in conduits	B1	Single-core cables in a conduit on a wall
	B2	Multi-core cable or multi-core sheathed cable in a conduit on a wall
Direct installation	C	Single or multi-core cable or single or multi-core sheathed cable on a wall
Installation in ground	D	Multi-core cable or multi-core sheathed cable in a conduit or in a cable vault in the ground
Installation in free air	E	Multi-core cable or multi-core sheathed cable in free air at a distance of at least 0.3 x diameter d from the wall
	F	Single-core cables or single-core sheathed cables, with contact, in free air at a distance of at least 1 x diameter d from the wall
	G	Single-core cables or single-core sheathed cables, with a spacing of d, in free air at a distance of at least 1 x diameter d from the wall
Ambient conditions		
Installation in air	Sufficiently large and ventilated rooms in which the ambient temperature is not noticeably increased by the heat dissipation of the cable or conductors	
	Ambient temperature ϑ_u = 30 °C	
	Installation under the ceiling for installation methods B1, B2, and C	
	Protection against direct irradiation of heat from the sun, etc.	
Installation in ground	Ambient temperature ϑ_u = 20 °C	
	Thermal resistivity of soil ς = 2.5 k · m / W	
	Installation depth t = 0.7 m	

To build powerful cable networks in industry, the cables are usually installed in free air on cable trays and cable ladders. When calculating the permissible current-carrying capacity of cables that are installed in free air on cable trays and cable ladders, rating factors for differing air temperatures and cable grouping must be taken into account. The following applies:

$$I_{perm} = I_{rL} \cdot f_{\vartheta} \cdot f_H \quad (11.2.1)$$

I_{rL} rated value of the current-carrying capacity of the cable for the reference method of installation E, F or G (Tables C11.23 to C11.26)

f_{ϑ} correction factor for differing air temperatures (Table C11.27)

f_H reduction factor for the grouping of cables on trays and ladders (Table C11.28 for reference method of installation E (multi-core cables) and Table C11.29 for reference method of installation F (single-core cable))

Table C11.23 Rated values for the current-carrying capacity of PVC cables and lines with a permissible operating temperature on the conductor of 70 °C (conductor material copper)

Standardized nominal cross-sectional area A_n [mm²]	Reference method of installation																		
	In thermally insulated walls				In conduits				On a wall		In ground		In free air, i. e. unhindered heat dissipation						
	A1		A2		B1		B2		C		D		E		F			G	
	Number of loaded conductors																		
	2	3	2	3	2	3	2	3	2	3	2	3	2	3	2	3	3	3	3
	Ratings I_r for the current-carrying capacity [1)] in A																		
1.5	14.5	13.5	14	13	17.5	15.5	16.5	15	19.5	17.5	18.5	15.5	22	18.5	--	--	--	--	--
2.5	19.5	18	18.5	17.5	24	21	23	20	27	24	25	21	30	25	--	--	--	--	--
4	26	24	25	23	32	28	30	27	36	32	32	27	40	34	--	--	--	--	--
6	34	31	32	29	41	36	38	34	46	41	40	34	51	43	--	--	--	--	--
10	46	42	43	39	57	50	52	46	63	57	54	45	70	60	--	--	--	--	--
16	61	56	57	52	76	68	69	62	85	76	69	59	94	80	--	--	--	--	--
25	80	73	75	68	101	89	90	80	112	96	88	76	119	101	131	114	110	146	130
35	99	89	92	83	125	110	111	99	138	119	106	91	148	126	162	143	137	181	162
50	119	108	110	99	151	134	133	118	168	144	126	108	180	153	196	174	167	219	197
70	151	136	139	125	192	171	168	149	213	184	156	133	232	196	251	225	216	281	254
95	182	164	167	150	232	207	201	179	258	223	184	161	282	238	304	275	264	341	311
120	210	188	192	172	269	239	232	206	299	259	209	183	328	276	352	321	308	396	362
150	240	216	219	196	300	262	258	225	344	299	236	205	379	319	406	372	356	456	419
185	273	245	248	223	341	296	294	255	392	341	265	231	434	364	463	427	409	521	480
240	321	286	291	261	400	346	344	297	461	403	307	266	514	430	546	507	485	615	569
300	367	328	334	298	458	394	394	339	530	464	347	298	593	497	629	587	561	709	659
400	--	--	--	--	--	--	--	--	--	--	--	--	--	--	754	689	656	852	795
500	--	--	--	--	--	--	--	--	--	--	--	--	--	--	868	789	749	982	920
630	--	--	--	--	--	--	--	--	--	--	--	--	--	--	1,005	905	855	1,138	1,070

1) According to DIN VDE 0298-4 (VDE 0298-4): 2003-08 [11.39], IEC 60364-5-52: 2009-10 [11.40] and E DIN IEC 60364-5-52 (VDE 0100-520): 2004-07 [11.45]

Table C11.24 Rated values for the current-carrying capacity of PVC cables and lines with a permissible operating temperature on the conductor of 70 °C (conductor material aluminium)

Standardized nominal cross-sectional area A_n [mm²]	Reference method of installation																		
	In thermally insulated walls				In conduits				On a wall		In ground		In free air, i. e. unhindered heat dissipation						
	A1		A2		B1		B2		C		D		E		F			G	
	Number of loaded conductors																		
	2	3	2	3	2	3	2	3	2	3	2	3	2	3	2	3	3	3	3
	Ratings I_r for the current-carrying capacity [1)] in A																		
1.5	--	--	--	--	--	--	--	--	--	--	--	--	--	--	--	--	--	--	--
2.5	15	14	14.5	13.5	18.5	16.5	17.5	16.5	21	18.5	19	17	--	--	--	--	--	--	--
4	20	18.5	19.5	17.5	25	22	24	21	28	25	25	22	--	--	--	--	--	--	--
6	26	24	25	23	32	28	30	27	36	32	32	28	--	--	--	--	--	--	--
10	36	32	33	31	44	39	41	36	49	44	41	36	--	--	--	--	--	--	--
16	48	43	44	41	60	53	54	48	66	59	54	46	--	--	--	--	--	--	--
25	63	57	58	53	79	70	71	62	83	73	68	59	89	78	98	87	84	112	99
35	77	70	71	65	97	86	86	77	103	90	82	71	111	96	122	109	105	139	124
50	93	84	86	78	118	104	104	92	125	110	96	83	135	117	149	133	128	169	152
70	118	107	108	98	150	133	131	116	160	140	119	103	173	150	192	173	166	217	196
95	142	129	130	118	181	161	157	139	195	170	141	124	210	183	235	212	203	265	241
120	164	149	150	135	210	186	181	160	226	197	161	141	244	212	273	247	237	308	282
150	189	170	172	155	234	204	201	176	261	227	181	158	282	245	316	287	274	356	327
185	215	194	195	176	266	230	230	199	298	259	204	180	322	280	363	330	315	407	376
240	252	227	229	207	312	269	269	232	352	305	235	208	380	330	430	392	375	482	447
300	289	261	263	237	358	306	308	265	406	351	266	237	439	381	497	455	434	557	519
400	--	--	--	--	--	--	--	--	--	--	--	--	--	--	600	552	526	671	629
500	--	--	--	--	--	--	--	--	--	--	--	--	--	--	694	640	610	775	730
630	--	--	--	--	--	--	--	--	--	--	--	--	--	--	808	746	711	900	852

1) According to DIN VDE 0298-4 (VDE 0298-4): 2003-08 [11.39], IEC 60364-5-52: 2009-10 [11.40] and E DIN IEC 60364-5-52 (VDE 0100-520): 2004-07 [11.45]

Table C11.25 Rated values for the current-carrying capacity of XLPE cables and lines with a permissible operating temperature on the conductor of 90 °C (conductor material copper)

Standardized nominal cross-sectional area A_n [mm²]	In thermally insulated walls				In conduits				On a wall		In ground		In free air, i. e. unhindered heat dissipation						
	A1		A2		B1		B2		C		D		E		F			G	
	Number of loaded conductors																		
	2	3	2	3	2	3	2	3	2	3	2	3	2	3	2	3	3	3	3
	Ratings I_r for the current-carrying capacity 1) in A																		
1.5	19	17	18.5	16.5	23	20	22	19.5	24	22	22	19	26	23	--	--	--	--	--
2.5	26	23	25	22	31	28	30	26	33	30	29	25	36	32	--	--	--	--	--
4	35	31	33	30	42	37	40	35	45	40	37	32	49	42	--	--	--	--	--
6	45	40	42	38	54	48	51	44	58	52	47	39	63	54	--	--	--	--	--
10	61	54	57	51	75	66	69	60	80	71	62	53	86	75	--	--	--	--	--
16	81	73	76	68	100	88	91	80	107	96	81	69	115	100	--	--	--	--	--
25	106	95	99	89	133	117	119	105	138	119	103	89	149	127	161	141	135	182	161
35	131	117	121	109	164	144	146	128	171	147	124	107	185	158	200	176	169	226	201
50	158	141	145	130	198	175	175	154	209	179	147	126	225	192	242	216	207	275	246
70	200	179	183	164	253	222	221	194	269	229	181	156	289	246	310	279	268	353	318
95	241	216	220	197	306	269	265	233	328	278	214	187	352	298	377	342	328	430	389
120	278	249	253	227	354	312	305	268	382	322	244	213	410	346	437	400	383	500	454
150	318	285	290	259	393	342	334	300	441	371	275	240	473	399	504	464	444	577	527
185	362	324	329	295	449	384	384	340	506	424	309	271	542	456	575	533	510	661	605
240	424	380	386	346	528	450	459	398	599	500	356	312	641	538	679	634	607	781	719
300	486	435	442	396	603	514	532	455	693	576	403	349	741	621	783	736	703	902	833
400	--	--	--	--	--	--	--	--	--	--	--	--	--	--	940	868	823	1,085	1,008
500	--	--	--	--	--	--	--	--	--	--	--	--	--	--	1,083	998	946	1,253	1,169
630	--	--	--	--	--	--	--	--	--	--	--	--	--	--	1,254	1,151	1,088	1,454	1,362

1) According to DIN VDE 0298-4 (VDE 0298-4): 2003-08 [11.39], IEC 60364-5-52: 2009-10 [11.40] and E DIN IEC 60364-5-52 (VDE 0100-520): 2004-07 [11.45]

Table C11.26 Rated values for the current-carrying capacity of XLPE cables and lines with a permissible operating temperature on the conductor of 90 °C (conductor material aluminium)

Standardized nominal cross-sectional area A_n [mm²]	Reference method of installation																		
	In thermally insulated walls				In conduits				On a wall		In ground		In free air, i. e. unhindered heat dissipation						
	A1		A2		B1		B2		C		D		E		F			G	
	Number of loaded conductors																		
	2	3	2	3	2	3	2	3	2	3	2	3	2	3	2	3	3	3	3
	Ratings I_r for the current-carrying capacity [1] in A																		
1.5	--	--	--	--	--	--	--	--	--	--	--	--	--	--	--	--	--	--	--
2.5	20	19	19.5	18	25	22	23	21	26	24	23	20	--	--	--	--	--	--	--
4	27	25	26	24	33	29	31	28	35	32	29	26	--	--	--	--	--	--	--
6	35	32	33	31	43	38	40	35	45	41	37	32	--	--	--	--	--	--	--
10	48	44	45	41	59	52	54	48	62	57	48	42	--	--	--	--	--	--	--
16	64	58	60	55	79	71	72	64	84	76	62	54	--	--	--	--	--	--	--
25	84	76	78	71	105	93	94	84	101	90	79	69	108	97	121	107	103	138	122
35	103	94	96	87	130	116	115	103	126	112	95	83	135	120	150	135	129	72	153
50	125	113	115	104	157	140	138	124	154	136	112	97	164	146	184	165	159	210	188
70	158	142	145	131	200	179	175	156	198	174	139	120	211	187	237	215	206	271	244
95	191	171	175	157	242	217	210	188	241	211	164	143	257	227	289	264	253	332	300
120	220	197	201	180	281	251	242	216	280	245	187	164	300	263	337	308	296	387	351
150	253	226	230	206	307	267	261	240	324	283	212	184	346	304	389	358	343	448	408
185	288	256	262	233	351	300	300	272	371	323	237	210	397	347	447	413	395	515	470
240	338	300	307	273	412	351	358	318	439	382	274	244	470	409	530	492	471	611	561
300	387	344	352	313	471	402	415	364	508	440	309	278	543	471	613	571	547	708	652
400	--	--	--	--	--	--	--	--	--	--	--	--	--	--	740	694	663	856	792
500	--	--	--	--	--	--	--	--	--	--	--	--	--	--	856	806	770	991	921
630	--	--	--	--	--	--	--	--	--	--	--	--	--	--	996	942	899	1,154	1,077

1) According to DIN VDE 0298-4 (VDE 0298-4): 2003-08 [11.39], IEC 60364-5-52: 2009-10 [11.40] and E DIN IEC 60364-5-52 (VDE 0100-520): 2004-07 [11.45]

Table C11.27 Correction factors f_ϑ for differing air temperatures

Type of construction	Permissible conductor temperature	Air temperature ϑ_{air}										
		10 °C	15 °C	20 °C	25 °C	30 °C	35 °C	40 °C	45 °C	50 °C	55 °C	60 °C
		Correction factors f_ϑ										
PVC cable [1)]	70 °C	1.22	1.17	1.12	1.06	1.00	0.94	0.87	0.79	0.71	0.61	0.50
XLPE cable [2)]	90 °C	1.15	1.12	1.08	1.04	1.00	0.96	0.91	0.87	0.82	0.76	0.71

1) e. g. NYY, NYCY, NAYY, NAYCY
2) e. g. N2XY, N2X2Y, NI2XY

Table C11.28 Reduction factors f_H for the grouping of multi-core cables on trays and ladders (reference method of installation E)

Arrangement of cable systems		Number of trays or ladders	Number of multi-core cables					
			1	2	3	4	6	9
			Reduction factors f_H					
Non-perforated cable trays	Touching, ≥ 300 mm, ≥ 20 mm	1	0.97	0.84	0.78	0.75	0.71	0.68
		2	0.97	0.83	0.76	0.72	0.68	0.63
		3	0.97	0.82	0.75	0.71	0.66	0.61
		6	0.97	0.81	0.73	0.69	0.63	0.58
Perforated cable trays	Touching, ≥ 300 mm, ≥ 20 mm	1	1.00	0.88	0.82	0.79	0.76	0.73
		2	1.00	0.87	0.80	0.77	0.73	0.68
		3	1.00	0.86	0.79	0.76	0.71	0.66
		6	1.00	0.84	0.77	0.73	0.68	0.64
	Spaced, d, ≥ 20 mm	1	1.00	1.00	0.98	0.95	0.91	--
		2	1.00	0.99	0.96	0.92	0.87	--
		3	1.00	0.98	0.95	0.91	0.85	--
	Touching, ≥ 225 mm	1	1.00	0.88	0.82	0.78	0.73	0.72
		2	1.00	0.88	0.81	0.76	0.71	0.70
	Spaced, d, ≥ 225 mm	1	1.00	0.91	0.89	0.88	0.87	--
		2	1.00	0.91	0.88	0.87	0.85	--
Cable ladders	Touching, ≥ 300 mm, ≥ 20 mm	1	1.00	0.87	0.82	0.80	0.79	0.78
		2	1.00	0.86	0.81	0.78	0.76	0.73
		3	1.00	0.85	0.79	0.76	0.73	0.70
		6	1.00	0.83	0.76	0.73	0.69	0.66
	Spaced, d, ≥ 20 mm	1	1.00	1.00	1.00	1.00	1.00	--
		2	1.00	0.99	0.98	0.97	0.96	--
		3	1.00	0.98	0.97	0.96	0.93	--

Table C11.29 Reduction factors f_H for the grouping of single-core cables on trays and ladders (reference method of installation F)

Arrangement of cable systems		Number of trays or ladders	Number of three-phase circuits comprising single-core cables		
			1	2	3
			Reduction factors f_H		
Perforated cable trays	Touching; ≥300 mm; ≥20 mm	1	0.98	0.91	0.87
		2	0.96	0.87	0.81
		3	0.95	0.85	0.78
	Touching; ≥225 mm	1	0.96	0.86	--
		2	0.95	0.84	--
Cable ladders	Touching; ≥300 mm; ≥20 mm	1	1.00	0.97	0.96
		2	0.98	0.93	0.89
		3	0.97	0.90	0.86
Perforated cable trays	Spaced; d; ≥2d; ≥300 mm; ≥20 mm	1	1.00	0.98	0.96
		2	0.97	0.93	0.89
		3	0.96	0.92	0.86
	Spaced; ≥225 mm; ≥2d; d	1	1.00	0.91	0.89
		2	1.00	0.90	0.86
Cable ladders	Spaced; d; ≥2d; ≥300 mm; ≥20 mm	1	1.00	1.00	1.00
		2	0.97	0.95	0.93
		3	0.96	0.94	0.90

By installation of single-core cables in free air at a distance of at least 1 × diameter *d* from the wall and a spacing of *d* (reference method of installation G), cable grouping is automatically ruled out. For that reason, reduction factors f_H for cable grouping only have to be considered for reference method of installation E and F but not for reference method of installation G. The necessary rating factors for site operating conditions in designing distribution circuits and final circuits with cables and lines with reference method of installation A1, A2, B1, B2, C or D are provided in DIN VDE 0298-4 (VDE 0298-4): 2003-08 [11.39] or IEC 60364-5-52: 2009-10 [11.40].

The conductor cross-sectional area according to the permissible current-carrying capacity must be chosen such that the following inequality is fulfilled:

$$I_{perm} \geq I_{load} \tag{11.3}$$

I_{perm} permissible current-carrying capacity under site operating conditions
I_{load} maximum load current

The load current in inequality (11.3) corresponds to the current in the line conductors of the three-phase system under a symmetrical load. If the neutral conductor is subjected to a load although the line conductors are not relieved of the corresponding load, the neutral conductor must also be included in the calculation of the current-carrying capacity of the cable [11.39, 11.40]. High neutral conductor currents arise due to phase currents with high harmonic contents that do not cancel one another out in the neutral conductor. The largest current harmonic that is not cancelled out is usually that of the third-order (150-Hz current). A high third-order harmonic content in the load current mainly occurs in the three-phase circuits of energy-efficient lighting systems (Section 10.1.1.5).

Tables C11.30 and C11.31 can be used for dimensioning four-core and five-core cables according to the 150-Hz current content in the load current.

Table C11.30 Dimensioning of four-core and five-core cables and lines according to the 150-Hz current content (3rd-order harmonic current) in the load current

Harmonic load due to 150-Hz current I_3	Dominant-loaded conductor of the cable	Selection of the cross-sectional area according to the		Dimensioning conditions to be complied with
		phase current $I_{load\text{-}L}$	neutral conductor current $I_{load\text{-}N}$	
$I_3 \leq 0.15 \cdot I_{load}$	Line conductor	I_{load}	--	$I_r > I_{load\text{-}L}$ ($A_L \Rightarrow A_N$)
$0.15 \cdot I_{load} < I_3 \leq 0.33 \cdot I_{load}$		$I_{load} / 0.86$ 1)	--	
$0.33 \cdot I_{load} < I_3 \leq 0.45 \cdot I_{load}$	Neutral conductor	--	$\frac{1}{0.86}$ 1) $\cdot 3 \cdot I_{load} \cdot p'_3$	$I_r > I_{load\text{-}N}$ ($A_N \Rightarrow A_L$)
$I_3 > 0.45 \cdot I_{load}$ 2)		--	$3 \cdot I_{load} \cdot p'_3$ 2)	

I_3 3rd-order harmonic current (150-Hz current)
I_{load} Load current of the three-phase system
I_r Rated value of the load current-carrying capacity (Tables C11.23 to C11.26)
p'_3 Percentage proportion of the 3rd harmonic current in the load current ($p'_3 = p_3 / 100$)
A_L Cross-sectional area of the line conductor
A_N Cross-sectional area of the neutral conductor

1) Reduction factor for 4- and 5-core cables according to DIN VDE 0298-4 (VDE 0298-4): 2003-08 [11.39] or IEC 60364-5-52: 2009-10 [11.40]

2) In the case of 150-Hz current content of more than 45 %, the neutral conductor current $I_{load\text{-}N}$ is more than 135 % of the phase current $I_{load\text{-}L}$. Based on this neutral conductor current for the cable rating, the three line conductors are not fully loaded. The reduction in heat generated by the line conductors offsets the heat generated by the neutral conductor to the extent that it is not necessary to apply any reduction factor to the current-carrying capacity for three loaded conductors [11.39, 11.40].

Example C7

Dimensioning of a four-core PVC cable (copper conductor, reference method of installation E) for 150-Hz current contents of 10 %, 20 %, 40 % and 50 % for a postulated load current of $I_{load} = 110$ A (Table C11.31).

Table C11.31 Example of cable dimensioning according to the 150-Hz current content (3rd-order harmonic current) in the load current (Example C7)

Postulated harmonic load due to 150-Hz current I_3	Postulated load current of the 3-phase circuit I_{load}	Selection of the cross-sectional area according to the		Dimensioning conditions to be complied with	Necessary cross-sectional area of the PVC cable for reference installation method E (Table C11.23)
		phase current $I_{load\text{-}L}$	neutral conductor current $I_{load\text{-}N}$		
$I_3 = 0.10 \cdot I_{load}$	110 A	110 A	---	$I_r > 110\,A$	$A_L = A_N = 35\,mm^2$ ($I_r = 126\,A$)
$I_3 = 0.20 \cdot I_{load}$		$\frac{110\,A}{0.86} = 127.9\,A$	---	$I_r > 127.9\,A$	$A_L = A_N = 50\,mm^2$ ($I_r = 153\,A$)
$I_3 = 0.40 \cdot I_{load}$		---	$\frac{1}{0.86} \cdot 3 \cdot 110\,A \cdot 0.4 = 153.5\,A$	$I_r > 153.5\,A$	$A_N = A_L = 70\,mm^2$ ($I_r = 196\,A$)
$I_3 = 0.50 \cdot I_{load}$		---	$3 \cdot 110\,A \cdot 0.5 = 165\,A$	$I_r > 165\,A$	$A_N = A_L = 70\,mm^2$ ($I_r = 196\,A$)

All the rules stated above for determining the nominal cross-sectional area only take account of the current-carrying capacity of the cable. Further criteria determining the cross-sectional area are considered below.

11.3.2 Protection against overload

For protection against overload, protective devices must be provided to break any overload current flowing in the conductors before such a current could cause a temperature rise detrimental to insulation, joints, terminations and surroundings of the cables and busbars [11.42, 11.43]. When assigning protective devices for the protection of cables and lines against overload, the dimensioning and tripping rules must be followed.

- *Conditions of the dimensioning rule to be met:*
  a) non-adjustable protective devices (e.g. fuses, miniature circuit-breakers (MCBs))

$$I_{load} \leq I_n \leq I_{perm} \qquad (11.4.1)$$

  b) adjustable protective devices (e.g. circuit-breakers (ACBs, MCCBs))

$$I_{load} \leq I_R \leq I_{perm} \qquad (11.4.2)$$

- *Condition of the tripping rule to be met:*

$$I_2 \leq 1.45 \cdot I_{perm} \qquad (11.5)$$

The terms of the inequalities mean

I_{load} load current of the cable or circuit,

I_n nominal current of the non-adjustable protective device,

I_{perm} permissible current-carrying capacity of the cable,

I_R setting value of the protective device (inverse-time-delay overload release L),

I_2 current that causes tripping of the protection device under conditions defined in the device standards (conventional tripping or fusing current).

Fig. C11.32 illustrates standard-compliant coordination of the characteristic values for the overload protection of cables and lines.

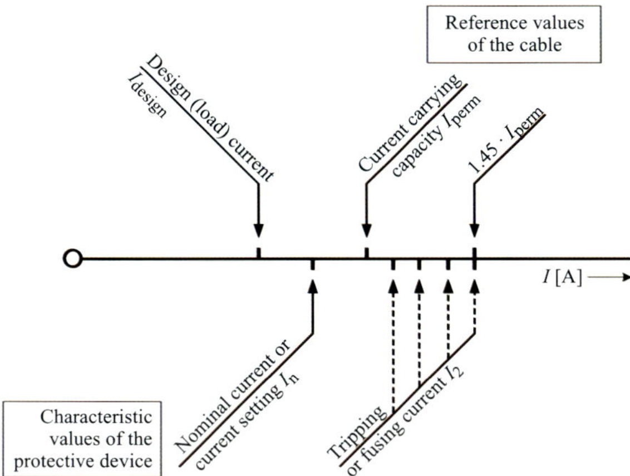

Fig. C11.32 Coordination of the characteristic values for the overload protection of cables and lines [11.42, 11.43]

The dimensioning rule for non-adjustable protective devices (Eq. 11.4.1) states that the nominal current I_n of the fuse or miniature circuit-breaker must be between the maximum load current I_{load} and the permissible current-carrying capacity I_{perm} of the cable. In the case of adjustable protective devices, the dimensioning rule is not applied with the nominal current I_n of the switching device but with the real setting value I_R of the overload release (circuit-breaker) or overload relay (motor-starter combination) (Eq. 11.4.2). The tripping rule (Eq. 11.5) includes the tripping tolerance of the protective device, which is always present, and a temporary overload of the cable increased by a factor of 1.45 in the protection dimensioning.

The current ensuring effective operation of the protective device is called the conventional tripping or fusing current I_2. It states from which multiple of the nominal current the protective device reliably operates within a defined time on an overload considering the tolerances. Where as the conventional tripping/fusing current I_2 defines the upper tolerance range for the operating current, the conventional non-tripping/non-fusing current I_1 defines the upper tolerance range for the non-operating current. A summary of the standardized conventional currents I_1 and I_2 for fuses of utilization category gG, miniature circuit-breakers (MCBs) and circuit-breakers (ACBs, MCCBs) is given in Table C11.33. For the maximum permissible current load of cables and lines, the tripping rule (Eq. 11.5) yields the following:

$$I_{perm} \geq \frac{I_2}{1.45} = k_u \cdot I_n \qquad (11.6)$$

k_u factor expressing the ratio of the tripping or fusing current to the nominal current of the protective device

Table C11.33 Conventional currents I_1 and I_2 for fuses of the utilization category gG, miniature circuit-breakers and circuit-breakers

Protective device				Conventional non-tripping/non-fusing current I_1	Conventional tripping/fusing current I_2	Conventional test duration
Fuses of utilization category gG	Nominal currents		$I_n \leq 4$ A	$1.5 \cdot I_n$	$2.1 \cdot I_n$	t = 1 h
			4 A < I_n < 16 A	$1.5 \cdot I_n$	$1.9 \cdot I_n$	t = 1 h
			16 A ≤ I_n ≤ 63 A	$1.25 \cdot I_n$	$1.6 \cdot I_n$	t = 1 h
			63 A < I_n ≤ 160 A	$1.25 \cdot I_n$	$1.6 \cdot I_n$	t = 2 h
			160 A < I_n ≤ 400 A	$1.25 \cdot I_n$	$1.6 \cdot I_n$	t = 3 h
			I_n > 400 A	$1.25 \cdot I_n$	$1.6 \cdot I_n$	t = 4 h
Miniature circuit-breaker (MCB)	Tripping characteristics	B	Nominal currents: I_n ≤ 63 A	$1.13 \cdot I_n$	$1.45 \cdot I_n$	t ≤ 1 h
			I_n > 63 A			t ≤ 2 h
		C	I_n ≤ 63 A	$1.13 \cdot I_n$	$1.45 \cdot I_n$	t ≤ 1 h
			I_n > 63 A			t ≤ 2 h
		D	I_n ≤ 63 A	$1.13 \cdot I_n$	$1.45 \cdot I_n$	t ≤ 1 h
			I_n > 63 A			t ≤ 2 h
Circuit-breaker (ACB, MCCB) with integrated overload release	Nominal currents		I_n ≤ 63 A	$1.05 \cdot I_R$	$1.30 \cdot I_R$ $1.20 \cdot I_R$ 1)	t = 1 h
			I_n > 63 A	$1.05 \cdot I_R$	$1.30 \cdot I_R$ $1.20 \cdot I_R$ 1)	t = 2 h

1) Valid for circuit-breakers for protecting motor circuits and motor-starter combinations with contactor and overload relay according to DIN EN 60947-4-1 (VDE 0660-102): 2006-04 [11.26] or IEC 60947-4-1: 2009-09 [11.27]

Table C11.34 k_u factors for fuses of the utilization category gG, miniature circuit-breakers and circuit-breakers

Protective device	Nominal current	k_u factor
Fuses of utilization category gG	I_n ≤ 4 A	1.31
	4 A < I_n < 16 A	1.21
	I_n ≥ 16 A	1.10
Miniature circuit-breaker (MCB)	I_n ≤ 63 A	1.0
	I_n > 63 A	
Circuit-breaker (ACB, MCCB) with integrated overload release	I_n ≤ 63 A	1.0
	I_n > 63 A	

The factor k_u can be determined from the conventional tripping/fusing current I_2 and the factor 1.45. The k_u factors to be used for non-adjustable and adjustable protective devices in the overload protection of cables and lines are stated in Table C11.34.

In the case of fuses that have passed the additional test of DIN EN 60269-1 (VDE 0636-1): 2010-03 [11.49] or IEC 60269-1: 2009-07 [11.50] stated in Section 8.4.3.5, the k_u factor is set to 1, that is, only the condition $I_{perm} \geq I_n$ has to be met. However, this procedure only applies to fuses of utilization category gG with nominal currents $I_n \geq 16$ A.

For miniature circuit-breakers according to DIN EN 60898-1 (VDE 0641-11): 2006-03 [11.51] or IEC 60898-1: 2003-07 [11.52], the conventional tripping current is $I_2 = 1.45 \cdot I_n$ (Table C11.33). This results in a k_u factor of 1.0.

For circuit-breakers according to DIN EN 60947-2 (VDE 0660-101): 2010-04 [11.53] or IEC 60947-2: 2009-05 [11.54], the conventional tripping current I_2 is less than $1.45 \cdot I_n$ (Table C11.33). It is $I_2 = 1.30 \cdot I_R$ (circuit-breakers for line protection) or $I_2 = 1.20 \cdot I_R$ (circuit-breakers for motor protection). With a sufficient rating reserve, a k_u factor of 1.0 can therefore be selected for circuit-breakers.

11.3.3 Protection against short circuit

Protection against short circuit must be ensured by the protective device disconnecting the short-circuit current before the permissible short-circuit temperature of the cable is reached. To ensure short-circuit protection, the following condition must be met:

$$k^2 \cdot A_n^2 \geq I_k^2 \cdot t_k \tag{11.7}$$

A_n	nominal conductor cross-sectional area in mm²
I_k	effective short-circuit current (RMS value) in A
t_k	short-circuit duration in sec
k	material-related value of the cable type in $A \cdot \sqrt{sec}/mm^2$

The k value is a factor by which the resistance, temperature coefficient and thermal capacity of the conductor material and the corresponding initial and final temperatures can be taken adequately into account. Table C11.35 lists the k values of common types of conductor insulation. The k value can also be replaced by the rated short-circuit current density J_{thr}. The rated short-time current densities J_{thr} for cables and lines with copper and aluminium conductors are stated in Table B4.7 or DIN VDE 0298-4 (VDE 0298-4): 2003-08 [11.39]/IEC 60364-5-52: 2009-10 [11.40].

Table C11.35 k values for live conductors [11.42, 11.43]

Conductor		Type of conductor insulation				
		PVC		Heat-resistant PVC		EPR / XLPE
		≤ 300 mm²	> 300 mm²	≤ 300 mm²	> 300 mm²	
Initial temperature ϑ_a		70 °C	70 °C	90 °C	90 °C	90 °C
Final temperature ϑ_e		160 °C	140 °C	160 °C	140 °C	250 °C
k value for the conductor material in $A \cdot \sqrt{sec} / mm^2$	Cu	115	103	100	86	143
	Al	76	68	66	57	94

EPR Ethylene-propylene monomer (EPM) or ethylene-propylene-diene monomer (EPDM)
PVC Polyvinyl chloride
XLPE Crosslinked polyethylene

The short-circuit withstand capability condition (11.7) applies to short-circuit durations in the range 0.1 sec $\leq t_k \leq$ 5 sec. For short circuits in this time range, the permissible disconnecting time $t_{a\text{-perm}}$ of the protective device must be calculated by the following formula [11.42, 11.43]:

$$t_{a\text{-perm}} \leq \left(k \cdot \frac{A_n}{I_k}\right)^2 \tag{11.7.1}$$

The clearing time t_a that can be achieved by the protective device must not exceed the permissible time according to Eq. (11.7.1). For circuit-breakers with an integrated overcurrent release system (LSI release system), the following must apply:

$$t_{a\text{-perm}} \geq \begin{cases} t_{sd} + \Delta T_0 \text{ for } I_{k\text{-min}} & \text{at the end of the circuit} \\ t_i + \Delta T_0 \text{ for } I_{k\text{-max}} & \text{at the beginning of the circuit} \end{cases} \tag{11.7.2}$$

t_{sd} tripping time of the short-time-delay overcurrent release (S release)
t_i tripping time of the instantaneous overcurrent release (I release)
ΔT_0 upper tolerance range of the tripping time

For very short clearing times ($t_a < 0.1$ sec), the DC component of the short-circuit current is no longer negligible. This especially applies to the use of current-limiting protective devices. In such cases, the permissible Joule heat value of the cable ($k^2 \cdot A_n^2$) must be greater than or equal to the let-through energy value of the protective device ($I^2 t_a$) during a short circuit (Eq. 11.8).

$$k^2 \cdot A_n^2 \geq I^2\, t_a \tag{11.8}$$

It is usually not sufficient to check the short-circuit withstand capability according to Eq. (11.7) or (11.8) only for the maximum short-circuit current of the circuit. It must be ensured that the cable is also short-circuit-proof in a short circuit at any point in the circuit. The quantity that determines the thermal short-circuit rating of a cable is its Joule heat value $k^2 \cdot A_n^2$. Converting the Joule heat value $k^2 \cdot A_n^2$ into equivalent $I^2 t$ values yields the "cable damage curve". In double logarithmic representation, the "cable damage curve" is a falling straight line. To ensure that the cable is short-circuit-proof in the event of all fault currents occurring in the circuit, this straight line must be above the tripping curve of the overcurrent release system (LSI release system).

Example C8

Comparison of the LSI tripping curve of a 630-A 3WL circuit-breaker with the "damage curve" of a PVC cable 3 × 1 × 150/70 mm² (Fig. C11.36).

With very large short-circuit currents at the beginning of a circuit, the short-circuit withstand capability of a cable can be exceeded despite the presence of an I release. As the let-through energy diagram in Fig. C11.37 shows, this would be the case for the PVC cable 3 × 1 × 150/70 mm² Cu under consideration if the short-circuit current in the cross-section-reduced conductor were to exceed the value $I_k = 36$ kA and, in the conductor with the full cross-sectional area, the value $I_k = 77$ kA.

For modern building installations, the use of cables with cross-section-reduced PEN or N conductors is obsolete due to possible harmonic loads (150-Hz currents). Today, only PE conductors are considered for conductors with a reduced nominal cross-sectional area A_n.

The short-circuit protection of the PE conductor is based exclusively on the maximum and minimum line-to-earth short-circuit current. For the short-circuit protection of the line conductors, on the other hand, both the maximum and the minimum short-circuit currents can be decisive in case of line-to-earth and three-phase faults (see Tables C9.1, C9.2 and C9.4).

Unlike fuseless functional units (circuit-breakers with integrated overcurrent release systems, fuseless motor-starter combinations), the relation between the let-through

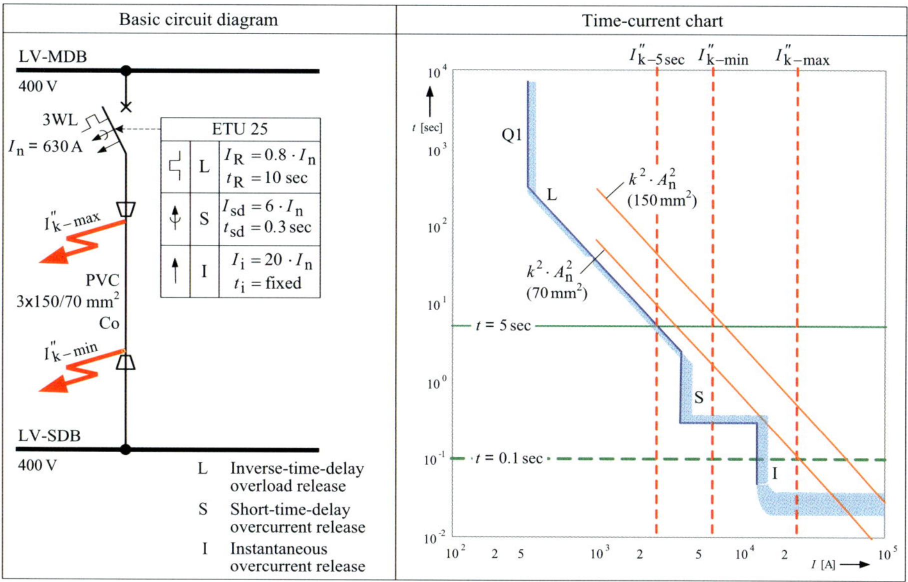

Fig. C11.36 Example comparison of the LSI tripping curve of a circuit-breaker with the damage curve of a PVC cable (Example C8)

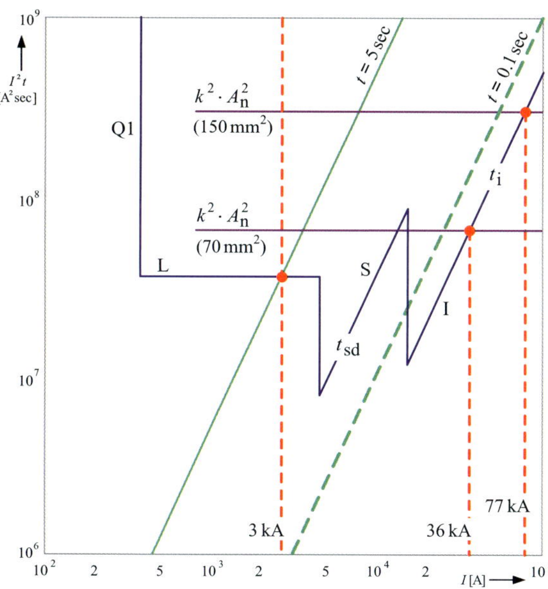

Fig. C11.37 Let-through energy diagram of the 630-A 3WL circuit-breaker in Example C8

energy of fuses and the fault current is antiproportional. Owing to the tripping characteristic of fuses, the let-through energy decreases as the fault current increases until a constant I^2t value is reached.

For short-circuit protection with fuses, therefore, the minimum short-circuit current at the end of the cable feeder is decisive.

For short-circuit proof dimensioning of fused circuits, Table C11.38 provides let-through energy values of Siemens 3NA LV HRC fuses of utilization category gG for a disconnecting time of $t_a = 5$ sec and the case that the fault disconnection is performed using the full short-circuit breaking capacity.

A comparison of the let-through energy values $I^2 t_a$ in Table C11.38 with the Joule heat values $k^2 \cdot A_n^2$ of cables that were selected according to the dimensioning rules for overload protection with fuses of utilization category gG shows that cables dimensioned for the requirements of overload protection automatically meet the conditions for sufficient short-circuit protection. The Joule heat values $k^2 \cdot A_n^2$ for the most common LV cables and lines are stated in Table C11.39.

Instead of the Joule heat value $k^2 \cdot A_n^2$ the rated short-circuit withstand current I_{cw} and the rated short time t_{cw} are used as the short-circuit strength quantity for busbar trunking systems.

Nominal current of the fuse I_n in A	$I^2 t_a$ let-through energy values in A^2sec for	
	$U_{nN} = 400$ V AC	$t_a = 5$ sec
6	$0.11 \cdot 10^3$	$2.72 \cdot 10^3$
10	$0.27 \cdot 10^3$	$7.81 \cdot 10^3$
16	$0.75 \cdot 10^3$	$1.61 \cdot 10^4$
20	$1.37 \cdot 10^3$	$2.52 \cdot 10^4$
25	$2.34 \cdot 10^3$	$4.04 \cdot 10^4$
32	$4.55 \cdot 10^3$	$6.08 \cdot 10^4$
35	$6.75 \cdot 10^3$	$8.72 \cdot 10^4$
40	$8.70 \cdot 10^3$	$1.20 \cdot 10^5$
50	$1.16 \cdot 10^4$	$2.05 \cdot 10^5$
63	$1.90 \cdot 10^4$	$3.56 \cdot 10^5$
80	$3.07 \cdot 10^4$	$6.45 \cdot 10^5$
100	$5.62 \cdot 10^4$	$1.16 \cdot 10^6$
125	$9.13 \cdot 10^4$	$1.76 \cdot 10^6$
160	$1.58 \cdot 10^5$	$2.95 \cdot 10^6$
200	$2.85 \cdot 10^5$	$6.10 \cdot 10^6$
224	$3.92 \cdot 10^5$	$8.10 \cdot 10^6$
250	$5.51 \cdot 10^5$	$1.06 \cdot 10^7$
300	$9.01 \cdot 10^5$	$1.84 \cdot 10^7$
315	$9.01 \cdot 10^5$	$1.84 \cdot 10^7$
355	$1.06 \cdot 10^6$	$2.33 \cdot 10^7$
400	$1.52 \cdot 10^6$	$3.25 \cdot 10^7$
425	$1.86 \cdot 10^6$	$4.71 \cdot 10^7$
500	$2.26 \cdot 10^6$	$6.56 \cdot 10^7$
630	$4.34 \cdot 10^6$	$1.09 \cdot 10^8$
800	$8.51 \cdot 10^6$	$1.93 \cdot 10^8$
1,000	$1.62 \cdot 10^7$	$3.60 \cdot 10^8$
1,250	$2.91 \cdot 10^7$	$6.25 \cdot 10^8$

Table C11.38
Let-through I^2t values of Siemens 3NA LV HRC fuses of utilization category gG [11.55]

Table C11.39 Joule heat values $k^2 \cdot A_n^2$ of common LV cables and lines

Nominal cross-sectional area of the conductor A_n [mm²]	$k^2 \cdot A_n^2$ Joule heat values [A^2 sec]					
	Copper conductor			Aluminium conductor		
	PVC insulation		EPR / XLPE insulation	PVC insulation		EPR / XLPE insulation
	≤ 300 mm²	> 300 mm²		≤ 300 mm²	> 300 mm²	
1.5	$2.98 \cdot 10^4$	-	$4.60 \cdot 10^4$	-	-	-
2.5	$8.26 \cdot 10^4$	-	$1.28 \cdot 10^5$	$3.61 \cdot 10^4$	-	$5.52 \cdot 10^4$
4	$2.11 \cdot 10^5$	-	$3.27 \cdot 10^5$	$9.24 \cdot 10^4$	-	$1.41 \cdot 10^5$
6	$4.76 \cdot 10^5$	-	$7.36 \cdot 10^5$	$2.08 \cdot 10^5$	-	$3.18 \cdot 10^5$
10	$1.32 \cdot 10^6$	-	$2.04 \cdot 10^6$	$5.77 \cdot 10^5$	-	$8.83 \cdot 10^5$
16	$3.38 \cdot 10^6$	-	$5.23 \cdot 10^6$	$1.48 \cdot 10^6$	-	$2.26 \cdot 10^6$
25	$8.26 \cdot 10^6$	-	$1.28 \cdot 10^7$	$3.61 \cdot 10^6$	-	$5.52 \cdot 10^6$
35	$1.62 \cdot 10^7$	-	$2.50 \cdot 10^7$	$7.07 \cdot 10^6$	-	$1.08 \cdot 10^7$
50	$3.30 \cdot 10^7$	-	$5.11 \cdot 10^7$	$1.44 \cdot 10^7$	-	$2.21 \cdot 10^7$
70	$6.48 \cdot 10^7$	-	$1.00 \cdot 10^8$	$2.83 \cdot 10^7$	-	$4.33 \cdot 10^7$
95	$1.19 \cdot 10^8$	-	$1.84 \cdot 10^8$	$5.21 \cdot 10^7$	-	$7.97 \cdot 10^7$
120	$1.90 \cdot 10^8$	-	$2.94 \cdot 10^8$	$8.32 \cdot 10^7$	-	$1.27 \cdot 10^8$
150	$2.98 \cdot 10^8$	-	$4.60 \cdot 10^8$	$1.30 \cdot 10^8$	-	$1.99 \cdot 10^8$
185	$4.52 \cdot 10^8$	-	$7.00 \cdot 10^8$	$1.98 \cdot 10^8$	-	$3.02 \cdot 10^8$
240	$7.62 \cdot 10^8$	-	$1.18 \cdot 10^9$	$3.33 \cdot 10^8$	-	$5.09 \cdot 10^8$
300	$1.19 \cdot 10^9$	-	$1.84 \cdot 10^9$	$5.20 \cdot 10^8$	-	$7.95 \cdot 10^8$
400	-	$1.70 \cdot 10^9$	$3.27 \cdot 10^9$	-	$7.40 \cdot 10^8$	$1.41 \cdot 10^9$
500	-	$2.65 \cdot 10^9$	$5.11 \cdot 10^9$	-	$1.15 \cdot 10^9$	$2.21 \cdot 10^9$

By equating $k^2 \cdot A_n^2 = I_{cw}^2 \cdot t_{cw}$, it is possible to derive the calculation formula from Eqs. (11.7) and (11.8) for determining the necessary rated short-time withstand current of the busbar trunking system. The following applies:

$$I_{cw} \geq \begin{cases} I_k \cdot \sqrt{\dfrac{t_k}{t_{cw}}} & \text{for } 0.1\ \text{sec} \leq t_k \leq 5\ \text{sec} \\ \sqrt{\dfrac{I^2 t_a}{t_{cw}}} & \text{for } t_a < 0.1\ \text{sec} \end{cases} \quad (11.9)$$

I_{cw} required rated short-circuit withstand current of the busbar trunking system
I_k prospective short-circuit current (RMS value) in case of a solid short circuit
$I^2 t_a$ let-through or breaking energy of the protection device when the prospective short-circuit current I_k is flowing
t_k short-circuit duration
t_{cw} rated short time (t_{cw} = 1 sec for the SIVACON 8PS busbar trunking system)

The busbar trunking systems have to be not only thermally but also mechanically short-circuit-proof. The mechanical short-circuit rating must be verified as follows:

$$I_{pk} \geq \begin{cases} i_p = \sqrt{2} \cdot \kappa \cdot I_k'' & \text{using protective devices without a current-limiting effect} \\ I_D = f(\text{device}) & \text{using protective devices with a current-limiting effect} \end{cases} \tag{11.10}$$

I_{pk} required rated peak withstand current
i_p peak short-circuit current
κ asymmetrical current peak factor
I_k'' initial symmetrical short-circuit current
I_D let-through current of the protective device (e.g. current-limiting MCCB or fuse)

Data about the thermal and mechanical short-circuit rating of the switchboards and busbar trunking systems from the Siemens type program are provided in Tables C11.9, C11.14, C11.15 and C11.17. With the standardized, graduated short-circuit strength quantities I_{cw} and I_{pk} from these tables, it is possible to optimize busbar trunking systems for the short-circuit current conditions prevailing in the industrial LV network.

11.3.4 Protection against electric shock

The preferred type of system earthing for designing industrial LV networks is the TN system (Section 10.2.1.3). In LV networks that are implemented as a TN system, protection against electric shock is effected by automatic disconnection. According to DIN VDE 0100-410 (VDE 0100-410): 2007-06 [11.20] or IEC 60364-4-41: 2005-12 [11.21], cables and protective devices of TN distribution circuits must be coordinated in such a way that on a fault with a connection to earth (line to PE or PEN conductor, line to exposed conductive parts of stationary equipment connected to the PE or PEN conductor), the power supply is disconnected in $t_a \leq 5$ sec. The line-to-earth short-circuit current must reach the following value:

$$I_{k1\text{-min}} \geq I_a \tag{11.11}$$

I_a operating current of the protective device (Tables C10.30 and C10.31 for miniature circuit-breakers and LV HRC fuses of utilization category gG; for circuit-breakers, $I_a = 1.2 \cdot I_{sd}$ (short-time-delay overcurrent release) or $I_a = 1.2 \cdot I_i$ (instantaneous overcurrent release) can be selected [11.56])
$I_{k1\text{-min}}$ minimum line-to-earth short-circuit current at the end of the feeder cable

For calculation of the minimum line-to-earth short-circuit current $I_{k1\text{-min}}$, the following conditions apply:

- use of the voltage factor c, that according to DIN EN 60909-0 (VDE 0102): 2002-07 [11.57] or IEC 60909-0: 2001-07 [11.58] must be applied to calculate the smallest short-circuit current,
- choice of the system supply or the circuit state that results in the smallest values of the short-circuit current at the fault location (e.g. system supply with a diesel generator set as the emergency power supply for the loads),
- non-consideration of short-circuit currents contributed by motors,
- use of resistance values for cables and busbar trunking systems that arise at the conductor temperature at the end of the short-circuit duration.

Taking these conditions into account, the minimum line-to-earth short-circuit current $I''_{k1\text{-min}}$ can be calculated by means of the short-circuit impedance at the fault location (positive-sequence short-circuit impedance $\underline{Z}_1$ and zero-sequence short-circuit impedance $\underline{Z}_0$) or the impedance of the entire fault loop (loop impedance $\underline{Z}_S$). Alternatively, the following calculations are possible:

- $$I''_{k1\text{-min}} = \frac{c_{min} \cdot \sqrt{3} \cdot U_{nN}}{\sqrt{(2 \cdot R_1 + R_0)^2 + (2 \cdot X_1 + X_0)^2}} \quad (11.12)$$

$$R_1 = R_{1\text{-N}} + R_{1\text{-T}} + R_{1\text{-L}} \quad (11.12.1)$$

$$X_1 = X_{1\text{-N}} + X_{1\text{-T}} + X_{1\text{-L}} \quad (11.12.2)$$

$$R_0 = R_{0\text{-T}} + R_{0\text{-L}} \quad (11.12.3)$$

$$X_0 = X_{0\text{-T}} + X_{0\text{-L}} \quad (11.12.4)$$

- $$I''_{k1\text{-min}} = \frac{c_{min} \cdot U_{nN}}{\sqrt{3} \cdot \sqrt{R_s^2 + X_s^2}} \quad (11.13)$$

$$R_s = \frac{2 \cdot R_1 + R_0}{3} \quad (11.13.1)$$

$$X_s = \frac{2 \cdot X_1 + X_0}{3} \quad (11.13.2)$$

c_{min}	minimum voltage factor ($c_{min} = 0.95$ at $U_{nN} \leq 1{,}000$ V)
R_1, X_1	positive-sequence short-circuit resistance or reactance up to the fault location
R_0, X_0	zero-sequence short-circuit resistance or reactance up to the fault location (Using MV/LV transformers of vector group Dyn for the incoming supply the zero system of the MV network is decoupled from the LV network. Therefore only the positive-sequence system of the upstream network is included in calculation of the line-to-earth short-circuit current.)
$R_{1\text{-N}}, X_{1\text{-N}}$	short-circuit resistance or reactance of the upstream MV network in the positive-sequence system
$R_{1\text{-T}}, X_{1\text{-T}}$	short-circuit resistance or reactance of the supplying MV/LV transformer(s) in the positive-sequence system
$R_{1\text{-L}}, X_{1\text{-L}}$	short-circuit resistance or reactance of the cable, line or busbar trunking system in the positive-sequence system
$R_{0\text{-T}}, X_{0\text{-T}}$	short-circuit resistance or reactance of the supplying MV/LV transformer(s) in the zero-sequence system
$R_{0\text{-L}}, X_{0\text{-L}}$	short-circuit resistance or reactance of the cable, line or busbar trunking system in the zero-sequence system
R_S	resistance of the entire fault loop
X_S	reactance of the entire fault loop

The short-circuit impedances of the equipment (upstream MV network, MV/LV transformers, cables/busbars) must be calculated using the specified formulas in DIN EN 60909-0 (VDE 0102): 2002-07 [11.57] or IEC 60909-0: 2001-07 [11.58]. Relevant equipment parameters for calculation can be taken from Tables C11.8a/b (GEAFOL cast-resin transformers) and C11.42 to C11.47 (LV cables).

To provide a way of checking that the disconnection condition (11.11) has been met, the calculation of the minimum line-to-earth short-circuit current $I''_{k1\text{-min}}$ can be simplified. This is done by summating the loop impedances of the serially connected equipment arithmetically rather than geometrically. By way of simplification, it may be stated:

$$I''_{k1\text{-min}} \approx \frac{c_{min} \cdot U_{nN}}{\sqrt{3} \cdot \sqrt{(Z_{s\text{-}N} + Z_{s\text{-}L})}} \qquad (11.14)$$

$Z_{s\text{-}N}$ loop impedance of the upstream network (Tables C11.40 and C11.41)

$Z_{s\text{-}L}$ loop impedance of the cable (loop impedances per unit length for single-core and multi-core cables, see Tables C11.42 to C11.47)

To ensure that the simplified short-circuit calculations according to Eq. (11.14) are with safe values, the following conventions are used:

- The system supply with Dyn transformers is based on a minimum short-circuit power of the upstream network of $S''_{k\text{-min}} = 100$ MVA. Transformers with an impedance voltage at rated current of $u_{rZ} = 6$ % are used.
- The system supply with LV generators is based on a steady-state line-to-earth short-circuit current of $I_{k1} = 5 \cdot I_{rG}$. If generators with a steady-state short-circuit current of $I_{k1} = 3 \cdot I_{rG}$ are used, the Z_s values in Table C11.41 must be multiplied by a factor of 5/3 = 1.67.
- The loop impedance values $Z_{s\text{-}N}$ of the upstream network apply to distances of $l \leq 20$ m between the supplying transformer and generator and the low-voltage main distribution board.
- The loop impedance values $Z_{s\text{-}L}$ of the single-core and multi-core cables are based on a conductor temperature of $\vartheta = 80$ °C.

Table C11.40 Loop impedance of the upstream network with respective transformer supply variants

Rated individual power of the transformers T1 – T3 (u_{rZ} = 6%) S_{rT} [kVA]	Variants of supply		
	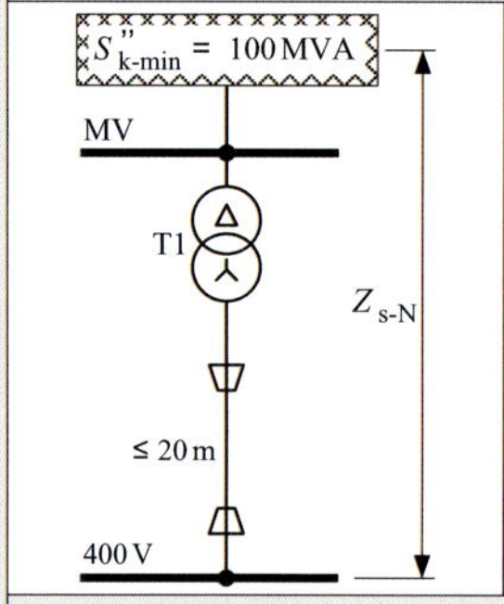 $S''_{k\text{-min}}$ = 100 MVA; MV; T1; ≤ 20 m; 400 V; $Z_{s\text{-}N}$	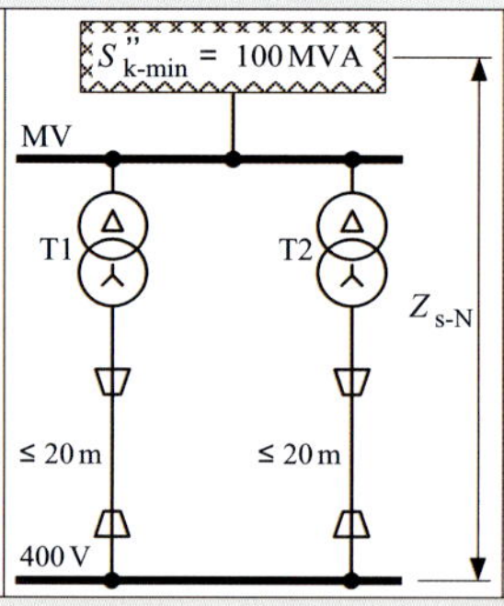 $S''_{k\text{-min}}$ = 100 MVA; MV; T1, T2; ≤ 20 m, ≤ 20 m; 400 V; $Z_{s\text{-}N}$	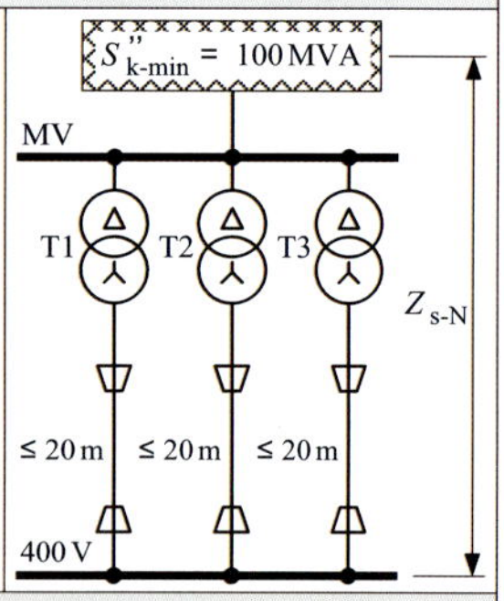 $S''_{k\text{-min}}$ = 100 MVA; MV; T1, T2, T3; ≤ 20 m, ≤ 20 m, ≤ 20 m; 400 V; $Z_{s\text{-}N}$
	Maximum loop impedance $Z_{s\text{-}N}$ [mΩ]		
630	17.902	8.951	5.967
800	14.517	7.259	4.839
1,000	11.726	5.863	3.909
1,250	9.579	4.790	3.193
1,600	7.760	3.880	2.587
2,000	6.379	3.190	--
2,500	5.314	--	--

Table C11.41 Loop impedance of the upstream network with supply from one or more generators

Variant of supply	Maximum loop impedance $Z_{s\text{-}N}$ if			
	S_{rG} < 1,000 kVA		S_{rG} > 1,000 kVA	
	S_{rG} [kVA]	Z_s [1)] [mΩ]	S_{rG} [kVA]	Z_s [1)] [mΩ]
S_{rG}; G 3~; $I_{k1} = 5 \cdot I_{rG}$; $Z_{s\text{-}N}$; ≤ 20 m; 400 V	250	125.367	1,040	30.186
	280	110.917	1,250	25.085
	315	98.604	1,350	23.285
	365	85.433	1,450	21.602
	410	76.284	1,600	19.638
	455	68.797	1,860	16.892
	500	62.720	2,050	15.307
	550	57.193	2,160	14.553
	600	52.100	2,250	13.936
	660	47.385	2,500	12.542
	725	43.205	--	--
	775	40.501	--	--
	880	35.825	--	--
	890	35.154	--	--

1) For $n \geq 2$ parallel supplying generators, the Z_s values must be divided by n

The maximum permissible loop impedance of a circuit can also be used for fast and reliable verification of the disconnection condition (11.11). The following applies for this loop impedance:

$$Z_{s\text{-}L} \leq \frac{c_{min} \cdot U_{nN}}{\sqrt{3} \cdot I_a} - Z_{s\text{-}N} \tag{11.15}$$

The geometric data of XLPE or EPR-insulated multi-core cables with enhanced behaviour in case of fire (e.g. NHXHX) that are decisive for their electrical characteristics differ only slightly from PVC-insulated multi-core cables (e.g. NYY) [11.60]. If no data can be obtained from the manufacturer for these cables, the impedances and loop impedances per unit length specified in Tables C11.42 and C11.44 provide an initial approximation.

11.3.5 Permissible voltage drop

The present codes and standards do not stipulate mandatory rules for the permissible voltage drop. Only recommendations are made. DIN IEC 60038 (VDE 0175): 2002-11 [11.46] or IEC 60038: 2009-06 [11.47] recommends that the voltage at the point of coupling between the distribution network and the consumer's installation should not deviate from the nominal voltage by more than ±10 %. Within the consumer's installation, the voltage drop according to DIN VDE 0100-520 (VDE 0100-520): 2003-06 [11.44] or IEC 60364-5-52: 2009-10 [11.40] should not be larger than 4 % of the nominal voltage of the power system.

In the case of the 110-kV/MV transfer transformers with automatic voltage regulation commonly used in the incoming supply of large industrial plants, a supply voltage corresponding to the full value of the nominal system voltage can be assumed at the nodes of the MV distribution network. Accordingly, the voltage drops between the MV-side ter-

Table C11.42 Impedances and loop impedances per unit length of multi-conductor cables with PVC insulation (conductor material copper, type NYY)

Conductor cross-sectional area	Multi-core cable NYY (Cu conductor, PVC insulation)						
	Impedances per unit length 1)				Loop impedances per unit length 2)		
A_n [mm²]	R'_{20} [mΩ/m]	X' [mΩ/m]	R'_0/R'_1 [1]	X'_0/X'_1 [1]	R'_s [mΩ/m]	X'_s [mΩ/m]	Z'_s [mΩ/m]
3x25/16	0.7270	0.0860	5.75	4.79	2.321	0.195	2.33
3x35/16	0.5240	0.0820	7.58	5.21	2.068	0.197	2.08
3x50/25	0.3870	0.0830	6.64	4.77	1.377	0.187	1.39
3x70/35	0.2680	0.0800	6.87	5.14	0.979	0.190	1.00
3x95/50	0.1930	0.0800	7.02	5.08	0.717	0.189	0.74
3x120/70	0.1530	0.0790	6.26	4.66	0.521	0.175	0.55
3x150/70	0.1240	0.0780	7.48	4.94	0.484	0.180	0.52
3x185/95	0.0990	0.0780	6.84	4.81	0.361	0.177	0.40
3x240/120	0.0750	0.0770	7.09	4.85	0.282	0.176	0.33
3x300/150	0.0600	0.0770	7.19	4.87	0.228	0.176	0.29
4x10	1.8300	0.0951	4.00	4.00	4.523	0.190	4.53
4x16	1.1500	0.0894	4.00	4.00	2.842	0.179	2.85
4x25	0.7270	0.0878	4.00	4.00	1.797	0.176	1.81
4x35	0.5240	0.0850	4.00	4.00	1.295	0.170	1.31
4x50	0.3870	0.0846	4.00	4.00	0.957	0.169	0.97
4x70	0.2680	0.0824	4.00	4.00	0.662	0.165	0.68
4x95	0.1930	0.0820	4.00	4.00	0.477	0.164	0.50
4x120	0.1530	0.0805	4.00	4.00	0.378	0.161	0.41
4x150	0.1240	0.0805	4.00	4.00	0.306	0.161	0.35
4x185	0.0991	0.0803	4.00	4.00	0.245	0.161	0.29
4x240	0.0754	0.0799	4.00	4.00	0.186	0.160	0.25
4 x 300	0.0601	0.0798	4.00	4.00	0.149	0.160	0.22

1) According to DIN EN 60909-0 Bbl. 4 (VDE 0102 Bbl. 4): 2009-08 [11.59]
2) Applies to a conductor temperature of $\vartheta = 80\,°C$

Table C11.43 Impedances and loop impedances per unit length of multi-conductor cables with PVC insulation (conductor material aluminium, type NAYY)

Conductor cross-sectional area	Multi-core cable NAYY (Al conductor, PVC insulation)						
	Impedances per unit length 1)				Loop impedances per unit length 2)		
A_n [mm²]	R'_{20} [mΩ/m]	X' [mΩ/m]	R'_0/R'_1 [1]	X'_0/X'_1 [1]	R'_s [mΩ/m]	X'_s [mΩ/m]	Z'_s [mΩ/m]
4x35	0.8680	0.0859	4.00	4.00	2.156	0.172	2.16
4x50	0.6410	0.0847	4.00	4.00	1.592	0.169	1.60
4x70	0.4430	0.0822	4.00	4.00	1.100	0.164	1.11
4x95	0.3200	0.0820	4.00	4.00	0.795	0.164	0.81
4x120	0.2530	0.0804	4.00	4.00	0.628	0.161	0.65
4x150	0.2060	0.0802	4.00	4.00	0.512	0.160	0.54
4x185	0.1640	0.0805	4.00	4.00	0.407	0.161	0.44

1) According to DIN EN 60909-0 Bbl. 4 (VDE 0102 Bbl. 4): 2009-08 [11.59]
2) Applies to a conductor temperature of $\vartheta = 80\,°C$

Table C11.44 Impedances and loop impedances per unit length of multi-conductor cables with PVC insulation and concentric outer conductor (conductor material copper, type NYCWY)

Conductor cross-sectional area	Multi-core cable NYCWY (Cu conductor, PVC insulation, concentric outer conductor)						
	Impedances per unit length 1)				Loop impedances per unit length 2)		
A_n [mm^2]	R'_{20} [mΩ/m]	X' [mΩ/m]	R'_0/R'_1 [1]	X'_0/X'_1 [1]	R'_s [mΩ/m]	X'_s [mΩ/m]	Z'_s [mΩ/m]
3x25/16	0.7270	0.0807	5.62	1.56	2.282	0.096	2.28
3x35/16	0.5240	0.0780	7.41	1.18	2.031	0.083	2.03
3x35/35	0.5240	0.0778	3.92	1.21	1.278	0.083	1.28
3x50/25	0.3870	0.0778	6.53	1.14	1.360	0.081	1.36
3x50/50	0.3870	0.0778	3.77	1.16	0.919	0.082	0.92
3x70/35	0.2680	0.0747	6.71	1.21	0.962	0.080	0.96
3x70/70	0.2680	0.0747	3.85	1.24	0.646	0.081	0.65
3x95/50	0.1930	0.0747	6.55	1.18	0.680	0.079	0.68
3x95/95	0.1930	0.0747	3.92	1.21	0.471	0.080	0.48
3x120/70	0.1530	0.0731	6.00	1.07	0.504	0.075	0.51
3x120/120	0.1530	0.0731	3.92	1.10	0.373	0.076	0.38
3x150/70	1.1240	0.0734	7.17	1.10	0.468	0.076	0.47
3x150/150	1.1240	0.0734	3.88	1.13	0.300	0.076	0.31
3x185/95	0.0991	0.0733	6.69	1.10	0.355	0.076	0.36
3x185/185	0.0991	0.0733	3.88	1.14	0.240	0.077	0.25
3x240/120	0.0754	0.0725	6.93	1.09	0.277	0.075	0.29
3x300/150	0.0601	0.0725	6.94	1.06	0.221	0.074	0.23

1) According to DIN EN 60909-0 Bbl. 4 (VDE 0102 Bbl. 4): 2009-08 [11.59]

2) Applies to a conductor temperature of ϑ = 80 °C

Table C11.45 Impedances and loop impedances per unit length of multi-core cables with PVC insulation and concentric outer conductor (conductor material aluminium, type NAYCWY)

Conductor cross-sectional area	Multi-core cable NAYCWY (Al conductor, PVC insulation, concentric outer conductor)						
	Impedances per unit length 1)				Loop impedances per unit length 2)		
A_n [mm^2]	R'_{20} [mΩ/m]	X' [mΩ/m]	R'_0/R'_1 [1]	X'_0/X'_1 [1]	R'_s [mΩ/m]	X'_s [mΩ/m]	Z'_s [mΩ/m]
3x50/25	0.6410	0.0775	4.34	1.31	1.683	0.085	1.68
3x50/50	0.6410	0.0775	2.67	1.34	1.239	0.086	1.24
3x70/35	0.4430	0.0749	4.45	1.28	1.183	0.082	1.19
3x70/70	0.4430	0.0749	2.73	1.31	0.867	0.083	0.87
3x95/50	0.3200	0.0749	4.35	1.25	0.841	0.081	0.84
3x95/95	0.3200	0.0749	2.76	1.29	0.631	0.082	0.64
3x120/70	0.2530	0.0731	4.02	1.25	0.631	0.079	0.64
3x120/120	0.2530	0.0731	2.77	1.29	0.500	0.080	0.51
3x150/70	0.2060	0.0734	4.71	1.20	0.572	0.078	0.58
3x150/150	0.2060	0.0734	2.73	1.24	0.403	0.079	0.41
3x185/95	0.1640	0.0734	4.44	1.06	0.437	0.075	0.44
3x185/185	0.1640	0.0734	2.77	1.11	0.324	0.076	0.33

1) According to DIN EN 60909-0 Bbl. 4 (VDE 0102 Bbl. 4): 2009-08 [11.59]

2) Applies to a conductor temperature of ϑ = 80 °C

Table C11.46 Impedances and loop impedances per unit length of single-conductor cables with PVC insulation for various conductor arrangements (conductor material copper, type NYY)

Conductor cross-sectional area	Single-core cable NYY (Cu conductor, PVC insulation)						
	Conductor arrangement A1:						
	Impedances per unit length 1)				Loop impedances per unit length 2)		
A_n [mm^2]	R'_{20} [mΩ/m]	X' [mΩ/m]	R'_0/R'_1 [1]	X'_0/X'_1 [1]	R'_s [mΩ/m]	X'_s [mΩ/m]	Z'_s [mΩ/m]
4 x 1 x 10	1.830	0.136	4	5.15	4.523	0.323	4.53
4 x 1 x 16	1.150	0.126	4	5.24	2.842	0.303	2.86
4 x 1 x 25	0.727	0.112	4	5.39	1.797	0.276	1.82
4 x 1 x 35	0.524	0.106	4	5.48	1.295	0.263	1.32
4 x 1 x 50	0.382	0.102	4	5.53	0.944	0.257	0.98
4 x 1 x 70	0.268	0.095	4	5.65	0.662	0.242	0.71
4 x 1 x 95	0.193	0.091	4	5.71	0.477	0.235	0.53
4 x 1 x 120	0.153	0.090	4	5.74	0.378	0.231	0.44
4 x 1 x 150	0.124	0.090	4	5.73	0.306	0.232	0.38
4 x 1 x 185	0.099	0.088	4	5.77	0.245	0.229	0.34
4 x 1 x 240	0.075	0.086	4	5.81	0.186	0.225	0.29
4 x 1 x 300	0.060	0.083	4	5.87	0.149	0.219	0.26
4 x 1 x 400	0.046	0.083	4	5.88	0.114	0.218	0.25
4 x 1 x 500	0.037	0.082	4	5.90	0.091	0.216	0.23
Conductor cross-sectional area	Single-core cable NYY (Cu conductor, PVC insulation)						
	Conductor arrangement A2:						
	Impedances per unit length 1)				Loop impedances per unit length 2)		
A_n [mm^2]	R'_{20} [mΩ/m]	X' [mΩ/m]	R'_0/R'_1 [1]	X'_0/X'_1 [1]	R'_s [mΩ/m]	X'_s [mΩ/m]	Z'_s [mΩ/m]
4 x 1 x 10	1.830	0.143	4	4	4.523	0.286	4.53
4 x 1 x 16	1.150	0.133	4	4	2.842	0.266	2.85
4 x 1 x 25	0.727	0.119	4	4	1.797	0.238	1.81
4 x 1 x 35	0.524	0.113	4	4	1.295	0.226	1.31
4 x 1 x 50	0.382	0.110	4	4	0.944	0.220	0.97
4 x 1 x 70	0.268	0.102	4	4	0.662	0.204	0.69
4 x 1 x 95	0.193	0.099	4	4	0.477	0.198	0.52
4 x 1 x 120	0.153	0.097	4	4	0.378	0.194	0.43
4 x 1 x 150	0.124	0.097	4	4	0.306	0.194	0.36
4 x 1 x 185	0.099	0.096	4	4	0.245	0.192	0.31
4 x 1 x 240	0.075	0.094	4	4	0.186	0.188	0.26
4 x 1 x 300	0.060	0.091	4	4	0.149	0.182	0.23
4 x 1 x 400	0.046	0.090	4	4	0.114	0.181	0.21
4 x 1 x 500	0.037	0.089	4	4	0.091	0.179	0.20

1) According to DIN EN 60909-0 Bbl. 4 (VDE 0102 Bbl. 4): 2009-08 [11.59]

2) Applies to a conductor temperature of ϑ = 80 °C

Table C11.47 Impedances and loop impedances per unit length of single-conductor cables with PVC insulation for various conductor arrangements (conductor material copper, type NYY)

Conductor cross-sectional area	Single-core cable NYY (Cu conductor, PVC insulation)						
	Conductor arrangement A3						
	Impedances per unit length 1)				Loop impedances per unit length 2)		
A_n [mm²]	R'_{20} [mΩ/m]	X' [mΩ/m]	R'_0/R'_1 [1]	X'_0/X'_1 [1]	R'_s [mΩ/m]	X'_s [mΩ/m]	Z'_s [mΩ/m]
4 x 1 x 10	1.830	0.150	4	4.92	4.523	0.346	4.54
4 x 1 x 16	1.150	0.140	4	4.99	2.842	0.326	2.86
4 x 1 x 25	0.727	0.127	4	5.09	1.797	0.299	1.82
4 x 1 x 35	0.524	0.120	4	5.15	1.295	0.286	1.33
4 x 1 x 50	0.382	0.117	4	5.18	0.944	0.280	0.98
4 x 1 x 70	0.268	0.109	4	5.26	0.662	0.265	0.71
4 x 1 x 95	0.193	0.106	4	5.30	0.477	0.258	0.54
4 x 1 x 120	0.153	0.104	4	5.33	0.378	0.254	0.46
4 x 1 x 150	0.124	0.105	4	5.32	0.306	0.254	0.40
4 x 1 x 185	0.099	0.103	4	5.34	0.245	0.252	0.35
4 x 1 x 240	0.075	0.101	4	5.37	0.186	0.248	0.31
4 x 1 x 300	0.060	0.098	4	5.41	0.149	0.242	0.28
4 x 1 x 400	0.046	0.098	4	5.41	0.114	0.241	0.27
4 x 1 x 500	0.037	0.097	4	5.43	0.091	0.239	0.26
Conductor cross-sectional area	Single-core cable NYY (Cu conductor, PVC insulation)						
	Conductor arrangement A4 (d_a)						
	Impedances per unit length 1)				Loop impedances per unit length 2)		
A_n [mm²]	R'_{20} [mΩ/m]	X' [mΩ/m]	R'_0/R'_1 [1]	X'_0/X'_1 [1]	R'_s [mΩ/m]	X'_s [mΩ/m]	Z'_s [mΩ/m]
4 x 1 x 10	1.830	0.194	4	4.71	4.523	0.433	4.54
4 x 1 x 16	1.150	0.184	4	4.75	2.842	0.413	2.87
4 x 1 x 25	0.727	0.170	4	4.81	1.797	0.386	1.84
4 x 1 x 35	0.524	0.164	4	4.84	1.295	0,374	1.35
4 x 1 x 50	0.382	0.160	4	4.86	0.944	0.367	1.01
4 x 1 x 70	0.268	0.153	4	4.90	0.662	0.352	0.75
4 x 1 x 95	0.193	0.149	4	4.92	0.477	0.345	0.59
4 x 1 x 120	0.153	0.148	4	4.94	0.378	0.341	0.51
4 x 1 x 150	0.124	0.148	4	4.94	0.306	0.341	0.46
4 x 1 x 185	0.099	0.147	4	4.94	0.245	0.339	0.42
4 x 1 x 240	0.075	0.145	4	4.96	0.186	0.335	0.38
4 x 1 x 300	0.060	0.141	4	4.98	0.149	0.329	0.36
4 x 1 x 400	0.046	0.141	4	4.98	0.114	0.328	0.35
4 x 1 x 500	0.037	0.140	4	4.98	0.091	0.326	0.34

1) According to DIN EN 60909-0 Bbl. 4 (VDE 0102 Bbl. 4): 2009-08 [11.59]

2) Applies to a conductor temperature of $\vartheta = 80\,°C$

minals of the distribution transformers and the LV-side terminals of the load-consuming equipment can be summated to a value of $\Delta u = 14\,\%$ that is permissible according to the standards. This standardized permissible value is reliably undershot if the voltage drop per spur cable run is limited to $\Delta u = 2...3\,\%$. The limitation of the voltage drop to $\Delta u = 2...3\,\%$ has proven advantageous above all in the dimensioning of supply cables for connection of LV subdistribution boards in power stations and industrial plants [11.56]. To ensure safe motor starting, dimensioning of the motor circuits should normally be based on a voltage drop of $\Delta u \leq 2\,\%$ occurring with the load current.

The rule is that, if the permissible voltage drop at the end of the cable is selected, reliable operation of all connected loads will be ensured. To supply power to sensitive production processes, it may be necessary to choose voltage drops below the guidance values recommended by the standards for conventional consumers‘ installations.

The voltage drop in power systems can be calculated as follows according to DIN VDE 0100 Bbl. 5 (VDE 0100 Bbl. 5): 1995-11 [11.61]:

$$\Delta U = \sqrt{3} \cdot l \cdot I_{\mathrm{load}} \cdot (R_{\mathrm{L}}' \cdot \cos\varphi + X_{\mathrm{L}}' \cdot \sin\varphi) \qquad (11.16)$$

$$\Delta u = \frac{\Delta U}{U_{\mathrm{nN}}} \cdot 100\,\% \qquad (11.17)$$

ΔU absolute voltage drop
Δu relative voltage drop
U_{nN} nominal system voltage
I_{load} maximum load current
R_{L}' resistance per unit length of the conductor (calculated acc. to Eq. 11.18)
X_{L}' reactance per unit length of the conductor (Tables C11.42 to C11.47)
l cable length

Because the resistance per unit length R_{L}' is a temperature-dependent quantity, the resistance increase that would occur with a conductor temperature deviating from 20 °C should be taken into account in the calculation of a real voltage drop. To obtain a realistic result for normal operating conditions, the resistance calculation can be based on an appropriate conductor cross-section temperature rather than on the final conductor temperature ($\vartheta_{\mathrm{e}} = 70$ °C for a PVC and $\vartheta_{\mathrm{e}} = 90$ °C for an XLPE cable).

An average conductor temperature of $\vartheta_{\mathrm{e}} = 55$ °C is considered appropriate. The temperature-dependent resistance increase can be calculated as follows:

$$R_{\mathrm{L}\vartheta_{\mathrm{e}}}' = R_{\mathrm{L20}}' \cdot [1 + \alpha_{20} \cdot (\vartheta_{\mathrm{e}} - 20\,°\mathrm{C})] \qquad (11.18)$$

$R_{\mathrm{L}\vartheta_{\mathrm{e}}}'$ resistance per unit length of the conductor at temperature ϑ_{e}
R_{L20}' resistance per unit length of the conductor at $\vartheta = 20$ °C (Tables C11.42 to C11.47)
ϑ_{e} conductor temperature deviating from $\vartheta = 20$ °C
α_{20} temperature-coefficient according to IEC 60228: 2004-11 [11.62] ($\alpha_{20} = 0.00393\,\frac{1}{°\mathrm{C}}$ for copper and $\alpha_{20} = 0.00403\,\frac{1}{°\mathrm{C}}$ for aluminium conductors at $\vartheta = 20$ °C)

Compliance with the voltage drops in a power system can be simply and conveniently checked based on Tables C11.48 (for multi-core cables) and C11.49 (for single-core cables). These tables contain normalized values for the permissible cable lengths for a voltage drop of $\Delta u = 1\,\%$. The length data that applies to $I_{\mathrm{norm}} = 1$ A and $U_{\mathrm{norm}} = 1$ V takes account of a temperature-dependent resistance increase of the impedances per

Table C11.48 Normalized cable lengths for multi-core cables for conversion to any load currents and nominal voltages based on the voltage drop

Normalized voltage U_{norm} = 1 V					
Nominal cross-sectional area of the conductor	Normalized design current	Multi-core cable			
		PVC insulation		XLPE/EPR insulation	
		Cu	Al	Cu	Al
A_n [mm^2]	I_{norm} [A]	Normalized permissible cable length ℓ_{norm} in m for Δu = 1 %			
1.5	1	0.40	--	0.37	--
2.5		0.65	--	0.61	--
4		1.06	--	0.99	--
6		1.59	--	1.49	--
10		2.63	--	2.47	--
16		4.19	--	3.93	--
25		6.61	--	6.20	--
35		9.13	5.52	8.58	5.17
50		12.27	7.45	11.54	6.99
70		17.46	10.72	16.46	10.07
95		23.60	14.70	22.33	13.82
120		28.94	18.38	27.48	17.31
150		34.31	22.23	32.72	20.98
185		40.46	27.19	38.82	25.75
240		48.10	32.88	46.52	31.27
300		53.96	--	52.56	--

unit length from Tables C11.42 to C11.47, which are based on a final conductor temperature of ϑ_e = 70 °C (PVC cable) or ϑ_e = 90 °C (XLPE cable).

Conversion to any load currents, nominal system voltages and voltage drops is possible with Eq. (11.19).

$$l_{perm} = l_{norm} \cdot \frac{U_{nN}/V}{I_{load}/A} \cdot \Delta u_{perm}/\% \quad (11.19)$$

l_{perm} permissible cable length for compliance with the permissible voltage drop
l_{norm} normalized permissible cable length (Tables C11.48 and C11.49)
U_{nN} nominal system voltage to convert to
I_{load} load or rated current to convert to
Δu_{perm} permissible relative voltage drop for reliable operation of the loads

For single-phase AC circuits, the values calculated with Eq. (11.19) for the permissible cable length l_{perm} must be halved.

The l_{norm} values in Tables C11.48 and C11.49 yield the shorter cable length l_{perm} which is within safe values. The normalized length data are based on the assumption that the phase angle $\cos\varphi_{load}$ of the rated or load current and the impedance angle of the cable φ_L are equal, that is,

$$\cos\varphi_{\text{load}} = \cos\varphi_{\text{L}} = \cos\left(\arc\tan\frac{X'_{\text{L}}}{R'_{\text{L}}}\right) \tag{11.20}$$

The usually deviating $\cos\varphi_{\text{load}}$ of the load current I_{load} results in a longer cable length and is therefore a "length safety margin". If the calculated cable length l_{perm} is complied with, the permissible voltage drop Δu_{perm} will definitely not be exceeded. However, the permissible cable length l_{perm} does not provide any information about compliance with the necessary voltage stability during operation of motor loads. This must be checked based on the calculation method explained in Section 10.1.1.2.

Table C11.49 Normalized cable lengths for PVC-insulated single-core cables for conversion to any load currents and nominal voltages based on the voltage drop

Normalized voltage U_{norm} = 1 V					
Nominal cross-sectional area of the conductor A_{n} [mm²]	Normalized design current I_{norm} [A]	Single-core cable NYY (Cu conductor, PVC insulation)			
		Conductor arrangement			
		A1	A2	A3	A4
		Normalized permissible cable length ℓ_{norm} in m for Δu = 1 %			
10	1	2.63	2.63	2.63	2.63
16		4.18	4.18	4.18	4.16
25		6.58	6.58	6.57	6.52
35		9.09	9.07	9.05	8.92
50		12.34	12.29	12.25	11.94
70		17.29	17.18	17.07	16.30
95		23.30	23.03	22.79	21.07
120		28.41	27.95	27.51	24.65
150		34.37	32.69	31.92	27.64
185		39.18	37.99	36.91	30.75

11.3.6 Dimensioning example

Example C9

Let us assume the following: In an industrial plant, there is a small 400-V subnetwork supplied through a 1,000-kVA transformer from the upstream 20-kV factory power system. To handle an outage of the normal power supply, a 775-kVA diesel generator is connected to the low-voltage main distribution board. In normal operation, the diesel generator is switched off. Only on a power outage does it automatically start and provide an emergency power supply to the loads.

A 110-kW pump motor to boost the performance of a sprinkler system is also to be connected to the low-voltage main distribution board of the subnetwork as an additional load. Connection must be chosen and dimensioned according to the valid criteria for the overload and short-circuit protection, protection against electric shock and voltage drop.

The following supply conditions are prescribed:

- use of a moulded-case circuit-breaker (MCCB) with integrated overcurrent release for motor protection,
- connection of the motor through a 90-m long PVC-insulated multi-core cable,
- installation of the PVC-insulated multi-core cable on a perforated cable tray with three further cable systems, touching,
- consideration of an ambient temperature of $\vartheta = 40\,°\mathrm{C}$,
- compliance with a voltage drop of $\Delta u = 2\,\%$ when the load current is applied to the motor connection cable.

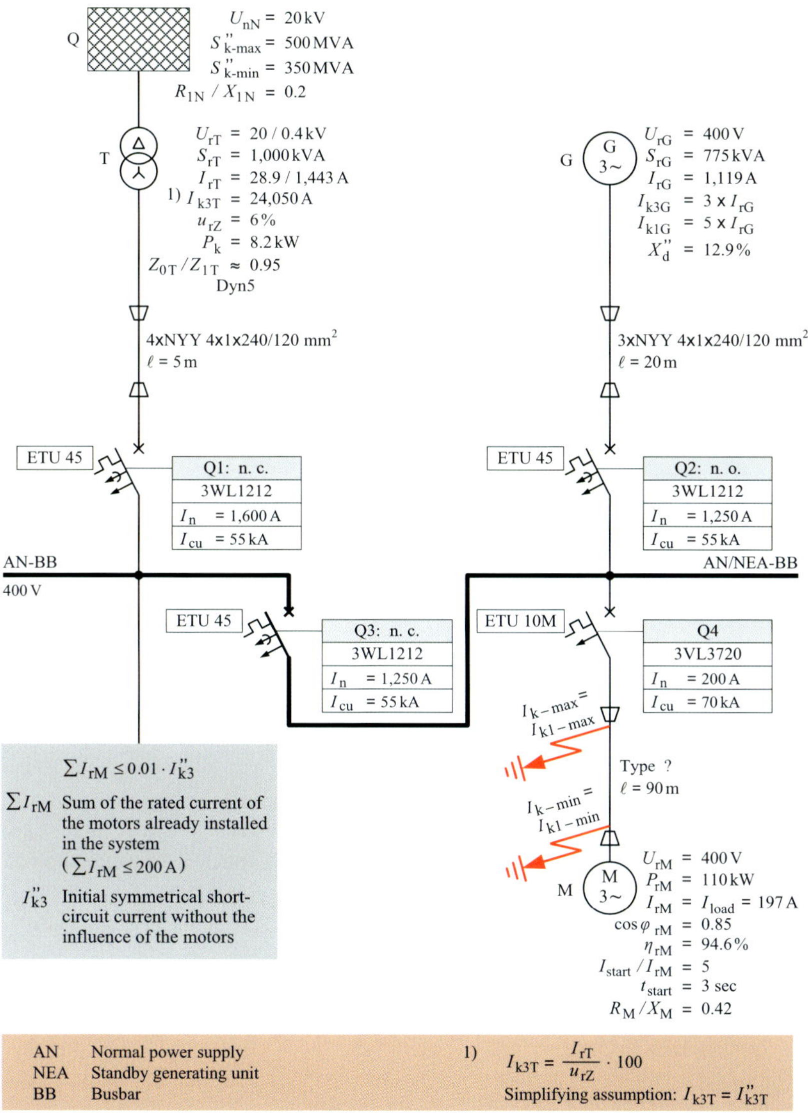

Fig. C11.50 400-V subnetwork for example of cable dimensioning (Example C9)

In addition to the supply conditions, the equipment data are required for network calculation. Fig. C11.50 shows the 400-V subnetwork with the equipment data required for the calculation.

The cable for the example of motor connection must be dimensioned as follows:

a) *Selection of the cable according to the permissible current-carrying capacity (Section 11.3.1)*

The following rating factors result for deviating ambient temperatures and grouping of cables on trays and ladders:

- $f_\vartheta = 0.87$ according to Table C11.27 for a PVC cable installed in the air at an ambient temperature of $\vartheta = 40\,°C$,
- $f_H = 0.79$ according to Table C11.28 for grouping on a perforated cable tray with $n = 4$ multi-core cables, touching.

At a maximum load current of $I_{load} = I_{rM} = 197$ A, the following rated value can be calculated for the current-carrying capacity of the PVC cable with these rating factors:

$$I_{rL} \geq \frac{I_{load}}{f_\vartheta \cdot f_H} = \frac{197\text{ A}}{0.87 \cdot 0.79} = 287\text{ A}$$

According to Table C11.23, this rated value with reference method of installation E (multi-core cable installed in free air) requires cable type NYY 3 × 150/70 mm². With this PVC-insulated multi-core cable, the load current condition $I_{perm} \geq I_{load}$ (Eq. 11.3) is reliably met ($I_{perm} = I_{rL} \cdot f_\vartheta \cdot f_H = 319\text{ A} \cdot 0.87 \cdot 0.79 = 219$ A; 219 A > 197 A).

b) *Setting and checking the overload protection (Section 11.3.2)*

According to the maximum load current of the 110-kW motor of $I_{load} = I_{rM} = 197$ A, a 200-A 3VL moulded-case circuit-breaker for motor-protection with integrated overcurrent release system of type ETU 10M is required for connection of the cable NYY 3 × 150/ 70 mm² to the low-voltage main distribution board. The inverse-time-delay overload release (L release) of the ETU 10M electronic release system is set to $I_R = 197$ A. With this setting, it is possible to comply both with the dimensioning rule and the tripping rule.

- The dimensioning rule according to Eq. (11.4.2): $I_{load} \leq I_R \leq I_{perm}$ is complied with for $I_{load} = I_R = 197$ A and $I_{perm} = 219$ A.
- The tripping rule according to Eq. (11.5): $I_2 \leq 1.45 \cdot I_{perm}$ is complied with for $I_2 = 1.2 \cdot I_R = 236$ A (circuit-breaker for motor protection) and $1.45 \cdot I_{perm} = 317$ A.

c) *Setting and checking the short-circuit protection (Section 11.3.3)*

To set the instantaneous overcurrent release (I release) of the electronic release system ETU 10M and to check the short-circuit withstand capability of the motor connection cable, the maximum and minimum short-circuit current must be known.

- Maximum short-circuit current $I''_{k\text{-}max}$

  The maximum short-circuit current results in normal operation when the 400-V subnetwork is supplied from the 20-kV in-plant network through the 1,000-kVA transformer. If the maximum short-circuit current is calculated according to DIN EN 60909-0 (VDE 0102): 2002-07 [11.57] or IEC 60909-0: 2001-07 [11.58], the contribution of the LV motors is negligible if the condition $\sum I_{rM} \leq 0.01 \cdot I''_{k3}$ applies. This condition is met in the 400-V subnetwork because the sum of the LV motors already installed is only $\sum I_{rM} = 200\text{ A}$.

The Dyn5 transformer supplying the 400-V subnetwork exhibits an impedance ratio of $Z_{0T}/Z_{1T} \approx 0.95$. Based on this impedance ratio, without considering a contribution by motors, the line-to-earth short-circuit current I''_{k1} at the LV main distribution board is larger than the three-phase short-circuit current I''_{k3} (see Table C9.2, column (T)). The line-to-earth short-circuit current I''_{k1} can be calculated as a decimal multiple of the three-phase short-circuit current I''_{k3}. Without considering the very short LV supplying cable, the following simplified short-circuit calculation can be performed:

$$I''_{k3\text{-max}} = I''_{k3T} \cdot \frac{S''_{k\text{-max}}}{S''_{k\text{-max}} + 1.1 \cdot \dfrac{S_{rT} \cdot 100}{u_{rZ}}} \tag{11.21}$$

$$I''_{k3\text{-max}} = 24{,}050\ \text{A} \cdot \frac{500 \cdot 10^6\ \text{VA}}{500 \cdot 10^6\ \text{VA} + 1{,}1 \cdot \dfrac{1.000 \cdot 10^3\ \text{VA} \cdot 100}{6\ \%}} = 23{,}200\ \text{A}$$

$$I''_{k\text{-max}} = I''_{k1\text{-max}} = \frac{3}{2 + \dfrac{Z_{0T}}{Z_{1T}}} \cdot I''_{k3\text{-max}} \tag{11.22}$$

$$I''_{k\text{-max}} = \frac{3}{2 + 0.95} \cdot 23{,}200\ \text{A} = 23{,}590\ \text{A}$$

- Minimum short-circuit current $I''_{k\text{-min}}$

  The minimum short-circuit current occurs at the end of the motor connection cable in operation under fault conditions when loads are powered by the 775-kVA diesel generator. Because of the impedance ratios $R_{0L}/R_{1L} = 7.48$ and $X_{0L}/X_{1L} = 4.94$ prevailing in the motor circuit (see Table C11.42), the line-to-earth short-circuit current I''_{k1} at the end of the connecting cable is the smallest of all short-circuit currents. For calculation of the minimum line-to-earth short-circuit current $I''_{k1\text{-min}}$ according to Eq. (11.14), the loop impedance values $Z_{s\text{-N}} = 40.501\ \text{m}\Omega$ from Table C11.41 and $Z_{s\text{-L}} = Z'_S \cdot l = 0.52\ \text{m}\Omega/\text{m} \cdot 90\ \text{m} = 46.8\ \text{m}\Omega$ from Table C11.42 can be used. With these Z_s values, the following minimum line-to-earth short-circuit current results:

$$I''_{k1\text{-min}} = \frac{c_{min} \cdot U_{nN}}{\sqrt{3} \cdot (Z_{s\text{-N}} + Z_{s\text{-L}})} = \frac{0.95 \cdot 400\ \text{V}}{\sqrt{3} \cdot (40.501 + 46.8) \cdot 10^{-3}\ \Omega} = 2{,}513\ \text{A}$$

The instantaneous overcurrent release (I release) of the ETU 10M release system must be set taking the minimum short-circuit current $I''_{k\text{-min}} = I''_{k1\text{-min}} = 2{,}513$ A and the motor starting current $I_{start} = 5 \cdot I_{rM} = 985$ A into account. The following starting condition must be met:

$$I_{start} < I_i < I''_{k\text{-min}} \tag{11.23}$$

$$985\ \text{A} < I_i < 2{,}513\ \text{A}$$

To meet the starting condition (11.23), the instantaneous overcurrent release is set to $I_i = 8 \cdot I_n = 1{,}600$ A. With this setting, the upper tripping tolerance of the I release of $1.2 \cdot I_i = 1{,}920$ A is still clearly below the minimum line-to-earth short-circuit current of $I''_{k1\text{-min}} = 2{,}513$ A, so that both the maximum short-circuit cur-

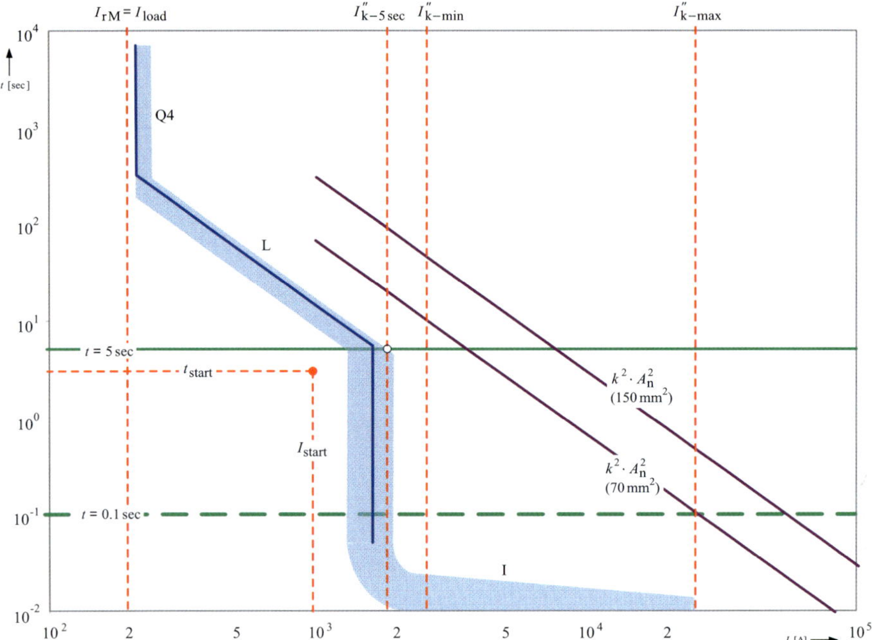

Fig. C11.51 Overload and short-circuit protection of the example motor circuit

rent at the beginning of the cable and the minimum short-circuit current at the end of the cable can be disconnected instantaneously ($t_i < 0.1$ sec) (Fig. C11.51).

With disconnecting times $t_i < 0.1$ sec, the DC component of the short-circuit current is no longer negligible. Based on the Joule heat value $k^2 \cdot A_n^2$ it is therefore necessary to check whether the connecting cable NYY 3 × 150/70 mm² is short-circuit-proof on a short-circuit at any point in the motor circuit.

According to Table C11.39, the conductors of the motor connecting cable NYY 3 × 150/70 mm² have the following Joule heat values $k^2 \cdot A_n^2$:

- 150-mm² line conductor: $2.98 \cdot 10^8$ A²sec,
- 70-mm² protective conductor: $6.48 \cdot 10^7$ A²sec.

Converting these Joule heat values into equivalent $I^2 \cdot t$ figures yields the "cable damage curves" represented in Fig. C11.51. As Fig. C11.51 shows, the line conductors and the PE conductor of the motor connecting cable are thermally short-circuit-proof in the entire fault current range 2,513 A $\leq I''_k \leq$ 23,590 A.

d) Checking the protection against electric shock (Section 11.3.4)

The minimum line-to-earth short-circuit current of the motor circuit is $I''_{k1\text{-min}}$ = 2,513 A. With this short-circuit current, the disconnection condition (11.11) for protection against electric shock is reliably met (2,513 A > I_a; $I_a = 1.2 \cdot I_i = 1.2 \cdot 1{,}600$ A = 1,920 A). Disconnection is performed instantaneously ($t_i < 0.1$ sec).

Protection against electric shock can also be checked using Eq. (11.15).

$$Z_{\text{s-L}} \leq \frac{c_{\min} \cdot U_{\text{nN}}}{\sqrt{3} \cdot I_{\text{a}}} - Z_{\text{s-N}} \leq \frac{0.95 \cdot 400\ \text{V}}{\sqrt{3} \cdot 1{,}920\ \text{A}} - 40.501\ \text{m}\Omega \leq 73.766\ \text{m}\Omega$$

$Z_{\text{s-N}}$ loop impedance of the upstream network (in this case, supply from a 775-kVA generator (see Table C11.41))

$Z_{\text{s-L}}$ loop impedance of the cable used ($Z_{\text{s-L}} = Z_{\text{s}}' \cdot l$; for Z_{s}' for cable type NYY 3×150/70 mm² see Table C11.42)

The maximum loop impedance of the motor connection cable can be $Z_{\text{s-L}} = 73.766\ \text{m}\Omega$. For a loop impedance per unit length of the motor connecting cable NYY 3 × 150/70 mm² of $Z_{\text{s}}' = 0.52\ \text{m}\Omega/\text{m}$, the impedance value $Z_{\text{s-L}} = 73.766\ \text{m}\Omega$ corresponds to a maximum cable length of $l_{\max} = 142$ m. The connecting cable of the motor circuit is only $l = 90$ m long. The motor circuit therefore provides sufficient protection against indirect contact.

e) *Checking the voltage drop (Section 11.3.5)*

Conventionally, when the motor connection cable NYY 3 × 150/70 mm² is loaded with the rated motor current, a voltage drop of $\Delta u = 2\,\%$ must be complied with. According to Table C11.42, the cable NYY 3 × 150/70 mm² has a resistance per unit length of $R_{\text{L20}}' = 0.124\ \text{m}\Omega/\text{m}$ and a reactance per unit length of $X_{\text{L}}' = 0.078\ \text{m}\Omega/\text{m}$. To calculate a realistic voltage drop, the resistance increase occurring when the connecting cable is loaded with the full load current must be taken into account. This resistance increase is based on a conductor temperature of $\vartheta_{\text{e}} = 55\,°\text{C}$. According to Eq. (11.18), the following resistance per unit length results for a realistic voltage drop calculation:

$$R_{\text{L55}}' = R_{\text{L20}}' \cdot [1 + \alpha_{20} \cdot (\vartheta_{\text{e}} - 20\,°\text{C})] =$$

$$0.124\ \text{m}\Omega/\text{m} \cdot [1 + 0.00393\ {}^{1}\!/_{°\text{C}} \cdot (55\,°\text{C} - 20\,°\text{C})] = 0.141\ \text{m}\Omega/\text{m}$$

The resistance and reactance values per unit length $R_{\text{L55}}' = 0.141\ \text{m}\Omega/\text{m}$ and $X_{\text{L}}' = 0.078\ \text{m}\Omega/\text{m}$ must be inserted into Eq. (11.16).
The following voltage drop then occurs under normal operating conditions ($I_{\text{load}} = 195$ A and $\cos\varphi_{\text{M}} = 0.85$):

$$\Delta U = \sqrt{3} \cdot l \cdot I_{\text{load}} \cdot (R_{\text{L55}}' \cdot \cos\varphi_{\text{M}} + X_{\text{L}}' \cdot \sin\varphi_{\text{M}}) =$$

$$\sqrt{3} \cdot 90\ \text{m} \cdot 195\ \text{A} \cdot \left(0.141 \cdot 10^{-3}\frac{\Omega}{\text{m}} \cdot 0.85 + 0.078 \cdot 10^{-3}\frac{\Omega}{\text{m}} \cdot 0.527\right) = 4.89\ \text{V}$$

Eq. (11.17) can be used to convert the absolute voltage drop ΔU into the relative voltage drop.

$$\Delta u = \frac{\Delta U}{U_{\text{nN}}} \cdot 100\,\% = \frac{4.89\ \text{V}}{400\ \text{V}} \cdot 100\,\% = 1.22\,\%$$

The calculated relative voltage drop is smaller than 2 % and therefore permissible.
The length-related check of the voltage drop according to Eq. (11.19) has proven especially convenient. For this check , the length normalized at $I_{\text{norm}} = 1$ A and $U_{\text{norm}} = 1$ V of the motor connection cable NYY 3 × 150/70 mm² is required. Accord-

ing to Table C11.48, a normalized permissible cable length of l_{norm} = 34.31 m can be assigned to a PVC-insulated multi-core cable with 150-mm^2 line conductors made of copper. For the voltage drop of Δu = 2 % to be complied with at I_{load} = 195 A in the 400-V motor circuit, the following maximum permissible cable length results:

$$l_{perm} = l_{norm} \cdot \frac{U_{nN}/V}{I_{load}/A} \cdot \Delta u/\% = 34.31\ \text{m} \cdot \frac{400}{195} \cdot 2\,\% = 141\ \text{m}$$

The real cable length is only l = 90 m.

As the result of both checks shows, compliance with permissible voltage drop can also be ensured with the PVC multi-core cable type NYY 3 × 150/70 mm^2. Consequently, all criteria of cable dimensioning are fulfilled.

12 Reactive-power compensation

12.1 Technical and economic reasons for compensation

Inductive linear equipment and loads (e.g. transformers, reactors, asynchronous motors) and inductive non-linear loads (e.g. static converters for variable-speed drives, welding machines, arc furnaces, rectifiers, electronic valves, thyristors, brightness and temperature controllers, AC power controllers and three-phase AC power controllers, frequency converters, gas discharge lamps with ballasts) require reactive power to produce a magnetic field. Because the magnetic fields of the inductive loads increase and decrease with the energy and at the frequency of the supplying power system, the reactive power moves back and forth between the power source and the load-consuming equipment. This movement back and forth is effected by means of the reactive current that places an extra load on the equipment in addition to the current for the active power.

For technical and economic reasons, it is convenient to compensate for the reactive power as close as possible to the load, that is, usually on the LV side. These reasons are:

- to save investment costs by economic utilization of the equipment capacity,
- to provide high quality of supply due to better voltage stability,
- to provide high energy efficiency due to reduction in power system losses.

Noteworthy is the contribution that reactive-power compensation makes to climate protection. Calculations by the ZVEI (German electrical and electronics industry association) in 2007 for Germany show that reactive-power compensation has the potential to reduce system losses by about 3.5 billion kWh from 2007's level. That would equate to a reduction in CO_2 emissions of about 1.8 million t per year [12.1]. The greatest potential for CO_2 reduction with reactive-power compensation can be achieved by improving the power factor by an average of $\cos\varphi$-Ø = 0.70 to $\cos\varphi$-target = 0.95 in industrial power supplies [12.2]. For this reason, reactive-power compensation in industrial networks is beneficial for reasons both of cost-efficiency and of environmental protection.

12.2 Compensation when supplying linear loads

Linear inductive loads draw from the system approximately sinusoidal current, which lags behind the system voltage by an angle of φ. The current lags due to linear loads requiring reactive power to set up a magnetic field. Because the reactive power is not useful energy, the absorbed apparent power is greater than the required active power in all inductive loads.

The ratio of active power to apparent power is termed the power factor $\cos\varphi$. Fig. C12.1 shows the relation between active, reactive and apparent power on which reactive-power compensation for the operation of linear loads is based. Linear inductive loads do not produce harmonic-distortion (non-linear) reactive power. The reactive power Q_{tot} is equal to the fundamental-frequency reactive power Q. For reactive-power compensation in networks in which linear inductive loads prevail, capacitor units without reactors can therefore generally be used.

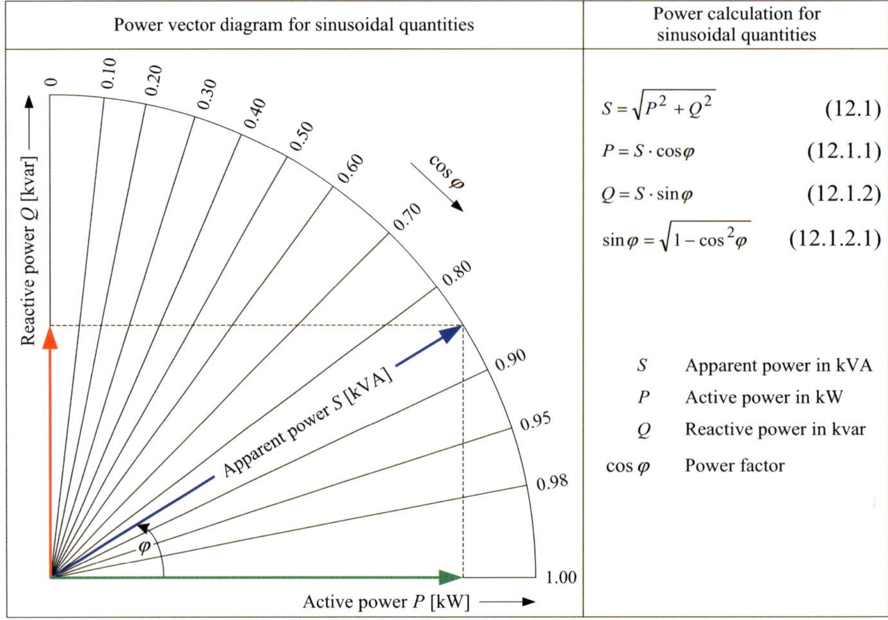

Fig. C12.1 Relation between sinusoidal power quantities

12.2.1 Determining the necessary capacitive power

The necessary capacitive power Q_c is determined differently for a new installation that is being planned and for an existing installation that is already in operation.

Determining the power at the planning and project engineering stage

The capacitive power for an installation that is still at the planning or project engineering stage can be determined with sufficient precision applying the multiple coefficient method that makes use of the mathematical relationship between the demand factor, coincidence factor and level of utilization factor.

The following applies:

$$P_{max} = \begin{cases} g \cdot \sum_i P_{pr_i} \cdot a_i & \text{with } \sum_i P_{pr_i} \cdot a_i = \sum_i P_{max_i} \\ \sum_i P_{pr_i} \cdot c_i & \text{with } c_i = g \cdot a_i \\ b \cdot \sum_i P_{pr_i} & \text{with } b = \dfrac{g \cdot \sum_i P_{pr_i} \cdot a_i}{\sum_i P_{pr_i}} \end{cases} \quad (12.2)$$

P_{max}	maximum active power demand of the installation
P_{pr}	power rating (for motors $P_{pr} = P_{rM}/\eta$)
a	utilization factor (see Table C12.2)
b	demand factor (see Tables A2.3 and C12.2)
c	maximum load component factor
g	coincidence factor (see Table A2.3)
i	incrementing index for loads and load groups ($i = 1(1)n$)

To improve the power factor of an installation with the calculated active power demand P_{max} from $\cos \varphi_1$ to $\cos \varphi_2$, the following capacitive power is required:

$$Q_c = P_{max} \cdot (\tan \varphi_1 - \tan \varphi_2) \quad (12.3)$$

$$\tan \varphi = \sqrt{\frac{1 - \cos^2 \varphi}{\cos^2 \varphi}} \quad (12.3.1)$$

The average power factor $\cos \varphi_1$ of a planned new installation has to be either derived or estimated from the results of measurements conducted on similar installations. Table C12.2 contains a selection of average power factors $\cos \varphi_1$ that can be used as rough guidance values. The $(\tan \varphi_1 - \tan \varphi_2)$ values for determining the capacitive power Q_c are listed in Table C12.3.

Table C12.2 Guidance values for the utilization factor a, power factor b and average power factor $\cos \varphi_1$

Industry or consumer load	Factors for power demand calculation according to [12.3 to 12.7]		
	Utilization factor a	Demand factor b	Average power factor $\cos \varphi_1$
Metal processing and machining			
Machine tools for small batch production	0.14 ··· 0.20	0.16 ··· 0.20	0.40 ··· 0.50
Machine tools for large series production	0.18	0.23	0.50 ··· 0.60
Welding machines	0.20	0.30	0.60 ··· 0.65
Welding transformers	0.20	0.30	0.40 ··· 0.50
Cranes	---	---	0.50 ··· 0.60
Large machine tools, incl. presses	---	0.23	0.65 ··· 0.70
Water pumps	---	---	0.80 ··· 0.85
Mechanical workshops	0.20	0.15 ··· 0.30	0.50 ··· 0.60
Fans	---	---	0.70 ··· 0.80
Compressors	---	---	0.70 ··· 0.80
Foundries	---	0.40 ··· 0.50	0.60 ··· 0.70
Motor vehicle repair shops	---	0.15 ··· 0.28	0.70 ··· 0.80
Wood-working industry			
Sawmill	0.42	0.80	0.60 ··· 0.70
Cabinetmaker's and joiner's workshop	0.36	0.25 ··· 0.40	0.60 ··· 0.70
Plywood factories	---	0.15 ··· 0.30	0.60 ··· 0.70
Food and beverages industry			
Meat processors	0.50	0.60	0.60 ··· 0.70
Bakeries	0.35	0.50	0.60 ··· 0.70
Sugar production	0.60	0.52	0.80 ··· 0.85
Breweries	0.50	0.40 ··· 0.50	0.60 ··· 0.70
Textile industry			
Cotton spinning mills	0.60 ··· 0.85	0.80 ··· 0.95	---
Weaving mills	0.60 ··· 0.75	0.70 ··· 0.80	---

Table C12.3 ($\tan\varphi_1 - \tan\varphi_2$) values for determining the capacitive power Q_c for correcting $\cos\varphi_1$ to $\cos\varphi_2$

$\cos\varphi_1$ \ $\cos\varphi_2$	Target power factor										
Actual power factor	0.70	0.75	0.80	0.85	0.90	0.92	0.94	0.95	0.96	0.98	1.00
0.40	1.27	1.41	1.54	1.67	1.81	1.87	1.93	1.96	2.00	2.09	2.29
0.45	0.96	1.10	1.23	1.36	1.50	1.56	1.62	1.66	1.69	1.78	1.98
0.50	0.71	0.85	0.98	1.11	1.25	1.31	1.37	1.40	1.44	1.53	1.73
0.55	0.50	0.64	0.77	0.90	1.03	1.09	1.16	1.19	1.23	1.32	1.52
0.60	0.31	0.45	0.58	0.71	0.85	0.91	0.97	1.00	1.04	1.13	1.33
0.65	0.15	0.29	0.42	0.55	0.68	0.74	0.81	0.84	0.88	0.97	1.17
0.70	---	0.14	0.27	0.40	0.54	0.59	0.66	0.69	0.73	0.82	1.02
0.75	---	---	0.13	0.26	0.40	0.46	0.52	0.55	0.59	0.68	0.88
0.80	---	---	---	0.13	0.27	0.32	0.39	0.42	0.46	0.55	0.75
0.85	---	---	---	---	0.14	0.19	0.26	0.29	0.33	0.42	0.62
0.90	---	---	---	---	---	0.06	0.12	0.16	0.19	0.28	0.48

Example C10

Determining the necessary capacitive power at the planning stage

- Problem to be solved:

  Car body production is to be expanded to boost the annual output of a car factory. The planning documents show that to expand car body production an electrical power rating of $P_{pr} = \sum_i P_{pr_i} = 4{,}700\ \text{kW}$ results. Most new loads are welding machines with an average power factor of $\cos\varphi_1 = 0.60$. To permit energy-efficient production of car bodies, the power factor is to be improved to $\cos\varphi_2 = 0.98$. The capacitive power required to improve the power factor from $\cos\varphi_1 = 0.60$ to $\cos\varphi_2 = 0.98$ must be determined.

- Solution:

  The maximum active power demand of the new car body production plant can be calculated with the equation (12.2) and the demand factor b for welding machines from Table C12.2.

$$P_{max} = b \cdot \sum_i P_{pr_i} = 0.30 \cdot 4{,}700\ \text{kW} = 1{,}410\ \text{kW}$$

Based on this active power demand and the corresponding ($\tan\varphi_1 - \tan\varphi_2$) value from Table C12.3, the following capacitive power is required to improve the power factor from $\cos\varphi_1 = 0.60$ to $\cos\varphi_2 = 0.98$:

$$Q_c = P_{max} \cdot (\tan\varphi_1 - \tan\varphi_2) = 1{,}410\ \text{kW} \cdot 1.13 =$$

$$1{,}593.3\ \text{kvar} \approx 1{,}600\ \text{kvar}$$

Because the LV welding network of the new car body production plant is to be supplied from the upstream MV network using $n = 4$ decentrally installed load-centre substations, a PF correction system with a capacitor rating of $Q_c = 400$ kvar must be provided for each load-centre substation.

Power calculation for installations already in operation

For power distribution installations that are already in operation, the necessary capacitive power can be determined by measurements. If active and reactive power meters are installed, calculation can be performed as follows:

$$Q_c = \frac{W_{reactive} - W_{active} \cdot \tan\varphi_2}{t} \tag{12.4}$$

$W_{reactive}$ measured reactive energy consumption in kvarh
W_{active} measured active energy consumption in kWh
t operating time or number of hours of use in h

The basis for the capacitive power calculated according to Eq. (12.4) is formed by average values. The use of Eq. (12.4) therefore assumes that the equipment is operated with approximately constant load. If operation includes alternation between phases with mainly resistive load and phases with extremely inductive load, measurements should be aimed at determining the precise instantaneous values for current, voltage and $\cos\varphi$ in the periods of the inductive peak load.

Example C11

Determining the required capacitive power for an installation that is already in operation

- Problem to be solved:

A small industrial plant with an average active power demand of $P = 500$ kW, an average power factor of $\cos\varphi_1 = 0.70$ and an annual number of operating hours of $t = 4{,}000$ h intends to reduce its reactive energy costs. The power system operating company sent the industrial plant the following power bill (statement of the annual values) [12.1]:

- active energy consumption (day rate): $W_{active} = 2{,}000{,}000$ kWh
- reactive energy consumption (day rate): $W_{reactive} = 2{,}040{,}408$ kvarh
- reactive energy consumption (free): $W_{reactive\text{-}free} = 1{,}000{,}000$ kvarh
- reactive energy consumption (remainder): $W_{reactive\text{-}rem} = 1{,}040{,}408$ kvarh
- reactive energy costs: $K_{reactive} = 1{,}040{,}408$ kvarh $\cdot$ 0.013 €/kvarh = € 13,525

The cost efficiency of a reactive-power compensation system that improves the power factor from $\cos\varphi_1 = 0.70$ to $\cos\varphi_2 = 0.90$ is to be demonstrated. To demonstrate the cost efficiency, the capacitive power required to reduce the reactive power costs for $\cos\varphi_2 = 0.90$ must be known.

- Solution:

The capacitive power required to reduce the reactive energy costs for a target power factor of $\cos\varphi_2 = 0.90$ can be calculated based on the measured active and reactive energy consumption using Eq. (12.4). A PF correction system with the following capacitor rating is required:

$$Q_c = \frac{W_{reactive} - W_{active} \cdot \tan \varphi_2}{t} =$$

$$\frac{(2{,}040{,}408 - 2{,}000{,}000 \cdot 0.4843)\ \text{kvar}}{4.000\ \text{h}} = 268\ \text{kvar}$$

According to the calculated capacitive power, a PF correction system with a capacitor rating of $Q_c = 300$ kvar must be installed. Installation of a 300-kvar PF correction system incurs investment costs of approximately € 8,000 [12.1]. Considering annual reactive energy costs of $K_{reactive} =$ € 13,525, the investment costs for the PF correction system are already recouped after about 7 months.

12.2.2 Types of reactive-power compensation

It is basically possible to compensate for inductive loads individually, in groups or centrally. In larger power distribution plants, it is also common to use two or all three types of reactive-power compensation together (mixed compensation). The choice of the type of compensation is influenced by economic and installation engineering considerations.

12.2.2.1 Individual compensation

In the case of the individual compensation shown in Fig. C12.4, a magnitude of inductive reactive power that depends on the agreed target cos φ is compensated for directly where it arises. For this purpose, an appropriately dimensioned capacitor or a capacitor bank is directly connected to the terminals of the inductive load. The capacitors are connected to and disconnected from the load by a common switching device. The main advantages of individual compensation are:

- relief of the load feeder cables from reactive current,
- reduction of the power loss of the cable (saving of active energy costs through reduced conduction losses),
- saving on switching devices for the capacitor,

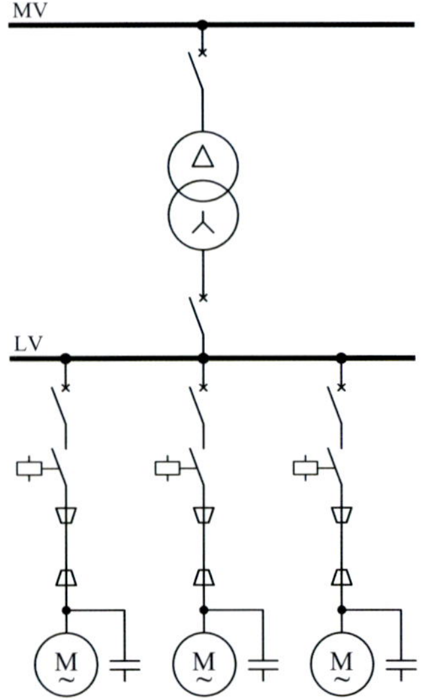

Fig. C12.4
Individual reactive-power compensation

- no need for a special PF correction control unit due to common switching of the capacitor with the load.

Individual reactive-power compensation should be used for

- large inductive loads with
- a constant power demand during
- continuous operation (duty ratio at least 50 to 70 %).

Typical applications for individual compensation include separate fixed compensation of asynchronous motors and distribution transformers (see Section 12.2.4).

12.2.2.2 Group compensation

In a group compensation system, one capacitor unit is associated with a specific group of loads, usually located close to each other. This load group may, for example, consist of motors and/or fluorescent lamps connected to the LV network via a common contactor or switch.

To calculate the group compensation capacity, the reactive power actually occurring during operation is required. It is therefore necessary to check whether the load of a group has to be considered in calculation of the demand factor b (see Section 12.2.1). Considering the demand factor b in group compensation results in a saving on compensation capacity that would not be possible with individual compensation. Group compensation is generally used for a total nominal motor rating of $\sum P_{rM} \geq 10$ kW. Fig. C12.5 shows implementation of group reactive-power compensation.

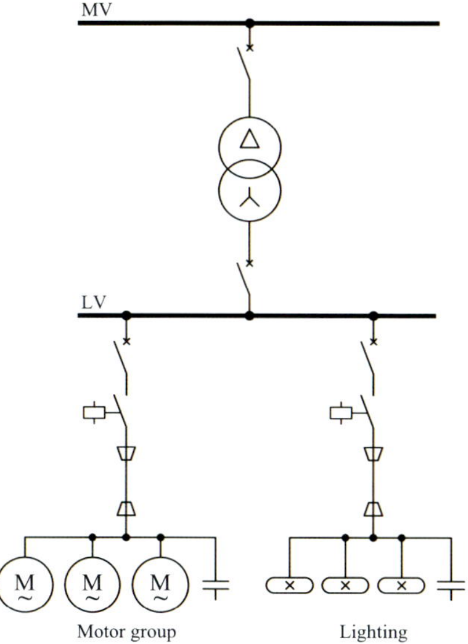

Fig. C12.5
Group reactive-power compensation

12.2.2.3 Centralized compensation

Centralized compensation is preferably used in large power distribution plants with constantly varying loads. For centralized compensation of load-induced reactive power, automatic reactive-power compensation systems are mainly used, which are assigned to a specific load-centre substation, main distribution board or subdistribution board (Fig. C12.6). With this type of compensation, the desired power factor $\cos\varphi_2$ can be

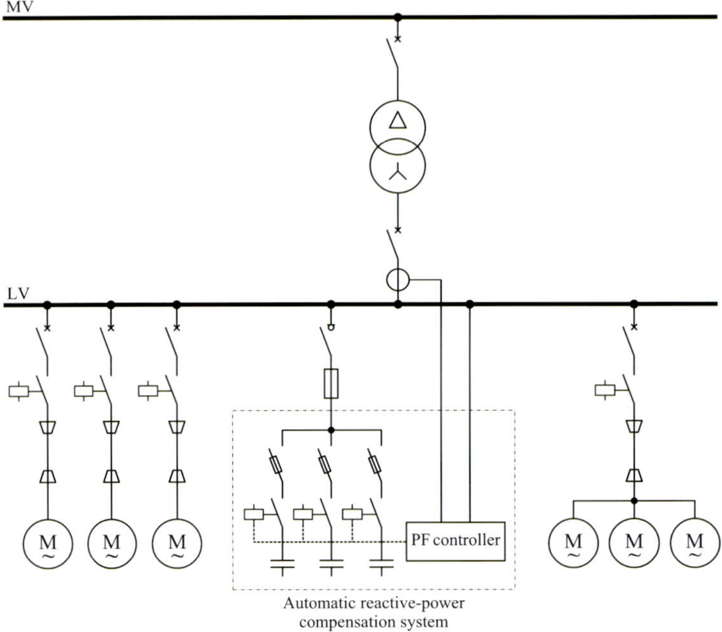

Fig. C12.6 Centralized compensation of load-induced reactive power

maintained during the consumption of electrical power by automatic, stepped connection and disconnection of the necessary or excess capacitor power.

A centralized reactive-power compensation system should preferably be installed if operation of the LV network the power factor of which is to be improved is dominated by

- many small loads with
- different active and reactive power demands and
- varying duty cycles.

Under these operational conditions, use of centralized compensation has the following economic and technical advantages:

- easy checking of the compensation equipment due to its centralized arrangement,
- simple subsequent installation or expansion,
- accurate adjustment of the capacitive power to the varying reactive power demand of the loads,
- saving on capacitor power to be installed by considering the demand factor (at a demand factor in the range $0.7 \leq b \leq 0.8$, the cost efficiency of individual and centralized compensation is generally identical).

Given its importance for energy-efficient power supplies in industry, centralized reactive-power compensation using an automatic PF correction system is explained in more detail in Section 12.2.5.

12.2.2.4 Hybrid or mixed compensation

In the case of loads being only a short time in operation, exclusively pursuing the aim of compensating for reactive power directly where it arises by way of individual compensation results in the sum of the individual compensation power being much higher than the average compensation power actually required. For this reason, mixed compensation is often used in industrial networks. In this solution, large loads with a constant power demand in continuous operation are compensated individually.

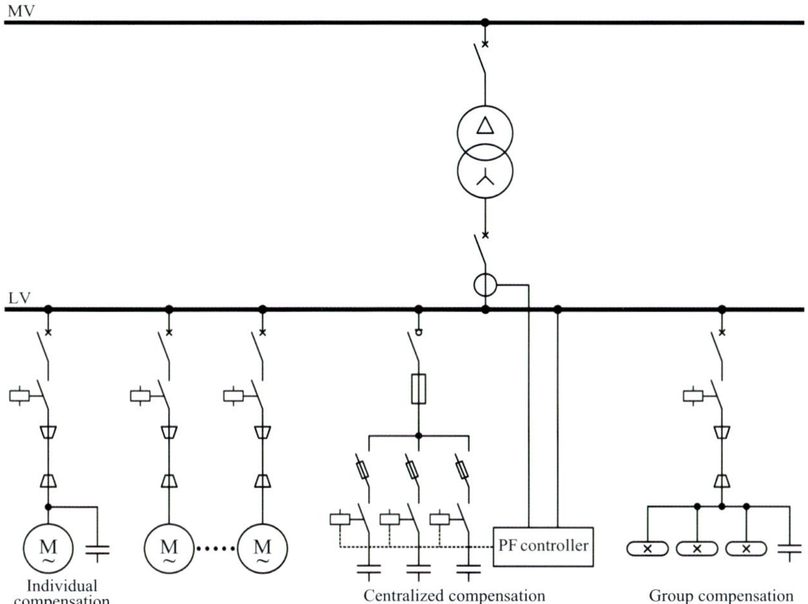

Fig. C12.7 Principle of mixed reactive-power compensation

Whereas group compensation is used in reactive load centres, the time-variable loads are compensated centrally and controllably. Fig. C12.7 shows the principle of mixed compensation as used, for example, in large industrial plants.

12.2.3 Choosing the most advantageous type of compensation

To make it easier to decide on the best way of compensating for the reactive power, Table C12.8 provides information about the specific operation conditions and the economic and installation engineering-related advantages and disadvantages of each type of compensation.

A common feature of industrial networks (especially in the metal-processing industry) are numerous loads with different power demands and varying duty ratios. To operate industrial networks with this load structure energy-efficiently, centralized compensation systems are mainly used. For energy-efficient operation of radial networks in an interconnected cable system (Section 10.3.1.3) and radial networks interconnected through busbar trunking systems (Section 10.3.1.5), they are integrated into the load-centre substations (Section 11.2.5). By integrating automatic reactive-power compensation systems into the load-centre substations, it is usually possible to match the actual cos φ of the network configurations stated above to the agreed target cos φ with ade-

Table C12.8 General evaluation of the compensation types according to [12.8]

Type of reactive-power compensation	Operation conditions / characteristics	Advantages	Disadvantages
Individual compensation	• Use for large loads with constant power demand and high duty ratio • A separate capacitor with the correct rating is allocated to each load	• Compensation of the inductive reactive power directly where it arises • Relief of the load feeder cables from reactive current • Reduced conduction losses • Saving on a separate switching device for the capacitors	• Installed capacitor power consisting of very small capacitors results in higher costs than a larger capacitor that is adapted to the real reactive power demand of the LV system
Group compensation	• Use for groups of inductive loads that are usually in close physical proximity • The PF correction system is assigned to one load group each • The capacitor is switched together with the load group	• Reduction of capacitor costs • Saving on an additional switching device for the capacitor unit • Relief of the distribution cables from reactive current	• Reactive currents remain in the incoming feeder cables
Centralized compensation	• Use for many smaller loads with different power demands and varying duty ratio • Utilization of the capacitive power at one point in the system (e. g. a load-centre substation)	• Better utilization of the installed capacitor power • Better adaptation of the capacitor power to the reactive power demand • General improvement in the voltage stability of the system • General reduction of the system losses • Automatic power factor correction (APFC) by a microprocessor-based PFC controller possible • Simple subsequent installation or expansion	• Residual transmission of reactive power in the LV system due to utilization of the capacitive power at just one separate point
Hybrid or mixed compensation	• Individual PF correction of the loads for which it is intended and group and/or centralized PF correction of the other loads		

quate precision. It is still necessary to decide on a case-by-case basis whether the installation of automatic PF correction systems oriented towards time-variable reactive power demand in load-centre substations can be cost-efficiently supplemented by individual compensation of relatively large loads with a constant power demand and a high duty ratio.

12.2.4 Reactive-power compensation of three-phase asynchronous motors and distribution transformers

Separate compensation of the reactive power of three-phase asynchronous motors and distribution transformers is one of the typical applications of individual compensation. These two applications for compensation of reactive power directly where it arises are described below.

12.2.4.1 Three-phase asynchronous motors

To avoid self-excitation of a decelerating motor that results in hazardous overvoltage at the motor and capacitor terminals, the capacitor power Q_c to be installed must never exceed the no-load reactive power of the motor. For that reason, a capacitor power is chosen that is no more than 90 % of the inductive no-load reactive power of the motor to be compensated. The compensation rating equation for three-phase asynchronous motors is therefore:

$$Q_c = 0.9 \cdot \sqrt{3} \cdot U_{nN} \cdot I_{0M} \tag{12.5}$$

I_{0M} no-load current of the motor

If the no-load current is missing, it is possible to calculate the capacitor power Q_c to be installed approximately based on the motor rating plate data as follows:

$$Q_c \approx 0.9 \cdot \frac{P_{rM}}{\eta_{rM}} \cdot \left(\frac{1 - \cos\varphi_{rM}}{\cos\varphi_{rM} \cdot \sin\varphi_{rM}} \right) \tag{12.6}$$

P_{rM} rated motor power output
η_{rM} motor efficiency
$\cos\varphi_{rM}$ power factor of the motor at rated load

In addition to the compensation rating equations given, Table C12.9 contains guidance values for assigning capacitor powers Q_c to rated motor power outputs P_{rM} at a synchronous motor speed of $n_{syn} = 1{,}500\ \text{min}^{-1}$ (rpm). The guidance values stated in Table C12.9 for Q_c must be converted as follows at speeds differing from 1,500 revolutions per minute [12.8]:

- $n_{syn} = 750\ \text{min}^{-1}$: $1.15 \cdot Q_c$
- $n_{syn} = 1{,}000\ \text{min}^{-1}$: $1.05 \cdot Q_c$
- $n_{syn} = 3{,}000\ \text{min}^{-1}$: $0.90 \cdot Q_c$

For dimensioning the fixed compensation of the motor, the starting method of the motor (direct-on-line (DOL) or star-delta) is also relevant. In the case of DOL starting of squirrel-cage or wound-rotor motors, the delta-arranged capacitors can be connected directly to the motor terminals U, V, W.

The motor and capacitor unit are switched together (Fig. C12.10). For the capacitors connected according to Fig. C12.10, no special short-circuit protection is usually necessary. The short-circuit protection of the capacitors is taken over from the motor fuses (fuse protection) or instantaneous short-circuit release (fuseless protection).

When parameterizing the overload protection, it is important to note that the overload relay (Fig. C12.10a) or overload release (Fig. C12.10b) must be set to a reduced current. The setting current for the overload protection of individually compensated motors must be calculated as follows:

$$I_R = \sqrt{I_{active\text{-}M}^2 + (I_{reactive\text{-}M} - I_{rC})^2} \tag{12.7}$$

$$I_{active\text{-}M} = I_{rM} \cdot \cos\varphi_{rM} \tag{12.7.1}$$

$$I_{reactive\text{-}M} = \sqrt{I_{rM}^2 - I_{active\text{-}M}^2} \tag{12.7.2}$$

$$I_{rC} = \frac{Q_{rC}}{\sqrt{3} \cdot U_{nN}} \tag{12.7.3}$$

I_R current setting of the overload relay or release (L release, see Section 13.1.2)
$I_{active\text{-}M}$ active component of the rated motor current
$I_{reactive\text{-}M}$ reactive component of the rated motor current
I_{rC} rated capacitor current
I_{rM} rated motor current
Q_{rC} rated capacitor power for operation at nominal voltage U_{nN}

Rated motor power output P_{rM} [kW]	Capacitor power Q_c [kvar]
1.0 ··· 3.9	≈ 55 % of P_{rM}
4.0 ··· 4.9	2
5.0 ··· 5.9	2.5
6.0 ··· 7.9	3.0
8.0 ··· 10.9	4.0
11.0 ··· 13.9	5.0
14.0 ··· 17.9	6.0
18.0 ··· 21.9	8.0
22.0 ··· 29.9	10.0
30.0 ··· 39.9	≈ 40 % of P_{rM}
≥ 40	≈ 35 % of P_{rM}

Table C12.9
Guidance values for fixed reactive-power compensation of motors with a rotational speed of n_{syn} = 1,500 min^{-1} (rpm) [12.8]

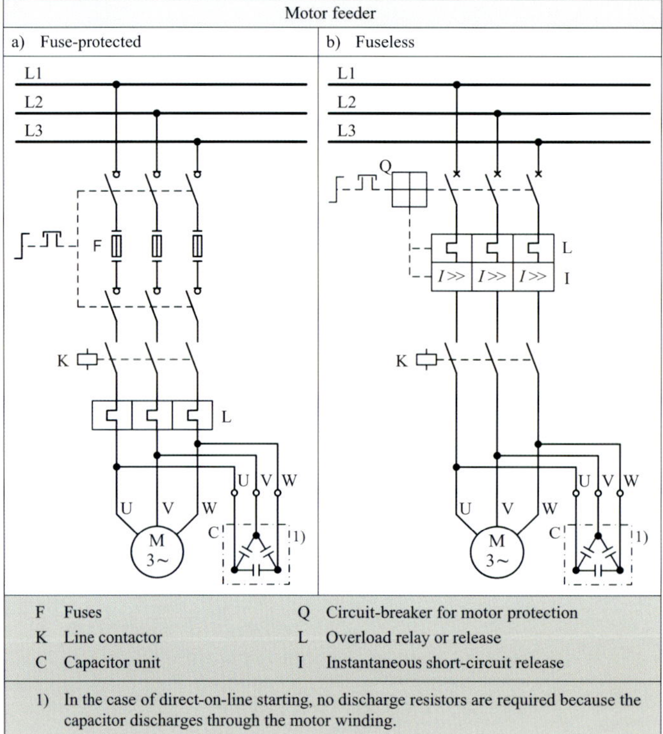

Fig. C12.10 Fixed reactive-power compensation for direct-online starting of squirrel-cage and wound-rotor motors

If the motor is started using star-delta starters, the capacitor unit must not be connected directly to its terminals U, V, W.

In the case of motors that start using a star-delta starter and whose capacitor remains on the star-connected winding during the changeover to delta connection, a dangerous self-excitation occurs in the dead interval. The motor acts as a generator and draws its excitation current from the capacitor unit. Thus an overvoltage of up to twice the value

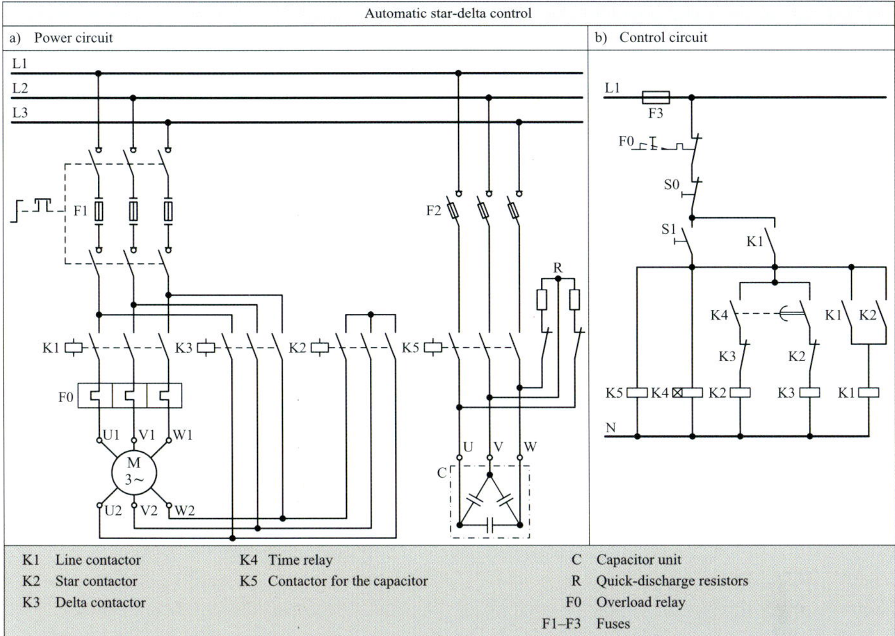

Fig. C12.11 Individual compensation with an automatic star-delta starting of motors

of the system voltage can occur at the terminals of the motor, driven by the flywheel effect of a coupled production machine [12.9].

A further risk that can arise is that the charged capacitor may be connected in phase opposition to the system voltage during the de-energized switchover interval between star and delta connection and the associated short interruption of the circuit. The subsequent potential equalization would result in high compensating currents that subject the motor and capacitor to heavy stress and, above all, would cause great contact wear on the power contactors. The contactor connection shown in Fig. C12.11 is used to avoid self-excitation, series resonance and connection in phase opposition.

In the contactor circuit diagram illustrated in Fig. C12.11, the capacitor unit (C) is separately connected to the power system by an additional contactor (K5). The capacitor unit is discharged through low-resistance resistors (R) when the contactor has dropped out. If discharge resistors are used, it must be ensured that the discharge voltage remaining on the capacitor unit has decreased to a value smaller than $0.1 \cdot U_{nN}$ before reconnection.

12.2.4.2 Distribution transformers

When under load, transformers absorb a certain reactive power $Q_{load\text{-}T}$ that is composed of the no-load reactive power Q_{0T} and the stray-field reactive power of the short-circuit reactance. The following applies:

$$Q_{load\text{-}T} = Q_{0T} + \frac{u_{rZ}}{100} \cdot \left(\frac{S_{load}}{S_{rT}}\right)^2 \cdot S_{rT} \tag{12.8}$$

$$Q_{0T} \approx S_{0T} = \frac{i_{0T} \cdot S_{rT}}{100} \tag{12.8.1}$$

u_{rZ} impedance voltage at rated current as a percentage
S_{rT} rated power of the transformer in kVA
S_{0T} no-load apparent power of the transformer in kvar
S_{load} load in kVA
Q_{0T} no-load reactive power of the transformer in kvar
i_{0T} no-load current of the transformer as a percentage

To avoid overcompensation in times of low load, the capacitor power Q_c must not be determined according to the maximum reactive power demand $Q_{load\text{-}T}$. It must be based exclusively on the magnetizing reactive power Q_{0T} of the transformer under no load.

To calculate the no-load reactive power Q_{0T} of 4GB GEAFOL cast-resin transformers, the no-load currents i_0 from Table C11.8a/b must be inserted into Eq. (12.8.1). The no-load reactive power Q_{0T} must preferably be compensated for by installation of a fixed capacitor unit on the secondary side of the transformer.

In the case of distribution transformers with reduced no-load losses (see Table C11.8a/b), installation of a non-load-dependent fixed capacitor unit does not usually have any considerable economic advantages. For such transformers, fixed-value individual compensation only makes sense if part of the reactive power of the connected loads is to be compensated for. In this application of transformer fixed-value compensation, it is important to ensure that no resonance problems with harmonics occur. To avoid such resonance problems, the installed capacitor power Q_c of the fixed compensation unit without reactors must not exceed a certain value. Exceeding this value would result in an impermissible voltage rise and amplification of harmonics. The permissible capacitive power Q_c that

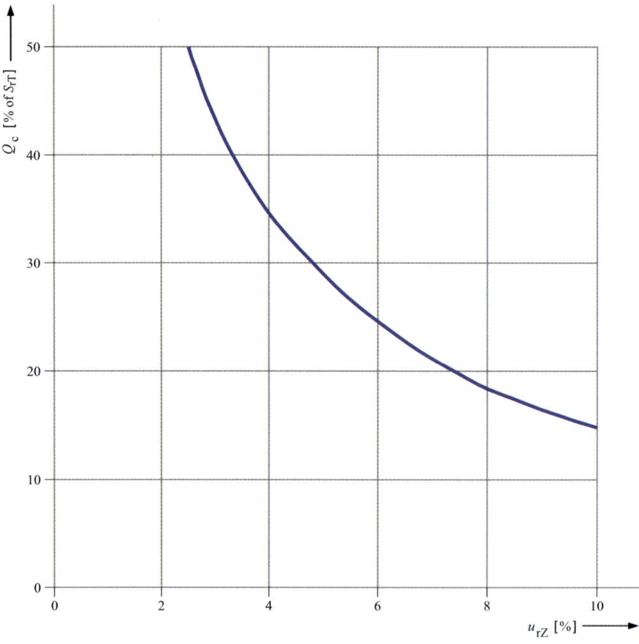

Fig. C12.12 Permissible capacitive power downstream of a no-load transformer for $h = \{5, 7\}$

can be installed downstream of a no-load transformer without amplification of the 5th-order and 7th-order harmonics, normally contained in the wave of the supply voltage ($\underline{U} = \underline{U}_1 + \underline{U}_5 + \underline{U}_7$), is shown in Fig. C12.12 as a function of the impedance voltage at rated current.

An approximate value of the maximum permissible capacitive power Q_c can also be determined by using the following equation:

$$Q_c < \frac{S_{rT} \cdot 100}{h^2 \cdot u_{rZ}} \tag{12.9}$$

S_{rT} rated power of the transformer in kVA

u_{rZ} impedance voltage at rated current as a percentage

h order of the highest critical harmonic ($h = 5, 7, 11, 13, \ldots$)

Example C12

Determining of the capacitor power for exemplary fixed compensation of a transformer

For a 1,000-kVA distribution transformer ($u_{rZ} = 6\,\%$), fixed reactive-power compensation on the secondary side is preferable. Resonance should be avoided up to the 13th-order harmonic. For this purpose, the permissible capacitive power Q_c must be defined. Applying Eq. (12.9) yields the following result:

$$Q_c < \frac{S_{rT} \cdot 100}{h^2 \cdot u_{rZ}} = \frac{1{,}000\ \text{kVA} \cdot 100}{13^2 \cdot 6\,\%} \approx 99\ \text{kvar}$$

The capacitor power to be installed on the secondary-side in a fixed compensation unit must be limited to less than 99 kvar.

Fig. C12.13 shows the schematic diagram of non-load-dependent secondary-side fixed compensation of a transformer. Fuses (F1) are used for short-circuit protection of the capacitor unit (C). Because fuses may operate, discharge resistors (R1, R2) must be connected to the capacitor terminals.

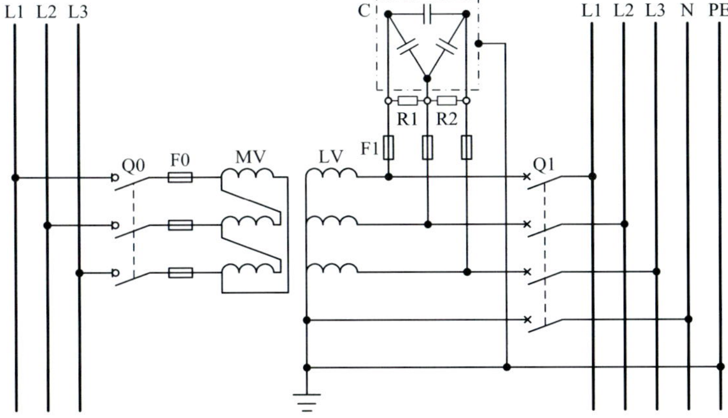

Fig. C12.13 Non-load-dependent secondary-side fixed reactive-power compensation of a distribution transformer

12.2.5 Connecting and operating automatic compensation systems

For energy-efficient operation of decentralized multiply-fed industrial networks, each load-centre substation must have its own centralized reactive-power compensation system for automatic PF correction. The automatic reactive-power compensation systems used for load-related control of the reactive power comprise a PF controller and a power section. The power section includes:

- fuse-switch-disconnector and connecting cable,
- power capacitors without reactors (for linear loads) or with reactors (for non-linear loads) with parallel discharge resistors,
- contactors for switching the power capacitors,
- fuses for the capacitor branch circuits.

Fig. C12.14 shows the schematic diagram of a centralized reactive-power compensation system for automatic PF correction. Because the automatic compensation units are nowadays supplied already wired and ready to connect, planning and dimensioning of centralized compensation systems should focus on the electrical operating conditions (current transformer for connection of the PF controller, number of control steps and step power, *C*/*k* response value of the PF controller, fuse protection and cross-sectional area of the connection cable, rated voltage and overload capability of the power capacitors). Meeting the most important electrical conditions for safe and reliable operation of automatic compensation systems is explained below.

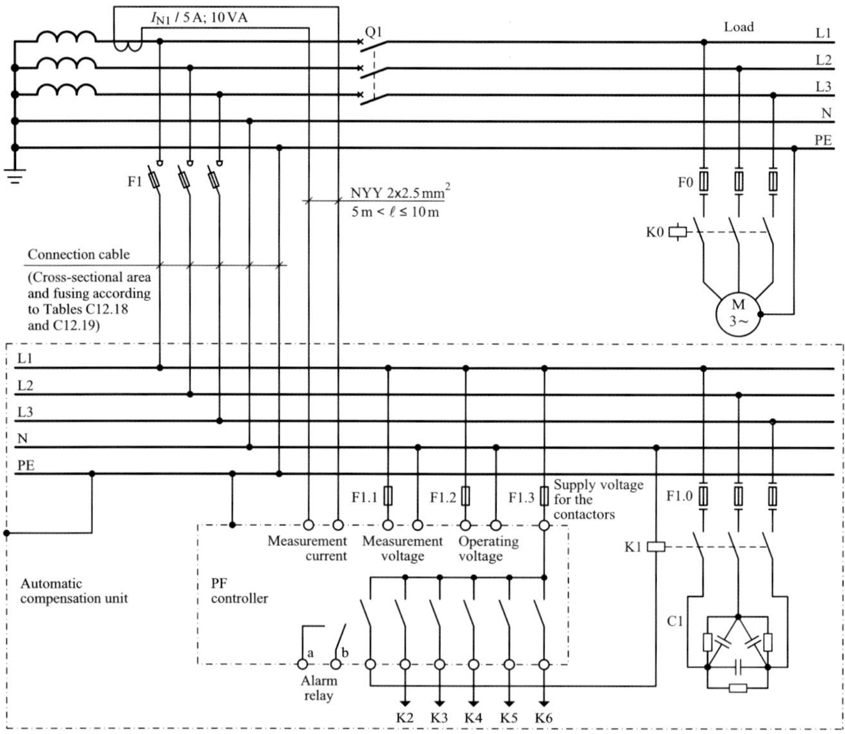

Fig. C12.14 Schematic diagram of a centralized reactive-power compensation system for automatic PF correction

12.2.5.1 Selecting a current transformer for the PF controller

Connection of a PF controller with a measurement current input of $I_{N2} = 5$ A (normal type) or $I_{N2} = 1$ A makes installation of a current transformer necessary. The nominal primary current I_{N1} of the current transformer must be selected according to the maximum load current of the supplying distribution transformer. In the case of 4GB GEAFOL cast-resin transformers (Section 11.1) that are installed in load-centre substations with forced air circulation (AF-cooling with radial-flow fans), selection must be made according to the AF load current. The following applies:

$$I_{N1} \geq I_{T\text{-}AF} = \frac{k_{AF} \cdot S_{rT}}{\sqrt{3} \cdot U_{rT}} \tag{12.10}$$

I_{N1} nominal primary current of the current transformer

$I_{T\text{-}AF}$ maximum load current of the transformer with radial-flow fans switched on (AF mode)

S_{rT} rated power of the transformer

U_{rT} rated voltage of the transformer

k_{AF} factor for the overload capability in AF mode ($k_{AF} \leq 1.4$)

The following data are relevant to current transformers provided for connecting the PF controller [12.8]:

- type of current transformer (e.g. bushing type, core balance or bar-primary type),
- highest voltage for equipment U_m ($U_m = 660$ V or $U_m = 800$ V is usual in the normal version),
- rated transformation ratio $k = I_{N1}/I_{N2}$ (e.g. $k = 2500$ A/5 A $= 500$),

Table C12.15 Current change caused in the PF controller of an automatic compensation unit

Nominal primary current I_{N1} of the current transformer [A]	Capacitor power Q_c [kvar] and resulting current change ΔI [A] for I_{N1} / 5 A current transformers												
	5	10	15	20	25	30	40	45	50	60	75	100	150
50	0.144	0.289	0.433	0.577	0.722	0.866	1.155	1.299	1.443	1.732	2.165	2.887	4.330
75	0.096	0.192	0.289	0.385	0.481	0.577	0.770	0.866	0.962	1.155	1.443	1.925	2.887
100	0.072	0.144	0.217	0.289	0.361	0.433	0.577	0.650	0.722	0.866	1.083	1.443	2.165
150	--	0.096	0.144	0.192	0.241	0.289	0.385	0.433	0.481	0.577	0.722	0.962	1.443
200	--	0.072	0.108	0.144	0.180	0.217	0.289	0.325	0.361	0.433	0.541	0.722	1.083
250	--	0.058	0.087	0.115	0.144	0.173	0.231	0.260	0.289	0.346	0.433	0.577	0.866
300	--	--	0.072	0.096	0.120	0.144	0.192	0.217	0.241	0.289	0.361	0.481	0.722
400	--	--	0.054	0.072	0.090	0.108	0.144	0.162	0.180	0.217	0.271	0.361	0.541
500	--	--	--	0.058	0.072	0.087	0.115	0.130	0.144	0.173	0.217	0.289	0.433
600	--	--	--	--	0.060	0.072	0.096	0.108	0.120	0.144	0.180	0.241	0.361
700	--	--	--	--	0.052	0.062	0.082	0.093	0.103	0.124	0.155	0.206	0.309
800	--	--	--	--	--	0.054	0.072	0.081	0.090	0.108	0.135	0.180	0.271
1,000	--	--	--	--	--	--	0.058	0.065	0.072	0.087	0.108	0.144	0.217
1,200	--	--	--	--	--	--	--	0.054	0.060	0.072	0.090	0.120	0.180
1,500	--	--	--	--	--	--	--	--	--	0.058	0.072	0.096	0.144
2,000	--	--	--	--	--	--	--	--	--	--	0.054	0.072	0.108
2,500	--	--	--	--	--	--	--	--	--	--	--	0.058	0.087
3,000	--	--	--	--	--	--	--	--	--	--	--	--	0.072
4,000	--	--	--	--	--	--	--	--	--	--	--	--	0.054

- rated power S_{rb} (S_{rb} = 10 VA is generally chosen),
- accuracy class (classes 1 to 3).

When selecting the current transformer, the sensitivity of the PF controller must also be taken into account. A specific current change ΔI is required to start the control process. For example, a Modl PF controller [12.11, 12.12] in conjunction with a 5-A current transformer only initiates control as from a current change of $\Delta I \geq 50$ mA.

Table C12.15 provides information about the current changes ΔI caused in PF controllers when deployed in automatic compensation units with different sizes of current transformer and capacitor step powers.

The values stated in Table C12.15 apply to a nominal system voltage of $U_{nN} = 400$ V. They can also be used for current transformers with a transformation ratio of I_{N1}/1A because, if an I_{N1}/1A current transformer is installed, not only the current change ΔI but also the sensitivity of the controller (see Eq. 12.11) is reduced by a factor of 5. If an automatic compensation unit is used in 690-V networks, the values from Table C12.15 must be multiplied by a factor of 0.58 (400 V / 690 V). Only current-transformer step-power combinations are permissible that still ensure a current change of $\Delta I \geq 0.050$ A after this multiplication.

Correct installation of the current transformer is essential for correct power-factor correction (PFC). Seen in the direction of power flow, the current transformer must always be installed upstream of the branch circuit to the compensation unit (Fig. C12.14). Installation downstream of the branch circuit to the compensation unit would prevent detection of the compensation effect.

12.2.5.2 Defining the number of steps and the step power

To implement a PFC system that is matched to the prevailing network conditions, the necessary total capacitor power must be divided into steps. The number of steps denotes the ratio of the total compensation power to the smallest switchable capacitor power. Good matching to the required power factor cos φ_2 can be achieved with just five switching steps [12.10].

In decentralized multiply-fed LV networks, no more than three to four capacitor control steps are required for each load-centre substation. For example, if $n = 4$ decentralized distribution transformers are supplying the LV network (see Example C10 in Section 12.2.1), division of the total capacitor power should be performed in 12 to 16 steps.

Only in industrial plants with a relatively large number of motors and small connected loads might it be convenient to provide more finely stepped PFC. Experience has shown that, to correct the power factor of cos $\varphi_1 = 0.7$ to cos $\varphi_2 = 0.9...0.95$, approximately 40 % of the transformer rating is required as capacitor rating ($Q_c \approx 0.4 \cdot S_{rT}$). The capacitor step power should be approximately 10 to 20 % of the required capacitor power ($Q_{c\text{-step}} \approx (0.1...0.2) \cdot Q_c$). The capacitor step power is therefore the smallest switchable capacitor power. Freely selectable step function ratios are characteristic of the allocation and gradation of the total capacitor power. A distinction is made between step function ratios based on arithmetic series (1:1:1:...), mixed series (1:2:2:...) and geometric series (1:2:4:...). The arithmetic series entails greater expense for components but permits use of contactors with a correspondingly low switching capacity.

Table C12.16a shows the arithmetic switching sequence of a 400-kvar PFC unit as a switching step diagram.

The equivalent switching step diagram for a mixed step function ratio is illustrated by Table C12.16b.

Today's PF controllers are capable of all three step function ratios described. For design and budget reasons, however, intelligent step function ratios that combine the classic

Table C12.16a Switching step diagram (Q_c = 400 kvar, arithmetic step function ratio)

Capacitor step	Switching step					
Arithmetic step function ratio: 1:1:1:1:1	0	1	2	3	4	5
80						
80						
80						
80						
80						
Power of the switching step [kvar]	0	80	160	240	320	400
Capacitor step:	switched off/ disconnected			switched on/ connected		

Table C12.16b Switching step diagram (Q_c = 400 kvar, mixed step function ratio)

Capacitor step	Switching step					
Mixed step function ratio: 1:2:2	0	1	2	3	4	5
80						
160						
160						
Power of the switching step [kvar]	0	80	160	240	320	400
Capacitor step:	switched off/ disconnected			switched on/ connected		

step function ratios (e.g. 1:1:2:2:2:4) are used. An “intelligent” controller attempts to switch capacitors of the same power equally frequently.

12.2.5.3 Setting the controller sensitivity (*C*/*k* response value)

The PF controller detects the reactive power at the incoming supply by measuring the current and voltage (see Fig. C12.14). If the reactive power deviates from the set target value, the controller outputs control commands to the capacitor contactors and connects or disconnects them in steps, as required.

The PF-controller requires a defined sensitivity value to initiate the control process. This sensitivity value is termed the C/k response value (C = capacitor with the smallest step power, k = transformation ratio of the current transformer). The C/k response value specifies the point from which control can be initiated. Taking account of all tolerances, that is, those of the controller, current transformer and the capacitors, a response threshold of 60 to 85 % of the smallest capacitor step power has emerged in practice [12.8]. In accordance with this response threshold, the C/k value must be calculated as follows:

$$C/k = (0.60...0.85) \cdot \frac{Q_c}{\sqrt{3} \cdot U_{nN} \cdot k} \qquad (12.11)$$

Q_c smallest capacitor step power
U_{nN} nominal system voltage
k transformation ratio of the current transformer ($k = I_{N1}/I_{N2}$)

The *C*/*k* response value calculated according to Eq. (12.11) ensures that temporary load peaks do not result in hunting of the automatic control. Modern controllers also permit non-hunting control by time-delayed connection and disconnection of the capacitor step power. For centralized compensation of the reactive power, controllers with the following *C*/*k* value settings are now available on the market:

- manual *C*/*k* value setting (setting of the PF controller according to the result calculated with Eq. (12. 11)),
- semi-automatic *C*/*k* value setting (entry of the transformation ratio *k* of the current transformer and the smallest capacitor step power Q_c),
- fully automatic *C*/*k* value setting.

In PF controllers with fully automatic *C*/*k* value setting, it is neither necessary to enter the CT transformation ratio *k* nor the capacitor step power Q_c. The compensation effect achieved by each capacitor is stored and monitored, including the response threshold of 60 to 85 %. Because of the "intelligence" of the PF controller, very adaptable reactive-power compensation can be achieved without an invariably predefined switching step program.

12.2.5.4 Requirements, connection and fuse protection of the power capacitors

Power capacitors for improving the power factor cos φ of LV networks must meet the requirements defined in DIN EN 60831-1 (VDE 0560-46): 2003-8 [12.13] or IEC 60831-1: 2002-11 [12.14]. The values specified in Table C12.17 must be complied with for the electrical strength of the power capacitors for PF correction.

The decisive quantity for the electrical strength of the capacitor is its rated voltage U_{rC}. When selecting the rated voltage U_{rC}, both the nominal system voltage U_{nN} and the reaction of the capacitors on the power system must be taken into account.

Table C12.17 Permissible voltage values for operation of power capacitors for PF correction [12.13, 12.14]

<table>
<tr><th>Frequency</th><th>Voltage
(RMS value)</th><th>Maximum
duration</th><th>Remarks</th></tr>
<tr><td rowspan="5">System frequency</td><td>$1.00 \cdot U_{rC}$</td><td>Continuous</td><td>Highest average value during any period of capacitor energization</td></tr>
<tr><td>$1.10 \cdot U_{rC}$</td><td>8 h per day</td><td rowspan="2">Low-voltage fluctuations</td></tr>
<tr><td>$1.15 \cdot U_{rC}$</td><td>30 min per day</td></tr>
<tr><td>$1.20 \cdot U_{rC}$</td><td>5 min</td><td rowspan="2">Voltage rise at light load</td></tr>
<tr><td>$1.30 \cdot U_{rC}$</td><td>1 min</td></tr>
<tr><td colspan="4">U_{rC} Rated voltage (RMS value of the sinusoidal alternating voltage for which the capacitor has been designed. For polyphase capacitors with internal electrical connections between the phases and for polyphase capacitor banks, U_{rC} refers to the line-to-line voltage.)</td></tr>
</table>

In industrial networks, there can be a considerable difference between the nominal system voltage and the rated voltage of the capacitor. Capacitors for compensating for the fundamental-frequency (linear) reactive power in decentralized multiply-fed networks with a nominal voltage of $U_{nN} = 400$ V, for example, have a rated voltage in the range 440 V < $U_{rC} \leq 480$ V. At a rated voltage in this range, the voltage increase caused by capacitor units without reactors is usually reliably handled. In decentralized multi-end-fed networks, capacitors cause the following permanent voltage increase:

$$\Delta U \approx U \cdot \frac{\sum_{i=1}^{n} Q_{c_i}}{S_k''} \tag{12.12}$$

ΔU	voltage rise in V
U	voltage before connection of the compensation power installed in the load-centre substation in V
S_k''	characteristic short-circuit power of the LV network at the decentralized installation locations of the load-centre substations in MVA
Q_{c_i}	available compensation power of a decentralized load-centre substation i ($i = 1, 2, ..., n$) in Mvar
i	incrementing index for decentralized load-centre substations, $i = 1(1)n$
n	number of decentralized load-centre substations

For compensation of harmonic-distortion (non-linear) reactive power in 400-V power systems, a rated voltage in the range 525 V ≤ $U_{rC} \leq 690$ V is recommended for capacitor units with reactors.

The capacitive power Q_c (Section 12.2.1) always refers to operation at the nominal system voltage U_{nN}. At a rated capacitor voltage that is higher than the nominal system voltage, the compensation effect of the capacitors is diminished. To achieve the full capacitive power with an increased electrical strength of the capacitors, the following calculation must be made:

$$Q_{rC} = Q_c \cdot \left(\frac{U_{rC}}{U_{nN}}\right)^2 \tag{12.13}$$

Q_{rC}	capacitor rating required for the increased electrical strength in kvar
Q_c	capacitive power required to improve the power factor from $\cos\varphi_1$ to $\cos\varphi_2$ at the nominal system voltage in kvar
U_{rC}	rated capacitor voltage for the increased electrical strength in V
U_{nN}	nominal system voltage in V

In a 400-V network, for example, according to Eq. (12.13), dimensioning of a 250-kvar capacitor bank for the increased rated capacitor voltage $U_{rC} = 525$ V would necessitate installation of a rated power of $Q_{rC} = 430$ kvar ($Q_{rC} = 250$ kvar (525 V/400 V)^2 = 430 kvar).

In addition to the necessary electrical strength, the power capacitors must have a certain safety margin for thermal loads due to the fundamental current and harmonic currents. According to DIN EN 60831-1 (VDE 0560-46): 2003-08 [12.13] or IEC 60831-1: 2002-11 [12.14], the safety margin is limited to $1.3 \cdot I_{nC}$ in continuous operation with sinusoidal rated voltage. Considering the capacitance tolerance of $1.15 \cdot C_r$, the maximum permissible current can have values up to $1.5 \cdot I_{nC}$. These maximum values of the current-carrying capacity must not be exceeded during operation of capacitor units either with or without reactors (Section 12.3.2.1).

Taking the permissible current-carrying capacity of power capacitors into account, Tables C12.18 and C12.19 contain fuse protection recommendations for compensation units with and without reactors. In addition to the fuse protection recommendations, these tables also contain recommendations for dimensioning the connection cables of capacitor units and capacitor banks.

The recommendations for dimensioning the connection cables are based on the current-carrying capacity criteria of DIN VDE 0298-4 (VDE 0298-4): 2003-08 [12.15] or IEC 60364-5-52: 2009-10 [12.16].

Table C12.18 Recommended line fuses and connection cables for compensation units without reactors [12.12]

Power of the capacitor unit or bank Q_{rC} [kvar]	Nominal system voltage U_{nN}								
	400 V AC; 50 Hz			525 V AC; 50 Hz			690 V AC; 50 Hz		
	Nominal current I_{nC} [A]	Fuse per phase L1, L2, L3 $I_{n\text{-LVHRC}}$ [A]	Cable cross-section per phase L1, L2, L3 A_n [mm^2]	Nominal current I_{nC} [A]	Fuse per phase L1, L2, L3 $I_{n\text{-LVHRC}}$ [A]	Cable cross-section per phase L1, L2, L3 A_n [mm^2]	Nominal current I_{nC} [A]	Fuse per phase L1, L2, L3 $I_{n\text{-LVHRC}}$ [A]	Cable cross-section per phase L1, L2, L3 A_n [mm^2]
≤ 21	30.3	35	10	--	--	--	--	--	--
25	36.1	63	16	27.5	50	10	20.9	50	10
30	43.3	63	16	--	--	--	--	--	--
35	50.5	80	25	--	--	--	--	--	--
40	57.7	100	35	--	--	--	--	--	--
45	64.9	100	35	--	--	--	--	--	--
50	72.2	100	35	54.9	100	35	41.8	63	16
60	86.6	160	70	--	--	--	--	--	--
70	101	160	70	--	--	--	--	--	--
75	108	160	70	82.5	125	35	62.7	100	25
80	115	200	95	--	--	--	--	--	--
100	144	250	120	110	200	95	83.6	125	35
125	180	300	150	137	200	95	105	160	70
150	217	355	2 x 70	165	250	120	126	200	95
160	231	355	2 x 70	--	--	--	--	--	--
175	253	400	2 x 95	192	300	150	146	250	120
200	289	500	2 x 120	220	355	185	167	250	120
250	361	630	2 x 150	275	400	2 x 95	209	315	185
300	433	2 x 355 1)	2 x 185	330	500	2 x 120	251	400	2 x 95
350	505	2 x 400 1)	4 x 95	385	630	2 x 150	293	500	2 x 120
400	577	2 x 400 1)	4 x 120	440	2 x 355 1)	2 x 185	335	500	2 x 120
450	650	2 x 500 1)	4 x 120	495	4 x 400 1)	4 x 95	377	2 x 315 1)	2 x 185
500	722	2 x 630 1)	4 x 150	550	2 x 500 1)	4 x 120	418	2 x 315 1)	2 x 185

1) LV HRC fuses should only be recommended up to size 3 (for 500 V to 630 A and for 690 V to 500 A). For higher fuse currents, LV HRC fuse-switch-disconnectors of sizes 4 and 4a are required. Because their installation requires greater effort, two small LV HRC parallel fuse-switch-disconnectors are better in terms of installation engineering.

Table C12.19 Recommended line fuses and connection cables for compensation units with reactors [12.12]

Power of the capacitor unit or bank Q_{rC} [kvar]	Nominal system voltage U_{nN} = 400 V AC; 50 Hz — Detuning factor p [1) — 5.67 %			7 %			14 %		
	Rated current [2) I_{rC} [A]	Fuse per phase L1, L2, L3 $I_{n\text{-LVHRC}}$ [A]	Cable cross-section per phase L1, L2, L3 A_n [mm²]	Rated current [2) I_{rC} [A]	Fuse per phase L1, L2, L3 $I_{n\text{-LVHRC}}$ [A]	Cable cross-section per phase L1, L2, L3 A_n [mm²]	Rated current [2) I_{rC} [A]	Fuse per phase L1, L2, L3 $I_{n\text{-LVHRC}}$ [A]	Cable cross-section per phase L1, L2, L3 A_n [mm²]
≤ 21	39	50	10	36	50	10	34	50	10
25	47	63	16	43	63	16	40	63	16
30	56	63	16	52	63	16	48	63	16
35	66	80	25	60	80	25	56	80	25
40	75	100	35	69	100	35	64	100	35
45	84	100	35	78	100	35	72	100	35
50	94	100	35	86	100	35	80	100	35
60	112	160	70	104	160	70	96	160	70
70	131	160	70	121	160	70	112	160	70
75	140	160	70	130	160	70	120	160	70
80	150	200	95	138	200	95	128	160	70
100	187	250	120	173	200	95	160	200	95
125	234	300	150	216	250	120	200	250	120
150	281	355	2 x 70	260	300	150	240	300	150
160	300	355	2 x 70	277	355	2 x 70	256	300	150
175	328	400	2 x 95	303	355	2 x 70	281	355	2 x 70
200	374	500	2 x 120	346	500	2 x 120	321	500	2 x 120
250	468	630	2 x 150	433	500	2 x 120	401	500	2 x 120
300	562	2 x 355 [3)	2 x 185	519	630	2 x 185	481	630	2 x 185
350	656	2 x 400 [3)	4 x 95	606	2 x 355	3 x 120	560	630	2 x 185
400	750	2 x 400 [3)	4 x 95	692	2 x 400	3 x 120	640	2 x 400	3 x 120
450	845	2 x 500 [3)	4 x 120	780	2 x 500	4 x 120	721	2 x 500	4 x 120
500	938	2 x 630 [3)	4 x 185	866	2 x 500	4 x 120	801	2 x 500	4 x 120

1) The detuning factor is the ratio of 50-Hz power of the reactor preceding the capacitor to the 50-Hz power of the capacitor. The use of detuned capacitor units (capacitor units with reactors) is described in Section 12.3.2.1.

2) The nominal current I_{nC} stated in Table 12.18 refers to the fundamental-frequency nominal power. Because of the harmonics, the following safety margins arise for the rated current I_{rC} of detuned capacitor units (capacitor units with reactors):

$$I_{rC} = \begin{cases} 1.3 \cdot I_{nC} & \text{if } p = 5.67\% \\ 1.2 \cdot I_{nC} & \text{if } p = 7\% \\ 1.11 \cdot I_{nC} & \text{if } p = 14\% \end{cases}$$

At this design current, the circuit filter reactors are not yet disconnected due to excessive temperature [12.12].

3) LV HRC fuses should only be recommended up to size 3 (for 500 V to 630 A and for 690 V to 500 A). For higher fuse currents, LV HRC fuse-switch-disconnectors of size 4 and 4a are required. Because their installation requires greater effort, two small LV HRC parallel fuse-switch-disconnectors are better in terms of installation engineering.

The recommended cross-sectional areas apply to installation on a wall of four-core PVC power cables with copper conductors (NYY) in air at an ambient temperature of ϑ = +30 °C (reference method of installation C with three loaded cores according to Tables C11.22 and C11.23). Because of the limitation to reference method of installation C and an ambient temperature of $\vartheta = +30\,°C$, the cable cross-sectional areas provided in Tables C12.18 and C12.19 are non-mandatory values for guidance only. For other methods of installation, ambient temperatures and cable groupings, the cable must be precisely dimensioned (Section 11.3).

The power section of a reactive-power compensation system for automatic PF correction must be dimensioned such that all components (switching devices, protection equipment, connection cables) reliably withstand the thermal and dynamic stresses caused by the fundamental current and any harmonic or switching impulse currents.

12.2.5.5 Reactions affecting audio-frequency ripple control systems

Audio-frequency ripple control systems are installed in public power distribution networks to perform switching operations, such as tariff-rate switchover, switching street lighting and lit traffic control equipment on and off, etc. via connected receivers. This is done by feeding audio-frequency signals (AF signals) into the public power distribution network. The AF signals are transmitted by pulse trains in the range $166\text{ Hz} \leq f_{TRA} < 2{,}000\text{ Hz}$. The audio frequency chosen is very much dependent on the extent of the power distribution network. The BDEW (German Association of the Energy and Water Industries [12.17]) recommends audio frequencies of $f_{TRA} < 250\text{ Hz}$ for extensive public networks and $f_{TRA} > 250\text{ Hz}$ for power distribution networks of limited extent.

To ensure that the reliability of a ripple control system and the ripple control receivers connected in the power distribution network is not impaired, industrial power supply installations (customers‘ installations of the operating company of the public power distribution network) must neither impermissibly lower the control voltage of the audio-frequency pulses nor overload the transmission systems [12.18].

In assessing the reactions affecting the ripple-control frequency, it is important to observe whether the point of common coupling of the industrial power distribution plant (customer‘s installation) is in the public MV or in the LV network. The power supply concepts for industrial plants and high-tech businesses described in Section 5.4 feature a point of common coupling in the public MV or 110-kV network. To supply industrial plants through $n \geq 2$ dedicated distribution transformers from the public MV network, the impedance factor α must be used to assess the reactions on the ripple control. To avoid impermissible reactions, the following must apply [12.18]:

$$\alpha = \frac{Z_{TRA}}{Z_{A\text{-}50\,Hz}} \geq 0.4 \qquad (12.14)$$

α impedance factor of the entire industrial plant (customer‘s installation)

Z_{TRA} impedance of the entire industrial plant at the point of common coupling at a ripple-control frequency of f_{TRA}

$Z_{A\text{-}50Hz}$ connection impedance of the industrial or customer‘s installation at a system frequency of $f_N = 50\text{ Hz}$ ($Z_{A\text{-}50Hz} = U_{nN}^2 / P_{An}$)

Precise calculation of the impedances Z_{TRA} and $Z_{A\text{-}50Hz}$ requires detailed knowledge of the loads and equipment in the industrial plant and of how they work.

To assess the reactions in the case of a <u>single</u> dedicated distribution transformer, on the other hand, a simplified calculation according to Eq. (12.15) is usually sufficient [12.18].

$$\alpha^* = \frac{Z^*_{\mathrm{TRA}}}{U^2_{\mathrm{nN}}/S_{\mathrm{rT}}} \geq 0.5 \tag{12.15}$$

α^*	impedance factor of the entire industrial plant (customer's installation) without taking the load into account
Z^*_{TRA}	impedance of the supplying distribution transformer and the centralized reactive-power compensation system
$U^2_{\mathrm{nN}}/S_{\mathrm{rT}}$	impedance calculated from the transformer rating
S_{rT}	rated power of the supply distribution transformer (customer's transformer)
U_{nN}	nominal system voltage

Application of Eq. (12.15) requires that the reaction affecting the ripple control primarily emanate from the centralized reactive-power compensation system. In case of centralized reactive-power compensation with detuned capacitor units, special attention must be paid to reactions on the ripple-control frequencies $f_{\mathrm{TRA}} < 250$ Hz. To avoid impermissible reactions on ripple-control frequencies $f_{\mathrm{TRA}} < 250$ Hz, a certain detuning factor p is required.

Based on an impedance factor of $\alpha^* \geq 0.5$ at the point of common coupling in the public MV network, the following permissible detuning factor results [12.18]:

$$p \geq \frac{\dfrac{1}{n^2} + \dfrac{Q_{\mathrm{rC}}}{S_{\mathrm{rT}}} \cdot \left(\dfrac{1}{2 \cdot n} - u_{\mathrm{rZ}}\right)}{1 + \dfrac{Q_{\mathrm{rC}}}{S_{\mathrm{rT}}} \cdot \left(\dfrac{1}{2 \cdot n} - u_{\mathrm{rZ}}\right)} \tag{12.16.1}$$

Eq. (12.16.1) applies if the detuning frequency is <u>below</u> the ripple-control frequency ($p > 1/(f_{\mathrm{TRA}}/f_{\mathrm{N}})^2$). For very low ripple-control frequencies ($f_{\mathrm{TRA}} < 200$ Hz), a detuning frequency <u>above</u> the ripple-control frequency can also result in a sufficiently large impedance factor α^* [12.18].

For $p < 1/(f_{\mathrm{TRA}}/f_{\mathrm{N}})^2$ the required detuning factor must be calculated as follows:

$$p \leq \frac{\dfrac{1}{n^2} - \dfrac{Q_{\mathrm{rC}}}{S_{\mathrm{rT}}} \cdot \left(\dfrac{1}{2 \cdot n} + u_{\mathrm{rZ}}\right)}{1 - \dfrac{Q_{\mathrm{rC}}}{S_{\mathrm{rT}}} \cdot \left(\dfrac{1}{2 \cdot n} + u_{\mathrm{rZ}}\right)} \tag{12.16.2}$$

p	required detuning factor of the centralized reactive-power compensation system
n	ratio of the ripple-control frequency f_{TRA} to the system frequency f_{N} ($n = f_{\mathrm{TRA}}/f_{\mathrm{N}}$)
Q_{rC}	rated capacitive power (total capacitor rating)
S_{rT}	rated power of the supplying distribution transformer
u_{rZ}	impedance voltage at rated current of the supplying distribution transformer

Section 12.3.2.1 explains reactive-power compensation with detuned capacitor units (capacitor units with reactors) in more detail.

The selection scheme shown in Fig. C12.20 can be used for reactive-power compensation systems to avoid reactions affecting the ripple-control frequencies in public power distribution networks. In the case of points of common coupling in the public MV or LV network, this selection scheme does not release the planner from the obligation to plan each reactive-power compensation system to be installed in an industrial network in close coordination with the responsible operating company of the public power distribution network.

With the introduction of smart-grid technologies (e.g. smart metering) to public power supplies, the importance of audio-frequency ripple-control systems is expected to decline. The gradual development of the public power distribution network into a smart grid [12.19] will obviate the operation of audio-frequency ripple-control systems and simplify the planning of reactive-power compensation systems.

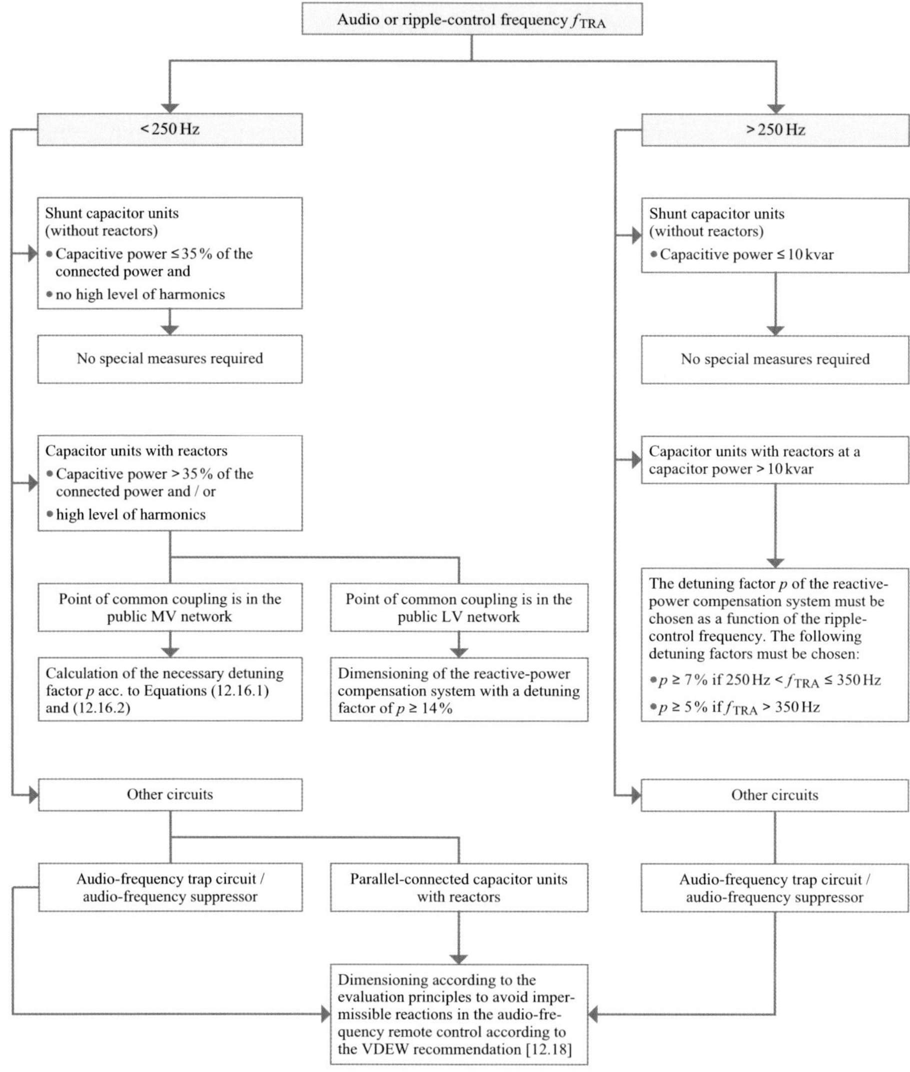

Fig. C12.20 Selection scheme for reactive-power compensation systems according to ripple-control frequencies in the public power distribution network [12.18]

12.3 Compensation when supplying non-linear loads

Typical non-linear loads in industrial networks include converter-fed drives. Due to their use as "electronic gear mechanism" for adjusting and controlling the speed, they cause system disturbances because they draw not only active power but also lagging reactive power and non-sinusoidal current from the network. This current can be broken down into a fundamental current with a system frequency and a series of harmonic currents whose frequencies are integer multiples of the system frequency (Fig. C12.21).

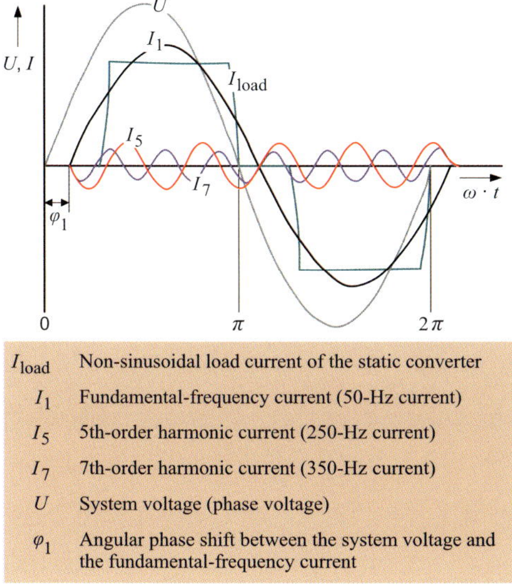

Fig. C12.21
Breakdown of the converter current into fundamental-frequency and harmonic components

The harmonic currents are imposed onto the three-phase system. This produces negative impacts which are eliminated by means of measures described below.

12.3.1 Negative effects of harmonics on the power system

To be able to evaluate the influence on a system by harmonic currents caused by converters and other non-linear loads, it is necessary to know their order h and magnitudes I_h.

Harmonic orders

$$h = p_z \cdot k \pm 1 \quad (k = 1, 2, 3, \ldots; \text{ pulse number of static converter } p_z > 1) \qquad (12.17)$$

Within the range $h \leq 49$ (Table A2.12), which is of interest to us here, according to Eq. (12.17) harmonic orders $h = 5$, 7, 11, 13, 17, 19, 23, 25, 29, 31, 35, 37, 41, 43, 47 and 49 occur in six-pulse converters ($p_z = 6$) and $h = 11$, 13, 23, 25, 35, 37, 47 and 49, in twelve-pulse converters ($p_z = 12$). Twelve-pulse rectification circuits are used only for powers above 1 MW. Depending on the power, use of twelve-pulse static converters is recommended in supply systems with an emergency power supply (standby diesel generator) (Fig. C11.50). The short-circuit power of such supply systems is much higher

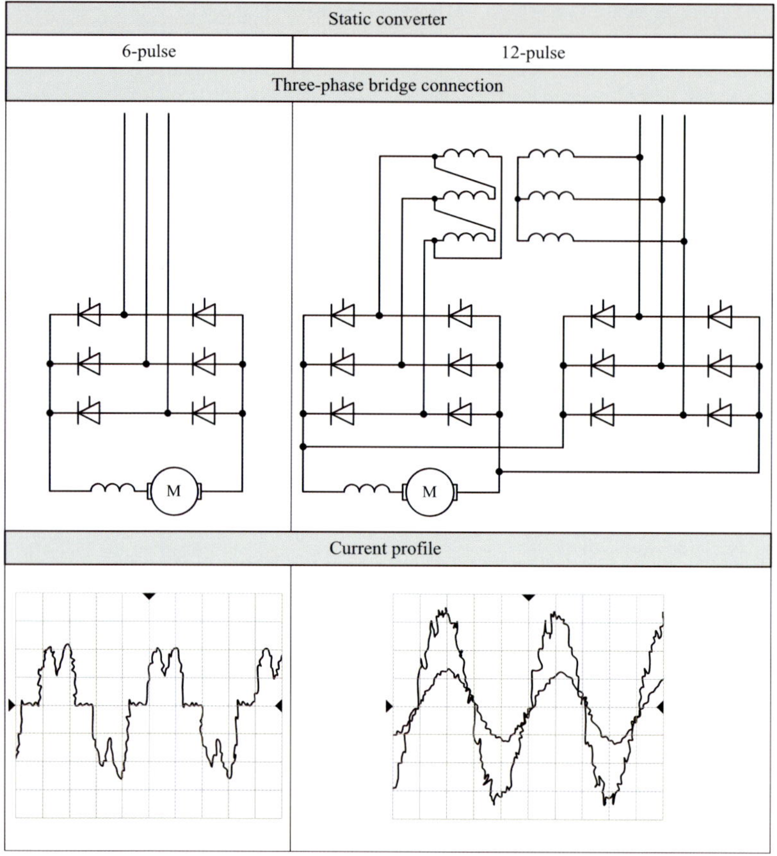

Fig. C12.22 Three-phase bridge circuit and current profile of a six-pulse and twelve-pulse static converter

during normal operation than in emergency operation with the diesel generator. For this reason, the same harmonic currents distort the generator voltage much more heavily than the system voltage [12.20].

By using static converters in a twelve-pulse bridge circuit, it may be possible to comply with the compatibility levels for the standardized voltage quality (Table A2.12) not only in normal but also in emergency power operation without additional measures to reduce harmonics.

Generally, however, the six-pulse bridge circuit is the converter circuit most frequently found in industrial networks. Fig. C12.22 shows the three-phase bridge circuit and current profile of a six-pulse and a twelve-pulse converter.

In addition to converter-fed drives, energy-efficient lighting systems with energy-saving lamps are especially obvious producers of harmonics. Energy-efficient lighting systems with energy-saving lamps produce harmonic currents of the following orders:

$$h = 3 + (k - 1) \cdot 6 \quad (k = 1, 2, 3, \ldots) \tag{12.18}$$

Current harmonics with orders according to Eq. (12.18) are divisible by three ($h = 3, 9, 15, 21, ..., 45$) and are summated in the N/PEN conductor of LV networks. For that reason, N/PEN conductors may have to be dimensioned according to the 150-Hz current (see Tables C10.23, C11.30 and C11.31). As a further remedial measure against excessive loads by 150-Hz current, active filters (Section 12.3.2.3) installed in the N conductors can be considered.

Installation in the PEN conductor is not permitted because PEN conductors according to DIN VDE 0100-460 (VDE 0100-460): 2002-08 [12.21] and IEC 60364-4-41: 2005-12 [12.22] must not be switched or isolated.

Magnitudes

$$I_{\mathrm{h}} \approx \frac{1}{h} \cdot I_1 \qquad (12.19)$$

I_1 fundamental current (to be calculated approximately from the power according to the formula $I_1 = S/(\sqrt{3} \cdot U_{\mathrm{nN}})$

h harmonic order

Because Eq. (12.19) only applies to a perfectly filtered direct current, the following values should be used for calculation in practice:

$I_3 = (0.33...0.77) \cdot I_1$

$I_5 = (0.25...0.30) \cdot I_1$

$I_7 = (0.12...0.13) \cdot I_1$

$I_{11} = (0.06...0.09) \cdot I_1$

$I_{13} = (0.05...0.07) \cdot I_1$

Due to the RMS value range of the current harmonics, all system components in the network are subjected to an additional thermal load. The following formula applies to the thermal load on capacitors by harmonic currents:

$$I_{\mathrm{th}} = \sqrt{I_1^2 + \sum_{\mathrm{h}} I_{\mathrm{h}}^2} \qquad (12.20)$$

I_1 fundamental-frequency current taken over by the capacitor

I_h hth-order harmonic current taken over by the capacitor

Based on the harmonic load according to Eq. (12.20), the dimensioning standard DIN EN 60831-1 (VDE 0560-46): 2003-08 [12. 13] or IEC 60831-1: 2002-11 [12.14] prescribes a 30 % margin for thermal overload of capacitors in compensation units. This thermal margin is usually only exceeded in extreme harmonic loads.

Resonance phenomena due to harmonics are much more dangerous than the additional thermal load from system components. Resonance phenomena with serious consequences can occur if the compensation unit forms a parallel-resonant circuit with the supplying system (see Fig. C12.23).

In this parallel-resonant circuit formed by the capacitance of the connected power capacitors and the inductance of the power system and transformer, the harmonics are generally amplified. If one of the harmonics caused by a static converter is near to the intrinsic frequency of the parallel-resonant circuit shown in Fig. C12.23b, very high currents occur that result in the overloading of equipment or tripping of protective de-

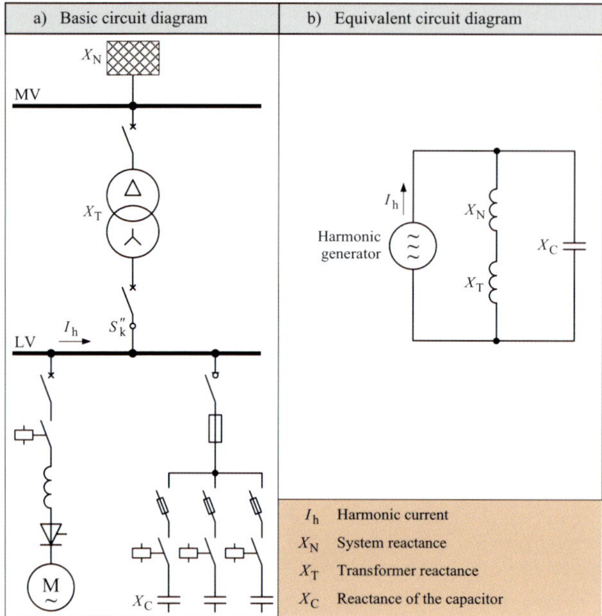

Fig. C12.23 Network configuration showing the possibility of parallel resonance due to converter load

vices. The resonance frequency f_R of such a parallel-resonant circuit can be calculated as follows with initial approximation:

$$f_R \approx f_N \cdot \sqrt{\frac{S''_k}{Q_c}} \tag{12.21}$$

f_R resonance frequency
f_N system frequency
S''_k short-circuit power at the respective system node (e.g. busbar on the supplying distribution transformer)
Q_c capacitive power of the compensation unit

Harmonics also result in additional voltage drops on the system inductances. Because the inductance rises proportionally with the frequency, for example, the current of the fifth-order harmonic on the system inductance causes a voltage drop that is five times greater than that caused by an equally large current at the fundamental frequency. The harmonic voltages caused by the harmonic currents at the system inductances are superimposed onto the originally sinusoidal system voltage (fundamental waveform), thus distorting it.

This influence on the quality and magnitude of the system voltage poses a risk that voltage-sensitive loads may be disturbed or even destroyed. The hazard posed by harmonics in networks with installed PFC systems can be roughly calculated based on the maximum permissible capacitive power Q_c according to Eq. (12.9).

The system perturbations caused by harmonics must be limited to the extent that the internal compatibility of all loads of a process and the external compatibility of the process with the public power distribution is assured. The compatibility levels for harmonics to be complied with by the network operating company in the supply of electrical power from the public distribution network are defined in DIN EN 50160 (EN 50160): 2008-04 [12.23] (see Table A2.10).

The system user must comply with the compatibility levels for harmonics in the consumption of electrical energy at the points of common coupling (PCCs) with the public power distribution network, which are defined in the following guidelines and standards:

- D-A-CH-CZ Guideline 2007 [12.24] (Technical rules for assessing system perturbations),
- DIN EN 61000-2-2 (VDE 0839-2-2): 2003-02 [12.25] or IEC 61000-2-2: 2002-03 [12.26] (Environment – Compatibility levels for low-frequency conducted disturbances and signalling in public low-voltage power supply systems),
- IEC/TR 61000-3-6: 2008-02 [12.27] (Assessment of emission limits for the connection of distorting installations to MV, HV and EHV power systems).

The compatibility levels stated in Tables C12.24 and C12.25 are taken from these standards. These compatibility and indicative planning levels can be used for planning industrial power distribution plants that produce harmonics and whose point of common coupling (PCC) is located in the public LV, MV or HV network.

When planning industrial power distribution plants, the compatibility levels must be observed not only at the external points of common coupling (PCCs) but also at the in-plant points of coupling (IPCs).

The compatibility levels for harmonics to be complied with to ensure a good-quality supply of power to processes at the in-plant points of coupling of an industrial plant are defined in DIN EN 61000-2-4 (VDE 0839-2-4): 2003-05 [12.28] or IEC 61000-2-4: 2002-06 [12.29] (Compatibility levels in industrial plants for low-frequency conducted disturbances) and can be found in Table A2.12.

Table C12.24 Compatibility levels for harmonics to be complied with for a PCC of an industrial plant located in the public LV network ($U_{nN} \leq 1$ kV)

Odd harmonics				Even harmonics	
Non-multiple of 3		Multiple of 3			
Harmonic order h	Permissible voltage harmonic content [%]	Harmonic order h	Permissible voltage harmonic content [%]	Harmonic order h	Permissible voltage harmonic content [%]
5	6.0	3	5.0	2	2.0
7	5.0	9	1.5	4	1.0
11	3.5	15	0.4	6	0.5
13	3.0	21	0.3	8	0.5
$17 \leq h \leq 49$	$2.27 \cdot \left(\frac{17}{h}\right) - 0.27$	$21 < h \leq 45$	0.2	$10 \leq h \leq 50$	$0.25 \cdot \left(\frac{10}{h}\right) + 0.25$
Comment: The compatibility level for the total harmonic distortion is $THD_{LV\text{-perm}} = 8\,\%$. The total harmonic distortion THD must be calculated according to Eq. (2.15). The reference value for the compatibility level of the voltage harmonic contents and of the total harmonic distortion stated as percentages is the fundamental frequency of the system voltage.					

Table C12.25 Indicative planning levels for harmonic voltages for a PCC of an industrial plant located in the public MV or HV network

Odd harmonics						Even harmonics		
Non-multiple of 3			Multiple of 3					
Harmonic order h	Permissible voltage harmonic content [%] MV 1)	HV 2)	Harmonic order h	Permissible voltage harmonic content [%] MV 1)	HV 2)	Harmonic order h	Permissible voltage harmonic content [%] MV 1)	HV 2)
5	5	2	3	4	2	2	1.8	1.4
7	4	2	9	1.2	1	4	1	0.8
11	3	1.5	15	0.3	0.3	6	0.5	0.4
13	2.5	1.5	21	0.2	0.2	8	0.5	0.4
$17 \le h \le 49$	$1.9 \cdot \left(\frac{17}{h}\right) - 0.2$	$1.2 \cdot \left(\frac{17}{h}\right)$	$21 < h \le 45$	0.2	0.2	$10 \le h \le 50$	$0.25 \cdot \left(\frac{10}{h}\right) + 0.22$	$0.19 \cdot \left(\frac{10}{h}\right) + 0.16$

1) $1\,\text{kV} < U_{\text{nN}} \le 35\,\text{kV}$, $\text{THD}_{\text{MV-perm}} = 6.5\,\%$

2) $35\,\text{kV} < U_{\text{nN}} \le 230\,\text{kV}$, $\text{THD}_{\text{HV-perm}} = 3\,\%$

Comment: The total harmonic distortion THD must be calculated according to Eq. (2.15). The reference value for the compatibility level of the voltage harmonic contents and of the total harmonic distortion stated as percentages is the fundamental frequency of the system voltage.

Compliance with the compatibility levels for harmonics in the consumption of electrical power must be verified with the Eqs. (2.14) and (2.16).

12.3.2 Measures to mitigate harmonics

To mitigate the harmonics produced in the LV network directly, capacitor units with reactors, passive tuned filter circuits or active filters are required. The use of passive tuned filter circuits, in particular, is limited to LV networks subject to extreme harmonic loads. Active filters are ideal for harmonics of non-linear loads that hardly require fundamental-frequency reactive power because their power factor is approximately $\cos\varphi = 1$.

12.3.2.1 Installation of capacitor units with reactors

The installation of capacitor units with reactors for reactive-power compensation is recommended wherever the proportion of harmonics-producing loads (static converters) is more than 15 % of the total load. Like capacitor units without reactors, capacitor units with reactors must be chosen in accordance with the reactive power demand of the loads (see Section 12.2.1). However, it is important to note that capacitor units with reactors must be dimensioned for a higher voltage stress. If a capacitor unit with reactors is used, a voltage higher than the system voltage arises on the capacitors. The voltage applied to a reactor-connected capacitor can be calculated as follows [12.30]:

$$U_{\text{c}} = \frac{U_{\text{nN}}}{(1 - p/100)} \tag{12.22}$$

U_{c} voltage applied to a reactor-connected capacitor
U_{nN} nominal system voltage
p detuning factor ($p = X_{\text{L}}/X_{\text{C}}$, see Fig. C12.26)

When used in 400-V networks, a rated capacitor voltage in the range 525 V ≤ U_{rC} ≤ 690 V is recommended to cope with the voltage rise occuring at the reactor-connected capacitors.

Reactor-connected capacitors form a series-resonant circuit that is tuned such that its resonance frequency f_{OR} is below the fifth-order harmonic, that is, 250 Hz. This makes the compensation unit for all harmonics ≥ 250 Hz occurring in the converter current inductive, so that no resonance is possible with the system inductance (see Figs. C12.26 and C12.27).

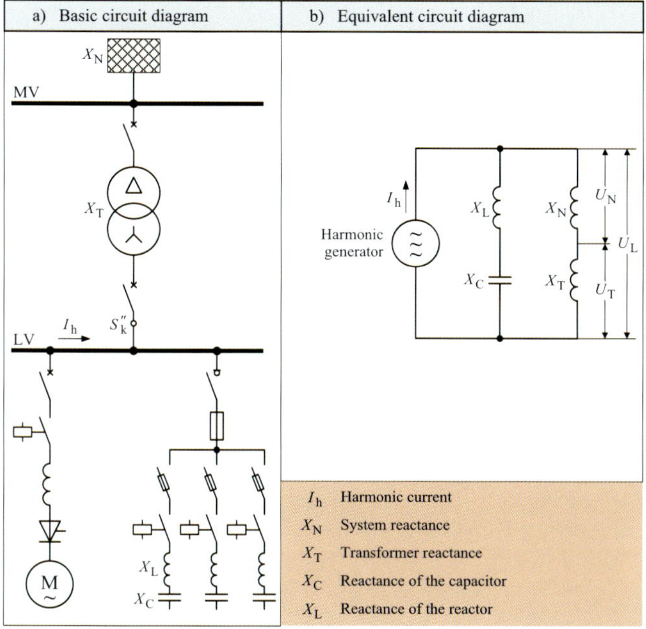

Fig. C12.26 Reactor-connected capacitors in a PFC application

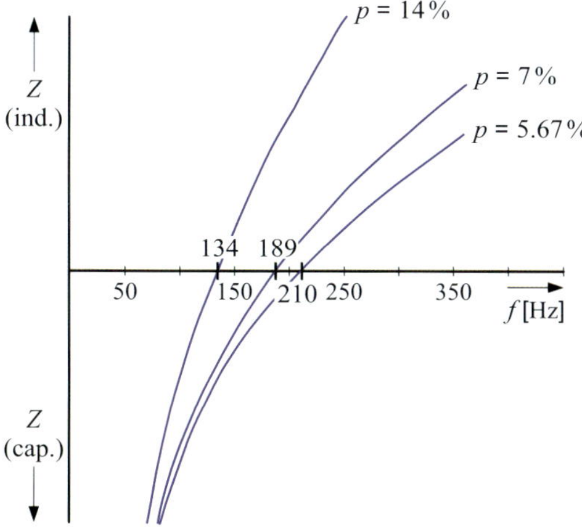

Fig. C12.27
Frequency-related impedance characteristic when using reactor-connected capacitor units for PFC

A further effect of using reactor-connected capacitors is that part of the harmonic currents produced by the converters, especially the fifth-order harmonic currents, flows into the compensation unit. This means that the network is relieved not only of lagging reactive power but also of harmonics. This, in turn, results in improvement of the voltage quality (reduced voltage distortions) in the network.

In practice, reactor-connected capacitor units with a detuning factor in the range between $5\,\% < p \leq 7\,\%$ are primarily used.

The detuning factor p is a percentage value for the size of the upstream reactor reactance X_L in relation to the downstream capacitor reactance X_C at system frequency f_N. Using the detuning factor p, it is possible to determine the tuning frequency of the series-resonant circuit formed by the reactor and capacitor.
The following applies:

$$f_{OR} = n_{OR} \cdot f_N = \frac{f_N}{\sqrt{p/100}} \tag{12.23}$$

$$n_{OR} = \sqrt{\frac{100}{p}} \tag{12.23.1}$$

f_{OR} series-resonance frequency or tuning frequency in Hz
f_N system frequency in Hz
n_{OR} frequency ratio of the series resonance frequency
p detuning factor of the compensation unit as a percentage

The impedance factor α (Eq. 12.14) or α^* (Eq. 12.15) can be used to assess whether the compensation unit with reactor-connected capacitors of a system user impermissibly influences the ripple control of the power system operating company.

Simplified assessment of the reaction affecting the ripple control is possible with the impedance factor α^* if the industrial plant is powered from the public MV network only through one dedicated distribution transformer and the detuning frequency of the compensation unit (series-resonance frequency f_{OR}) is below the audio frequency f_{TRA} applied. On this basis, the impedance factor α^* must be calculated as follows :

$$\alpha^* = \frac{u_{rZ}}{100} \cdot n + \frac{S_{rT}}{Q_{rC} \cdot \left(1 - \frac{p}{100}\right)} \cdot \left(\frac{p}{100} \cdot n - \frac{1}{n}\right) \tag{12.24}$$

u_{rZ} impedance voltage at rated current as a percentage
S_{rT} transformer power rating in kVA
Q_{rC} capacitor rating for the reactive-power compensation in kvar
p detuning factor of the compensation unit as a percentage
n ratio of the audio (ripple-control) frequency f_{TRA} to the system frequency ($n = f_{TRA}/f_N$)

If $\alpha^* \geq 0.5$, an impermissible reaction on the audio frequency is not usually to be expected [12.18].

Example C13

Assessment of the reaction caused by a centralized reactive-power compensation system with reactor-connected capacitor units on the AF ripple control system of the power-supply company (source network operator).

The reactive power of the loads in a very small industrial plant is to be compensated for by a centralized reactive-power compensation system with reactor-connected capacitor units. The industrial plant is supplied by a dedicated 1,250-kVA transformer ($u_{rZ} = 6\,\%$) from the public MV network. The centralized compensation system features a total capacitor rating of $Q_{rC} = 400$ kvar and a detuning factor of $p = 7\,\%$. It is necessary to check whether this compensation system can be installed without an audio-frequency suppressor because of the presence of an audio or ripple-control frequency of $f_{TRA} = 216.7$ Hz. According to Eq. (12.23), the compensation system to be installed exhibits the following series-resonance frequency:

$$f_{OR} = \frac{f_N}{\sqrt{p/100}} = \frac{50\text{ Hz}}{\sqrt{7\,\%/100}} \approx 189\text{ Hz}$$

Because the series-resonance frequency $f_{OR} = 189$ Hz is below the ripple-control frequency $f_{TRA} = 216.7$ Hz, the impedance factor α^* according to Eq. (12.24) can be used to assess the reaction affecting the AF ripple control system of the power supply company.

$$\alpha^* = \frac{u_{rZ}}{100} \cdot n + \frac{S_{rT}}{Q_{rC} \cdot \left(1 - \frac{p}{100}\right)} \cdot \left(\frac{p}{100} \cdot n - \frac{1}{n}\right) =$$

$$\frac{6\,\%}{100} \cdot \frac{216.7\text{ Hz}}{50\text{ Hz}} + \frac{1{,}250\text{ kVA}}{400\text{ kvar} \cdot \left(1 - \frac{7\,\%}{100}\right)} \cdot \left(\frac{7\,\%}{100} \cdot \frac{216.7\text{ Hz}}{50\text{ Hz}} - \frac{1}{\frac{216.7\text{ Hz}}{50\text{ Hz}}}\right) = 0.504$$

Calculation according to Eq. (12.24) yields the impedance factor $\alpha^* \geq 0.5$. According to this result, the reactive-power compensation system with reactor-connected capacitor units for centralized PF correction may be installed in the LV network of the industrial plant without an additional audio-frequency suppressor.

Compensation systems with the detuning factor of $p = 7\,\%$ ($f_{OR} = 189$ Hz) applied in Example C13 are to be used chiefly in networks with an average harmonic load because they only absorb about 30 % of the fifth-order harmonic. The greatest proportion of the harmonics is superimposed on the upstream network with this detuning factor.

Where in-plant production of harmonics is high, compensation systems with a detuning factor of $p = 5.67\,\%$ must be used. With a series-resonance frequency of $f_{OR} = 210$ Hz, about 50 % of the fifth-order harmonic can be suppressed.

In the presence of very low ripple-control frequencies ($f_{TRA} < 200$ Hz), it may be necessary to choose a detuning factor of $p = 14\,\%$ for the compensation system. 14% reactor-connected compensation systems with a series-resonance frequency of $f_{OR} = 134$ Hz achieve sufficient impedance to keep the load on the transmission signal low for all ripple-control frequencies. However, at this series-resonance frequency, only about 10 % of the fifth-order harmonic can be suppressed.

Fig. C12.28 shows, based on a practical example [12.31], that by PF correction of a static converter installation from $\cos \varphi_1 = 0.7$ to $\cos \varphi_2 = 0.9$ using reactor-connected capacitors (bar chart c), the current load of the equipment is reduced by 24 %. PF correction of the static converter installation without reactor-connected capacitors (bar chart b), on the other hand, is not subject to this limiting effect.

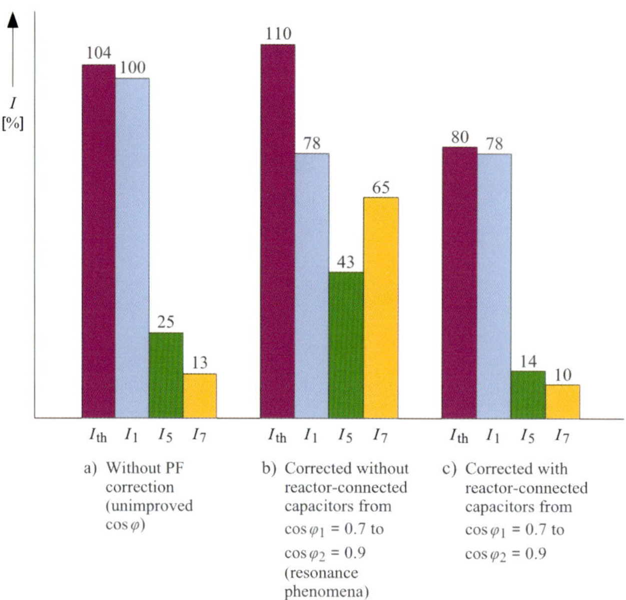

Fig. C12.28 Percentage share and distribution of harmonic currents in uncorrected, no-reactor-corrected and reactor-corrected operation of a converter installation [12.31]

12.3.2.2 Use of tuned filter circuits

A filter circuit is a series-resonant circuit comprising reactors and capacitors that is tuned to defined harmonic frequencies occurring in the network (Fig. C12.29). They are used both for compensation of reactive power and, above all, to prevent the propagation of harmonics to the network. For this reason, the resonant circuits are tuned in such a way that they exhibit impedances approximating to zero for the individual harmonic currents I_h.

Connection of the filter circuits tuned to the harmonics is performed from the step with the lower harmonic order to the step with the larger harmonic order. The filter circuits are disconnected in the reverse sequence.

By analogy with the impedance characteristic of series-resonant circuits with fixed detuning factors p, the impedance characteristic of tuned filter circuits $z = f(f_h)$ also causes harmonics to be suppressed. Fig. C12.30 shows the frequency-related impedance characteristic of filter circuits that are tuned to 250 Hz ($h = 5$), 350 Hz ($h = 7$) or 550 Hz ($h = 11$).

By tuning the filter circuits to the harmonic frequency, it was possible to suppress up to 90 % of the harmonic currents. Fig. C12.31 illustrates the filtering effect of passive tuned filter circuits.

The drawback in using passive tuned filter circuits is the requirement for leading reactive power. To suppress harmonics effectively, the capacitive power Q_{hc} is required. Modern frequency converters exhibit a power factor that is close to unity ($\cos\varphi = 1$). There is therefore no fundamental-frequency (linear) reactive power that has to be compensated for by switchable capacitor units. Tuned passive filters are not suitable for compensating for pure harmonic-distortion (non-linear) reactive power. The harmonic currents in networks with pure harmonic-distortion reactive power can only be mitigated using active filters (Section 12.3.2.3).

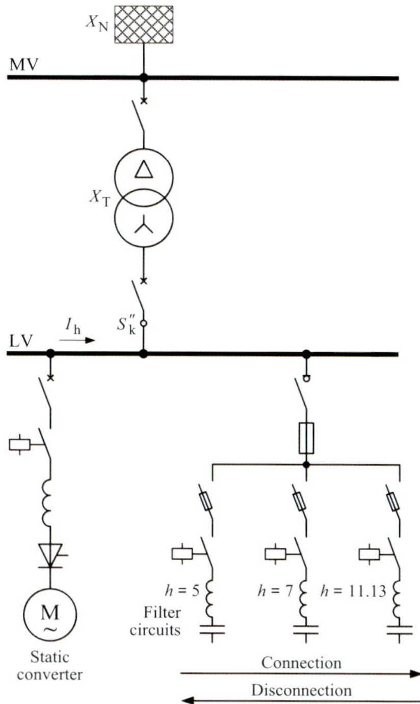

Fig. C12.29
Passive tuned filter circuits for mitigating harmonics

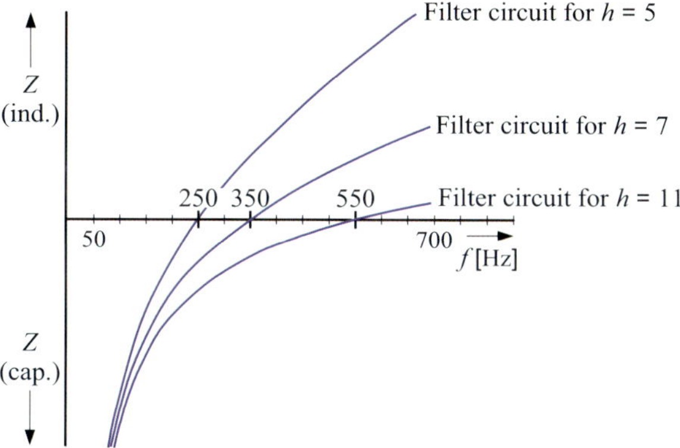

Fig. C12.30 Frequency-related impedance characteristic of passive tuned filter circuits

Filter circuits must be able to suppress the harmonic currents of the static converters in any operating condition. They should always be built up from the lowest harmonic order upwards.

The use of passive tuned filter circuits is useful for the 5th, 7th, 11th and 13th-order harmonics. A common series-resonant circuit is allocated to the 11th and 13th-order harmonics (Fig. C12.31).

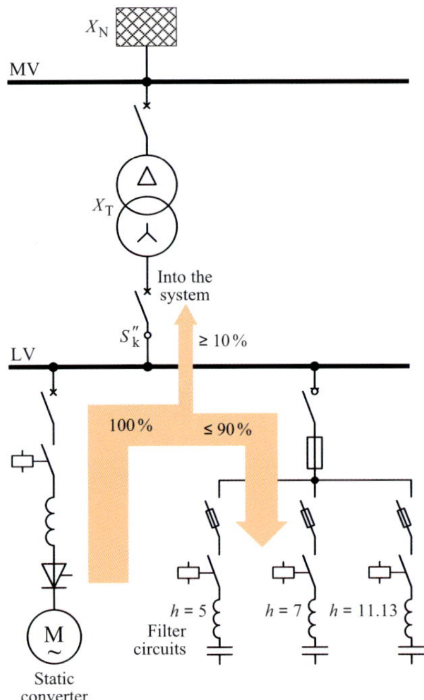

Fig. C12.31
Filtering of harmonic currents with passive tuned filter circuits

The fundamental compensation capacity of a unit is distributed as follows:

- 50 % to the 5th-order harmonic,
- 25 % to the 7th-order harmonic and
- 25 % to the 11th and 13th-order harmonic.

Generally, the dimensioning of filter and series-resonant circuits for harmonics is based on the following power system and installation engineering parameters [12.8]:

- system load voltage and system frequency,
- existing short-circuit power at the point of common coupling (PCC) or point of connection (PC),
- fundamental-frequency reactive power to be compensated for or intended fundamental-frequency compensation capacity,
- expected or actual harmonic currents,
- relevant audio frequency of the power supply company or source network operator.

It is also important to note that filter circuit systems or reactor-capacitor units basically should never be operated in parallel with non-reactor control units, terminated on the same busbar. If reactor-connected capacitor units and capacitor units without reactors are operated jointly on one busbar, parallel resonances may arise that are in the critical range of the harmonics occurring in the network.

12.3.2.3 Operation with active filters

While, with operation with passive filters, it is only possible to absorb harmonic currents of a certain frequency from the network, active filters enable an improvement in power quality aimed at "pure" sinusoidal current and voltage. As active filters, self-commutated converters are used that compensate for the harmonic-distortion reactive power by feeding in the negative (phase-shifted by 180°) harmonic spectrum.

Depending on the method of connection of the static converter, currents and voltages of almost any curve shape can be fed in [12.32]. Fig. C12.32 shows the mode of operation of an active filter with parallel and series coupling.

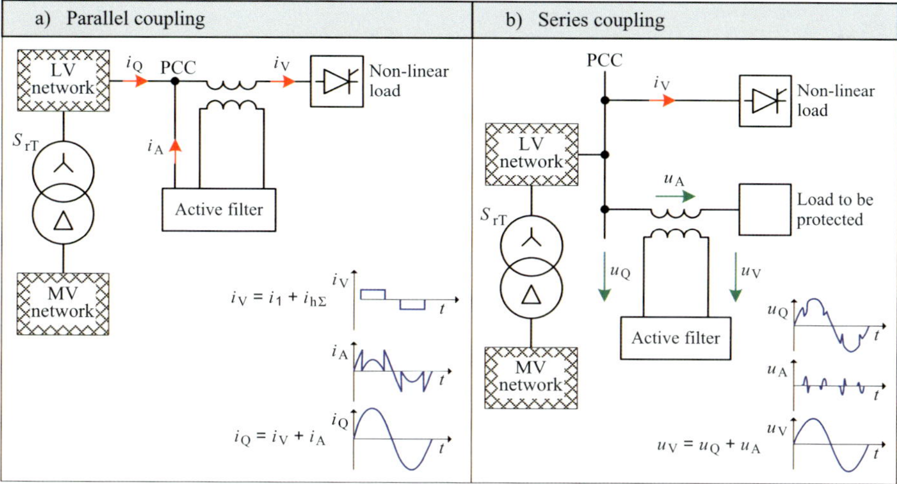

Fig. C12.32 Mode of operation of an active filter with parallel and series coupling [12.32]

The active filter with parallel coupling is a measure to improve the power quality nearby the harmonic source and the active filter in series coupling is a VQ measure for the load to be protected. If an active filter with parallel coupling (Fig. C12.32a) is used, the harmonic spectrum $i_{h\Sigma}$ is measured in the current i_V of the load and a current spectrum i_A in phase opposition to this is produced. The current spectrum i_A in phase opposition is added to the harmonic spectrum $i_{h\Sigma}$ of the current i_V of the load, resulting in an approximately sinusoidal system current i_Q.

Active filters in series coupling (Fig. C12.32b) are chiefly used for targeted improvement of the voltage quality of individual loads and load groups. To improve the voltage quality (VQ) for especially sensitive loads, harmonic voltages measured in the system voltage u_Q at the point of common coupling (PCC) are added to the complementary voltage spectrum u_A produced by the active filter using an in-phase booster. Ideally, with this addition, the purely sinusoidal voltage u_V is applied to the voltage-sensitive loads.

A perfectly pure sinusoidal wave shape of the voltage and current cannot be achieved even with active filters because the finite control time and the switching frequency in the kHz range of self-commutated converters place certain limits on the harmonics mitigation in the frequency range ($h \geq 50$) [12.32]. In measures to improve the voltage quality, however, it may already be sufficient to eliminate only the dominating frequencies of the harmonics, that is, the 5th, 7th, 11th and 13th-order harmonics.

The elimination of harmonics whose order *h* is divisible by three is becoming increasingly important. Harmonic currents of this order are mainly produced in energy-efficient lighting systems using energy-saving lamps. The harmonic currents produced by the energy-saving lamps are summated in the N conductor of the lighting circuits. With the installation of active filters in the lighting circuits, the 150-Hz currents only load the N conductor up to the point where they are installed. The N and PEN conductors installed downstream of the filter installation location are relieved of these harmonic currents. The current relief achieved with the active filters is especially important if N and/or PEN conductors with reduced cross-sectional areas have been installed in the main circuits of the wiring system.

It is important to note once again that no filters may be installed in the PEN conductors of wiring systems.

Use of active filters to eliminate harmonics is especially appropriate if

- non-linear loads produce high harmonic levels and require hardly any fundamental-frequency reactive power,
- changes to the load structure increase the harmonic level of the power system impermissibly (e.g. when conventional motor drives are replaced by converter-fed drives),
- high N conductor loads occur due to 150-Hz currents (e.g. in existing wiring systems when conventional lamps are replaced by energy-saving lamps).

12.4 Planning of compensation systems with products from Modl

To implement concepts for reactive-power compensation, the selection scheme shown in Fig. C12.33 for capacitors, reactive-power compensation units and filter circuits from Modl GmbH [12.11] can be used. The current Modl type range for compensation of fundamental-frequency (linear) and harmonic-distortion (non-linear) reactive power is comprised in Modl's product and system catalogue [12.12]. This catalogue contains the products summarized in Table C12.34 to C12.37 for reactive-power compensation with and without reactors and for passive and active filtering of harmonics.

The selection scheme (Fig. C12.33) for products from the Modl type range for reactive-power compensation and filtering of harmonics (Tables C12.34 to C12.37) is an easy-to-use planning aid for designing safe and energy-efficient power systems with high benefit for the distribution network operator (DNO) and the environment.

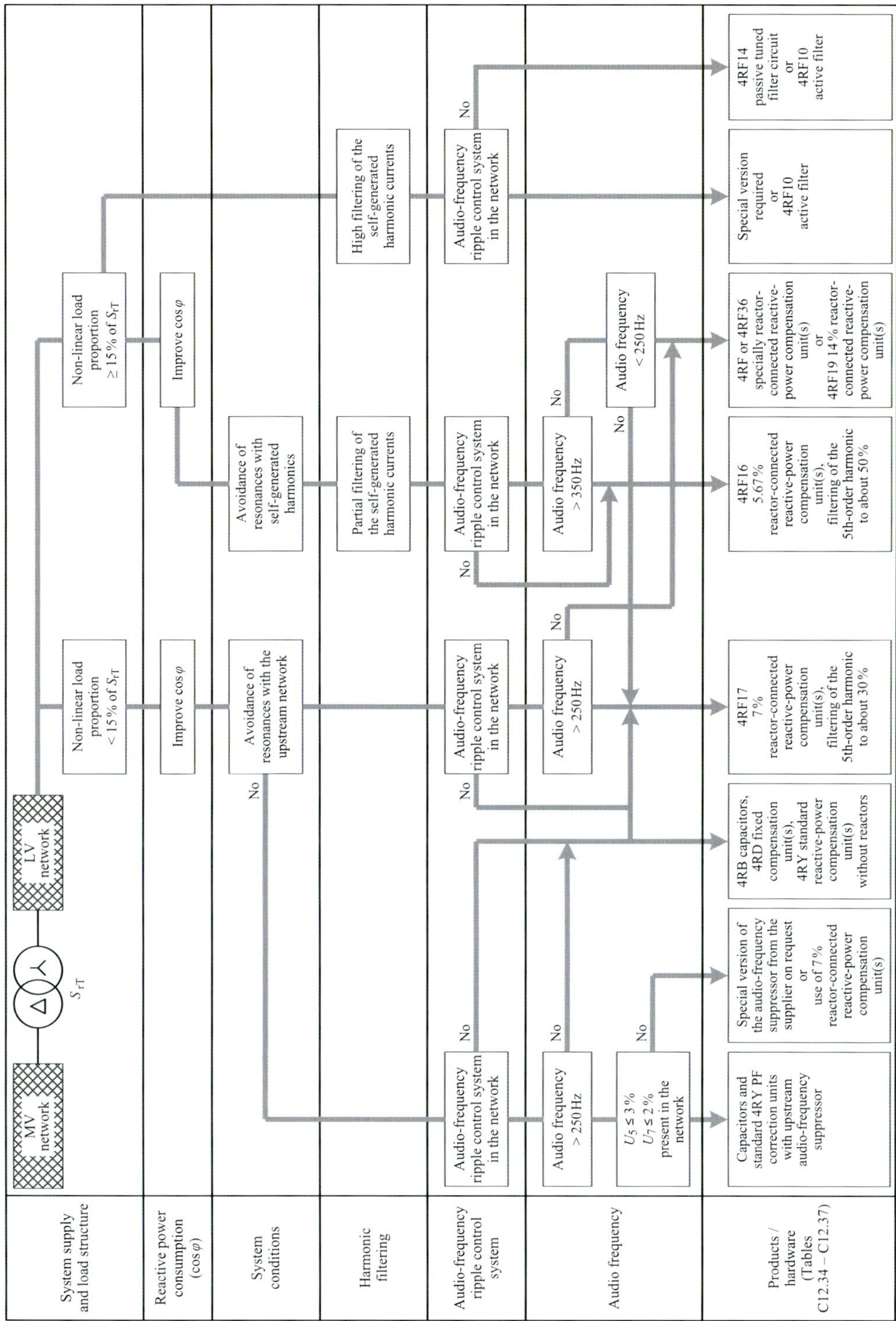

Fig. C12.33 Selection scheme for capacitors, reactive-power compensation units and filter circuits [12.12]

Table C12.34 Modl type range for PF correction without reactor-connected compensation units in LV networks with mainly linear loads [12.11, 12.12]

Type	Rated voltage U_{rC} in V (50 Hz)	Capacitor rating Q_{rC} in kvar
Fixed compensation units without reactors		
4RB	400	• 4.8; 7.5; 9.7; 12.5; 20.0; 25.0; 50.0
	440	• 10.4; 12.5; 14.2; 15.0; 20.0; 25.0; 28.1; 30.0; 33.0
	480	• 6.3; 10.4; 12.5; 15.0; 16.7; 20.0; 25.0; 30.0
	525	• 10.0; 12.5; 15.0; 20.0; 25.0; 30.0
	560	• 26.6
	690	• 12.5; 20.8; 25.0
	765	• 30.0
	800	• 12.5; 25.0; 28.0
4RD	400	• 5.0; 6.25; 10.0; 12.5; 15.0; 20.0; 25.0; 37.5; 50.0; 62.5; 75.0; 100
		• 5.0; 6.25; 10.0; 12.5; 15.0; 20.0; 25.0; 30.0; 37.5; 40.0; 50.0
	$400 < U_{rC} \leq 800$	• on request [12.11]
Reactive-power compensation units without reactors		
4RY	400	• 10.0; 17.5; 25.0; 45.0; 50.0
		• 15.0; 20.0; 25.0; 30.0; 40.0; 50.0
		• 25.0; 45.0; 50.0
		• 75; 100; 125; 150; 175; 200; 250; 300; 350; 400; 450; 500
	525; 690	• 75; 100; 125; 150; 175; 200; 250; 300; 350; 400; 450; 500

Table C12.35 Modl type range for PF correction with reactor-connected compensation units in LV networks with a high proportion (> 15 %) of non-linear loads [12.11, 12.12]

Type	Rated voltage U_{rC} in V (50 Hz)	Capacitor rating Q_{rC} in kvar
Fixed compensation units with reactors		
4RF66 [1] (p = 5.67 %)	400; 525; 690	• 25; 50
4RF67 [2] (p = 7 %)	400	• 5; 10; 15; 20; 25; 30; 40; 50
	525; 690	• on request [12.11]
4RF69 [3] (p = 14 %)	400	• 5; 10; 15; 20; 25; 30; 40; 50
	525; 690	• on request [12.11]
Reactive-power compensation units with reactors		
4RF16 [1] (p = 5.67 %)	400; 525; 690	• 150; 200; 250; 300; 350; 400; 450; 500; 600; 700; 800
4RF17 [2] (p = 7 %)	400; 525; 690	• 20; 25; 30; 40; 50; 75; 100; 125; 150; 200; 250; 300; 350; 400; 450; 500; 600; 700; 800
4RF19 [3] (p = 14 %)	400	• 20; 25; 30; 40; 50; 75; 100; 125; 150; 200; 250; 300; 350; 400; 450; 500; 600; 700; 800
	525; 690	• on request [12.11]

1) Filtering of approx. 50 ··· 60 % of the 5th-order harmonic
2) Filtering of approx. 30 % of the 5th-order harmonic
3) Filtering of approx. 10 % of the 5th-order harmonic
p Detuning factor of the PF correction equipment

Table C12.36 Modl type range for passive filtering of harmonics in LV power systems [12.11, 12.12]

<table>
<tr><th>Type</th><th>Rated voltage U_{rC}
in V (50 Hz)</th><th>Capacitor rating Q_{rC}
in kvar</th></tr>
<tr><td colspan="3">Passive tuned filter circuits</td></tr>
<tr><td rowspan="2">4RF14</td><td>400</td><td>• 29; 44; 58; 87; 115; 130; 175</td></tr>
<tr><td>525; 690</td><td>• on request [12.11]</td></tr>
</table>

Table C12.37 Modl type range for active filtering of harmonics [12.11, 12.12]

<table>
<tr><th>Type</th><th colspan="2">Nominal system voltage U_{nN}
in V (50 Hz)</th><th>Rated filtering current I_{rF}
in A</th></tr>
<tr><td colspan="4">Active filter circuits</td></tr>
<tr><td rowspan="3">Bluewave
4RF10</td><td rowspan="2">400</td><td>3-wire operation</td><td>• 30; 50; 100; 200; 250; 300</td></tr>
<tr><td>4-wire operation 1)</td><td>• 100; 200; 250; 300</td></tr>
<tr><td colspan="2">500; 690</td><td>• on request [12.11]</td></tr>
<tr><td rowspan="3">Cabinet
units 2)
4RF10</td><td colspan="2">400</td><td>• 170; 350; 630; 1,270</td></tr>
<tr><td colspan="2">500</td><td>• 135; 280; 500; 1,000; 1,400</td></tr>
<tr><td colspan="2">690</td><td>• 100; 200; 365; 730; 1,200</td></tr>
<tr><td colspan="4">1) To reduce the N-conductor current, i. e. the harmonic current, whose harmonic order h is divisible by 3</td></tr>
<tr><td colspan="4">2) 4-wire operation not possible</td></tr>
</table>

12.5 Demonstration of the economic and technical benefit of reactive-power compensation

The benefit provided by reactive-power compensation can be evaluated by the following three quantifiable indicators:

- reduction of power system losses,
- saving of rated transformer apparent power to cover the power demand,
- improved voltage stability.

An average power factor of $\cos\varphi_1 = 0.7$ is representative of the consumption of electrical power in industrial plants (Table C12.2). Evaluation of the benefit due to reactive-power compensation is therefore based on an example of improvement of the power factor from $\cos\varphi_1 = 0.7$ to $\cos\varphi_2 = 0.9$. With the indicators stated above, the following benefit can be quantified with this $\cos\varphi$ improvement.

Reduction of power system losses

Because the load-dependent power system losses are proportional to the square of the apparent current I_S, they can be considerably reduced by reactive-current compensation. This is illustrated by the function curve shown in Fig. C12.38 for the power loss P_{loss} and apparent current I_S with a variable power factor $\cos\varphi$ and constant active power consumption P.

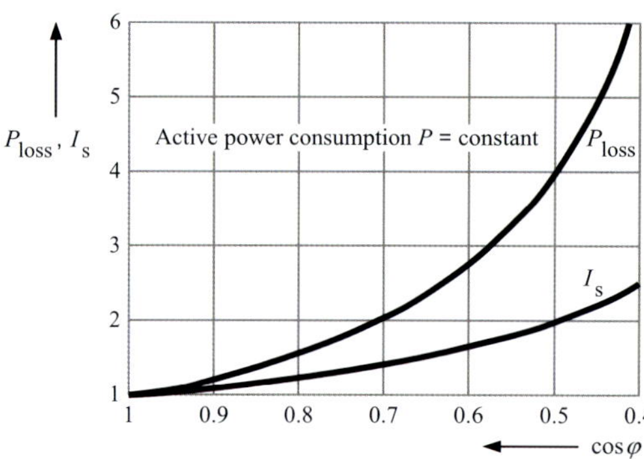

Fig. C12.38
Function curve of the power loss P_{loss} and apparent current I_S

With constant active power consumption *P*, the amount saved on power system losses by improving the power factor can be calculated as follows:

$$\Delta P_{loss} = \left(1 - \frac{P_{loss2}}{P_{loss1}}\right) \cdot 100 \qquad (12.25)$$

ΔP_{loss} saved amount of system losses in %
P_{loss1} magnitude of system losses at $\cos\varphi_1$ in kW
P_{loss2} magnitude of system losses at $\cos\varphi_2$ in kW

If the active power consumption is constant, the magnitude of the power system losses is calculated as follows:

$$P_{loss_i} = \frac{P^2}{U_{nN}^2 \cdot \cos^2\varphi_i} \cdot R \quad (i = 1, 2) \qquad (12.26)$$

By inserting Eq. (12.26) into Eq. (12.25), the magnitude of the power system losses ΔP_V saved by reactive-power compensation for PFC can be calculated as a function of $\cos\varphi$. The following applies:

$$\Delta P_{loss} = \left[1 - \left(\frac{\cos\varphi_1}{\cos\varphi_2}\right)^2\right] \cdot 100 \qquad (12.27)$$

$\cos\varphi_1$ power factor before correction
$\cos\varphi_2$ power factor after correction

According to Eq. (12.27), the following reduction in power system losses results from an improvement in the power factor from $\cos\varphi_1 = 0.7$ to $\cos\varphi_2 = 0.9$:

$$\Delta P_{loss} = \left[1 - \left(\frac{\cos\varphi_1}{\cos\varphi_2}\right)^2\right] \cdot 100 = \left[1 - \left(\frac{0.7}{0.9}\right)^2\right] \cdot 100 = 39.5\%$$

As is shown, with this improvement in $\cos\varphi$, power system losses can be reduced by 39.5 %. The same results are provided by the diagram shown in Fig. C12.39 for determining the loss reduction on correction from $\cos\varphi_1$ to $\cos\varphi_2$. By reducing the power system losses, the reactive-power compensation can make a lasting contribution to climate protection because low-loss power distribution networks help limit the worldwide CO_2 emissions from fossil-fuel power stations.

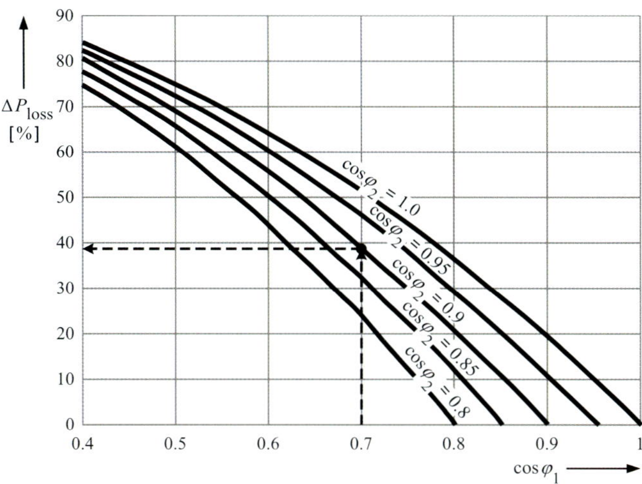

Fig. C12.39 Percentage reduction of power system losses on correction from cos φ_1 to cos φ_2

Saving on rated transformer apparent power

The amount of rated transformer apparent power that is additionally available or can be saved by correcting the power factor from cos φ_1 to cos φ_2 can be calculated as follows:

$$\Delta S_{rT} = S_{rT1} \cdot \left(1 - \frac{\cos\varphi_1}{\cos\varphi_2}\right) = S_{rT1} \cdot K \quad (12.28)$$

ΔS_{rT} magnitude of rated transformer apparent power that is additionally available or can be saved
S_{rT1} required rated transformer apparent power at cos φ_1
K reduction factor (Fig. C12.40)

The magnitude of rated transformer apparent power that is additionally available or can be saved with correction from cos φ_1 to cos φ_2 can also be determined graphically with the diagram in Fig. C12.41.

According to Fig. C12.41, improving the power factor from cos $\varphi_1 = 0.7$ to cos $\varphi_2 = 0.9$ reduces the rated apparent power of the transformer to be installed by about 22 %. The capacitor rating required for this reduction is 37 % of the rated apparent power of the transformer originally to be installed. Based on the relationships between sinusoidal quantities shown in Fig. C12.41, for example, use of a 1,600-kVA transformer produces a magnitude of the rated apparent power that can be saved of $\Delta S_{rT} = 0.22 \cdot S_{rT} = 352$ kVA.

Because of this, at cos $\varphi_2 = 0.9$ a 1,250-kVA transformer could also be used to cover the active power demand ($S_2 = S_{rT} - \Delta S_{rT} = 1{,}600$ kVA – 352 kVA = 1,248 kVA).

By reducing the rated apparent power of the transformer from $S_{rT} =$ 1,600 kVA to $S_{rT} = 1{,}250$ kVA, the investment costs could be reduced by about 18 %.

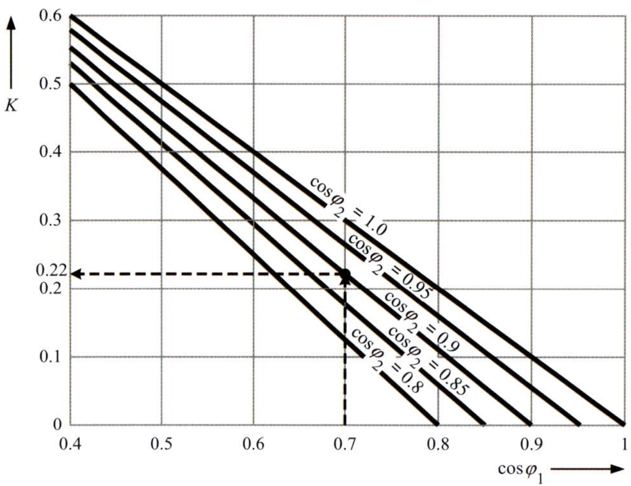

Fig. C12.40 Reduction factor K for determining the magnitude of rated transformer apparent power that is additionally available or can be saved by correction from cos φ_1 to cos φ_2

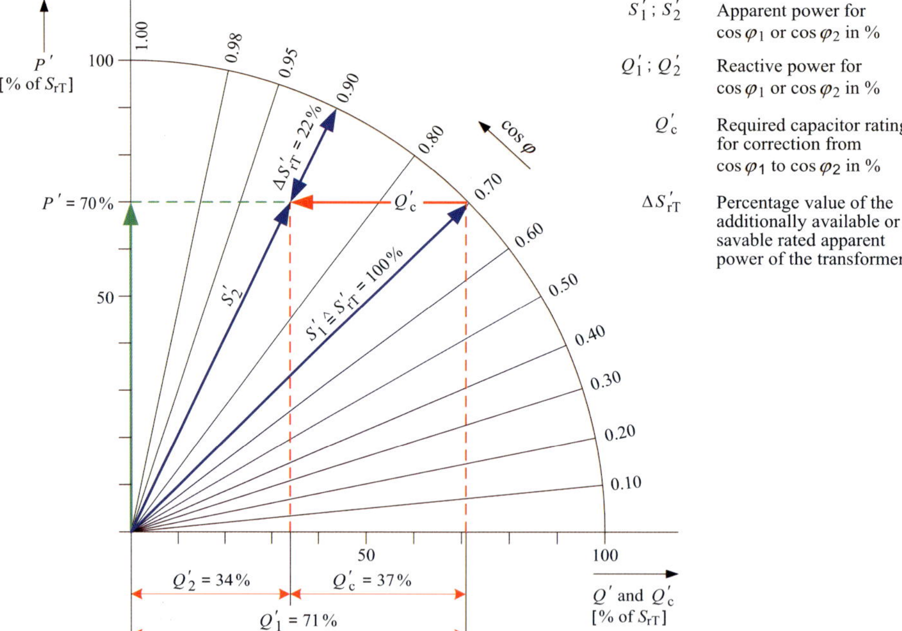

Fig. C12.41 Graphic determining of the rated transformer apparent power $\Delta S'_{rT}$ that is additionally available or can be saved and of the required capacitor rating Q'_c when correcting from cos φ_1 to cos φ_2

Improvement of the voltage stability

The voltage drop of a transformer can be determined with sufficient precision according to the following formula:

$$\Delta u \approx u_{rR} \cdot \cos\varphi + u_{rX} \cdot \sin\varphi \qquad (12.29)$$

$$u_{rX} = \sqrt{u_{rZ}^2 - u_{rR}^2} \qquad (12.29.1)$$

$$u_{rR} = \frac{P_k}{S_{rT}} \cdot 100 \qquad (12.29.1)$$

u_{rR} active component of u_{rZ} as a percentage
u_{rX} reactive component of u_{rZ} as a percentage
u_{rZ} impedance voltage at rated current as a percentage (Table C11.8a/b)
P_k short-circuit or copper losses (Table C11.8a/b)
S_{rT} standardized rated power (Table C11.8a/b)

Fig. C12.42 shows the curve for the voltage drop as a function of cos φ for a low-loss GEAFOL cast-resin transformer with a rated power of S_{rT} = 1,600 kVA and an impedance voltage at rated current of u_{rZ} = 6 %. As can be seen in the figure, the voltage drop Δu after correction from cos φ_1 = 0.7 to cos φ_2 = 0.9 is only 68 % of its original value. If the reduction of the apparent current due to the PF correction is also taken into account, the voltage drop is reduced to values < 60 %.

For safe and cost-efficient operation of industrial networks, the potential offered to improve voltage stability by reactive-power compensation should always be used to the full.

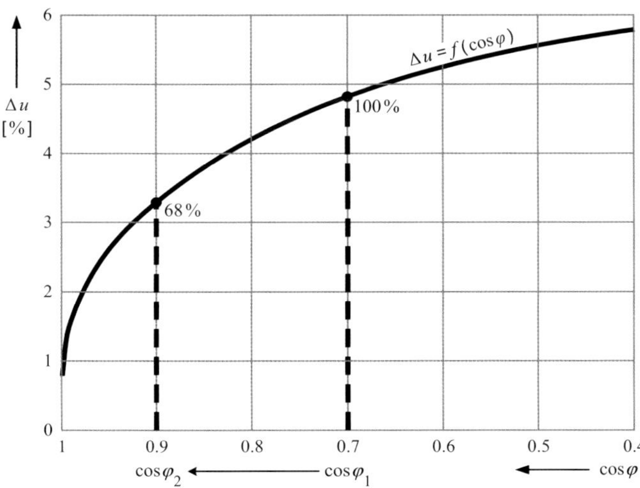

Fig. C12.42 Curve of the voltage drop as a function of cos φ for a GEAFOL cast-resin transformer with S_{rT} = 1,600 kVA and u_{rZ} = 6 %

Conclusion

In industrial networks, reactive-power compensation is an important measure to improve the energy efficiency and voltage quality. It generally pays for itself due to saved power system losses and use of equipment reserves.

13 Designing the LV power system protection

13.1 Fundamentals of protection engineering and equipment

LV system protection must be parameterized and coordinated in interaction with the MV system protection (Chapter 7).

The aim of the parameterization and coordination is to enable faulty parts of the system and installation to be selectively isolated from the distribution network by fast and reliable detection of the type and location of the fault. The LV-side system protection performs this task with the following functions:

- protection against overcurrent, that is, protection against
  - short circuit,
  - overload and/or
- protection against overtemperature.

Protection against short circuit is primarily achieved with overcurrent releases and relays with instantaneous and time-delay tripping characteristics as well as LV HRC fuses.

Protection against overload is essentially performed by thermal, inverse-time and electronic time-delay overload releases and relays.

The function of protection against overtemperature can be performed using thermistor protective devices.

Table C13.1 provides an overview of the protective devices used in industrial LV networks for protection against short circuit, overload and overtemperature, and indicates the valid standards.

Table C13.1 Protective equipment for use in LV networks

<table>
<tr><th rowspan="3">Protective equipment/devices</th><th rowspan="3">Valid standard</th><th colspan="3">Use for protection in case of</th></tr>
<tr><th colspan="2">overcurrent</th><th rowspan="2">over-temperature</th></tr>
<tr><th>short circuit</th><th>overload</th></tr>
<tr><td>Fuses</td><td>DIN EN 60269-1 (VDE 0636-1): 2010-03 [13.1]
or IEC 60269-1: 2009-07 [13.2]</td><td>X</td><td>X</td><td>---</td></tr>
<tr><td>Circuit-breakers with integrated overcurrent release</td><td>DIN EN 60947-2 (VDE 0660-101): 2010-04 [13.3]
or IEC 60947-2: 2009-05 [13.4]</td><td>X</td><td>X</td><td>---</td></tr>
<tr><td>Miniature circuit-breaker</td><td>DIN EN 60898-1 (VDE 0641-11): 2006-03 [13.5]
or IEC 60898-1: 2003-07 [13.6]</td><td>X</td><td>X</td><td>---</td></tr>
<tr><td>Switchgear assemblies comprising
• back-up fuse of utilization category gL/gG or aM and contactor with overload relay
or
• starter circuit-breaker and contactor with overload relay</td><td>DIN EN 60269-1 (VDE 0636-1): 2010-03 [13.1]
or IEC 60269-1: 2009-07 [13.2]
DIN EN 60947-4-1 (VDE 0660-102): 2006-04 [13.7]
or IEC 60947-4-1: 2009-09 [13.8]
DIN EN 60947-2 (VDE 0660-101): 2010-04 [13.3]
or IEC 60947-2: 2009-05 [13.4]
DIN EN 60947-4-1 (VDE 0660-102): 2006-04 [13.7]
or IEC 60947-4-1: 2009-09 [13.8]</td><td>X</td><td>X</td><td>---</td></tr>
<tr><td>Thermistor motor-protection devices</td><td>DIN EN 60947-8 (VDE 0660-302): 2007-07 [13.9]
or IEC 60947-8: 2006-11 [13.10]</td><td>---</td><td>---</td><td>X</td></tr>
</table>

Among the protective devices listed in Table C13.1, fuses, circuit-breakers and switchgear assemblies form the backbone of the LV system protection in industrial installations.

13.1.1 Fuses

Fuses are inverse-time-delay protective devices. With very high fault currents, the short pre-arcing time of the fuse greatly limits the peak short-circuit current i_p. The highest instantaneous value of the current that can be reached while the fuse is breaking the short-circuit current is known as the let-through current I_D. The let-through current diagrams for assessing the current limitation of Siemens LV HRC fuses are shown in the BETA Low-Voltage Circuit Protection catalog ET B1T [13.11].

Because of their current limitation and their very high breaking capacity (at least 120 kA for Siemens-produced LV HRC fuses), fuses provide especially effective protection against the thermal and dynamic effects of short circuits.

Low-voltage fuses are classified according to their function features. Table C13.2 provides information on the classification of low-voltage fuses.

Table C13.2 Classification of LV HRC fuses according to DIN EN 60269-1 (VDE 0636-1): 2010-03 [13.1] or IEC 60269-1: 2009-07 [13.2],

Function category	Rated continuous current up to	Rated breaking current	Utilization category	Protection of
Full-range breaking-capacity fuse-links				
g	I_n	$\geq I_f$ 3)	gL / gG 1)	Cables and lines
			gR / gS 2)	Semiconductors
			gB	Mining installations
			gTr	Transformers
Partial-range breaking-capacity fuse-links				
a	I_n	$\geq 4 \times I_n$	aM	Switchgear
		$\geq 2.7 \times I_n$	aR	Semiconductors

1) Utilization category gG (G for general use) will in future replace the characteristic gL

2) Fuses of utilization category gR operate faster than fuses of utilization category gS and have lower I^2t values

3) Conventional fusing current (s. Table C11.33)

Function category g refers to full-range fuses. Full-range fuses can carry their rated current continuously and break currents from the smallest fusing current up to the rated short-circuit breaking current. Fuses of utilization category gL for cable and line protection feature this capability.

Function category a contains accompanied fuses that can break currents above a multiple of their rated current up to the rated short-circuit breaking current. Accompanied fuses include switchgear fuses of utilization category aM. Because fuses of utilization category aM only break currents larger than four times the rated current, they are only used for short-circuit protection. For that reason, an additional protective device, for example a thermally delayed overload relay, must always be provided for overload protection.

Overload relays are used to equip switchgear assemblies. Fig. C13.3 shows an example of comparison of the pre-arcing time current characteristics ($t_s I$ characteristics) of 200-A LV HRC fuses of utilization categories gL and aM.

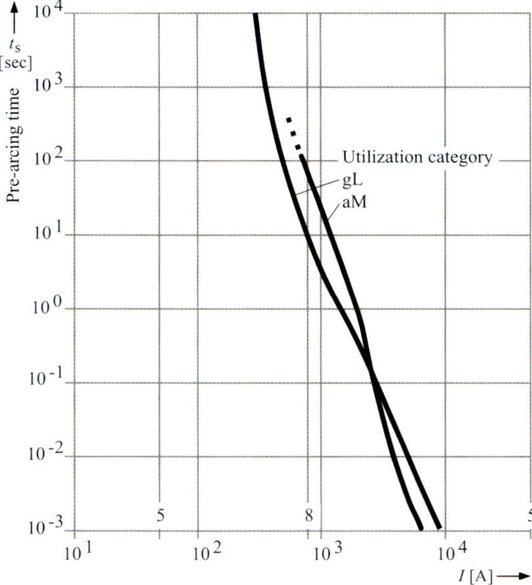

Fig. C13.3
Example of t_sI characteristics of LV HRC fuses of utilization categories gL and aM, $I_r = 200$ A

13.1.2 Circuit-breakers

Circuit-breakers are mainly used for protection against overload and short circuits. They can also be equipped with additional releases and relays for detecting earth-leakage currents (earth fault) and undervoltages. Whereas releases are permanent parts of the circuit-breaker, relays can only operate the mechanism of the circuit-breaker electrically via undervoltage and shunt releases.

Table C13.4 provides an overview of the releases used in Siemens circuit-breakers and their graphical symbols. The protection function of the circuit-breaker is determined by the choice of a particular release. We distinguish between thermomagnetic releases (TMs), previously also termed electromagnetic releases, and electronic trip units (ETUs). Thermomagnetic releases may be preset or adjustable. The electronic trip units of Siemens circuit-breakers are always adjustable.

Overcurrent releases can either be ready-installed in the circuit-breaker or supplied separately as modules for subsequent installation or replacement. See the manufacturers' data for possible exceptions [13.14].

Overload protection (L, formerly a)

The abbreviation for overload protection is L (Long-time delay). The L release is subject to an inverse-time delay and, depending on the type of release, exhibits the following characteristics:

- thermal bimetallic characteristic for TM releases.
- I^2t characteristic with or without additional I^4t characteristic for ETUs.

The operating current is designated I_R and a possible adjustable time delay is denoted by t_R. The time delay with a device-specific reference point defines the inverse-time curve of the target characteristic. In Siemens circuit-breakers, the reference point for the overload tripping is $6 \times I_R$ as standard, that is, at 6 times the set current tripping value I_R, the target characteristic of the overload release has the set tripping time t_R (t_R @ $6 \times I_R$). The complete curve of the set tripping characteristic results from the mathematical formula I^2t = constant or I^4t = constant.

Table C13.4 Releases of the circuit-breakers with their protection function and standardized graphical symbol

Protective function	Siemens identifi-cation symbol	Time-delay characteristics of release	Graphical symbols acc. to DIN EN 60617-7 (EN 60617-7): 1997-08 [13.12] or IEC 60617-DB-12M: 2001-11 [13.13]		
			Alternative graphical symbols for circuit diagram		Graphical symbol for block diagram
Overload protection	L	Inverse-time delay			
Selective overcurrent protection (delayed)	S [1]	Definite-time delay by timing element or inverse-time delay			$I>$
Earth fault protection (delayed)	G [1]	Definite-time delay or inverse-time delay			I ⏚
Overcurrent protection (undelayed)	I	Instantaneous			$I>>$

1) For SENTRON 3WL and SENTRON 3VL circuit-breakers, protection also includes "zone-selective interlocking" (ZSI)

The ETUs optionally feature overload protection for the neutral conductor (N conductor). This protection function is designated N (Neutral). The neutral conductors of LV networks with high harmonic loads (150-Hz currents) are reliably protected from overload by means of circuit-breakers with ETUs and integrated N conductor protection.

Short-time-delay overcurrent protection (S, formerly z)

The abbreviation for short-time-delay overcurrent protection is S. The S function of the overcurrent release permits time-selective short-circuit clearance in LV networks with multiple series-connected circuit-breakers.

The S release has a standard characteristic with a definite-time curve. According to this standard characteristic, the circuit-breaker interrupts a short circuit when a set current magnitude I_{sd} is exceeded after the associated delay time t_{sd} has elapsed. Optionally, the S release can exhibit an I^2t inverse-time tripping characteristic. The reference points for the tripping time t_{sd} for the I^2t characteristic depend on the basic type of switching or protective device (e.g. t_{sd} @ $12 \times I_n$ for 3WL and t_{sd} @ $8 \times I_R$ for 3VL).

When parameterizing the short-time-delay overcurrent protection (I_{sd}, t_{sd}), special attention must be paid to the rated short-time withstand current (I_{cw}, t_{cw}) of the circuit-breaker. If the load limit for the thermal short-circuit rating of the circuit-breaker is exceeded at the installation location because of the time delay t_{sd}, an I release that clears the short circuit instantaneously must be used.

Instantaneous overcurrent protection (I, formerly n)

The abbreviation for instantaneous overcurrent protection is the letter I. Depending on the application and type, I releases may have preset or adjustable operating currents or threshold values I_i. On air circuit-breakers (ACBs), the I releases can be deactivated in some cases because these switching devices normally have a very high thermal short-circuit rating I_{cw}. The I_{cw} values of moulded-case circuit-breakers (MCCBs) are lower because of their design and mode of operation. For that reason, an active I release must usually be installed for self-protection of the MCCBs.

Earth-fault protection (G, formerly g)

The abbreviation for earth-fault protection is G (Ground). The G release is used for clearing line-to-earth faults, that is, earth faults of one line conductor. Earth-fault protection can have definite-time (standard function) or inverse-time (optional I^2t function) characteristics. The adjustable operating current is designated I_g and the associated time delay is denoted by t_g.

Fig. C13.5 schematically illustrates the range of possible settings for Siemens circuit-breakers with integrated TM release and ETU.

According to their operating principle, air circuit-breakers (ACBs, e.g. SENTRON 3WL) are zero-current interrupters (not current limiting). The moulded-case circuit-breakers (MCCBs, e.g. SENTRON 3VL) and motor start protectors (MSPs, e.g. Siemens type 3RV), on the other hand, are constructed according to the current-limiting principle. The contacts of an MCCB or MSP open before the peak current of the prospective short-circuit current is reached. Thus the current-limiting circuit-breakers have a protective effect similar to that of fuses.

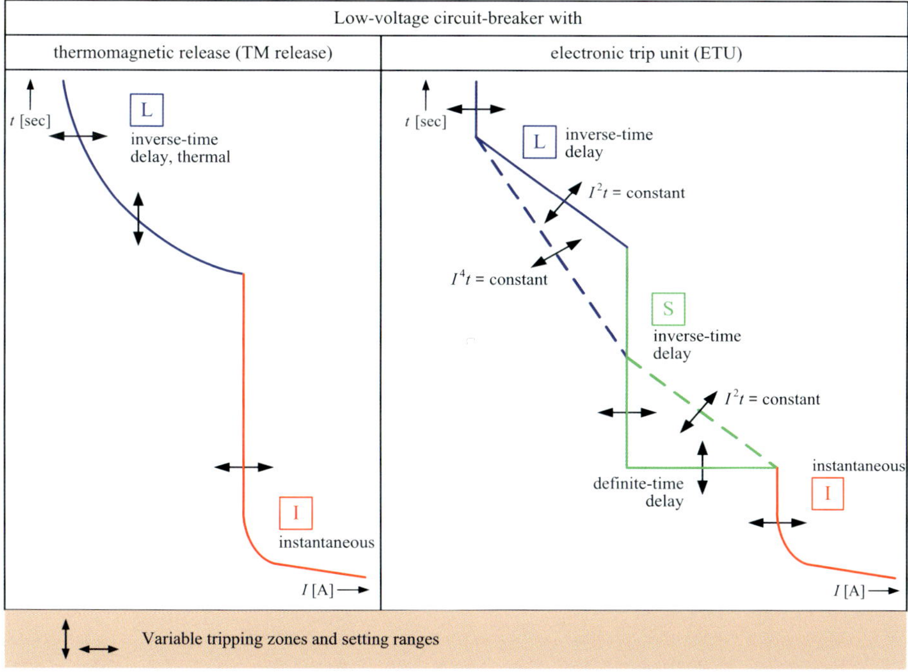

Fig. C13.5 Range of possible settings for Siemens circuit-breakers with TM releases or ETUs

To design selective LV networks without fuses, ACBs are preferably installed upstream and MCCBs downstream. Current-time selectivity can be achieved more simply with air circuit-breakers because the I release of these devices can usually be deactivated. The selective behaviour of an upstream moulded-case circuit-breaker, on the other hand, depends on its dynamic behaviour in the instantaneous short-circuit tripping range [13.15]. Selectivity tables provide information about the selective behaviour of a circuit-breaker combination in this range. The selectivity tables of a switchboard manufacturer contain the test results of the "limit current of selectivity" when series-connected circuit-breakers with instantaneous overcurrent releases are used. The "limit current of selectivity" is a measure of the magnitude of the short-circuit current that is required for the combination's behaviour to remain selective in the event of a fault.

When designing the protection, please note that no selectivity tables for the combination of circuit-breakers from different switchboard manufacturers exist.

Table C13.6 contains application examples for the use of Siemens circuit-breakers in LV networks and their most important technical features.

Table C13.6 Application examples for modern Siemens circuit-breakers and their most important technical features [13.16]

Circuit-breaker type	Rated current / release	Application example / technical features
SENTRON 3WL air circuit-breaker (ACB)	630 A to 6,300 A	As selective system protection up to 6,300 A for transformers, generators, PF correction equipment, busbar trunking systems, and cable connections • High rated short-time withstand current for current-time selectivity • Different switching capacity classes from ECO (N) to very high (C) for economical use (see Table C9.6) • Electronic, microprocessor-based overcurrent release not requiring external voltage with optional characteristics • Zone-selective interlocking (ZSI) with total delay time of 50 msec • Optional earth-fault protection
SENTRON 3VL current-limiting circuit-breaker (MCCB)	TM release: from 16 A to 630 A ETU release: from 63 A to 1,600 A	For system protection up to 1,600 A • Different switching power classes from ECO (N) to high (H) for economical use • Both with thermomagnetic and with electronic, microprocessor-based overcurrent release not requiring external voltage for optimal adaptation to the protection requirements • Optional earth-fault protection
	ETU release: from 63 A to 500 A	For motor protection up to 500 A • Electronic overload release with adjustable time-lag class for effective protection of motors
	M release: from 63 A to 500 A	For motor starter combinations up to 500 A • Unsusceptible to inrush peaks on direct-on-line starting of motors
	M release: from 100 A to 1,600 A	As isolating circuit-breaker (load interrupter) up to 1,600 A • Due to built-in short-circuit protection for self-protection no back-up fuse required
3RV1 current-limiting starter circuit-breaker (MSP)	TM release: from 0.16 A to 100 A	For motor protection in starter combinations up to 100 A • With overload and overcurrent protection • Can also be used for system protection (without neutral conductor protection)

13.1.3 Switchgear assemblies

Switchgear assemblies are series-connected switching and protective devices which perform specific tasks for protecting a part of a power system or installation; the first device (relative to flow of power) provides the short-circuit protection [13.16]. Because of the high making and breaking capacity of the moulded-case circuit-breakers, switchgear assemblies are now principally used as motor-starter combinations. When used as motor-starter combinations, they can be fused or fuseless.

Switchgear assemblies with fuses

Switchgear assemblies for motor starters in fused design comprise

- contactor for switching the motor on and off,
- (thermal or electronic) inverse-time-delay overload relay or motor management and control unit for protection against overload of the motor,
- fuse of utilization category gL/gG or aM for short-circuit protection.

For carefully coordinated interaction of the components used in the motor-starter combination, the following conditions must be met [13.16]:

- The time-current characteristic of the overload relay and the fuses must permit the motor to be run up to speed.
- The fuses must protect the overload relay against destruction by currents exceeding approximately 10 times the rated current of the relay.
- The fuses must interrupt overcurrents that the contactor cannot handle. These include currents larger than 10 times the rated load current of the contactor.

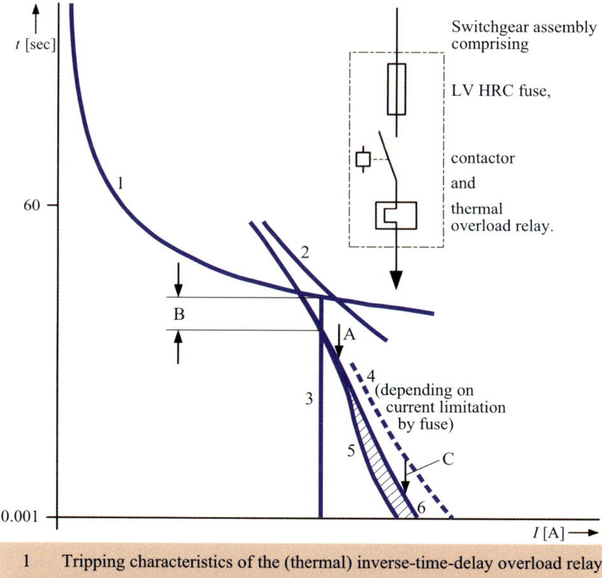

1	Tripping characteristics of the (thermal) inverse-time-delay overload relay
2	Destruction characteristic of the thermal overload relay
3	Rated breaking capacity of the contactor
4	Characteristics of the contactor for easily separable welding of the contacts
5	Pre-arcing time/current characteristic of the fuse, utilization category aM
6	Total clearing time characteristic of the aM fuse
A,B,C	Safety margins for reliable short-circuit protection

Fig. C13.7 Coordination of a fused switchgear assembly [13.16]

- The fuses must protect the contactor in the event of a short circuit so that destruction exceeding the agreed damage level (coordination type “1” or “2”, see p. 254) cannot occur.
- The contactor must withstand stresses due to the motor starting currents amounting to 8 to 12 times the rated load current without the contacts being welded.

Fig. C13.7 illustrates the correct protection coordination of a switchgear assembly (motor starter) comprising LV HRC fuses, contactor and thermal inverse-time-delay overload relay.

Switchgear assemblies without fuses

Switchgear assemblies for motor starters in fuseless design comprise

- contactor for switching the motor on and off,
- circuit-breaker with LI release for overload and overcurrent protection,

or

- contactor for switching the motor on and off,
- (thermal or electronic) inverse-time-delay overload relay or motor management and control unit for protection against overload of the motor,
- motor starter circuit-breaker with I release for instantaneous overcurrent protection.

The same conditions apply to coordination of a fuseless switchgear assembly comprising a circuit-breaker with an LI release and a contactor as apply to a fused switchgear assembly because the protection function of the fuse is performed by the I release of the circuit-breaker.

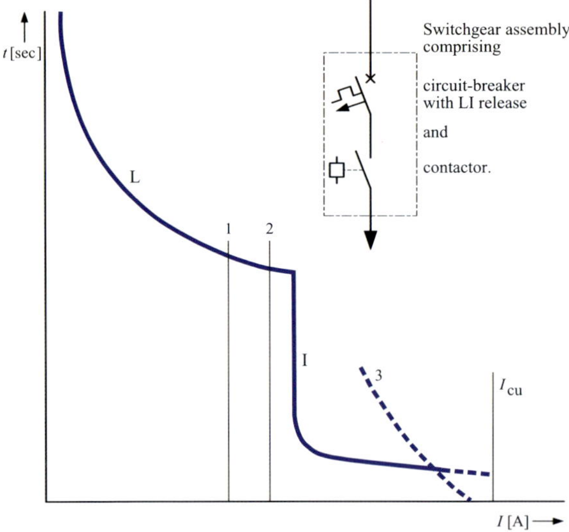

1 Rated breaking capacity of the contactor
2 Rated making capacity of the contactor
3 Characteristics of the contactor for easily separable welding of the contacts
L Characteristic curve of the inverse-time-delay overload release
I Characteristic curve of the instantaneous overcurrent release
I_{cu} Rated ultimate short-circuit breaking capacity of the circuit-breaker

Fig. C13.8
Coordination of a fuseless switchgear assembly

Fig. C13.8 shows an example of correct protection coordination for a fuseless switchgear assembly comprising a contactor and circuit-breaker with an LI release. The specific technical and selection data for use of fuseless switchgear assemblies with Siemens switchgear and protective devices are given in the brochure "SIRIUS Configuration" [13.17].

Switchgear assemblies with thermistor motor protective devices

When it is no longer possible to establish the winding temperature from the motor current, the limits of overload protection by the overload relay or release are reached. This is the case for

- high switching frequency,
- irregular intermittent duty,
- restricted cooling and
- high ambient temperature.

In these cases, switchgear assemblies with thermistor motor protective devices are deployed. Depending on the installation's configuration and protection concept, the switchgear assemblies are designed with or without fuses. The degree of protection that can be attained depends on whether the motor to be protected has a thermally critical stator or rotor. The operating temperature, coupling time constant and position of the temperature sensor in the motor winding are also crucial factors. These are usually specified by the motor manufacturer [13.18].

Motors with thermally critical stator can be adequately protected against overloads and overheating by means of thermistor motor protective devices without overload relays and releases (Fig. C13.9a and b).

Motors with thermally critical rotor, even if started with locked rotor, can only be provided with adequate protection if they are fitted with an additional overload relay or release (Fig. C13.9c and d) [13.18].

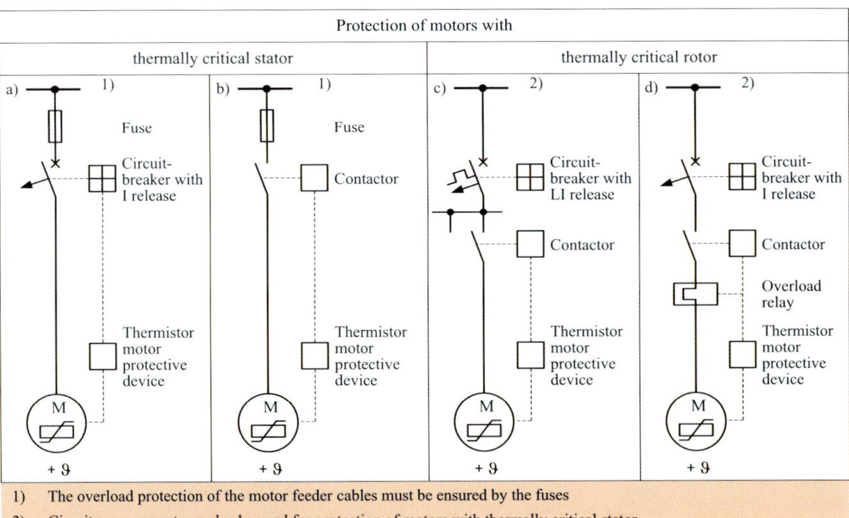

Fig. C13.9 Switchgear assemblies with thermistor motor protective device for protection of motors with thermally critical stator or rotor (Examples according to [13.16])

13.1.4 Comparative evaluation of the characteristics of protective devices

Tables C13.10 and C13.11 provide a comparative evaluation of the protective characteristics of fuses and circuit-breakers as well as fused and fuseless switchgear assemblies.

The consequences of the comparative evaluation of the protective characteristics for the devices and installation must be observed in designing a selective and reliable LV network. The design of selective LV networks is explained below.

Table C13.10 Comparative evaluation of the protective characteristics of fuses and circuit-breakers [13.16]

Characteristic	Fuse	Circuit-breaker
Rated breaking capacity for alternating voltage (AC)	> 100 kA, 690 V	$f(I_\mathrm{r}, U_\mathrm{e}, \text{type}^{1)})$
Current limitation	$f(I_\mathrm{r}, I_\mathrm{k})$	$f(I_\mathrm{r}, I_\mathrm{k}, U_\mathrm{e}, \text{type}^{1)})$
Additional arcing space	None	$f(I_\mathrm{r}, I_\mathrm{k}, U_\mathrm{e}, \text{type}^{1)})$
External indication of operability	Yes	No
Operational reliability	With some effort 2)	Yes
Remote switching	No	Yes
Automatic all-pole breaking	With some effort 3)	Yes
Indication facility	With some effort 4)	Yes
Interlocking facility	No	Yes
Readiness for reclosing after overload trip short-circuit clearing	 No No	 Yes f(system state)
Interrupted operation	Yes	f(system state)
Maintenance effort	No	f(number of operating cycles and system state)
Selectivity	Without effort	With some effort
Replaceability	Yes 5)	With unit of identical make
Short-circuit protection cable motor	 Very good Very good	 Good Good
Overload protection cable motor	 Adequate Not possible	 Good Good

1) The term “type“ embraces: current extinguishing method, short-circuit strength through inherent resistance, type of construction
2) For example, by means of shockproof fuse switch-disconnectors with snap-action closing
3) By means of fuse monitoring and the associated circuit-breaker
4) By means of fuse monitoring
5) Due to standardized pre-arcing time-current characteristics

Table C13.11 Comparative evaluation of the protective characteristics of switchgear assemblies [13.16]

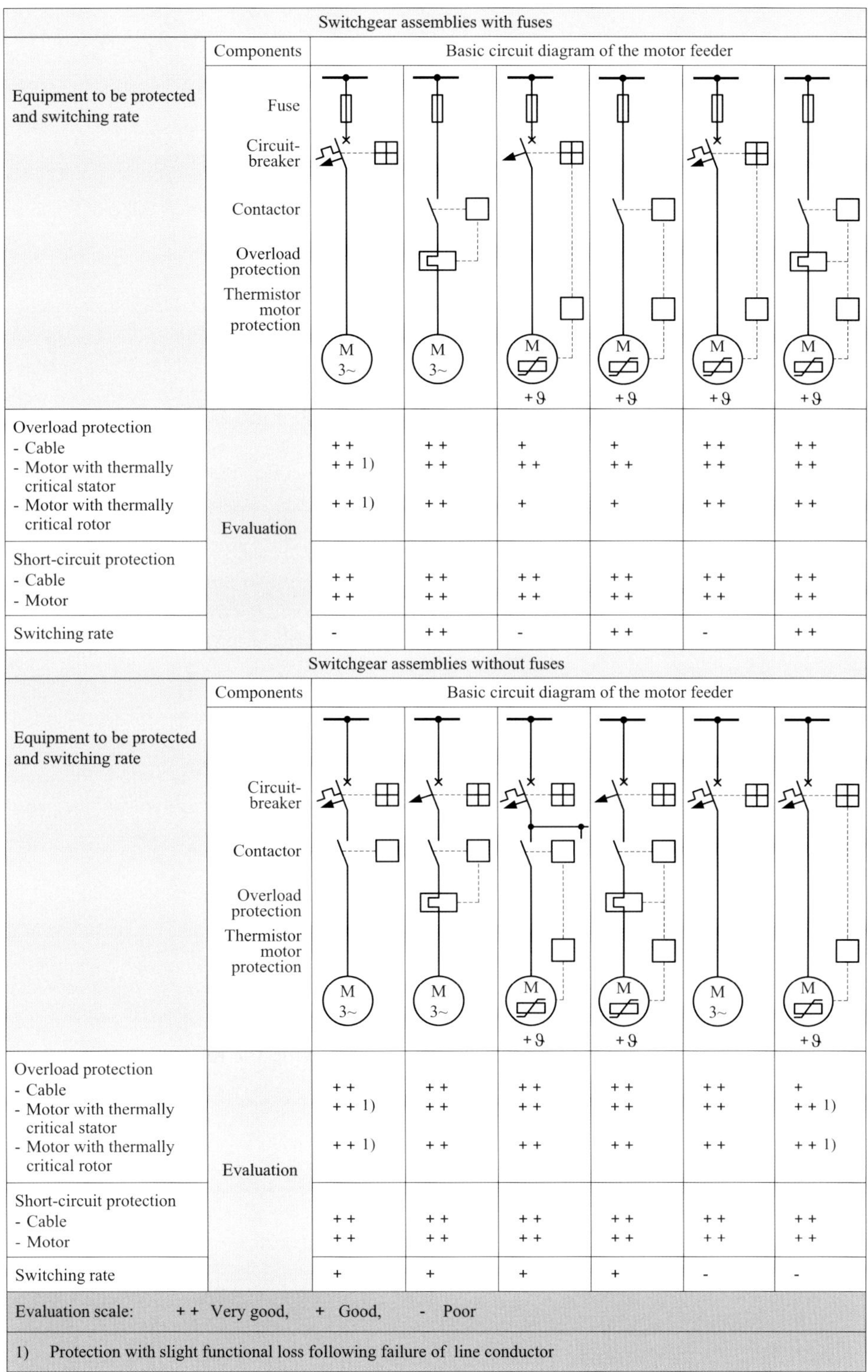

Switchgear assemblies with fuses							
Equipment to be protected and switching rate	Components	Basic circuit diagram of the motor feeder					
	Fuse Circuit-breaker Contactor Overload protection Thermistor motor protection	M 3~	M 3~	M +ϑ	M +ϑ	M +ϑ	M +ϑ
Overload protection - Cable - Motor with thermally critical stator - Motor with thermally critical rotor	Evaluation	+ + + + 1) + + 1)	+ + + + + +	+ + + +	+ + + +	+ + + + + +	+ + + + + +
Short-circuit protection - Cable - Motor		+ + + +	+ + + +	+ + + +	+ + + +	+ + + +	+ + + +
Switching rate		-	+ +	-	+ +	-	+ +
Switchgear assemblies without fuses							
Equipment to be protected and switching rate	Components	Basic circuit diagram of the motor feeder					
	Circuit-breaker Contactor Overload protection Thermistor motor protection	M 3~	M 3~	M +ϑ	M +ϑ	M 3~	M +ϑ
Overload protection - Cable - Motor with thermally critical stator - Motor with thermally critical rotor	Evaluation	+ + + + 1) + + 1)	+ + + + + +	+ + + + + +	+ + + + + +	+ + + + + +	+ + + 1) + + 1)
Short-circuit protection - Cable - Motor		+ + + +	+ + + +	+ + + +	+ + + +	+ + + +	+ + + +
Switching rate		+	+	+	+	-	-
Evaluation scale: + + Very good, + Good, - Poor							
1) Protection with slight functional loss following failure of line conductor							

13.2 Selectivity in LV networks

As defined in DIN VDE 0100-530 (VDE 0100-530): 2005-06 [13.19] or IEC 60364-5-53: 2002-06 [13.20], selectivity exists in an LV network if the operating/tripping characteristics of two or more overcurrent protective devices are coordinated in such a way that only the protective device connected directly upstream of the fault location trips when overcurrents occur. Whether selectivity is achieved by

- current grading (current selectivity),
- time grading (time selectivity),
- current/time grading (current-time selectivity) or
- specially trip-tested and selectivity-tested switchgear assemblies (dynamic/energy selectivity)

depends on the magnitude of the short-circuit currents at the network nodes. Tables C9.2, C9.4 and C9.5 provide information about the short-circuit currents occurring in industrial LV networks.

In LV networks that are implemented as a TN system and supplied with power through distribution transformers with the standard vector group Dyn5 or Dyn11 ($Z_{0T}/Z_{1T} \approx 0.95$)), the maximum fault current occurs on a near-to-infeed line-to-earth fault without a contribution to the short-circuit current by motors, or on a near-to-infeed three-phase fault with a contribution to the short-circuit current by motors. The lowest fault currents are always expected during far-from-infeed line-to-earth faults (e.g. line to exposed conductive part). While the maximum short-circuit currents form the basis for considering dynamic selectivity, mininum short-circuit currents are decisive for setting overcurrent tripping.

When the operating current of overcurrent releases is set according to the minimum short-circuit current, the contribution to the fault current by asynchronous motors is not taken into account. The maximum short-circuit currents, considering any asynchronous motors that may contribute to the short-circuit current in the case of a fault, determine the permissible disconnection times required to ensure reliable thermal short-circuit protection of the equipment used.

The short-circuit current distribution in simple radial networks can usually be calculated "manually" within a reasonable length of time. For complicated network configurations (meshed networks, radial networks in an interconnected cable system and radial networks interconnected through busbar trunking systems), use of the PSS™ SINCAL PC calculation program [13.21] is recommended.

13.2.1 Radial networks

Circuit-breakers and/or fuses can be connected in series in radial networks as seen in the direction of power flow, as follows:

- fuse with a downstream fuse,
- circuit-breaker with a downstream circuit-breaker,
- circuit-breaker with a downstream fuse,
- fuse with a downstream circuit-breaker,
- multiple parallel incoming feeders with or without bus couplers and with downstream circuit-breaker or fuse.

The selectivity requirements in networks incorporating the combinations of protective devices stated above are explained below.

13.2.1.1 Selectivity between LV HRC fuses

The pre-arcing time-current characteristics of series-connected LV HRC fuses with different rated currents and the same utilization category (e.g. gL/gG) are approximately parallel.

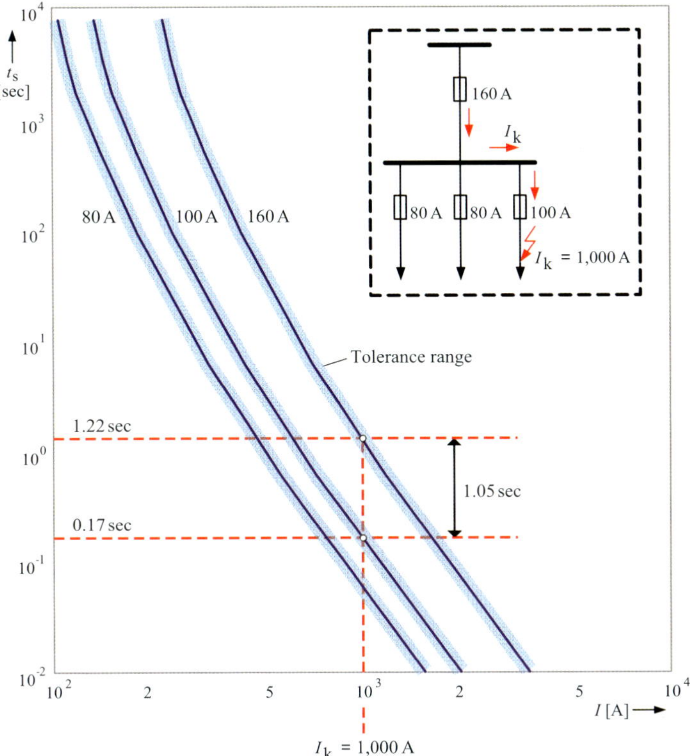

Fig. C13.12 Example of selectivity for series-connected 3NA LV HRC fuse-links of utilization category gL/gG

Absolute selectivity is when the I^2t_s pre-arcing (melting) value of the upstream fuse is greater than the I^2t_a operating value of the downstream fuse. To meet this condition, the rated currents of the series-connected LV HRC fuses must differ by a factor of 1.6 or more (Fig. C13.12). For grading the rated currents in the ratio 1/1.6, the characteristic comparison in the time-current diagram can be neglected for fuses in the same utilization category. Such a comparison is only required if fuses of different utilization categories (e.g. gL and aM) are connected in series. The data required for this are to be found in the BETA Low-Voltage Circuit Protection catalog ET B1T [13.11]. Moreover, the pre-arcing time-current characteristics of all Siemens fuses are stored in the database of the PC programs SIGRAGE [13.22] and SIMARIS® design (Section 13.3).

13.2.1.2 Selectivity between circuit-breakers

In the case of series-connected circuit-breakers with adjustable overload, short-time-delay and/or instantaneous overcurrent releases (LI or LSI releases), selectivity can be achieved by current and time grading. Circuit-breaker coordination (dynamic/energy selectivity) and zone-selective interlocking (ZSI) are further possible options.

a) Selectivity by current grading (grading of the operating currents of the instantaneous overcurrent releases (I releases))

Selective grading of the operating currents of the instantaneous overcurrent releases (I releases) for series-connected circuit-breakers with integrated LI releases is usually not possible. Selectivity by current grading could only be achieved if the operating current I_i of the instantaneous overcurrent release of the upstream circuit-breaker Q1 were set to a value above the maximum short-circuit current at the installation location of the downstream circuit-breaker Q2. Due to the time lag of the inverse-time-delay overload release (L-release), this setting does not ensure that all short circuits in the protection zone of the circuit-breaker Q1 are cleared within 5 sec (Fig. C13.13).

There are two possible solutions for ensuring reliable short-circuit protection (short-circuit duration $t_k \leq 5$ sec) and selectivity in the event of a fault with two series-connected circuit-breakers shown by way of example in Fig. C13.13.

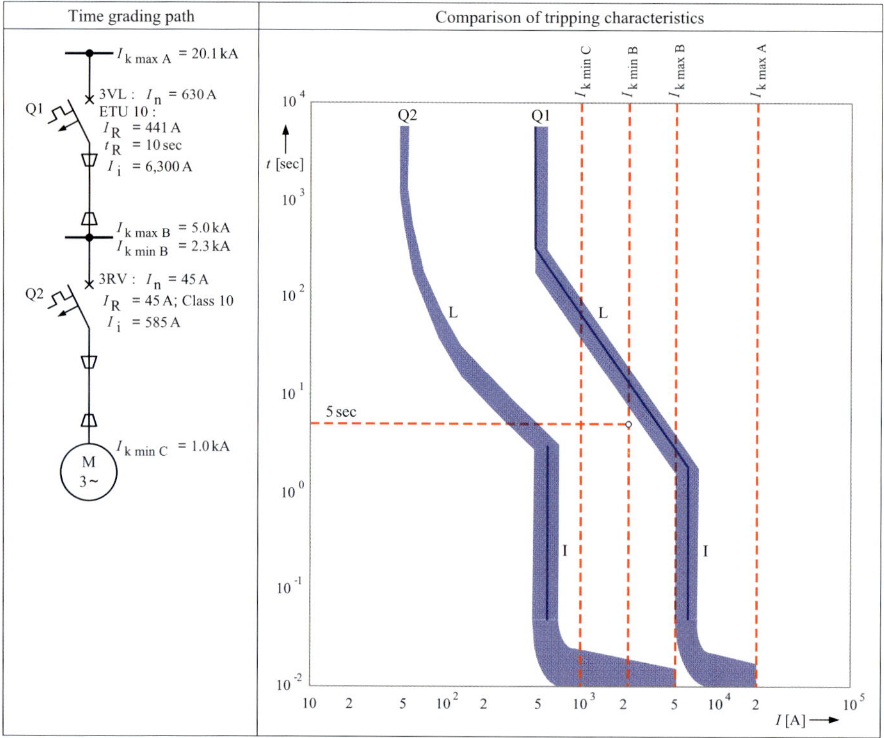

Fig. C13.13 Current selectivity of two circuit-breakers connected in series

- *Solution 1:*

  The operating current of the I release of Q1 must be set such that it reliably detects the minimum short-circuit current $I_{\text{k-min B}}$ at the end of its protection zone, taking tripping tolerances into account (setting e.g. to $I_i = 1{,}980$ A). If the maximum short-circuit current $I_{\text{k-max B}}$ flows with this setting in the event of a short circuit, both circuit-breaker Q1 and circuit-breaker Q2 trip.

  For evaluation of the selectivity that can be achieved with solution 1, the fault-current range of the dynamic selectivity (selectivity type b) is decisive (see Fig. C13.14).

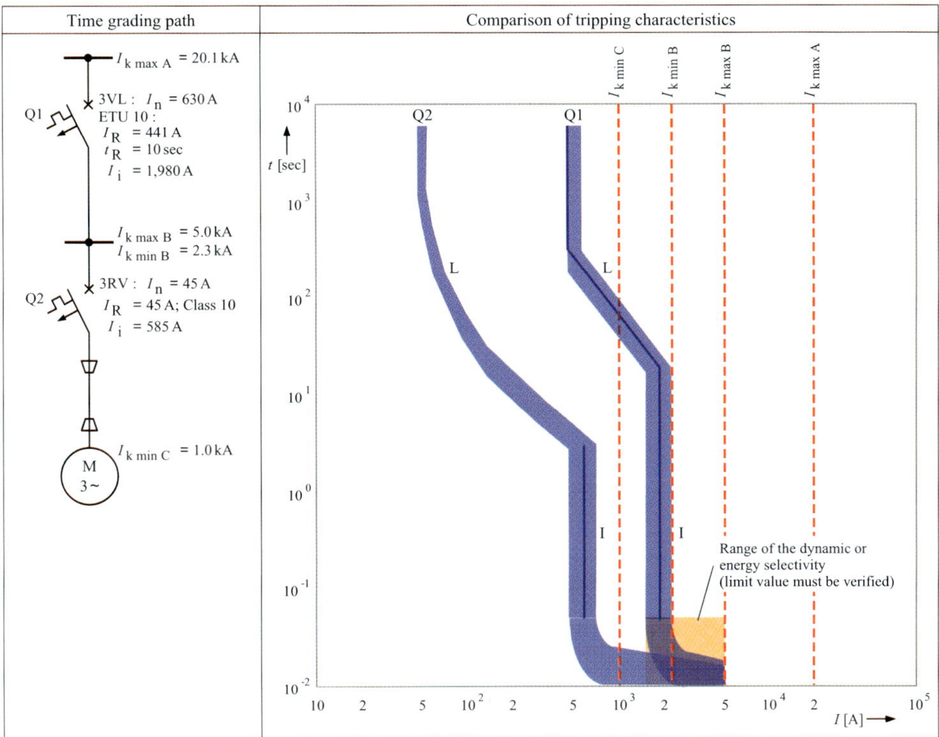

Fig. C13.14 Dynamic selectivity of two circuit-breakers connected in series

- *Solution 2:*

  Instead of a circuit-breaker with an LI release system, a circuit-breaker with an LSI release system is used for Q1. If a circuit-breaker with an LSI release system is used, short circuits in the protection zone of circuit-breaker Q1 are reliably and selectively cleared by coordinated setting of the S and I releases (setting of the S release e.g. to $I_{sd} = 1{,}980$ A, $t_{sd} = 0.1$ sec and of the I release to $I_i = 6{,}300$ A).

  Solution 2 ensures current-time selectivity (selectivity type c) for the example of a motor feeder with the two circuit-breakers Q1 and Q2.

b) Dynamic/energy selectivity (selectivity by coordination of the circuit-breakers connected in series)

In the case of series-connected circuit-breakers carrying the same short-circuit current, "natural" selectivity is provided simply by the difference in size. In this way, it is possible to determine a maximum value of the short-circuit current for each switch-switch combination up to which the downstream circuit-breaker will trip faster and alone. This current value is termed the selectivity-limit current I_{sel}. The response in the instantaneous short-circuit tripping range based on the let-through energy ($J_D^2 t$ value) of the downstream and the tripping energy ($J^2 t_a$ value) of the upstream circuit-breaker provides information about the selectivity-limit current I_{sel} (Fig. C13.14).

Full selectivity is ensured if the maximum short-circuit current at the installation location of the downstream circuit-breaker is less than the selectivity-limit current ($I_{k\text{-max}} < I_{sel}$) that applies to the selected switch-switch combination. If the value of the short-circuit current is above the determined selectivity-limit current, the upstream circuit-breaker will also trip.

The selectivity-limit current must be verified by testing according to DIN EN 60947-2 (VDE 0660-101): 2010-04 [13.3] or IEC 60947-2: 2009-05 [13.4]. Conducting a purely theoretical comparison of the let-through current (let-through energy) of the downstream and the tripping current (tripping energy) of the upstream protective device to determine the selectivity-limit current I_{sel} is not recommended, because the dynamic selectivity is determined by complicated electrodynamic and electrothermal processes.

The selectivity-limit currents tested on the hardware for Siemens circuit-breaker combinations (e.g. 3WL-3VL, 3VL-3RV) can be found in the selectivity tables of the Siemens application manual [13.16]. They are also stored in the database of the SIMARIS® design PC program (Section 13.3) .

c) Selectivity by time and current-time grading (grading of short-time-delay overcurrent releases (S releases))

If selectivity cannot be achieved either by current grading or by selecting circuit-breaker combinations specially tested for tripping and selectivity, time or time-current grading is used. For this type of grading, the upstream circuit-breakers are equipped with time-delay overcurrent releases (S releases). Both the tripping delays t_{sd} and the operating currents I_{sd} are graded.

In the case of the electronic trip units (ETUs) usual today, a grading time of Δt_{sd} = 70...100 msec is sufficient to take account of all possible variances of the mechanical delay of the switching device. The operating current of the time-delay overcurrent release should be set to at least 1.45 times the value of the downstream circuit-breaker. This setting takes account of a tripping tolerance of $\Delta I = -20...+20\,\%$. Because the Siemens 3WL circuit-breaker has a tripping tolerance of only of $\Delta I = 0...+20\,\%$, the grading factor for setting the current value I_{sd} can be reduced to 1.25. Both non-current-limiting (e.g. SENTRON 3WL) and current-limiting circuit-breakers (e.g. SENTRON 3VL) can be equipped with electronic LSI releases.

The principle of time and current-time grading is shown by way of example in Fig. C13.15 with four circuit-breakers connected in series. With short-circuit currents of approximately the same magnitude at the installation locations of the circuit-breakers (e.g. Q1 and Q2), only time grading (parameterization and coordination of I_{sd} and t_{sd}) is possible. The current-time grading (parameterization and coordination of I_{sd}, t_{sd} and I_i) requires an ability to set the I release without impairing the selectivity. In Fig. C13.15, circuit-breakers Q3 and Q4 have such a setting option. By setting the operating current I_i of the I release of Q3 to a value that is above the maximum short-circuit current at the installation location of Q4, the circuit-breaker Q3 only trips instantaneously on a "dead" short circuit in its protection zone ($I_{k\text{-max C}} < I_k \leq I_{k\text{-max B}}$).

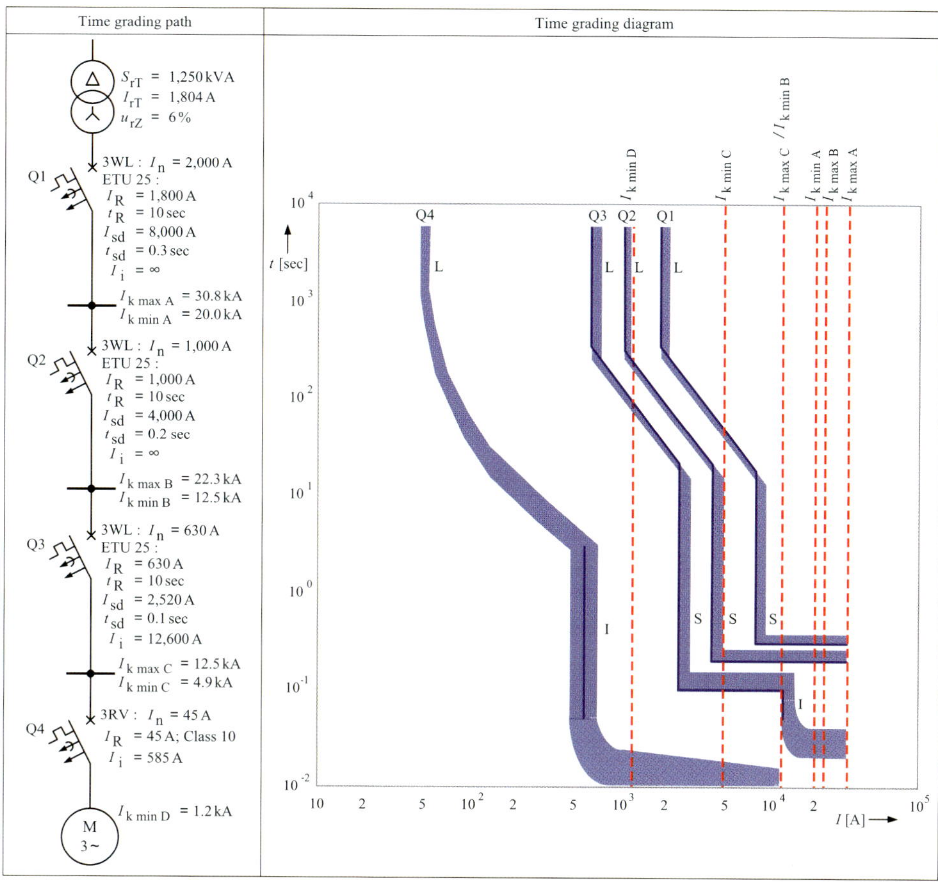

Fig. C13.15 Selective time and current-time grading for a feeder with four circuit-breakers connected in series

In protection zones in which selective current-time grading is possible, the instantaneous tripping of the I release greatly reduces the equipment stress in case of a “dead“ short circuit. With pure time grading, on the other hand, the short-circuit stress of the switching devices and equipment depends on the delay time t_{sd} set on the S release. As Fig. C13.15 shows, the short-circuit current at the installation location of the circuit-breaker is greatest with the longest delay time (circuit-breaker Q1). The short-circuit stresses occurring due to undesirably long delay times can be reduced using zone-selective interlocking (ZSI).

d) Zone-selective interlocking (ZSI)

Microprocessor-controlled zone-selective interlocking (ZSI) was developed to avoid an undesirably long total clearing time in LV networks with series-connected circuit-breakers. Fig. C13.16 shows the principle and method of operation of zone-selective interlocking. Zone-selective interlocking has the advantage that all short-circuits can be cleared in the network after no more than 50 msec, irrespective of the number of series-connected circuit-breakers. The shortest possible fault clearance times must be aimed at, especially with very high system short-circuit powers.

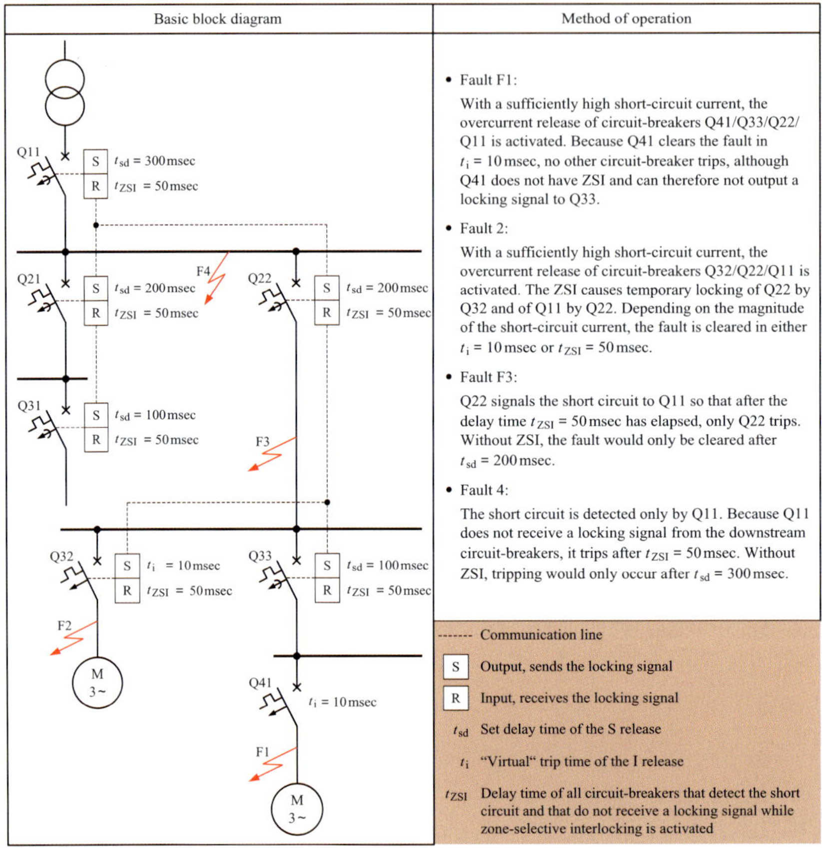

Fig. C13.16 Principle and method of operation of zone-selective interlocking (ZSI)

13.2.1.3 Selectivity between a circuit-breaker and LV HRC fuse

In LV networks, the circuit-breaker can be upstream or downstream of the LV HRC fuse. In selectivity considerations with upstream and downstream standard fuses, a permissible tolerance range of ±10 % in the direction of current flow must be taken into account in the current-time characteristics. If Siemens 3NA LV HRC fuses are used, the tolerance range is reduced to ±6 %.

a) Circuit-breaker with downstream LV HRC fuse

Circuit-breakers with LI or LSI releases act selectively in the overload range with respect to a downstream LV HRC fuse if there is a sufficient safety margin t_A between the upper tolerance range of the fuse characteristic and the lower tripping characteristic of the fully loaded L release (Fig. C13.17a). A safety margin of $t_A \geq 1$ sec is generally seen as sufficient.

At the operating temperature of a thermal overload release (type TM), a reduction of the tripping time of up to 25 % can be expected unless specified otherwise by the manufacturer.

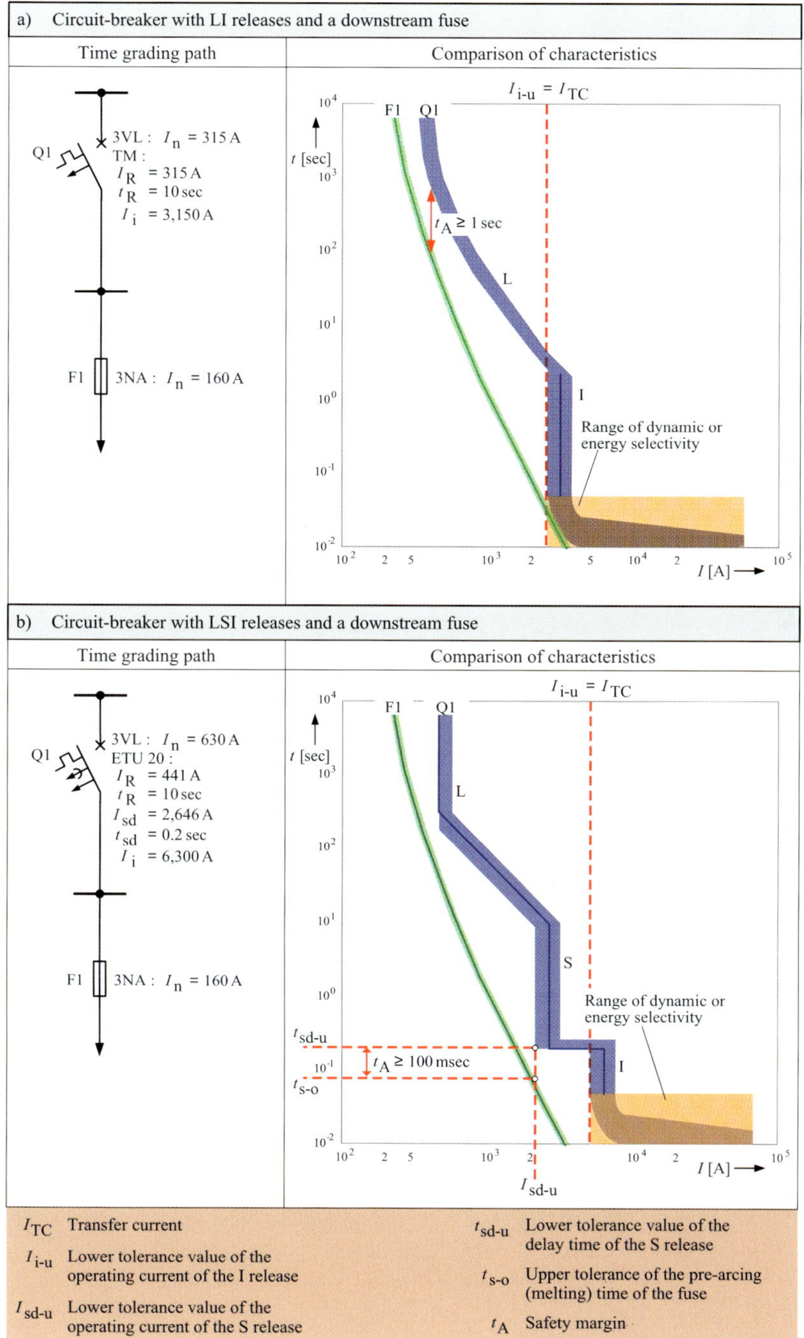

Fig. C13.17 Selectivity between a circuit-breaker and a downstream fuse

Absolute selectivity in the case of a short circuit is only ensured with use of circuit-breakers without a time-delay overcurrent release if the let-through current I_D of the LV HRC fuse (for let-through current diagrams for Siemens 3NA LV HRC fuses, see

BETA Low-Voltage Circuit Protection catalog ET B1T [13.11]) does not reach the lower tolerance value $I_{i\text{-}u}$ of the operating current of the I release. However, this can only be expected for a fuse whose rated current is very low compared with the rated continuous current of the circuit-breaker [13.16]. Otherwise, the principles for dynamic selec-

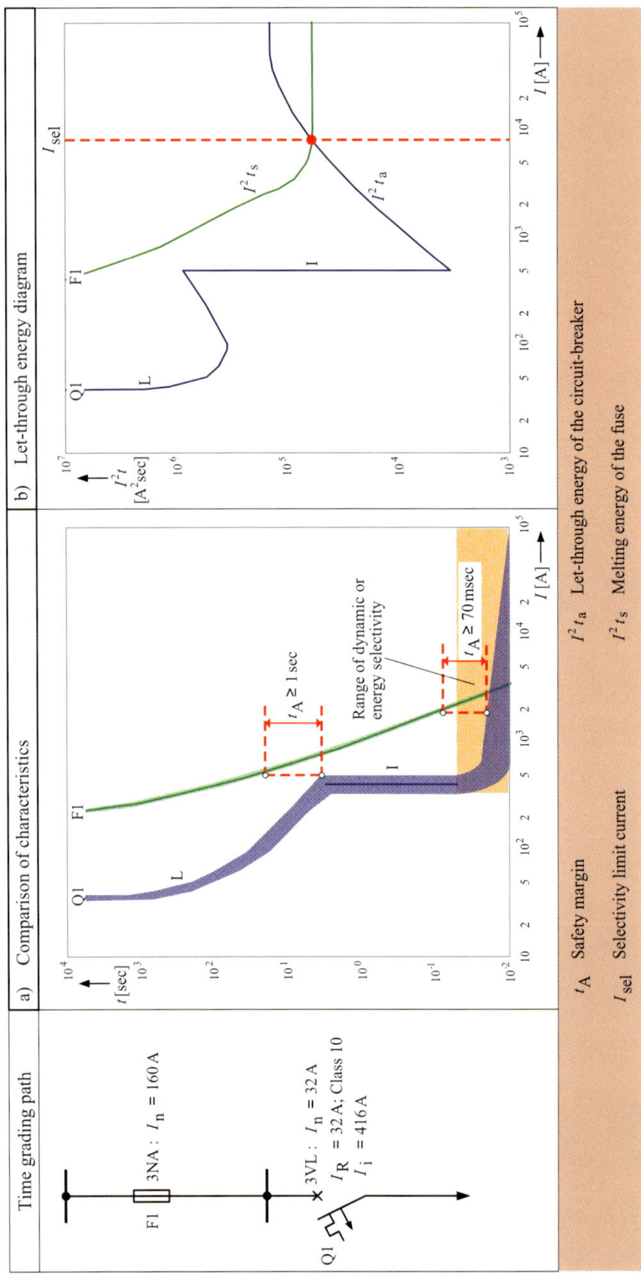

Fig. C13.18 Selectivity between a circuit-breaker and an upstream fuse

tivity must be observed. To clear short circuits selectively with the use of circuit-breakers with short-time-delay overcurrent releases, the safety margin between the lower tolerance value $t_{sd\text{-}u}$ of the delay time of the S release and the upper spread $t_{s\text{-}o}$ of the fuse characteristic must be at least 100 msec (Fig. C13.17b).

b) Circuit-breaker with upstream fuse

Selectivity exists in the overload range between the circuit-breaker and upstream fuse if a safety margin t_A of at least 1 sec is observed with the lower tolerance range of the fuse characteristic to the tripping characteristic of the inverse-time-delay overload release (L release) (Fig. C13.18a). For selectivity in the case of a short circuit, it must be considered that the fault current is already heating the LV HRC fuse during arc extinction as part of breaking. The selectivity limit is approximately where a safety margin between the lower tolerance range of the fuse characteristic and the upper tolerance value of the operating time of the instantaneous overcurrent release (I release) or the delay time of the short-time-delay overcurrent release (S release) of $t_A = 70$ msec is undershot (Fig. C13.18a). A safe and usually also a higher selectivity limit for the short-circuit range can be determined in the I^2t diagram (Fig. C13.18b). In this diagram, the maximum let-through I^2t_a value of the circuit-breaker is compared with the minimum pre-arcing I^2t_s value of the fuse. Because this is a comparison of extreme values, the tolerances are not considered. In case of a short circuit with selective clearance by the downstream circuit-breaker, it is not possible to state reliably whether selectivity of the fuse-switch combination will be retained, owing to of uncertainty about a possible pre-existing defect of the upstream fuse caused by the short-circuit current of unknown magnitude. To ensure full selectivity of the assembly after clearance of a short circuit by the downstream circuit-breaker, the upstream LV HRC fuses must always be replaced. This fuse replacement is often difficult in practical system operation (trained operating personnel, stocking of spare fuses, etc.).

13.2.1.4 Selectivity in case of incoming feeders connected in parallel

In parallel operation of multiple transformers on a common busbar, the partial short-circuit currents $I_{k\text{-}Ti}$ flowing in the individual incoming feeders add up to the total short-circuit current $I_{k\Sigma}$ in a faulty feeder in which a disturbance has occurred. This total short-circuit current is the basis for the current scale in the grading diagram. Using the current scale based on the total short-circuit current provides more favourable selectivity conditions for all protective devices downstream of the incoming feeder circuit-breakers. This applies to all types of fault.

a) Two identical incoming feeders

Fig. C13.19 details the distribution of short-circuit currents in case of a fault on an outgoing feeder with two parallel-connected transformers of the same power rating and the same length of incoming feeder cables.

The total short-circuit current $I_{k\Sigma}$ flowing through fuse F1 is composed of the partial short-circuit currents $I_{k\text{-}T1}$ and $I_{k\text{-}T2}$. Because the total short-circuit current $I_{k\Sigma}$ in the postulated ideal case (load feeder positioned exactly in the centre, disregarding the load current in other outgoing feeders) is divided equally among the incoming feeders, the tripping characteristic of circuit-breakers Q1 or Q2 can here permissibly be shifted by the characteristic displacement factor $K_{t\text{-}I} = 2$ to the right on the current scale as far as the perpendicular $I_{k\Sigma}$, the basis for this fault case. This shift results in both time selectivity and an additional gain in current selectivity (Fig. C13.19).

If the incoming and outgoing feeders are arranged and located asymmetrically on the busbar, the distribution of the short-circuit current is different depending on the impedance ratio of the transformers and incoming feeders. For the asymmetrical fault

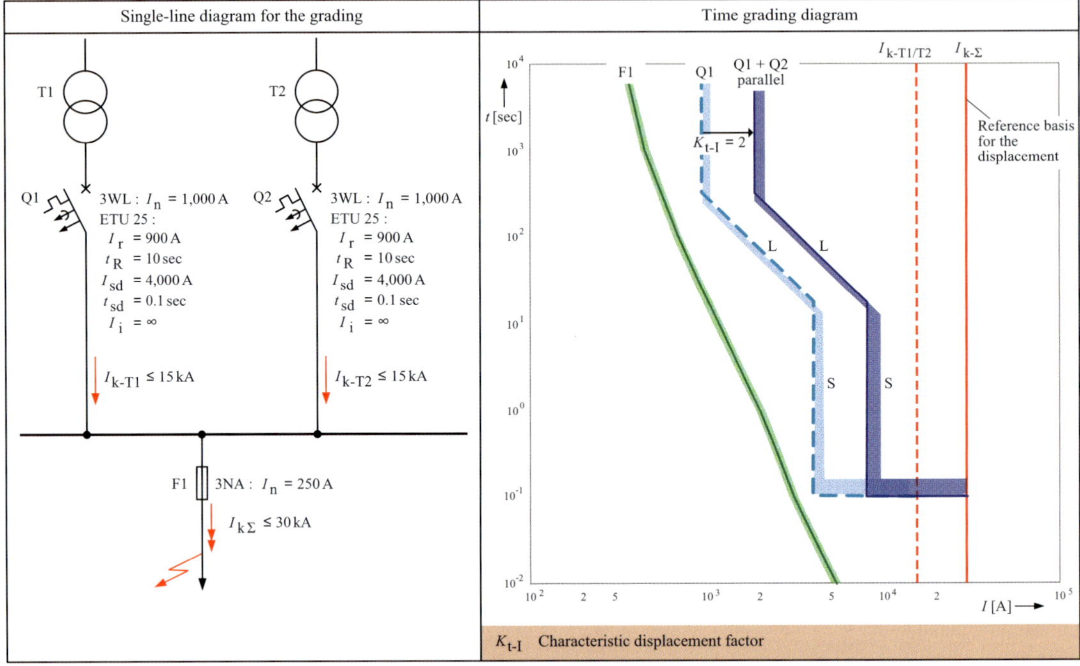

Fig. C13.19 Selectivity relations in the case of two parallel-connected transformers with the same power rating (example with faulty outgoing feeder positioned in the centre)

current distribution that usually occurs in practice, the characteristic displacement factor $K_{t\text{-}I}$ must be calculated as follows:

$$K_{t\text{-}I} = \frac{I_{k\Sigma}}{\max\limits_{i} I_{k\text{-}T_i}} \qquad (13.1)$$

b) Three identical incoming feeders

An even better current selectivity is achieved in the case of a short circuit on an outgoing feeder using $n = 3$ parallel-connected transformers, since the characteristic displacement factor for the incoming feeder circuit-breakers Q1, Q2 and Q3 is then in the range $2 < K_{t\text{-}I} \leq 3$ (for an idealized example, see Fig. C13.20a).

To clear faults between the secondary-side transformer terminals and incoming feeder circuit-breaker fast, reliably and selectively, LSI releases are used in incoming feeders. The operating current I_i of the instantaneous overcurrent release (I release) must be set such that its value is above the partial short-circuit current $I_{k\text{-}Ti}$ of a transformer. This means that only that incoming feeder circuit-breaker carrying the total short-circuit current $I_{k\Sigma}$ ever trips instantaneously in case of a secondary-side terminal short circuit on the transformer. The circuit-breakers in the faultless incoming feeders remain closed due to the delay time t_{sd} set on the S release (for an example, see Fig. C13.20b).

The selectivity example shown in Fig. C13.20b is based on a perfectly symmetrical fault current distribution, that is, for the transformer circuit-breakers Q2 and Q3 in the faultless incoming feeders, a characteristic displacement factor of $K_{t\text{-}I}$ = 30 kA/15 kA = 2 ideally applies. Owing to differing fault impedances among multiple incoming feeders,

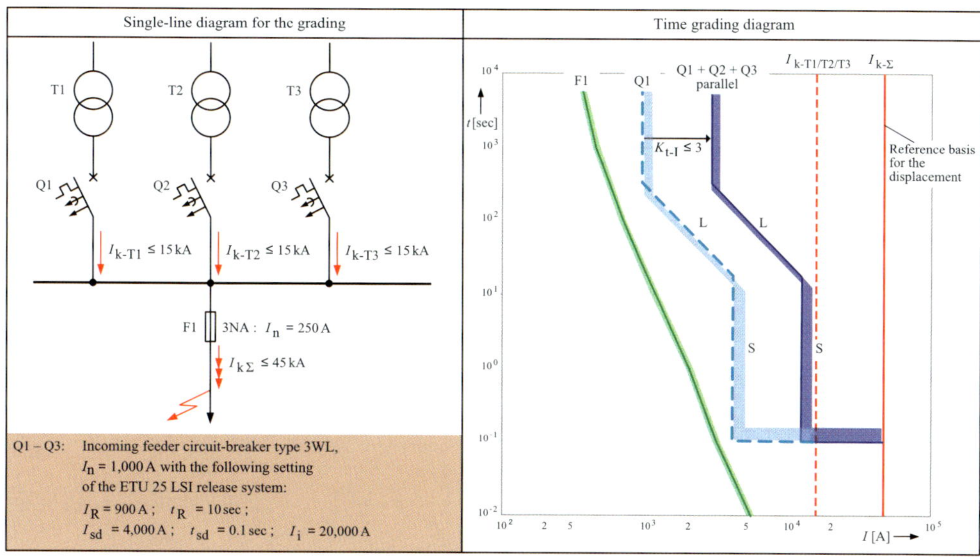

Fig. C13.20a Selectivity relations in the case of three parallel-connected transformers with the same power rating (short circuit on an outgoing feeder positioned in the centre)

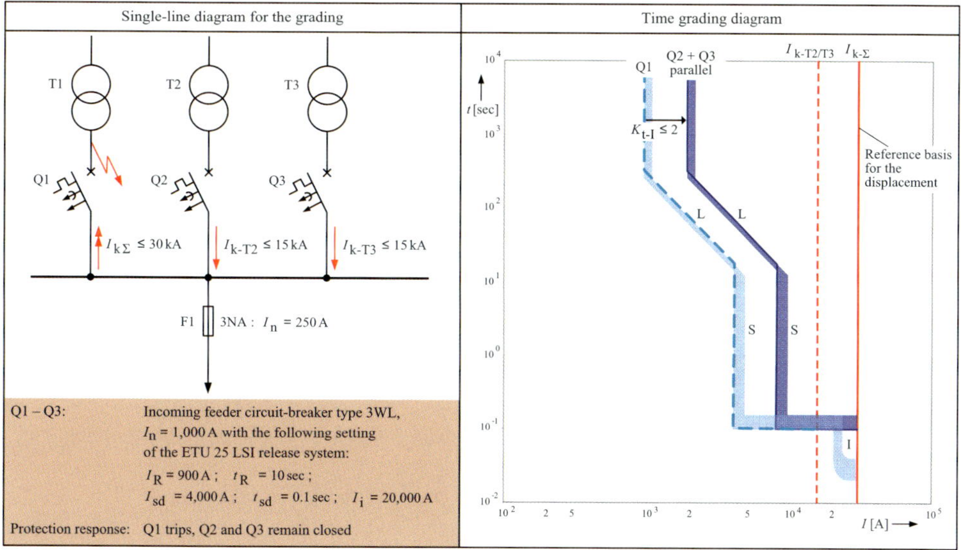

Fig. C13.20b Selectivity relations in the case of three parallel-connected transformers with the same power rating (secondary-side terminal short circuit on one transformer)

however, asymmetrical fault current distributions are the rule. With sufficient damping by the fault impedance, the total short-circuit current $I_{k\Sigma}$ can be kept so small that instantaneous short-circuit tripping of the transformer circuit-breaker in the incoming feeder in which the fault occurred is no longer certain. In such a case, all incoming feeders would trip simultaneously and non-selectively. To prevent non-selective tripping on strongly differing fault impedances, incoming feeder circuit-breakers must be used whose S releases exhibit an I^2t-dependent tripping characteristic. The gain in selectivity that can be achieved using S releases with an I^2t-dependent tripping characteristic is shown in Fig. C13.20c.

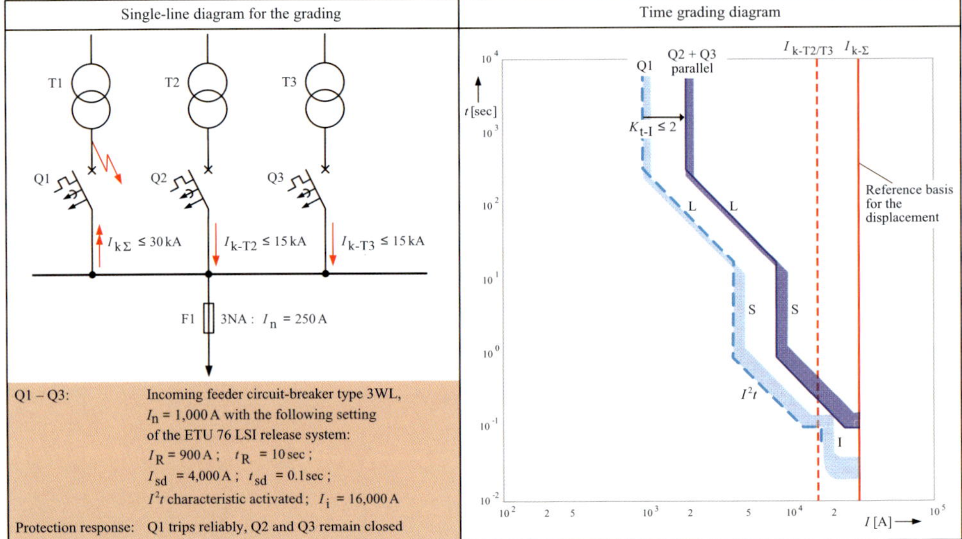

Fig. C13.20c Selectivity relations in the case of three parallel-connected transformers with the same power rating (secondary-side terminal short circuit on one transformer, I^2t-dependent fault clearance)

As a comparison of Fig. C13.20c with Fig. C13.20b shows, S releases with I^2t-dependent tripping characteristics are notable for much enhanced current-time selectivity.

c) Incoming feeders connected in parallel through bus coupler circuit-breakers

Circuit-breakers with overcurrent releases are used as bus couplers to meet the following objectives:

- fastest possible clearance of busbar faults,
- fault restriction to the respective busbar section,
- relieving the feeders of the effects of high total short-circuit currents.

With $n = 2$ parallel incoming feeders, only the partial short-circuit current of one transformer ever flows through the bus coupler. In this way, selectivity can only be achieved between the incoming feeder circuit-breaker and the coupler circuit-breaker by time grading. Only as from $n \geq 3$ parallel incoming feeders are the partial short-circuit currents flowing through the bus couplers dependent on the fault location (fault in the outer or central busbar section). Figs. C13.21 and C13.22 show the protection response of the transformer and bus coupler circuit-breakers in the case of three parallel incoming feeders and faults (fault on the outgoing feeder, fault on the incoming feeder) in the outer and central busbar section.

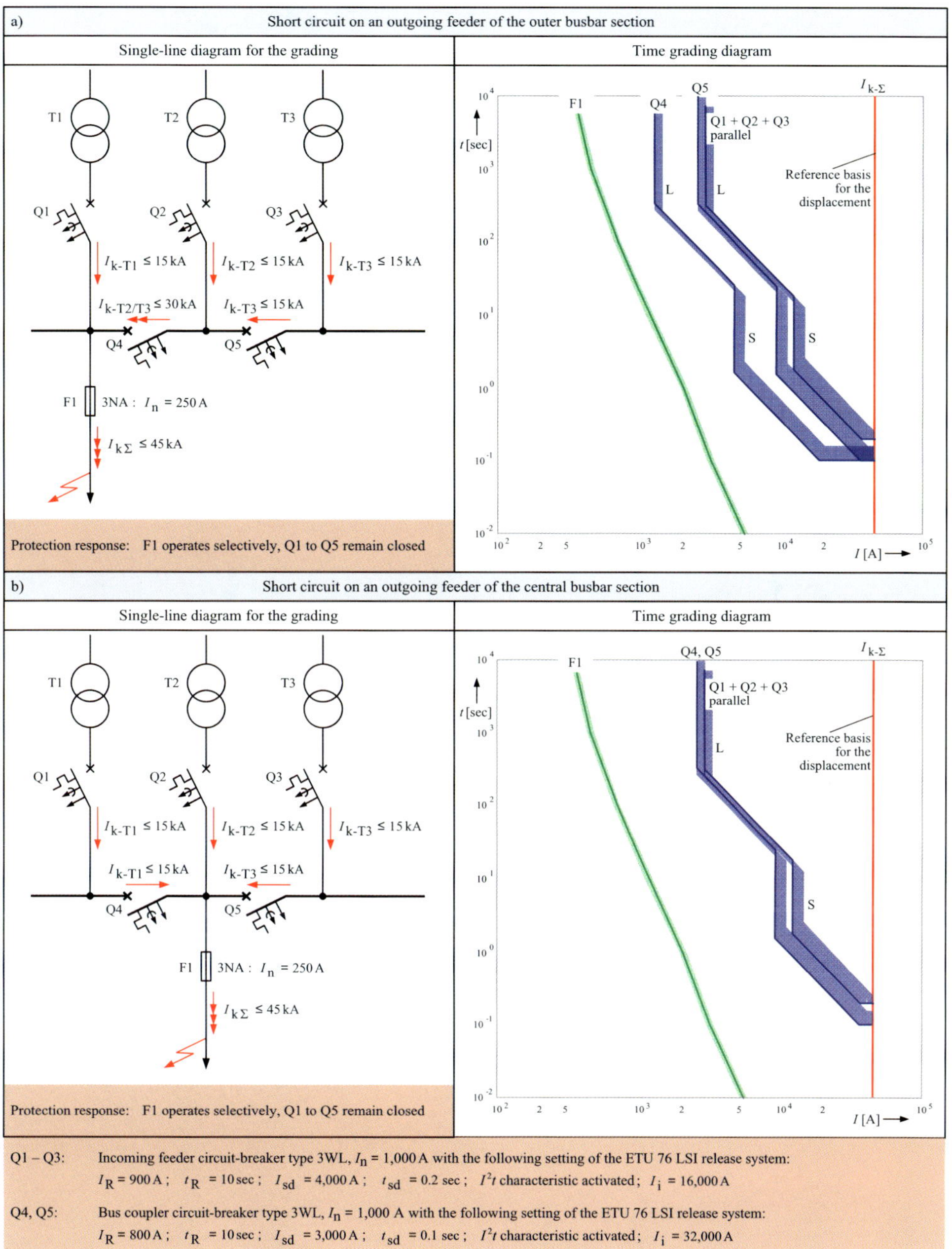

Fig. C13.21 Selectivity relations in the case of three identical transformers connected in parallel through bus coupler circuit-breakers (short circuit on an outgoing feeder)

Usually, further overcurrent protective devices are connected in series with the transformer and bus coupler circuit-breakers present in the multiple incoming supply of industrial LV networks. For this reason, inclusion of the bus coupler circuit-breakers in the selective time grading may inappropriately lengthen the disconnecting time for near-to-infeed short circuits.

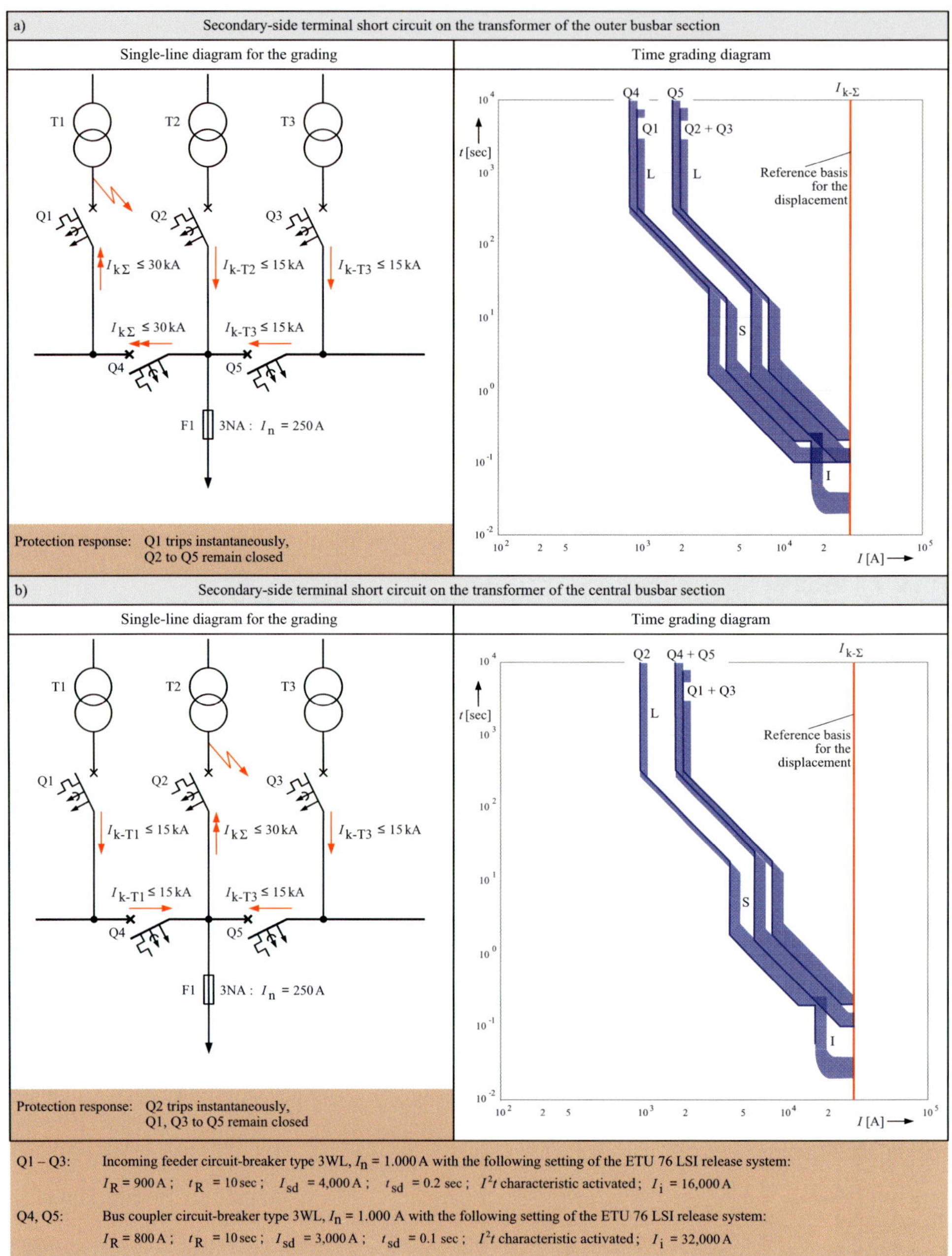

Fig. C13.22 Selectivity relations in case of three identical transformers connected in parallel through bus coupler circuit-breakers (terminal short circuit on one transformer)

Because there are often economic reservations against the use of zone-selective interlocking (ZSI), the S releases of the transformer and bus coupler circuit-breakers are no longer time-graded instead. In addition to setting identical time delays t_{sd}, it is also usual to dispense with overcurrent releases in the busbar couplings altogether.

13.2.1.5 Selectivity and undervoltage protection

On a short circuit the system voltage at the short-circuit location collapses, falling to a residual voltage that depends on the fault resistance. In a "dead" short circuit, that is, without resistance, the voltage at the short-circuit location is practically zero.

Generally, however, arcs occur during short circuits that experience has shown to have arc-drop voltages in the range 30 V $\leq U_{arc} \leq$ 70 V [13.18]. The residual voltage remaining during a short circuit rises against the direction of power flow proprotionately to the impedances occurring between the fault location and the power source. The residual voltage at the low-voltage main distribution board (LV-MDB) during a far-from-infeed short circuit is therefore greater than during a near-to-infeed short circuit (Fig. C13.23). The protection response of circuit-breakers equipped with an undervoltage release can be assessed based on the residual voltage occurring at the network nodes (LV-MDB, LV-SDB).

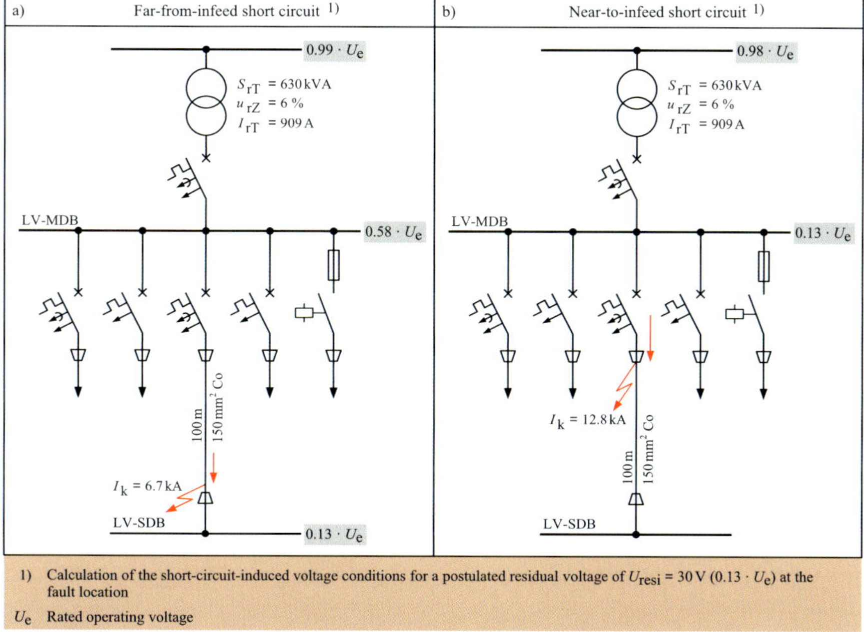

Fig. C13.23 Example of voltage conditions in a short-circuited LV network

If the rated operational voltage U_e collapses down to a residual voltage in the range $0.35 \cdot U_e \leq U_{resi} \leq 0.70 \cdot U_e$ and the voltage dip lasts for longer than $t_{\Delta u'}$ = 20 msec, all circuit-breakers equipped with undervoltage releases interrupt the supply of electrical power to the loads. Short-circuit-induced voltage dips also cause contactors to drop out. Contactors can be expected to drop out if the control supply voltage falls below 75 % of its rated value for longer than 5 msec to 30 msec [13.18].

Fig. C13.23 shows an example of the voltage conditions at the nodes of a short-circuited LV network. Comparing these voltage conditions with the limits for undervoltage-induced tripping, instantaneous overcurrent protection tripping must be possible within $t_a \leq$ 20 msec both for the near-to-infeed short circuit and for the far-from-infeed short circuit. Otherwise, there is a risk that the circuit-breakers equipped with instantaneous undervoltage releases in the network will also trip faultless feeders. It is also possible that all contactors may drop out. If an undervoltage-induced total clearing time of $t_{a\text{-total}} \leq 20$ msec cannot be achieved for all relevant short-circuit locations in the network by means of standard protection equipment, the use of undervoltage releases and contactors with a tripping delay must be examined. However, this tripping delay must also be in keeping with the immunity of voltage-sensitive loads to voltage dips.

13.2.2 Meshed and closed ring-operated networks

Meshed and closed ring-operated networks include

- radial networks in an interconnected cable system (Section 10.3.1.3),
- multi-end-fed meshed networks (Section 10.3.1.4),
- radial networks interconnected through busbar trunking systems (Section 10.3.1.5).

In implementing these network configurations, the following two selectivity tasks must be ensured:

a) Only the faulty connecting cable or the faulty component of the busbar trunking system may be isolated from the network.

b) In the case of faults occurring between an LV incoming feeder circuit-breaker and an MV-side switch-fuse combination (Section 7.3.1) or circuit-breaker-relay combination (Section 7.3.2), the respective supplying transformer must be disconnected on the primary and secondary sides.

13.2.2.1 Selectivity in meshed networks with node fuses

Selectivity in meshed networks can only be achieved with fuses of identical type (same rated current, same utilization category (Table C13.2)). To construct meshed networks, only cables with a uniform cross-sectional area may therefore be used. Fig. C13.24 shows the fault current distribution resulting from a short circuit at the node of a meshed network. Selectivity at this node is ensured if none of the partial short-circuit currents I_{ki} is greater than 80(63) % of the total short-circuit current $I_{k\Sigma}$.

In this current ratio, only the fuse carrying the total short-circuit current operates. The ratio $I_{ki}/I_{k\Sigma}$ is also termed "meshed network factor". A meshed network factor of 0.63 must be applied if commercial-type gL/gG fuses of different manufacturers are used [13.23].

The meshed network factors 0.80 or 0.63 also apply to extremely high short-circuit currents because the rated breaking capacity and the rated current of the fuse are mutally independent.

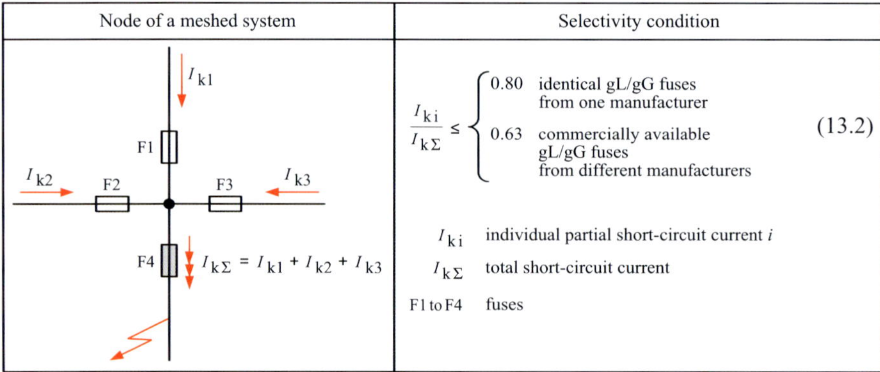

Fig. C13.24 Selective fault clearing in a meshed network with node fuses

13.2.2.2 Selectivity in operation of radial networks in an interconnected cable system

Reliable operation of radial networks in an interconnected cable system (Section 10.3.1.3) is ensured only if, on faults between the LV incoming feeder circuit-breaker and the MV-side switch-fuse combination, power backfeeding from the LV system to the fault location is prevented. To solve this selectivity problem with protection equipment, previously a network master relay (reverse-power relay) was used on the low-voltage side of the supplying transformers. Today, instead of network master relays, circuit-breakers with electronic release systems are used (e.g. ETU 45, 55, 76). The S release of such a trip system has an I^2t characteristic that can be used to implement selective short-circuit protection in the reverse direction.

Fig. C13.25 shows the protection principle used today for operation of radial networks in an interconnected cable system. With this protection principle, selective fault clearance is performed as follows:

Fault F1 (fault between switch-fuse combination and primary-side transformer terminals)

Short circuits and line-to-earth faults on the primary side of the supplying transformer are cleared by the switch-fuse combination. After the switch-fuse combination has operated, the transfer intertripping causes the associated LV incoming feeder circuit-breaker to open.

Fault F2 (fault between LV circuit-breaker and secondary-side transformer terminals)

On a short circuit or line-to-earth fault on the secondary-side transformer terminals, the total short-circuit current of all faultless transformers flows through the respective LV incoming feeder circuit-breaker. The circuit-breaker of the multiple incoming supply carrying the total short-circuit current trips faster than the circuit-breakers carrying the partial short-circuit currents owing to the I^2t-dependent response of the S releases (see Fig. C13.20c). In the case of non-simultaneous clearance of the fault F2 by the switch-fuse combination (e.g. in the event of a single-phase terminal short circuit, see Eq. 7.27.2), the transfer tripping causes the associated switch to open.

Fault F3 (fault on an LV MDB connecting cable)

A short circuit or line-to-earth fault on an LV-MDB connecting cable is always fed from two sides. In the case of a meshed network factor of ≥ 0.8 (Eq. 13.2) to be observed, the total overcurrent flowing to the fault location F3 from both sides is so large that only the node fuses of the faulty connecting cable operate. This selectively isolates the faulty connecting cable from the closed ring-main network. After disconnection of the faulty connecting cable, the network is operated as an open ring-main.

Selective isolation of LV-MDB connecting cables with a fault can also be achieved if LV circuit-breakers are used instead of LV HRC fuses. Like the S release of the incoming feeder circuit-breakers, the S release of these circuit-breakers must have an I^2t characteristic.

Radial networks in an interconnected cable system provide an instantaneous reserve or "hot standby" redundancy if a transformer incoming feeder fails. The available instantaneous reserve is dimensioned such that transformers remaining in operation after isolation of the fault location (e.g. F1 or F2 in Fig. C13.25) by the protection equipment are not overloaded. Table C11.7 gives information about the permissible load of GEAFOL cast-resin transformers in AF operation (radial-flow fans switched on). According to Table C11.7, the L releases of the LV circuit-breakers in the transformer incoming feeders must be parameterized for possible operation under fault conditions

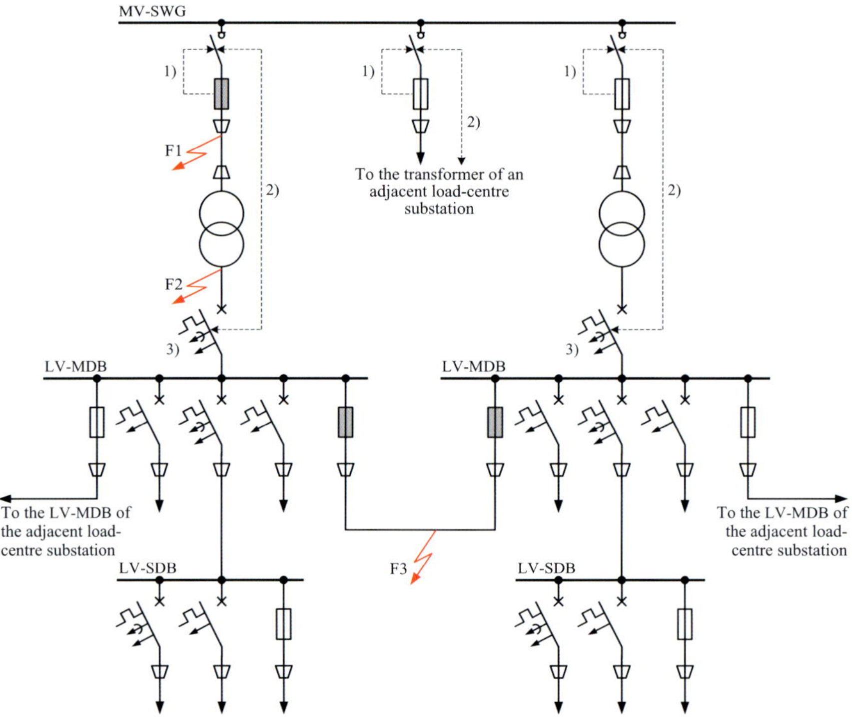

1) Switch-fuse combination (for selection and dimensioning, see Section 7.3.1)

2) Transfer tripping (tripping that causes the associated LV incoming feeder circuit-breaker to open when the MV switch opens and vice versa)

3) LV circuit-breaker with electronic LSI release system (The S release has an I^2t inverse-time tripping characteristic. The circuit-breaker receives the TRIP command on fault F1 through the transfer trip system and on fault F2 through the I release.)

Fig. C13.25 Protection principle for operation of radial networks in an interconnected cable system

(AF operation) of $n-1$ GEAFOL cast-resin transformers ($I_R > I_{T\text{-}AF}$). The characteristic displacement factor $K_{t\text{-}I}$ for the selectivity proof for decentralized multiple incoming supply can be calculated according to Eq. (13.1).

For faults not located between the LV incoming feeder circuit-breaker and the switch-fuse combination, the selectivity conditions explained in Section 13.2.1 apply.

13.2.2.3 Selectivity in operation of radial networks interconnected through busbar trunking systems

In radial networks interconnected through busbar trunking systems (Section 10.3.1.5), the low-voltage main distribution boards (LV-MDBs) and their connecting cables are replaced by a high-current busbar system (e.g. SIVACON 8PS (Section 11.2.4)). The high-current busbar system supplied from the decentralized dry-type transformers transmits the power to the busbar trunking systems. The busbar trunking systems perform the function of low-voltage subdistribution boards (LV-SDBs) for supplying electrical power to loads.

Purely in terms of protection equipment, therefore, the selectivity criteria that apply to radial networks in an interconnected cable system can also be applied to radial networks interconnected through busbar trunking systems. On occurrence of fault F1 (short circuit or line-to-earth fault between switch-fuse combination and primary-side transformer terminals) and F2 (short circuit or earth fault between LV incoming feeder circuit-breaker and secondary-side transformer terminals), selective fault clearance is performed in a way analogous to the radial networks in an interconnected cable system (Fig. C13.25). Because today‘s high-current busbar systems do not usually have coupler circuit-breakers (Fig. C10.50), all transformer incoming feeders would trip on fault F3 (busbar short circuit or earth fault). Unlike with cables, faults on high-current busbar systems with an inherently short-circuit-proof design and installation (e.g. SIVACON 8PS), however, can be excluded with very high probability. For that reason, the protection principle for radial networks interconnected through busbar trunking systems (general disconnection of the multiple incoming supply on line-to-earth and multi-phase faults on the high-current busbar system) is also appropriate from the point of view of decision criteria (acceptability criterion, see Fig. A2.21).

13.3 Example of selective protection coordination with SIMARIS® design

SIMARIS® design [13.24] is not a conventional network calculation tool; it is a tool for holistic dimensioning of low-voltage radial networks. Based on the results of network calculation, which is always performed first, that is

- load flow,
- short circuit (1-/2-/3-pole minimum and 1-/3-pole maximum) and
- energy balance,

SIMARIS® design automatically dimensions the network components required for a safe power supply to the loads that meets all the requirements. The network components and modules stored in the internal product database of SIMARIS® design include

- power supply sources (transformer(s), generator(s), neutral system infeed(s)),
- cables and busbars,
- switching and protective devices for all levels of the circuit hierarchy.

The SIMARIS® design PC program is characterized by the following performance and function features and more:

- general avoidance of expensive system overdimensioning by creation of sufficiently verified energy balances (considering utilization, coincidence and demand factors in calculation of the power demands),
- automatic dimensioning of the circuits for protection against overload, protection against short circuit, protection against electric shock and compliance with the permissible voltage drop,
- automatic device selection based on a relevant, coordinated Siemens device range,
- toleranced visualization of the characteristic device curves and tripping characteristics for worst-case selectivity scenarios,
- automatic evaluation of the selectivity, in particular dynamic selectivity (energy selectivity) of device combinations (more than 100,000 tested combinations of Siemens switching and protective devices in the time range < 100 msec are stored in the program‘s database),
- output selectivity diagrams and characteristic device curve comparisons as evidence of selectivity that is accepted by TÜV (German technical inspection agency).

Fig. C13.26 shows the unit circuit drawing for an example of a 20/0.4-kV radial network calculated and dimensioned using SIMARIS® design.

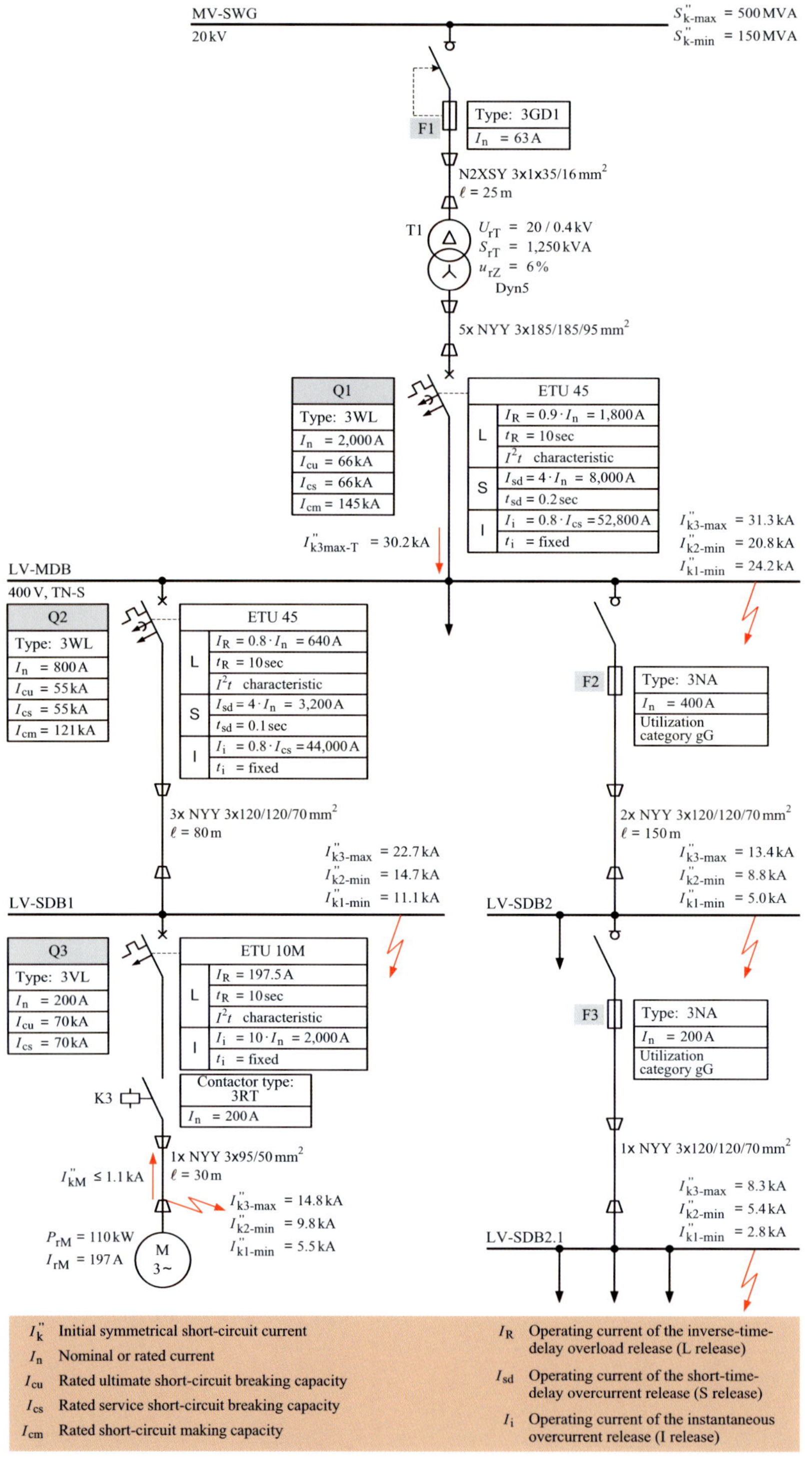

Fig. C13.26 Equipment selection and protection coordination for a simple radial network with SIMARIS® design (example of a unit circuit drawing)

SIMARIS® design provides the following proof that the protective devices parameterized according to Fig. C13.26 selectively trip single-phase and multi-phase faults occurring in the network:

- *Fig. C13.27*
  with the selectivity diagram for circuit-breaker Q1 (grading path F1-Q1-Q2/F2)

  The selectivity diagram in Fig. C13.27 visualizes the

  - protection setting of the incoming feeder circuit-breaker Q1 including tolerances,
  - lower tolerance curve of the upstream HV HRC fuse F1 and the
  - upper aggregate envelope tolerance limit for the tripping response of the downstream protection (circuit-breaker Q2 and LV HRC fuse F2).

  The maximum short-circuit current flowing through circuit-breaker Q1 is $I''_{\text{k-max-Q1}} = 30.2$ kA ($I''_{\text{k-max-Q1}} = I''_{\text{k3-max}} - I''_{\text{kM}} = 31.3\text{ kA} - 1.1\text{ kA} = 30.2\text{ kA}$). The minimum short-circuit current of $I''_{\text{k2-min}} = 20.8$ kA is broken by circuit-breaker Q1 after no less than 200 msec and no more than 270 msec. Selectivity is thus assured between Q1 and Q2 and F2.

  The lack of selectivity between F1 and Q1 in the high short-circuit current range can be accepted because clearance of a busbar fault concerns the same circuit (transformer incoming feeder).

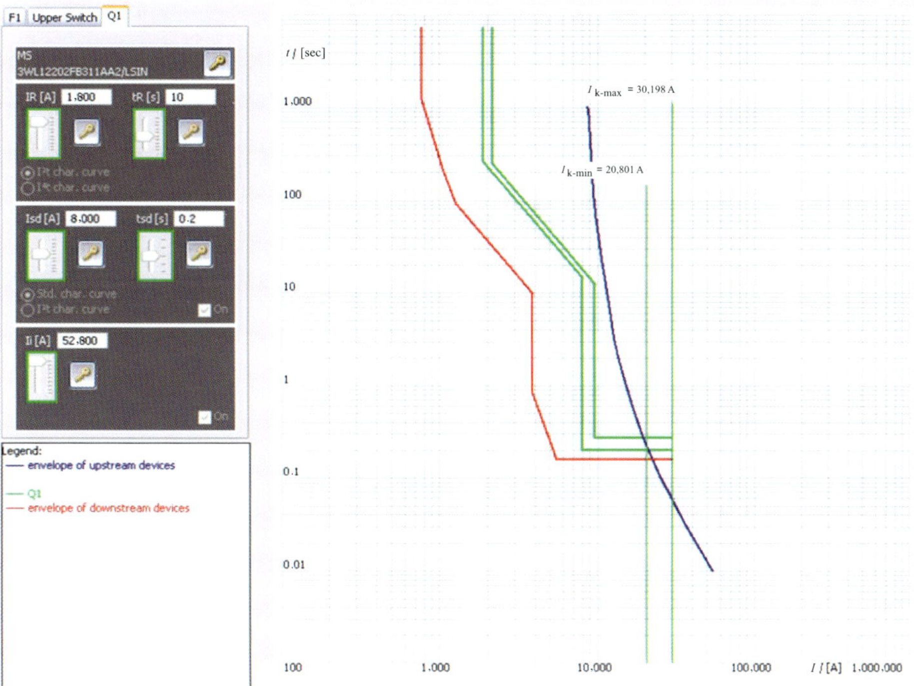

Fig. C13.27 Selectivity diagram for circuit-breaker Q1

- *Fig. C13.28a*
  with the selectivity diagram for circuit-breaker Q2 (grading path F1-Q1-Q2-Q3)

  The selectivity diagram in Fig. C13.28a visualizes the

  - protection setting of the outgoing feeder circuit-breaker Q2 including tolerances,
  - lower aggregate envelope tolerance limit for the tripping response of the upstream protection (circuit-breaker Q1 and HV HRC fuse F1) and the
  - upper tolerance curve of the downstream circuit-breaker Q3.

  The maximum short-circuit current flowing through circuit-breaker Q2 is $I''_{\text{k-max-Q2}}$ = 30.2 kA ($I''_{\text{k-max-Q2}} = I''_{\text{k3-max}} - I''_{\text{kM}}$ = 31.3 kA – 1.1 kA = 30.2 kA). The minimum short-circuit current of $I''_{\text{k1-min}}$ = 11.1 kA is broken by circuit-breaker Q2 after no less than 100 msec and no more than 160 msec on a single-phase fault at the end of the circuit. Selectivity is thus assured between Q2 and Q3. Selectivity is lacking between Q2 and F1 in the case of a dead three-phase fault at the beginning of the circuit. Dead faults, which are completely without resistance, can be ruled out with a very high probability in practical operation.

  Because short circuits always occur damped, the lack of selectivity in the high short-circuit current range can be accepted. To achieve full selectivity, a circuit-breaker relay combination would have to be used, for example, instead of the switch-fuse combination. Expensive modification of the transformer protection is not justified by the slight restriction on selectivity.

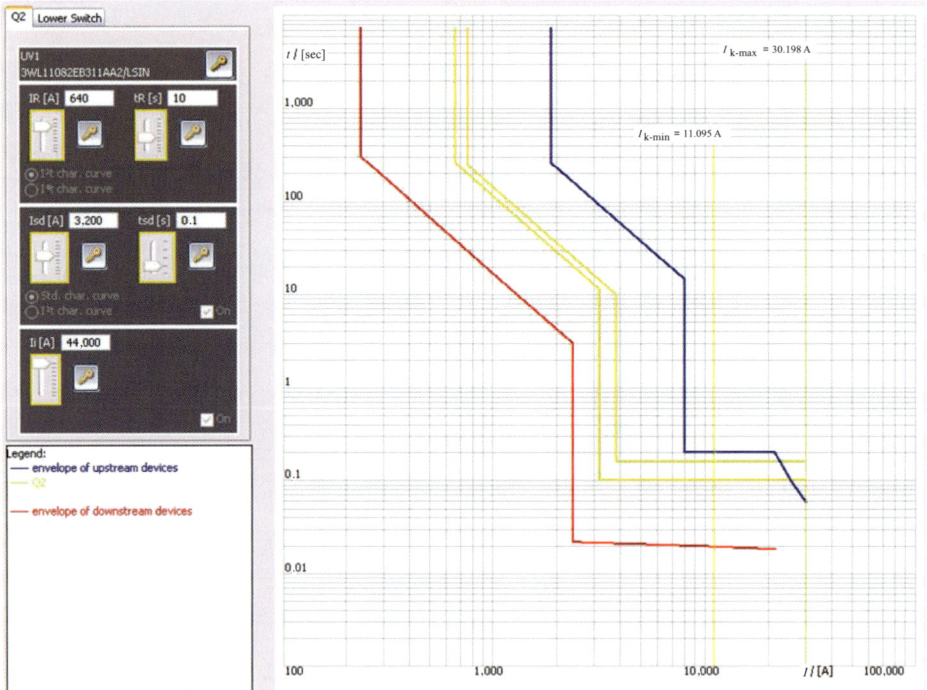

Fig. C13.28a Selectivity diagram for circuit-breaker Q2

- *Fig. C13.28b*
  with the selectivity diagram for circuit-breaker Q3 (grading path F1-Q1-Q2-Q3)

  The selectivity diagram in Fig. C13.28b visualizes the

  - protection setting of the outgoing feeder circuit-breaker Q3 including tolerances and the
  - lower aggregate envelope tolerance limit for the tripping response of the up-stream protection (circuit-breaker Q2, Q1 and HV HRC fuse F1).

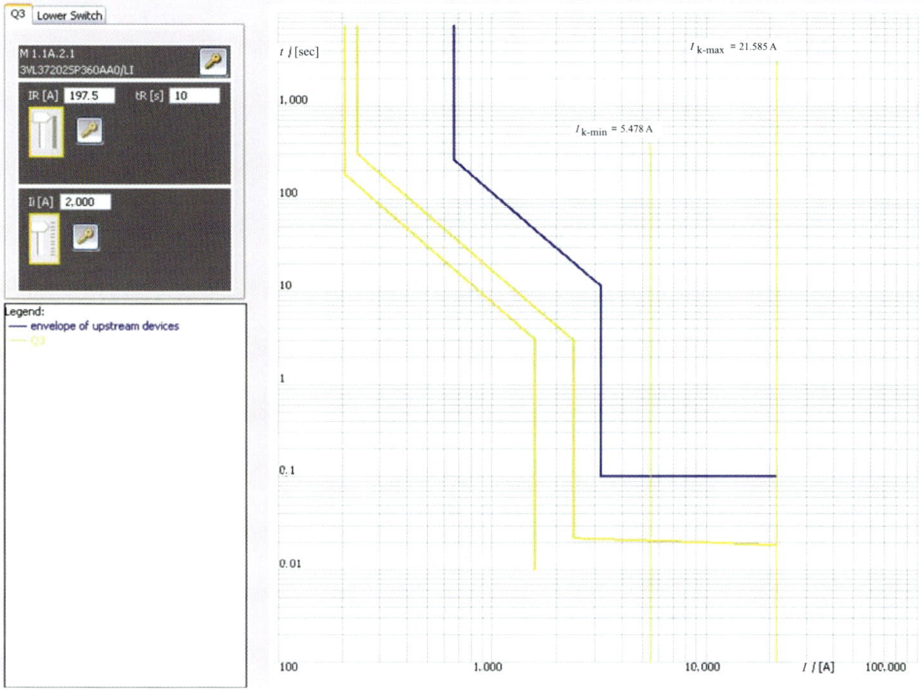

Fig. C13.28b Selectivity diagram for circuit-breaker Q3

Circuit-breaker Q3 breaks both the minimum short-circuit current $I''_{k1\text{-min}}$ = 5.5 kA and the maximum short-circuit current $I''_{k\text{-max-Q3}}$ = 21.6 kA ($I''_{k\text{-max-Q3}} = I''_{k3\text{-max}} - I''_{kM}$ = 22.7 kA − 1.1 kA = 21.6 kA) of the motor circuit to be protected instantaneously ($t_i \leq 20$ msec).

Selectivity exists in full with respect to the upstream protective devices (Q2, Q1, F1).

- *Fig. C13.29a*
  with the selectivity diagram for outgoing feeder fuse F2 (grading path F1-Q1-F2-F3)

  The time grading diagram in Fig. C13.29a visualizes the

  - pre-arcing time-current characteristic of the 400-A LV HRC fuse F2 with its toler-ances,
  - lower aggregate envelope tolerance limit for the tripping response of the up-stream protection (circuit-breaker Q1 and HV HRC fuse F1) and the
  - lower tolerance curve of the upstream LV HRC fuse F3.

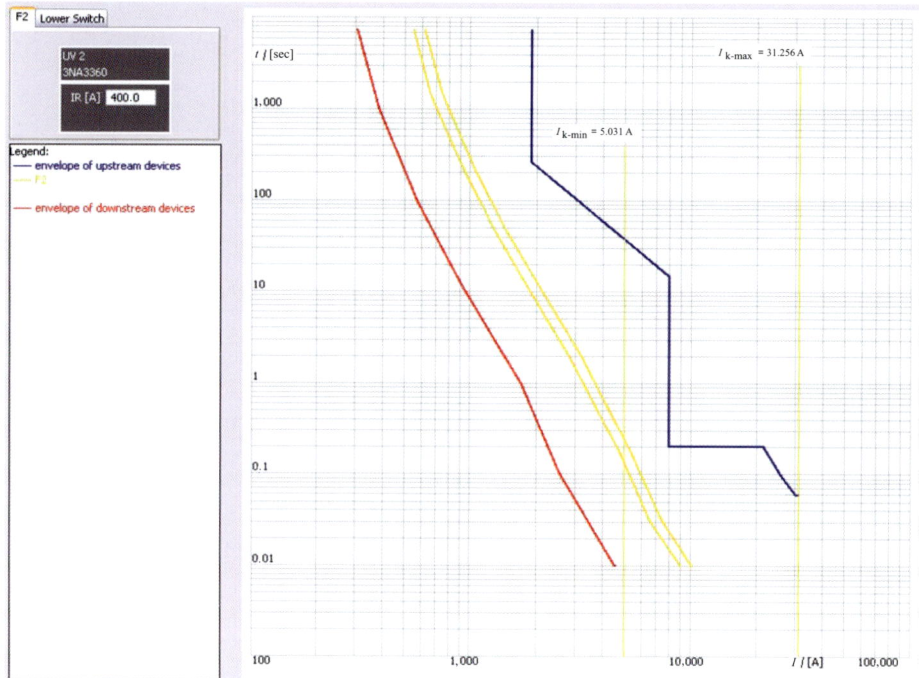

Fig. C13.29a Selectivity diagram for the LV HRC fuse F2

The maximum short-circuit current flowing through the LV HRC fuse F2 is $I''_{k3\text{-max}} = 31.3$ kA.

The minimum short-circuit current of $I''_{k1\text{-min}} = 5.0$ kA is interrupted by fuse F2 after no more than 250 msec in the case of a single-phase fault at the end of the circuit.

Full selectivity exists with respect to the protective devices (Q1, F1) over the entire current-time range.

- *Fig. C13.29b*
  with the selectivity diagram for outgoing feeder fuse F3 (grading path F1-Q1-F2-F3)

  The selectivity diagram in Fig. C13.29b visualizes the

  - pre-arcing time-current characteristic of the 200-A LV HRC fuse F3 with its tolerances and the
  - lower aggregate envelope tolerance limit for the tripping response of the upstream protection (LV HRC fuse F2, circuit-breaker Q1, HV HRC fuse F1).

  Fuse F3 breaks both the minimum short-circuit current $I''_{k1\text{-min}} = 2.8$ kA and the maximum short-circuit current $I''_{k3\text{-max}} = 13.4$ kA of the circuit to be protected instantaneously ($t_s \leq 70$ msec). Selectivity exists in full with respect to the upstream protective devices (F2, Q1, F1).

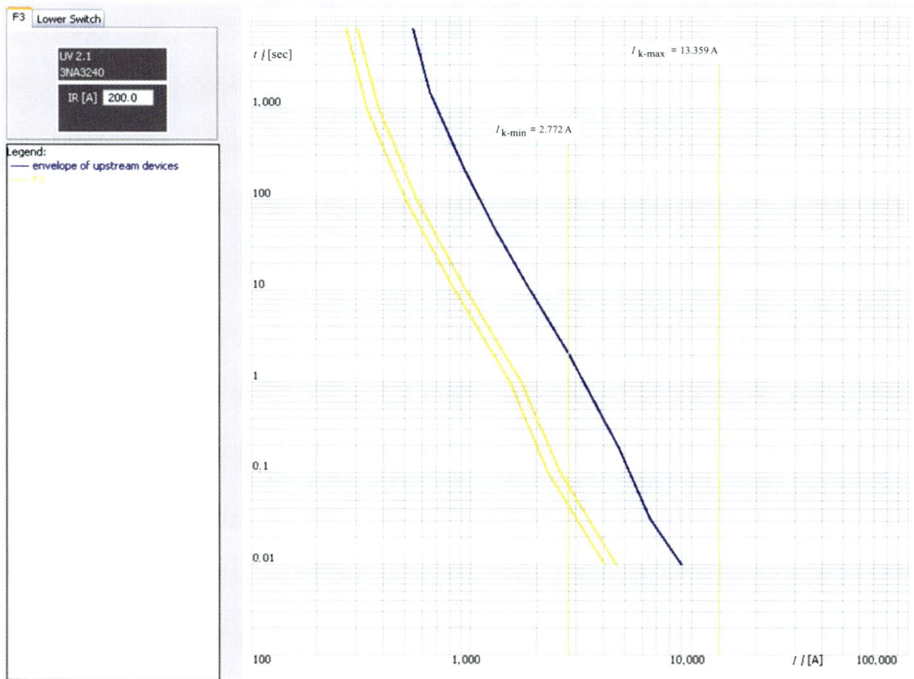

Fig. C13.29b Selectivity diagram for the LV HRC fuse F3

14 List of acronyms, abbreviations, symbols and subscripts used

14.1 Acronyms and abbreviations

AC	Alternating current
ACB	Air circuit-breaker
AF	Air-forced cooling
AF	Audio frequency
ALF	Accuracy limiting factor
AN	Air-natural cooling
AN	Allgemeine Netzversorgung (German for: normal power supply)
ANSI	American National Standards Institute
APFC	Automatic power factor correction
ASC	Arc-suppression coil (earth-fault neutralizer)
AVR	Automatic voltage regulator
BB	Busbar
BB-IE	Busbar for imported electricity
BB-IP	Busbar for in-plant electricity generation
BDEW	Bundesverband der Energie- und Wasserwirtschaft (German Federal Association of the Energy and Water Industries)
BF (protection)	Breaker failure (protection)
BGV	Berufsgenossenschaftliche Vorschriften (Rules of the employer‘s liability insurance association)
BH relay	Buchholz relay
CB	Circuit-breaker
CBCT	Core-balance current transformer (zero-sequence current transformer)
CBEMA	Computer and Business Equipment Manufacturers‘ Association
CEE	Commission on the Rules for the Approval of the Electrical Equipment
CEP	Central earthing point
CHP	Combined heat and power
CIRED	Congrès International des Réseaux Electriques de Distribution (International Conference on Electricity Distribution)
CO_2	Carbon dioxide
CT	Current transformer
Cu	Copper
DC	Direct current
DDUPS	Dynamic diesel uninterrupted power supply system
DIN	Deutsches Institut für Normung e. V. (German institute for standardization)
DNO	Distribution network operator
DOL	Direct on-line starting
DTL	Definite time-lag overcurrent (characteristic)
DVR	Dynamic voltage restorer
DySC	Dynamic sag corrector

EBA	Electronic ballast
EBDS	Electronic battery disconnecting switch
EBS	Electronic bypass switch
ECO	Economic
EDP	Electronic data processing
EMC	Electromagnetic compatibility
EN	Europa-Norm (European Standard)
EPR	Ethylene propylene rubber
ETU	Electronic trip unit
EU	European Union
EX	Expense, expenditure
FC	Forward conductor
Fe	Iron
FO	Fibre optic
GEAFOL	Cast-resin epoxy aluminium foil
GIS	Gas-insulated switchgear
HC-BB	High-current busbar(s)
HDLC	High-level data link control
HV	High voltage
HV HRC fuse	High-voltage high-rupturing-capacity fuse
ICEE	International Conference on Electrical Engineering
IDMTL	Inverse definite minimum time-lag overcurrent (characteristic)
IEC	International Electrotechnical Commission
IEEE	Institute of Electrical and Electronics Engineers
IGBT	Insulated gate bipolar transistor
IMD	Insulation monitoring device
IP-BB	In-plant electricity generation busbar
IPC	In-plant point of coupling/connection
IP (code)	Ingress protection (code)
ISO	International Organization for Standardization
IT	Information technology
ITIC	Information Technology Industry Council
KNOSPE	Kurzzeitige Niederohmige Sternpunkterdung (German for: short-time low-impedance neutral earthing)
LI	Lightning impulse
LPQI	Leonardo Power Quality Initiative
LV	Low voltage
LV HRC fuse	Low-voltage high-rupturing-capacity fuse
LV-MDB	Low-voltage main distribution board
LV-SDB	Low-voltage subdistribution board
MCB	Miniature circuit-breaker
MCC	Motor control centre
MCCB	Moulded-case circuit-breaker
MSP	Motor starter protector
MV	Medium voltage

N conductor	Neutral conductor
n.c.	Normally closed
NEA	Netzersatzanlage (German for: standby generating unit/system)
n.o.	Normally open
NOP	Normal operation
NOSPE	Niederohmige Sternpunkterdung (German for: low-impedance neutral earthing)
NP	Nodal point
NTr	Neutral earthing transformer
OHL	Overhead power line
ONAF	Oil-natural/air-forced cooling
ONAN	Oil-natural/air-natural cooling
OPFC	Operation under fault conditions
OSPE	Ohne Sternpunkterdung (German for: isolated neutral)
PC	Personal computer
PC	Point of connection
PCC	Point of common coupling/connection
PE	Polyethylene
PE conductor	Protective earth conductor
PEHLA	Prüfstelle für elektrische Hochleistungsapparate (Association of high-power testing laboratory owners in Germany and Switzerland)
PELV	Protective extra low voltage
PEN conductor	Combined protective and neutral conductor
PF	Power factor
PFC	Power factor correction
PI	Power inverter
PLC	Power limiting control
PQ	Power quality
PR	Power rectifier
PSS™ NETOMAC	Power System Simulator Network Torsion Machine Control
PSS™ SINCAL	Power System Simulator Siemens Network Calculation
PTTA	Partially type-tested low-voltage switchgear and controlgear assembly
PVC	Polyvinyl chloride
QS	Quality of supply
RC	Return conductor
RCBO	Residual current-operated circuit-breaker with integral overcurrent protection
RCCB	Residual current-operated circuit-breaker without integral overcurrent protection
RCD	Residual current-operated (protective) device
RCM	Residual-current monitor
RESPE	Resonanzsternpunkterdung (German for: resonant neutral earthing)
RMS	Root mean square (value)
RMU	Ring main unit

RPS	Redundant power supply
Schuko	Schutz-Kontakt (German for: earthing contact)
SDB	Subdistribution board
SELV	Safety extra low voltage
SF_6	Sulphur hexafluoride
SIL	Switching impulse level
SIPROTEC relay	Siemens protection relay
SITRABLOC	Siemens transformer block
SPE	Sternpunkterdung (German for: neutral earthing)
SR	Supply reliability
SSt	Substation
SVC	Static var compensator
SwD	Switch-disconnector
SWG	Switchgear
TM	Thermomagnetic
TRA	Tonfrequenz-Rundsteuer-Anlage (German for: audio-frequency remote control system)
TRBS	Technische Regeln für die Betriebssicherheit (German technical regulations for safety at the workplace)
TRV	Transient recovery voltage
TTA	Type-tested low-voltage switchgear and controlgear assembly
TÜV	Technischer Überwachungs-Verein (German technical inspection agency)
UPS	Uninterruptible power supply
VDE	Verband der Elektrotechnik Elektronik Informationstechnik e. V. (Association for Electrical, Electronic and Information Technologies)
VFD	Variable-frequency drive
VQ	Voltage quality
VT	Voltage transformer
XLPE	Cross-linked polyethylene
ZSI	Zone selective interlocking
ZVEI	Zentralverband der Elektrotechnik- und Elektronikindustrie e. V. (German electrical and electronics industry association)

14.2 Symbols

All symbols used in formulas are explained in the text the first time they appear. Symbols that are especially common or occur frequently in the text are also listed below. For symbols that are less common or appear less frequently in the text, please refer to their first occurrence in the text.

14.2.1 Currents

I current (general symbol)
I_{ASC} arc-suppression coil current
I_a operating current of a disconnecting protective device
I_b symmetrical short-circuit breaking current
I_{cE} capacitive earth-fault current
I_{cm} rated short-circuit making capacity
I_{cs} rated service short-circuit breaking capacity
I_{cu} rated ultimate short-circuit breaking capacity
I_{cw} rated short-time withstand current
I_D let-through current (cut-off current)
I_{design} design current
I_E inrush current (transformers)
I_F fault current
I_h harmonic current
I_k steady-state short-circuit current
I_k'' initial symmetrical short-circuit current
I_{load} load current
I_{ma} rated short-circuit making current (MV switchgear)
I_n nominal current
I_{pk} rated peak withstand current
I_r rated current
I_{resi} residual earth-fault current
I_{sc} rated short-circuit breaking current (MV switchgear)
I_{start} starting current (motors)
I_{TC} transfer current (switch-fuse combination)
I_{th} thermal equivalent short-circuit current
I_{thr} rated short-time withstand current (electrical equipment)
J_{thr} rated short-time current density
i_p peak short-circuit current
i_0 no-load current

14.2.2 Voltages

U voltage (general symbol)
U_E earthing voltage
U_e rated operational voltage or highest voltage for equipment (LV systems)
$U_{EN, en}$ neutral-point displacement voltage
U_F fault voltage
U_h harmonic voltage
U_L line (phase) voltage
U_{LE} line-to-earth voltage
U_{LL} line-to-line voltage

U_{LN} line-to-neutral voltage
U_m rated operational voltage or highest voltage for equipment (MV systems)
U_{nN} nominal network (system) voltage
U_r rated voltage
u_{rZ} impedance voltage of a transformer at rated current, expressed as a percentage
U_S step voltage
U_T touch voltage
U_0 nominal alternating voltage line-to-earth
Δu relative voltage drop caused by the load current
$\Delta u'$ relative voltage dip or relative voltage change

14.2.3 Resistances

R resistance (general symbol for ohmic resistance)
R_A resistance of the protection earth electrode
R_B resistance of the system earth electrode
R_D damping resistor against relaxation oscillations (ferroresonance)
R_E neutral earthing resistance
X reactance (general symbol for inductive reactance)
X_{ASC} reactance of an arc-suppression coil
X_d'' subtransient reactance of a synchronous machine (generator)
X_E neutral earthing reactance
Z impedance as an absolute value ($Z = \sqrt{R^2 + X^2}$)
$\underline{Z}$ impedance as a complex value ($\underline{Z} = R + jX$)
Z_E neutral earthing impedance
Z_k short-circuit impedance of a three-phase AC system
Z_s loop impedance
Z_1 positive-sequence short-circuit impedance
Z_2 negative-sequence short-circuit impedance
Z_0 zero-sequence short-circuit impedance

14.2.4 Powers and energy

P active power (general symbol)
P' per-unit active power, load per unit area
P_k nominal load losses (winding losses of a transformer)
P_{loss} network (system) losses
P_{pr} power rating of load-consuming apparatus
P_{rM} rated output of an asynchronous motor ($P_{rM} = S_{rM} \cdot \cos \varphi_{rM} \cdot \eta_{rM}$)
P_V dissipated power loss of HV HRC fuses
P_0 nominal no-load losses of a transformer
Q reactive power (general symbol)
Q_c capacitive power or capacitor power
$Q_{load\text{-}T}$ reactive power of a loaded transformer
Q_{0T} no-load reactive power of a transformer
S apparent power (general symbol)
S_k'' initial symmetrical short-circuit power (or more simply short-circuit power)
S_r rated apparent power of an electrical equipment item
S_{thr} rated short-time withstand power
ΔS dynamic load change (impulsive or fluctuating)

W_{active} active energy or active energy consumption
W_{arc} arc energy
$W_{reactive}$ reactive energy or reactive energy consumption

14.2.5 Time/duration

ED duty ratio of a welding machine (ratio of welding to cycle time)
$K_{T0.5}$ time to half value for the decay of the inrush current
T cycle time
T_u interruption duration
ΔT grading time or selective time interval
t time (general symbol)
t_a disconnecting or clearing time
$t_{a\text{-}perm}$ permissible clearing time
$t_{a\text{-}total}$ total clearing time
t_{cw} rated short-time (LV equipment)
t_k short-circuit duration
t_m melting or pre-arcing time
t_{min} minimum time delay (IEC/EN 60909-0)
t_s pre-arcing time (LV HRC fuses)
t_{start} starting time of the motor or generator
t_{thr} rated short-time (MV equipment)
t_w welding time

14.2.6 Factors

a capacity utilization factor
b demand factor
c voltage factor (IEC/EN 60909-0)
$\cos \varphi$ power factor
f factor (general symbol)
g coincidence factor
K correction factor for impedances (IEC/EN 60909-0)
K'_{ALF} actual accuracy limiting factor
k_u unbalance factor of the voltage
m factor for the heat effect of the DC component of the short-circuit current (IEC/EN 60909-0)
n factor for the heat effect of the AC component of the short-circuit current (IEC/EN 60909-0)
p detuning factor (PF correction with reactor-connected capacitors)
q interest rate factor for calculation of the present value (cash equivalent)
q_M factor for calculating the symmetrical short-circuit breaking current of asynchronous motors (IEC/EN 60909-0)
THD total harmonic distortion (value)
v detuning factor (resonant neutral earthing)
α impedance factor for assessing reactions on the AF ripple control
α_{20} temperature coefficient for Cu and Al conductors where $\vartheta = 20\,°C$
κ factor for calculating the peak short-circuit current (asymmetrical current peak factor, IEC/EN 60909-0)
μ_M factor for calculating the symmetrical short-circuit breaking current (IEC/EN 60909-0)

τ	system time constant ($\tau = X/(R \cdot \omega)$)
ς	resistivity of a conductor

14.2.7 Other quantities

A	area, cross-sectional area of a conductor (general symbol)
A_n	nominal cross-sectional area of a conductor
C_E	line-to-earth capacitance
E_m	mean illuminance
EX	expense
f	frequency (general symbol)
f_N	system frequency (50 Hz or 60 Hz)
f_R	resonance frequency
f_{TRA}	audio or ripple-control frequency
l	length
n_{syn}	synchronous speed
P_{lt}	long-term flicker intensity
P_{st}	short-term flicker intensity
$p_k(n)$	binomial probability that, out of *n* welding machines, exactly *k* machines are welding simultaneously
p_M	number of pole pairs of a motor ($p_M = f \cdot 60/n_{syn}$)
p_z	pulse number of a static converter
η	effficiency (general symbol)
ϑ	temperature (general symbol)
λ	failure rate for voltage-induced welding errors
ω	angular frequency ($\omega = 2 \cdot \pi \cdot f$)

14.3 Subscripts and superscripts

C	cable, capacitor
D	damping
E	earth
F	fault, failure
FC	forward conductor
G	generator
I	investments
L	line, phase conductor
M	motor
N	network, power system (general designation)
Q	network, power system (IEC/EN 60909-0)
RC	return conductor
T, Tr	transformer
b	breaking
c	capacitive
fluc	fluctuating
h	order of harmonics
i	incrementing index
j	incrementing index
k	short circuit

k1	line-to-earth or single-phase short circuit
k2	line-to-line or two-phase short circuit
k3	three-phase short circuit
load	load, on-load
lt	long term
max	maximum
min	minimum
n	nominal
norm	normalized
p	primary
perm	permissible
r	rated
ref	reference
req	required
resi	residual
s	secondary
sel	selective
st	short term, short time
start	starting
syn	synchronous
td	transient dimensioning
th	thermal
tot	total
w	welding
0	zero-sequence component
1	positive-sequence component
2	negative-sequence component

References and further reading

Chapter 1

[1.1] Bochanky, L.: Planung öffentlicher Elektroenergiesorgungsnetze. Leipzig: Deutscher Verlag für Grundstoffindustrie, 1985

[1.2] Leimgruber, H. u. Wolfrath, H.: Gesichtspunkte zum Aufbau elektrischer Gesamtanlagen in der metallverarbeitenden Industrie. Teil 1: Stromversorgung Mittelspannung. Siemens-Bericht: E335-2-A-01, Erlangen, 1978

[1.3] Leimgruber, H. u. Wolfrath, H.: Gesichtspunkte zum Aufbau elektrischer Gesamtanlagen in der metallverarbeitenden Industrie. Teil 2: Stromversorgung Niederspannung. Siemens-Bericht: E335-2-A-02, Erlangen, 1978

[1.4] Buchholz, B. M. u. a.: Planungsinhalte und Power Quality von Industrienetzen. Energiewirtschaftliche Tagesfragen 53 (2003) H. 11, S. 725–731

[1.5] Ziegler, G.: Digitaler Schutz für Industrieanlagen. etz 118 (1997) H. 18, S. 30–35

[1.6] Sämann, D.: Leistungsschalter in Industrieanlagen. etz 117 (1996) H. 12, S. 38–42

[1.7] Bauschbach, W.: Versorgungszuverlässigkeit in Verteilungsnetzen. ew 98 (1999) H. 23, S. 25–28

[1.8] Siemens AG, Erlangen: Die Qualität muss stimmen. Siemens-EV-Report (2000) H. 1, S. 6–9

[1.9] Scheifele, J.: Energieversorgung in der Industrie. EVU-Betriebspraxis 32 (1993) H. 7–8, S. 214–218

[1.10] Ehmcke, B. u. Schaller, F.: Erst planen, dann verlegen. Sonderdruck aus dem Siemens-EV-Report (1991) H. 4, S. 6–10

[1.11] Janotta, Ch.: Elektrische Verteilungsnetze der Industrie im Wandel. etz 131 (2010) H. 1, S. 30–32

[1.12] Bochanky, L.: Gesamtheitliche Planung der Verteilungsnetze. Energiewirtschaftliche Tagesfragen 46 (1996) H. 10, S. 624–629

[1.13] Balzer, G. u. Neunstöcklin, V.: Zu erwartende Trends bei MS-Schaltanlagen in industriellen Netzen. etz 122 (2001) H. 20, S. 16–21

[1.14] Bauhofer, P.: Netzplanung ist Denken im Kundennutzen. VEÖ-Journal 4 (1997) H. 10, S. 29–32

[1.15] Dommann, D.: Alle reden von Dienstleistungen, aber keiner will dienen. ew 96 (1997) H. 17, S. 867–870

[1.16] Fehling, H.: Elektrische Starkstromanlagen – Stromsysteme, Netze, Leitungen, Kurzschlußschutz. Berlin · Offenbach: VDE-Verlag, 1983

[1.17] Fettback-Heyden, K.-H.: Netzplanung – Tools, Training und Consulting im liberalisierten Markt. ew 101 (2002) H. 17–18, S. 58–63

[1.18] Flosdorff, R. u. Hilgarth, G.: Elektrische Energieverteilung. Stuttgart: B.G. Teubner-Verlag, 1986

[1.19] Friesenecker, W. u. a.: Bezugspunkte in elektrischen Netzen. VEÖ-Journal 11 (2004) H. 11–12, S. 28–41

[1.20] Fruth, W. u. a.: Totally Integrated Power – durchgängige Energieversorgungsanlagen für Zweckbau und Industriebau. ew 102 (2003) H. 13, S. 32–36

[1.21] Gester, J.: Starkstromleitungen und Netze. Berlin: Verlag Technik, 1979

[1.22] Heuck, K. u. Dettmann, K.-D.: Elektrische Energieversorgung. Braunschweig: Vieweg-Verlag, 1991

[1.23] Kopatz, M. u. a.: Zusammenarbeit von Industrie und Elektrizitätsversorgungsunternehmen bei der Grundsatzplanung elektrischer Netze. ew 78 (1979) H. 11, S. 373–374

[1.24] Lorenz, H. u. Meyer, B.: Begutachtung bei Anlagenerneuerung und Retrofit – Expertenwissen steht im Mittelpunkt. ew 97 (1998) H. 12, S. 23–25

[1.25] Märtel, P. u. Siegmund, D.: Kostenreduzierung bei Netzkomponenten in Verteilungsnetzen. ew 97 (1998) H. 25, S. 23–25

[1.26] Petrossian, E.: Nicht nur für große Netze. Siemens-EV-Report (1994) H. 3, S. 22–26

[1.27] Weck, K.-H. u. Welßow, W.H.: Kostenreduzierung bei Netzkomponenten in Verteilungsnetzen – FGH-Ausspracheveranstaltung 1998. ew 97 (1998) H. 25, S. 8–13

[1.28] Planung und Betrieb von städtischen Mittelspannungsnetzen. Frankfurt a.M.: VDEW-Verlag, 1991

Chapter 2

[2.1] Brauner, G.: Systematische Netzplanung – Planungsmethodik, Systemtheorie, Planung als vernetzter Prozess. VEÖ-Journal 4 (1997) H. 10, S. 55–56

[2.2] Backes, J. u. a.: Strategische Netzplanung – Schlüssel zur Wirtschaftlichkeit. etz 119 (1998) H. 16, S. 16–19

[2.3] Felgenhauer, D. u. a.: Integrierte Netzanalyse in Planungs- und Betriebsführungssystemen. etz 115 (1994) H. 10, S. 564–569

[2.4] Sachs, U.: Das Netzplanungsprogramm im Umfeld der DV-Landschaft. ew 101 (2002) H. 21–22, S. 44–49

[2.5] Floerke, H.: Leistungsbedarf elektrischer Anlagen. etz 104 (1983) H. 12, S. 586–589

[2.6] Bätz, H. u. a.: Elektroenergieanlagen. Berlin: Verlag Technik, 1989

[2.7] Berechnung des elektrischen Leistungs- und Energiebedarfs für Stromversorgungsanlagen von Industrienetzen. VEM-Projektierungsvorschrift, Ordnungs-Nr. 1.9/3.80. Berlin: Institut für Elektro-Anlagen, 1980

[2.8] Schroeder, K.-H.: Komplexe Energieversorgung von Territorien. Leipzig: Deutscher Verlag für Grundstoffindustrie, 1983

[2.9] Siemens AG, Erlangen/Germany: Electrical Equipment. Power Supply and Auxiliaries for the Metalworking Industry. Low-Voltage Systems. Brochure of Energy and Automation Group E375, Order No.: A19100-E375-A333-X-7600

[2.10] www.Tessag.com: Gigantische Datencenter benötigen neue Infrastruktur. ew 100 (2001), H. 5, S. 58–59

[2.11] Zimmermann, W.: Optimale Gestaltung von Industrienetzen. ELEKTRO-JAHR 1984, S. 55–59

[2.12] Backes, J. u. a.: Zuverlässigkeitsberechnung – Ein Verfahren wird praxistauglich. ew 97 (1998) H. 19, S. 30–36

[2.13] Börninck, S.: Ausfallkosten für Industrieunternehmen durch Stromversorgungsunterbrechungen. ew 101 (2002) H. 12, S. 36–39

[2.14] Jungbauer, J.: Wirtschaftliche Registrierung der Netzqualität im Nieder- und Mittelspannungsbereich. EVU-Betriebspraxis 38 (1999) H. 9, S. 27–29

[2.15] DIN EN 50160 (EN 50160): 2008-04 Merkmale der Spannung in öffentlichen Elektrizitätsversorgungsnetzen. Berlin · Wien · Zürich: Beuth Verlag

[2.16] DIN EN 61000-2-2 (VDE 0839-2-2): 2003-02 Elektromagnetische Verträglichkeit (EMV) - Teil 2-2: Umgebungsbedingungen – Verträglichkeitspegel für niederfrequente leitungsgeführte Störgrößen und Signalübertragung in öffentlichen Niederspannungsnetzen. Berlin · Offenbach: VDE Verlag

[2.17] IEC 61000-2-2: 2002-03 Electromagnetic compatibility (EMC) – Part 2-2: Environment – Compatibility levels for low-frequency conducted disturbances and signalling in public low-voltage power supply systems. Geneva/Switzerland: Bureau de la Commission Electrotechnique Internationale

[2.18] DIN EN 61000-2-4 (VDE 0839-2-4): 2003-05 Elektromagnetische Verträglichkeit (EMV) – Teil 2–4: Umgebungsbedingungen – Verträglichkeitspegel für niederfrequente leitungsgeführte Störgrößen in Industrieanlagen. Berlin · Offenbach: VDE Verlag

[2.19] IEC 61000-2-4: 2002-06 Electromagnetic compatibility (EMC) – Part 2–4: Environment – Compatibility levels in industrial plants for low-frequency conducted disturbances. Geneva/Switzerland: Bureau de la Commission Electrotechnique Internationale

[2.20] D-A-CH-CZ-Richtlinie. Technische Regeln zur Beurteilung von Netzrückwirkungen. 2. Auflage 2007. Frankfurt a.M.: VWEW Energieverlag

[2.21] Targosz, R. u. Manson, J.: Pan European LPQI Power Quality Survey. Paper 0263 in Conference Proceedings Part 1, 19th International Conference on Electricity Distribution (CIRED), 21–24 May 2007, Vienna/Austria

[2.22] VDEW-Messkampagne zu „Voltage Dips". EVU-Betriebspraxis 39 (2000) H. 1–2, S. 14–20

[2.23] McMichael, I. u. Barr, R.: Distribution Network Voltage Disturbances and Voltage Dip/Sag Compatibility. Paper 0325 in Conference Proceedings Part 1, 19th International Conference on Electricity Distribution (CIRED), 21–24 May 2007, Vienna/Austria

[2.24] DIN VDE 0276-1000 (VDE 0276-1000): 1995-06 Starkstromkabel. Strombelastbarkeit. Allgemeines. Umrechnungsfaktoren. Berlin · Offenbach: VDE Verlag

[2.25] DIN VDE 0265 (VDE 0265): 1995-12 Kabel mit Kunststoffisolierung und Bleimantel für Starkstromanlagen. Berlin · Offenbach: VDE Verlag

[2.26] DIN VDE 0271 (VDE 0271): 2007-01 Starkstromkabel. Festlegungen für Starkstromkabel ab 0,6/1 kV für besondere Anwendungen. Berlin · Offenbach: VDE Verlag

[2.27] DIN VDE 0276-620 (VDE 0276-620): 2009-05 Starkstromkabel mit extrudierter Isolierung für Nennspannungen von 3,6/6(7,2) kV bis einschließlich 20,8/36(42) kV. Berlin · Offenbach: VDE Verlag

[2.28] IEC 60502-2: 2005-03 Power cables with extruded insulation and their accessories for rated voltages from 1 kV (U_m = 1.2 kV) up to 30 kV (U_m = 36 kV) – Part 2: Cables for rated voltages from 6 kV (U_m = 7.2 kV) up to 30 kV (U_m = 36 kV). Geneva/Switzerland: Bureau de la Commission Electrotechnique Internationale

[2.29] DIN VDE 0298-4 (VDE 0298-4): 2003-08 Verwendung von Kabeln und isolierten Leitungen für Starkstromanlagen. Empfohlene Werte für die Strombelastbarkeit von Kabeln und Leitungen für feste Verlegung in und an Gebäuden und von flexiblen Leitungen. Berlin · Offenbach: VDE Verlag

[2.30] DIN EN 60076-12 (VDE 0532-76-12): 2008-01 Leistungstransformatoren - Teil 12: Belastungsrichtlinie für Trocken-Leistungstransformatoren. Berlin · Offenbach: VDE Verlag

[2.31] IEC 60076-12: 2008-11 Power transformers – Part 12: Loading guide for dry-type power transformers. Geneva/Switzerland: Bureau de la Commission Electrotechnique Internationale

[2.32] Siemens AG, Erlangen/Germany: PSS™SINCAL – Efficient planning software for electricity and pipe networks. Energy Sector, D SE PTI, Order No.: E50001-U610-A154-X-US00

[2.33] Siemens AG, Erlangen/Germany: www.siemens.com/sincal

[2.34] Bayer, W. u. a.: Netzrückwirkungen beim Schalten von Motoren und Generatoren. ETG-Fachbericht (1986) Bd. 17, S. 133–145

[2.35] Bochanky, L.: Entscheidungsproblematik in der Netztechnik. Elektrie 44 (1990) H. 5, S. 174–177

[2.36] Bochanky, L. u. Märtel, P.: Netzbemessung bei veränderten Anforderungsfestlegungen. ew 93 (1994) H. 19, S. 1144–1147

Chapter 3

[3.1] DIN IEC 60038 (VDE 0175): 2002-11 IEC-Normspannungen. Berlin · Offenbach: VDE Verlag

[3.2] IEC 60038: 2009-06 IEC standard voltages. Geneva/Switzerland: Bureau de la Commission Electrotechnique Internationale

[3.3] Leimgruber, H. u. Wolfrath, H.: Gesichtspunkte zum Aufbau elektrischer Gesamtanlagen in der metallverarbeitenden Industrie. Teil 1: Stromversorgung Mittelspannung. Siemens-Bericht: E335-2-A-01, Erlangen, 1978

[3.4] Balzer, G. u. Neunstöcklin, V.: Zu erwartende Trends bei MS-Schaltanlagen in industriellen Netzen. etz 122 (2001) H. 20, S. 16–21

[3.5] Tatsopoulos, P. u. Zimmermann, W.: Entwurf von Industrienetzen mit überwiegender Motorenlast – Einfluß auf Kurzschlußströme, Selektivität und Motorverhalten. ew 82 (1983) H. 9, S. 306–310

[3.6] Scheifele, J.: Energieversorgung in der Industrie. EVU-Betriebspraxis 32 (1993) H. 7–8, S. 214–218

[3.7] Janisch, O.: Industrienetze – Gesichtspunkte für die Netzgestaltung, speziell für die chemische und Mineralöl-Industrie. etz-B 22 (1970) H. 25, S. 601–602

[3.8] Petrossian, E. u. Steiniker, D.: Modernisierung der Stromversorgung von drei Autofabriken. ew 100 (2001) H. 9, S. 40–46

Chapter 4

[4.1] Bergauer, G.: Anforderungen an die Kurzschlussleistung für den sicheren Netzbetrieb im deregulierten Markt. VEÖ Journal 9 (2002) H. 9, S. 39–41

[4.2] DIN EN 60909-0 (VDE 0102): 2002-07 Kurzschlussströme in Drehstromnetzen. Berechnung der Ströme. Berlin · Offenbach: VDE Verlag

[4.3] IEC 60909-0: 2001-07 Short-circuit currents in three-phase a. c. systems - Part 0: Calculation of currents. Geneva/Switzerland: Bureau de la Commission Electrotechnique Internationale

[4.4] D-A-CH-CZ-Richtlinie. Technische Regeln zur Beurteilung von Netzrückwirkungen. 2. Auflage 2007. Frankfurt a.M.: VWEW Energieverlag

[4.5] Schlabbach, J.: Berechnung der Netzimpedanz zur Beurteilung von Netzrückwirkungen. ew 107 (2008) H. 6, S. 28–31

[4.6] DIN VDE 0101 (VDE 0101): 2000-01 Starkstromanlagen mit Nennwechselspannungen über 1 kV. Berlin · Offenbach: VDE Verlag

[4.7] IEC 61936-1: 2002-10 Power installations exceeding 1 kV a. c. - Part 1: Common rules. Geneva/Switzerland: Bureau de la Commission Electrotechnique Internationale

[4.8] DIN EN 60865-1 (VDE 0103): 1994-11 Kurzschlussströme – Berechnung der Wirkung, Begriffe und Berechnungsverfahren. Berlin · Offenbach: VDE Verlag

[4.9] IEC 60865-1: 1993-10 Short-circuit currents – calculation of effects – Part 1: Definitions and calculation methods. Geneva/Switzerland: Bureau de la Commission Electrotechnique Internationale

[4.10] DIN EN 60076-5 (VDE 0532-76-5): 2007-01 Leistungstransformatoren – Teil 5: Kurzschlussfestigkeit. Berlin · Offenbach: VDE Verlag

[4.11] IEC 60076-5: 2006-02 Power transformers - Part 5: Ability to withstand short circuit. Geneva/Switzerland: Bureau de la Commission Electrotechnique Internationale

[4.12] DIN VDE 0276-1000 (VDE 0276-1000): 1995-06 Starkstromkabel – Teil 1000: Strombelastbarkeit, Allgemeines, Umrechnungsfaktoren. Berlin · Offenbach: VDE Verlag

[4.13] Heinhold, L. u. Stubbe, R.: Kabel und Leitungen für Starkstrom. Grundlagen und Produkt-know-how für das Projektieren von Kabelanlagen. Erlangen: Publicis MCD Verlag, 1999

Chapter 5

[5.1] Leimgruber, H. u. Wolfrath, H.: Gesichtspunkte zum Aufbau elektrischer Gesamtanlagen in der metallverarbeitenden Industrie. Teil 1: Stromversorgung Mittelspannung. Siemens-Bericht: E335-2-A-01, Erlangen, 1978

[5.2] Pantenburg, N. u. Wolfrath, H.: Planung von Niederspannungsnetzen in der metallverarbeitenden Industrie. Siemens-Energietechnik 4 (1982) H. 4, S. 206–210

[5.3] Kiank, H. u. Pantenburg, N.: Optimum design and dimensioning of power supply systems in the automobile industry. Pages 3/1–3/5 in IEE Conference Publication No. 438. Part 1: Contribution. CIRED. 14th International Conference Exhibition on Electricity Distribution. 2–5 June 1997, Birmingham.

[5.4] Zentralverband Elektrotechnik- und Elektronikindustrie e.V. (ZVEI): Freiwillige Selbstverpflichtung der SF_6-Produzenten, Hersteller und Betreiber von elektrischen Betriebsmitteln >1 kV zur elektrischen Energieübertragung und -verteilung in der BR Deutschland zu SF_6 als Isolier- und Löschgas. Frankfurt a.M., Juli 2005

[5.5] Vereinigung Deutscher Elektrizitätswerke e.V. (VDEW): Bau und Betrieb von Übergabestationen zur Versorgung von Kunden aus dem Mittelspannungsnetz. Frankfurt a.M.: VDEW-Verlag, 1976

[5.6] Siemens AG, Erlangen: Schaltanlage Typ 8DH10 bis 24 kV, gasisoliert, anreihbar. Katalog HA 41.11 · 2008. Bestell-Nr.: E50001-K1441-A101-A7

[5.7] Siemens AG, Erlangen/Germany: Switchgear Type 8DH10 up to 24 kV, Gas-Insulated, Extendable. Catalog HA 41.11 · 2007. Order No.: E50001-K1441-A101-A7-7600

[5.8] Siemens AG, Erlangen: Schaltanlage Typ SIMOSEC bis 24 kV, luftisoliert, anreihbar. Katalog HA 41.21 · 2003. Bestell-Nr.: E50001-K1441-A211-A2

[5.9] Siemens AG, Erlangen/Germany: Switchgear Type SIMOSEC up to 24 kV, Air-Insulated, Extendable. Catalog HA 41.21 · 2007. Order No.: E50001-K1441-A211-A5-7600

[5.10] Siemens AG, Erlangen: Leistungsschalter-Festeinbauanlagen Typ NXPLUS C bis 24 kV, gasisoliert. Katalog HA 35.41 · 2009. Bestell-Nr.: E50001-K1435-A401-A9

[5.11] Siemens AG, Erlangen/Germany: Fixed-Mounted Circuit-Breaker Switchgear Type NXPLUS C up to 24 kV, Gas-Insulated. Catalog HA 35.41 · 2009. Order No.: E50001-K1435-A401-A9-7600

[5.12] Siemens AG, Erlangen: Leistungsschalteranlagen Typ NXAIR, NXAIR M und NXAIR P, bis 24 kV, luftisoliert. Katalog HA 25.71 · 2007. Bestell-Nr.: E50001-K1425-A811-A7

[5.13] Siemens AG, Erlangen/Germany: Circuit-Breaker Switchgear Types NXAIR, NXAIR M and NXAIR P, up to 24 kV, Air-Insulated. Catalog HA 25. 71 · 2007. Order No.: E50001-K1425-A811-A7-7600

[5.14] Siemens AG, Erlangen: Leistungsschalter-Festeinbauanlagen Typ 8DA und 8DB bis 40,5 kV, gasisoliert. Katalog HA 35.11 · 2010. Bestell-Nr.: E50001-K1435-A101-A10

[5.15] Siemens AG, Erlangen/Germany: Fixed-Mounted Circuit-Breaker Switchgear Type 8DA and 8DB up to 40,5 kV, Gas-Insulated. Catalog HA 35.11 · 2010. Order No.: E50001-K1435-A101-A10-7600

[5.16] DIN EN 62271-105 (VDE 0671-105): 2003-12 Hochspannungs-Schaltgeräte und -Schaltanlagen – Teil 105: Hochspannungs-Lastschalter-Sicherungs-Kombinationen. Berlin · Offenbach: VDE Verlag

[5.17] IEC 62271-105: 2002-08 High-voltage switchgear and controlgear -Part 105: Alternating current switch-fuse combinations. Geneva/Switzerland: Bureau de la Commission Electrotechnique Internationale

[5.18] DIN IEC 60076-7 (VDE 0532-76-7): 2008-02 Leistungstransformatoren – Teil 7: Leitfaden für die Belastung von ölgefüllten Leistungstransformatoren. Berlin · Offenbach: VDE Verlag

[5.19] IEC 60076-7: 2005-12 Power transformers - Part 7: Loading guide for oil-immersed power transformers. Geneva/Switzerland: Bureau de la Commission Electrotechnique Internationale

[5.20] Scheifele, J.: Energieversorgung in der Industrie. EVU-Betriebspraxis 32 (1993) H. 7-8, S. 214–218

[5.21] Siemens AG, Erlangen/Germany: PSS™ NETOMAC – Professional software for dynamic system analyses. Energy Sector, D SE PTI, Order No.: SWNM01-EN-200905

[5.22] Siemens AG, Erlangen/Germany: www.siemens.com/energy/power-technologies; Software Solutions

[5.23] Lerch, E. u. a.: Transientes Verhalten eines Industrienetzes in Verbund- und Inselbetrieb. etz 113 (1992) H. 3, S. 124–132

[5.24] DIN EN 61000-2-4 (VDE 0839-2-4): 2003-05 Elektromagnetische Verträglichkeit (EMV) – Teil 2–4. Umgebungsbedingungen – Verträglichkeitspegel für niederfrequente leitungsgeführte Störgrößen in Industrieanlagen. Berlin · Offenbach: VDE Verlag

[5.25] IEC 61000-2-4: 2002-06 Electromagnetic compatibility (EMC) - Part 2–4: Environment - Compatibility levels in industrial plants for low-frequency conducted disturbances. Geneva/Switzerland: Bureau de la Commission Electrotechnique Internationale

[5.26] Klinger, S. u. a.: Kundenspezifische Lösungen bei erhöhten Anforderungen an die Versorgungsqualität. ew 102 (2003) H. 24, S. 40–47

[5.27] Sezi, T.: Kompensation von Spannungseinbrüchen im MVA-Lastbereich. etz 122 (2001) H. 9, S. 20–23

[5.28] HITEC Power Protection BV, Almelo/Netherlands: www.hitecups.com

[5.29] Dipl.-Ing. Hitzinger Gesellschaft m.b.H., Linz/Austria: www.hitzinger. at

[5.30] Piller Power Systeme GmbH, Osterade/Germany: www.piller.com

[5.31] DIN VDE 0671-1 (VDE 0671-1): 2009-08 Hochspannungsschaltgeräte und -Schaltanlagen – Teil 1: Allgemeine Festlegungen. Berlin · Offenbach: VDE Verlag

[5.32] IEC 62271-1: 2007-10 High-voltage switchgear and controlgear - Part 1: Common specifications. Geneva/Switzerland: Bureau de la Commission Electrotechnique Internationale

Chapter 6

[6.1] Pundt, H.: Zur Wahl der Sternpunkterdung in Mittelspannungsnetzen. Elektrie 22 (1969) H. 3, S. 106–109

[6.2] Siemens AG, Erlangen: Der Erdschluß im Netzbetrieb. Bestell-Nr.: A 19100-E141-B166, 1984

[6.3] Körner, H. u. a.: Ein Netzleitsystem für die Betriebsführung des Hoch- und Mittelspannungsnetzes – Planung und Vorbereitung. ew 84 (1985) H. 4, S. 113–117

[6.4] Matthes, W. u. Schilling, K.: Erdschlußlöschung. Sonderdruck aus dem Siemens-EV-Report 1/92, S. 17–20

[6.5] Verband der Elektrizitätswerke Österreichs, Wien: Sternpunktbehandlung in Mittelspannungsnetzen. VEÖ-Bestell-Nr.: 202/110, ISBN-Nr.: 3-901411-19-4, 1996

[6.6] A. Eberle GmbH & Co. KG, Nuernberg/Germany: www.a-eberle.de; Download Center; Infoletters

[6.7] DIN VDE 0101 (VDE 0101): 2000-01 Starkstromanlagen mit Nennspannungen über 1 kV. Berlin · Offenbach: VDE Verlag

[6.8] Johannsen, A.: Einpoliger Fehler „Wischer“ – Kombinierte Sternpunktbehandlung in 10-kV-Mittelspannungsnetzen. etz 102 (1981) H. 7, S. 367–369

[6.9] Fiernkranz, K.: Mittelspannungsnetze mit isoliertem Sternpunkt oder Erdschlusskompensation. S. 64–78 in ETG-Fachbericht 24. Sternpunktbehandlung in 10- bis 110-kV-Netzen. Vorträge der ETG-Fachtagung vom 18.05.–19.05.1988 in Braunschweig. Berlin · Offenbach: VDE Verlag

[6.10] Hosemann, G.: Sternpunktbehandlung in Deutschland und im internationalen Vergleich. S. 7–19 in ETG-Fachbericht 24. Sternpunktbehandlung in 10- bis 110-kV-

Netzen. Vorträge der ETG-Fachtagung vom 18.05.–19.05.1988 in Braunschweig. Berlin · Offenbach: VDE Verlag

[6.11] Lehtonen, M.: Method for Distance Estimation of Single-Phase-to-Ground Faults in Electrical Distribution Network with an Isolated or Compensated Neutral. European Transaction on Electrical Power (ETEP). Volume 5, Issue 3, Mai/June 1995, Pages 193–198

[6.12] Hänninen, S. u. Lehtonen, M.: Method for Detection and Location of Very High Resitive Earth Faults. European Transaction on Electrical Power (ETEP). Volume 9, Issue 5, September/October 1999, Pages 285–291

[6.13] Schlabbach, J.: Sternpunktbehandlung. Anlagentechnik für elektrische Verteilungsnetze. Band 15. Berlin · Offenbach: VDE Verlag, 2002

[6.14] Andrä, W. u. Peiser, R.: Kippschwingungen in Drehstromnetzen ETZ-B (1966) H. 22, S. 825–832

[6.15] Kunckel, K.-H.: Bedämpfung subharmonischer Schwingungen in elektrischen Netzen. Elektrie 29 (1975) H. 10, S. 535–537

[6.16] Hoolmans, G. u. a.: Beitrag zur Lösung des Kippschwingungsproblems. EVU-Betriebspraxis 39 (2000) H. 12, S. 14–20

[6.17] DIN EN 60071-1 (VDE 0111-1): 2006-11 Isolationskoordination – Teil 1: Begriffe, Grundsätze und Anforderungen. Berlin · Offenbach: VDE Verlag

[6.18] IEC 60071-1: 2006-01 Insulation coordination – Part 1: Definitions, principles and rules. Geneva/Switzerland: Bureau de la Commission Electrotechnique Internationale

[6.19] DIN VDE 0276-620 (VDE 0276-620): 2009-05 Starkstromkabel mit extrudierter Isolierung für Nennspannungen von 3,6/6(7,2) kV bis einschließ-lich 20,8/36(42) kV. Berlin · Offenbach: VDE Verlag

[6.20] IEC 60502-2: 2005-03 Power cables with extruded insulation and their accessories for rated voltages from 1 kV (U_m = 1.2 kV) up to 30 kV (U_m = 36 kV) – Part 2: Cables for rated voltages from 6 kV (U_m = 7.2 kV) up to 30 kV (U_m = 36 kV). Geneva/Switzerland: Bureau de la Commission Electrotechnique Internationale

[6.21] Petersen, W.: Die Begrenzung des Erdschlussstromes und die Unterdrückung des Erdschlusslichtbogen durch die Erdschlussspule. ETZ 40 (1919) H. 1, S. 5–7 und H. 2, S. 17–19

[6.22] Mayer, G.: Die Brenndauer von Erdschlusslichtbögen in gelöschten Netzen. ETZ-A 52 (1931) H. 48, S. 1466–1469

[6.23] Erich, M. u. Heinze, H.: Löschung von Erdschlusslichtbögen in Mittelspannungsnetzen. ETZ 82 (1963) H. 5, S. 158–160

[6.24] Pundt, H.: Untersuchungen der Ausgleichsvorgänge bei Erdschluss in Energieversorgungsnetzen. Energietechnik 15 (1965) H. 10, S. 469–477

[6.25] Pietzsch, H.: Automatische Erdschlusskompensation in Mittelspannungsnetzen. Energietechnik 22 (1972) H. 5, S. 202–210

[6.26] Kahnt, R. u. Lindner, H.: Untersuchungen über die Sternpunktbehandlung in Mittelspannungsnetzen der ländlichen Versorgung. Elektrizitätswirtschaft 76 (1977) H. 13, S. 403–409

[6.27] Schwetz, P.: Ausgleichströme beim Erdschluss im gelöschten Netz. Elektrizitätswirtschaft 79 (1980) H. 22, S. 854–858

[6.28] Klockhaus, H.J. u. a.: Sternpunktbehandlung und Erdschlussfehlerortsuche im Mittelspannungsnetz. Elektrizitätswirtschaft 80 (1981) H. 22, S. 797–803

[6.29] Poll, J.: Löschung von Erdschlusslichtbögen. Elektrizitätswirtschaft 82 (1983) H. 7, S. 322–327

[6.30] Winkler, F.: Strombelastbarkeit von Mittelspannungskabeln bei Erdschluss. etz 105 (1984) H. 4, S. 178–182

[6.31] Heiß, W.: Einflüsse der Verkabelung auf Netzkomponenten und Netzbetrieb. etz 112 (1991) H. 8, S. 386–391

[6.32] Fiedler, H.: Anwendungsschranken der Resonanzsternpunkterdung in Kabelnetzen. Elektrie 46 (1992) H. 4, S. 124–127

[6.33] Lüke, E.: Sicherungsauslösung in Mittelspannungsnetzen bei Erdschlüssen. Elektrizitätswirtschaft 93 (1994) H. 11, S. 605–608

[6.34] Schäfer, H.-D. u. a.: Erhöhung der Verlagerungsspannung in Mittelspannungs-Kabelnetzen mit Erdschlusskompensation. Elektrizitätswirtschaft 93 (1994) H. 21, S. 1295–1302

[6.35] Bartels, H.: Erdschlussprobleme im Mittel- und Hochspannungsnetz. Die Erdschluß-spule (Petersenspule). EVU-Betriebspraxis 35 (1996) H. 10, S. 347–349

[6.36] Fickert, L.: Oberschwingungs-Relativmessung: Vorstellung eines neuartigen Erdschlussortungsverfahrens für gelöschte Strahlennetze. VEÖ Journal 3 (1996) H. 12, S. 58–62

[6.37] Fritzsche, W. u. Völker, R.: Verbesserung der wattmetrischen Erdschlussrichtungserfassung in digitalen Schutzrelais. Elektrizitätswirtschaft 99 (2000) H. 6, S. 80–82

[6.38] Molés y Biétrie, M. u. a.: Richtungsanzeiger für kompensierte Mittelspannungsnetze. EVU-Betriebspraxis 39 (2000) H. 10, S. 6–8

[6.39] Birkner, P. u. a.: Ortung von Erdschlüssen über Oberschwingungsanalyse. ew 101 (2002) H. 17–18, S. 50–56

[6.40] Stade, D.: Erdschlussströme in MS-Netzen mit Resonanzsternpunkterdung. etz 128 (2007) H. 11, S. 70–75

[6.41] Ulrich, P. u. a.: Berechnung oberschwingungsbehafteter Erdschlussrestströme. S. 17–21 in ETG-Fachbericht 116. Sternpunktbehandlung in Verteilungsnetzen – Stand, Herausforderungen, Perspektiven. Vorträge der ETG-Fachtagung vom 27.01.–28.01.2009 in Dresden.

[6.42] Obkircher, C.: Ausbaugrenzen und -möglichkeiten gelöschter Netze. S. 23–27 in ETG-Fachbericht 116. Sternpunktbehandlung in Verteilungsnetzen – Stand, Herausforderungen, Perspektiven. Vorträge der ETG-Fachtagung 27.01.–28.01.2009 in Dresden. Berlin · Offenbach: VDE Verlag

[6.43] Höpfner, S.: Untersuchungen von Einflüssen auf den Erdschlusslichtbogen. S. 29–34 in ETG-Fachbericht 116. Sternpunktbehandlung in Verteilungsnetzen – Stand, Herausforderungen, Perspektiven. Vorträge der ETG-Fachtagung 27.01.–28.01.2009 in Dresden. Berlin · Offenbach: VDE Verlag

[6.44] Druml, G.: Neue Methoden zur Erdschlusseingrenzung. S. 39–46 in ETG-Fachbericht 116. Sternpunktbehandlung in Verteilungsnetzen – Stand, Herausforderungen, Perspektiven. Vorträge der ETG-Fachtagung 27.01.–28.01.2009 in Dresden. Berlin · Offenbach: VDE Verlag

[6.45] Groß, G. u. a.: Technische Lösungen zur Kompensation der Oberschwingungen im Erschlussreststrom. S. 47–51 in ETG-Fachbericht 116. Sternpunktbehandlung in Verteilungsnetzen – Stand, Herausforderungen, Perspektiven. Vorträge der ETG-Fachtagung 27.01.–28.01.2009 in Dresden. Berlin · Offenbach: VDE Verlag

[6.46] Kaufmann, G. u. Druml, G.: Neue Methoden zur Resonanzabstimmung. S. 53–59 in ETG-Fachbericht 116. Sternpunktbehandlung in Verteilungsnetzen – Stand, Herausforderungen, Perspektiven. Vorträge der ETG-Fachtagung 27.01.–28.01.2009 in Dresden. Berlin · Offenbach: VDE Verlag

[6.47] Kahnt, R. u. Körner, H.: Niederohmige Sternpunkterdung in Mittelspannungs-Kabelnetzen. Elektrizitätswirtschaft 67 (1968) H. 13, S. 336–342

[6.48] Höppner, H. u. Lau, H.: Planungs- und Betriebsprobleme bei der Einführung der niederohmigen Sternpunkterdung in Mittelspannungskabelnetzen. Teil 1: Elektrie 28 (1975) H. 7, S. 349–353. Teil 2: Elektrie 28 (1975) H. 8, S. 426–429.

[6.49] Menzel, E. u. Remde, H.: Resistanz- und Reaktanzerdung als Varianten der niederohmigen Sternpunkterdung in einem städtischen Mittelspannungskabelnetz. Elektrizitätswirtschaft 75 (1976) H. 16, S. 400–403

[6.50] Lipken, H.: Sternpunkterdung über niederohmige Widerstände in einem städtischen 10-kV-Kabelnetz. . Elektrizitätswirtschaft 76 (1977) H. 21, S. 736–742

[6.51] Osman, Z. u. Pundt, H.: Optimierung der Erdkurzschlussströme in Übertragungsnetzen. Elektrie 34 (1980) H. 10, S. 507–510

[6.52] Lau, H.: NOSPE und KNOSPE. Der Elektro-Praktiker 35 (1981) H. 6, S. 185–188

[6.53] Pfeiler, V. u. a.: Einführung der niederohmigen Sternpunkterdung in einem 10-kV-Industrienetz. Elektrie 35 (1981) H. 12, S. 635–637

[6.54] Schuchardt, B.: Niederohmige Sternpunkterdung in Eigenbedarfsnetzen großer Blockkraftwerke. Elektrie 37 (1983) H. 5, S. 348–351

[6.55] Kaufmann, W.: Mittelspannungsnetze mit niederohmiger Sternpunktbehandlung. S. 111–119 in ETG-Fachbericht 24. Sternpunktbehandlung in 10- bis 110-kV-Netzen. Vorträge der ETG-Fachtagung vom 18.05.–19.05.1988 in Braunschweig. Berlin · Offenbach: VDE Verlag

[6.56] Höppner, H.: Resistanz- oder Reaktanzerdung zur niederohmigen Sternpunkterdung in städtischen 20-kV-Kabelnetzen. Elektrizitätswirtschaft 91 (1992) H. 26, S. 1726–1742

[6.57] Pühringer, M.: NOSPE/KNOSPE – Eine Alternative zur Resonanzsternpunkterdung? Elektrizitätswirtschaft 97 (1998) H. 4, S. 18–20

[6.58] Zeuschel, H. u. a.: Schutzfragen bei kurzzeitiger niederohmiger Sternpunkterdung (KNOSPE). Elektrizitätswirtschaft 98 (1999) H. 19, S. 40–44

[6.59] Erfahrungsbericht eines städtischen Netzbetreibers: Umstellung der Sternpunkterdung von erdschlusskompensiert auf niederohmig. S. 83–86 in ETG-Fachbericht 116. Sternpunktbehandlung in Verteilungsnetzen -Stand, Herausforderungen, Perspektiven. Vorträge der ETG-Fachtagung 27.01.–28.01.2009 in Dresden. Berlin · Offenbach: VDE Verlag

[6.60] Tenschert, W.: Sternpunktbehandlung in Drehstromnetzen der öffentlichen Stromversorgung. VEÖ Journal 4 (1997) H. 10, S. 40–42

[6.61] Fiedler, H.: Zuverlässigkeitsgrad- und Nutzeffektermittlung in Mittelspannungsnetzen in Abhängigkeit von der Sternpunktbehandlung. Teil 1: Elektrie 32 (1978) H. 12, S. 648–649. Teil 2: Elektrie 33 (1979) H. 2, S. 95–98.

[6.62] Fiedler, H.: Beispiele einer optimalen Sternpunktbehandlung in vorgegebenen Mittelspannungsnetzen. Elektrie 34 (1980) H. 8, S. 405–409

[6.63] Fiedler, H.: Zielfunktion zur Sternpunktbehandlung in Mittelspannungsnetzen. Elektrie 36 (1982) H. 8, S. 423–427

[6.64] DIN VDE 0228-2 (VDE 0228-2): 1987-12 Maßnahmen bei Beeinflussung von Fernmeldeanlagen. Beeinflussung durch Drehstromanlagen. Berlin · Offenbach: VDE Verlag

[6.65] Heinhold, L. u. Stubbe, R.: Kabel und Leitungen für Starkstrom. Grundlagen und Produkt-know-how für das Projektieren von Kabelanlagen. 5., wesentlich überarbeitete Auflage. Erlangen: Publicis MCD Verlag, 1999

[6.66] Wannow, K.: Störungen in Mittelspannungskabelnetzen verschiedener Sternpunktbehandlung. Elektrizitätswirtschaft 92 (1993) H. 21, S. 1296–1301

[6.67] Schlabbach, J.: Einsatz der Sternpunktbehandlung in Deutschland. ew 101 (2002) H. 4, S. 66–69

[6.68] Schubert, S.: Einfluss der Sternpunkterdung auf die Störungs- und Verfügbarkeitsstatistik. Download der Vortragspräsentationen der ETG-Fachtagung „Sternpunktbehandlung in Verteilungsnetzen“ vom 27.01.–28.01.2009 in Dresden. www.vde.com/ste2009

[6.69] Bielenberg, K.: CIRED 2007 – Sitzung 3: Betrieb, Steuerung und Schutz elektrischer Netze. ew 106 (2007) H. 23, S. 64–65

[6.70] Kiank, H.: Process-Depending Planning of Power Supply Systems for an Automobile Plant. Translation from “etz“, Edition 22/2003, Pages 30 to 34. Siemens Special Edition. Order No.: E50001-U229-A209-X-7600

[6.71] Heitbreder, R. u. Schmidt, St.: Verbesserung des Betriebsverhaltens eines Industrienetzes durch niederohmige Sternpunkterdung. ew 103 (2004) H. 15, S. 59–63

[6.72] Castor, W. u. a.: Niederohmige Sternpunkterdung bringt Vorteile. Umstellung der Sternpunkterdung im 20-kV-Kabelnetz. ew 108 (2009) H. 4, S. 54–57

[6.73] Gröber, R. u. Komurka, J.: Übertragung der Nullspannung bei zweiseitig geerdeten Transformatoren. Technische Mitteilung der Forschungsgemeinschaft für Hochspannungs- und Hochstromtechnik. Mannheim · Rheinau, 1973

[6.74] Balzer, G. u. Remde, H.: Probleme bei der beidseitigen Erdung von Transformatoren. Brown Boveri Technik 72 (1985) H. 7, S. 349–354

Chapter 7

[7.1] Siemens AG, Nuernberg/Germany: www.siprotec.com; SIPROTEC Download-Area

[7.2] Ziegler, G.: Digitaler Schutz für Industrieanlagen. etz 118 (1997) H. 18, S. 30–35

[7.3] Clemens, H. u. Rothe, K.: Relaisschutztechnik in Elektroenergiesystemen. Berlin: Verlag Technik, 1982

[7.4] Müller, L. u. Boog, E.: Selektivschutz elektrischer Anlagen. Frankfurt a.M.: VWEW-Verlag, 1990

[7.5] GEC ALSTHOM T&D Protection & Control Limited, Stafford: Protective Relays Application Guide. Third Edition, March 1995

[7.6] Doemeland, W.: Schutztechnik. Grundlagen, Schutzsysteme, Inbetriebsetzung. 6. aktualisierte und erweiterte Auflage. Berlin: Verlag Technik, 1997

[7.7] Herrmann, H.-J.: Digitale Schutztechnik. Grundlagen, Software, Ausführungsbeispiele. Berlin · Offenbach: VDE Verlag, 1997

[7.8] Schossig, W.: Netzschutztechnik. Reihe Anlagentechnik für elektrische Verteilungsnetze. Band 13. 2. Auflage. Frankfurt a.M.: VWEW-Verlag, 2001

[7.9] Siemens AG, Nuernberg/Germany: SIPROTEC Numerical Protection Relays. Catalog SIP-2008. Order No.: E50001-K4400-A101-A5-7600, 2008

[7.10] Siemens AG, Nürnberg: Multifunktionsschutz mit Steuerung. SIPROTEC 4, 7SJ61/62/63/64, 6MD63. Katalog SIP3.1-2007. Bestell-Nr.: E50001-K4403-A111-A5, 2007 (English version not available, catalog SIP 3.1 is part of [7.9])

[7.11] Döllinger, H. u. a.: Die neue Serie digitaler Schutzgeräte für Mittelspannungsanwendungen. Elektrie 46 (1992) H. 9, S. 428–432

[7.12] Meisberger, F.: Richtungsvergleichsschutz – ein alternatives Schutzkonzept. EVU-Betriebspraxis 39 (2000) H. 1/2, S. 28–31

[7.13] Siemens AG, Nuernberg/Germany: Optimum Motor Protection with SIPROTEC Protection Relays. Order No.: E50001-K4454-A101-A1-7600

[7.14] Schuster, N. u. Werben, St.: Adaptiver Differentialschutz für Industrienetze. etz 120 (1999) H. 13/14, S. 16–20

[7.15] Schiel, L. u. Schuster, N.: Umfassendes Konzept für den Transformatorschutz. etz 115 (1994) H. 9, S. 496–502

[7.16] Schuster, N. u. Schiel, L.: Multifunktionsschutz für Zweiwicklungs-Transformatoren. ew 100 (2001) H. 11, S. 40–44

[7.17] Schuster, N.: Differenzialschutz mit Transformatoren im Schutzbereich. etz 122 (2001) H. 13/14, S. 16–18

[7.18] Meisberger, F.: Leitungsdifferentialschutz mit digitaler Kommunikation über Drahtverbindungen. np 42 (2003) H. 1/2, S. 26–29

[7.19] Ziegler, G.: Numerical Differential Protection. Principles and Applications. Publisher: Publicis Corporate Publishing, Erlangen, 2005

[7.20] Bollmann, H. u. a.: Digitaler Abzweigschutz – Aufgaben, Einsatzplanung und Betriebserfahrungen. etz 109 (1988) H. 4, S. 148–153

[7.21] Lemmer, S. u. Meisberger, F.: Universeller Distanzschutz. etz 113 (1992) H. 17, S. 1076–1080

[7.22] Ziegler, G.: Numerical Distance Protection. Principles and Applications. Publisher: Publicis MCD, Munich and Erlangen, 1999

[7.23] Friemelt, N. u. a.: Mit koordiniertem Netzschutz zum optimierten Energieversorgungsnetz. ew 99 (2000) H. 13, S. 26–30

[7.24] Eßmann, St. u. Sachs, U.: Neue Hilfsmittel zur Bewertung und Dokumentation der Koordination von Distanzschutzgeräten in SINCAL. ew 99 (2000) H. 14, S. 17–24

[7.25] Spörl, K. u. a.: Verbesserte Erdschlussbehandlung im Distanzschutz 7SA6. np 46 (2007) H. 10, S. 30–33

[7.26] Granz, P. u. Driescher, F.: Konstruktion und Anwendung von HH-Sicherungen in EVU- und Industrie-Mittelspannungsnetzen bis 30/36kV. etz-b 30 (1978) H. 1, S. 11–17

[7.27] Dirks, R.: Die HH-Sicherung als Herausforderung an den Schaltanlagen-Konstrukteur. ew 87 (1988) H. 16/17, S. 781–782

[7.28] Bünger, St. u. a.: Zukunftsweisende HH-Ganzbereichsicherung als kompletter Schutz von Mittelspannungsgeräten. ew 99 (2000) H. 4, S. 31–37

[7.29] Dreischke, W.: Hochspannungs-Hochleistungssicherungen haben sich seit Jahrzehnten bewährt. Teil 1–6: EVU-Betriebspraxis 40 (2001) H. 11, S. 22–26. 40 (2001) H. 12, S. 20–22. 41 (2002) H. 4, S. 22–24. 41 (2002) H. 5, S. 28–29. 41 (2002) H. 6, S. 28. 41 (2002) H. 11–12, S. 34–35. Teil 7–9: np 42 (2003) H. 1–2, S. 40–41. np 42 (2003) H. 3, S. 34–35. np 42 (2003) H. 4, S. 32–33

[7.30] Wilhelm, D.: HH-Sicherungseinsätze des Typs SSK: Transformatorschutz bis 3.000 kVA. etz 128 (2007) H. 9, S. 69–74

[7.31] Siemens AG, Erlangen/Germany: www.energy.siemens.com; Power Distributions; IEC-Products and Solutions; Medium-Voltage Indoor Devices; Fuses

[7.32] DIN VDE 0670-402 (VDE 0670-402): 1988-05 Wechselstromschaltgeräte für Spannungen über 1 kV. Auswahl von strombegrenzenden Sicherungseinsätzen für Transformatorstromkreise. Berlin · Offenbach: VDE Verlag

[7.33] IEC/TR 60787: 2007-03 Application guide for the selection of high-voltage current-limiting fuse-links for transformer circuits. Geneva/Switzerland: Bureau de la Commission Electrotechnique Internationale

[7.34] FGH Engineering & Test GmbH, Mannheim: www.fgh-ma.com

[7.35] Siemens AG, Erlangen: 8DJ10 – Am sichersten mit Sicherungen. Bestell-Nr. E50001-U229-A32, 1992

[7.36] Hampel, W. u. a.: Schutz von Verteilungstransformatoren. ew 91 (1992) H. 25, S. 1660–1663

[7.37] DIN EN 60265-1 (VDE 0670-301): 1999-05 Hochspannungs-Lastschalter - Teil 1: Hochspannungs-Lastschalter für Bemessungsspannungen über 1 kV und unter 52 kV. Berlin · Offenbach: VDE Verlag

[7.38] IEC 60265-1: 1998-01 High-voltage switches – Part 1: Switches for rated voltages above 1 kV and less than 52 kV. Geneva/Switzerland: Bureau de la Commission Electrotechnique Internationale

[7.39] DIN EN 60282-1 (VDE 0670-4): 2006-12 Hochspannungssicherungen – Teil 1: Strombegrenzende Sicherungen. Berlin · Offenbach: VDE Verlag

[7.40] IEC 60282-1: 2009-10 High-voltage fuses – Part 1: Current-Limiting fuses. Geneva/Switzerland: Bureau de la Commission Electrotechnique Internationale

[7.41] DIN EN 62271-105 (VDE 0671-105): 2003-12 Hochspannungs-Schaltgeräte und -Schaltanlagen – Teil 105: Hochspannungs-Lastschalter-Sicherungs-Kombinationen. Berlin · Offenbach: VDE Verlag

[7.42] IEC 62271-105: 2002-08 High-voltages switchgear and controlgear – Part 105: Alternating current switch-fuse combinations. Geneva/Switzerland: Bureau de la Commission Electrotechnique Internationale

[7.43] Bessei, H.: „Ein Flugverbot für Hummeln?“ – Die neue VDE 0671 Teil 105 Hochspannungs-Lastschalter-Sicherungs-Kombinationen. np 43 (2004) H. 10, S. 30–36

[7.44] Siemens AG, Erlangen/Germany: www.siemens.com/sigrade

[7.45] Haas, H.-U. u. Wilhelm, D.: VDE 0671 Teil 105: Fortsetzung eines bewährten Konzepts. Sicherer Schutz von Verteiltransformatoren mit Lastschalter-Sicherungs-Kombination. np 44 (2005) H. 3, S. 14–19

[7.46] Vetter, Ch.: Lastschalter-Sicherungskombinationseinschub in luftisolierter Bauweise. etz 124 (2003) H. 7–8, S. 2–4

[7.47] DIN EN 62271-100 (VDE 0671-100): 2004-04 Hochspannungs-Schaltgeräte und Schaltanlagen – Teil 100: Wechselstrom-Leistungsschalter. Berlin · Offenbach: VDE Verlag

[7.48] IEC 62271-100: 2008-04 High-voltage switchgear and controlgear – Part 100: Alternating current circuit-breaker. Geneva/Switzerland: Bureau de la Commission Electrotechnique Internationale

[7.49] Siemens AG, Erlangen/Germany: Medium-Voltage Equipment. 3TL Vacuum Contactors. Catalog HG11.21 · 2008. Order No.: E50001-K1511-A211-A3-7600

[7.50] Heinhold, L. u. Stubbe, R.: Kabel und Leitungen für Starkstrom. Grundlagen und Produkt-know-how für das Projektieren von Kabelanlagen. 5., wesentlich überarbeitete Auflage. Erlangen: Publicis MCD Verlag, 1999

Chapter 8

[8.1] DIN IEC 60038 (VDE 0175): 2002-11 IEC-Normspannungen. Berlin · Offenbach: VDE Verlag

[8.2] IEC 60038: 2009-06 IEC standard voltages. Geneva/Switzerland: Bureau de la Commission Electrotechnique Internationale

[8.3] Schneider Electric GmbH, Ratingen/Germany: Planungskompendium Energieverteilung. Hüthig Verlag Heidelberg, 2007, ISBN 978-3-7785-4029-9

[8.4] Siemens AG, Nuernberg/Germany: IEC Squirrel-Cage Motors. Catalog D81.1-2008, Order No.: E86060-K5581-A111-A3-7600

[8.5] Siemens AG, Nuernberg/Germany: Low-Voltage Controls and Distribution. SIRIUS Configuration. Selection Data for Fuseless Load Feeders. Brochure - January 2009, Order No.: E86060-T1815-A101-A3-7600

[8.6] Siemens AG, Nuernberg/Germany: Vector Control. SIMOVERT MASTERDRIVES VC. Catalog DA 65.10-2003/2004, Order No.: E86060-K5165-A101-A3-7600

[8.7] Ose, K.: 660V Drehstrom für Netze hoher Energiedichte. ETZ-B 25 (1973) H. 5, S. 97–99

[8.8] Steiniger, E.: Spannungswahl und Netzaufbau von Industrienetzen. ETZ-A 96(1975) H. 10, S. 471–475

[8.9] Kloeppel, F. W. u. Voß, G.: Verluste senken durch Anhebung der (Nieder-)Spannungsebene. etz 122 (2001) H. 17, S. 18–27

[8.10] Albert, K. u. a.: Elektrischer Eigenbedarf. Energietechnik in Kraftwerken und Industrie. 2., korrigierte Auflage. Berlin · Offenbach: VDE Verlag, 1996

Chapter 9

[9.1] Lerch, E. u. a.: Beitrag von Niederspannungs-Asynchronmotoren zum dreipoligen Kurzschlussstrom. etz 110 (1989) H. 5, S. 222–229

[9.2] Lerch, E. u. a.: Kurzschlussströme in einem industriellen 660-V-Netz. etz 115 (1994) H. 5, S. 254–259

[9.3] DIN EN 60909-0 (VDE 0102): 2002-07 Kurzschlussströme in Drehstromnetzen. Berechnung der Ströme. Berlin · Offenbach: VDE Verlag

[9.4] IEC 60909-0: 2001-07 Short-circuit currents in three-phase a. c. systems – Part 0: Calculation of currents. Geneva/Switzerland: Bureau de la Commission Electrotechnique Internationale

[9.5] Noack, F. u. Pospiech, J.: Kurzschluss-Kenngrößen von Niederspannungsnetzen. etz 116 (1995) H. 5, S. 38–42

[9.6] Siemens AG, Nuernberg/Germany: Low-Voltage Controls and Distribution. SIRIUS-SENTRON-SIVACON. Catalog LV1-2009, Order No.: E86060-K1002-A101-A8-7600

[9.7] Siemens AG, Regensburg/Germany: BETA Low-Voltage Circuit Protection. Catalog ET B1-2009/2010, Order No.: E86060-K8220-A101-B1-7600

[9.8] Siemens AG, Regensburg/Germany: BETA Low-Voltage Circuit Protection. Technical Information on Catalog ET B1. Catalog ET B1T-2007, Order No.: E86060-K8229-A101-A8-7600

[9.9] Leimgruber, H. u. Wolfrath, H.: Gesichtspunkte zum Aufbau elektrischer Gesamtanlagen in der metallverarbeitenden Industrie. Teil 2: Stromversorgung Niederspannung. Siemens-Bericht: E335-2-A-02, Erlangen, 1978

[9.10] DIN VDE 0100-410 (VDE 0100-410): 2007-06 Errichten von Niederspannungsanlagen – Teil 4–41: Schutzmaßnahmen-Schutz gegen elektrischen Schlag. Berlin · Offenbach: VDE Verlag

[9.11] IEC 60364-4-41: 2005-12 Low-voltage installations – Part 4–41: Protection for safety – Protection against electric shock. Geneva/Switzerland: Bureau de la Commission Electrotechnique Internationale

Chapter 10

[10.1] Siemens AG, Erlangen/Germany: Electrical Equipment. Power Supply and Auxiliaries for the Metalworking Industry. Low-Voltage Systems. Brochure of Energy and Automation Group E375, Order No.: A19100-E375-A333-X-7600

[10.2] DIN EN 61000-2-4 (VDE 0839-2-4): 2003-05 Elektromagnetische Verträglichkeit (EMV) – Teil 2–4: Umgebungsbedingungen – Verträglichkeitspegel für niederfrequente leitungsgeführte Störgrößen in Industrieanlagen. Berlin · Offenbach: VDE Verlag

[10.3] IEC 61000-2-4: 2002-06 Electromagnetic compatibility (EMC) – Part 2–4: Environment – Compatibility levels in industrial plants for low-frequency conducted disturbances. Geneva/Switzerland: Bureau de la Commission Electrotechnique Internationale

[10.4] DIN EN 61000-2-2 (VDE 0839-2-2): 2003-02 Elektromagnetische Verträglichkeit (EMV) – Teil 2-2: Umgebungsbedingungen – Verträglichkeitspegel für niederfrequente leitungsgeführte Störgrößen und Signalübertragung in öffentlichen Niederspannungsnetzen. Berlin · Offenbach: VDE Verlag

[10.5] IEC 61000-2-2: 2002-03 Electromagnetic compatibility (EMC) – Part 2-2: Environment – Compatibility levels for low-frequency conducted disturbances and signalling in public low-voltage power supply systems. Geneva/Switzerland: Bureau de la Commission Electrotechnique Internationale

[10.6] D-A-CH-CZ-Richtlinie. Technische Regeln zur Beurteilung von Netzrückwirkungen. 2. Auflage 2007. Frankfurt a. M.: VWEW Energieverlag

[10.7] DIN EN 60909-0 (VDE 0102): 2002-07 Kurzschlussströme in Drehstromnetzen. Berechnung der Ströme. Berlin · Offenbach: VDE Verlag

[10.8] IEC 60909-0: 2001-07 Short-circuit currents in three-phase a. c. systems - Part 0: Calculation of currents. Geneva/Switzerland: Bureau de la Commission Electrotechnique Internationale

[10.9] Müller, K.-H.: Die Stromversorgung von Widerstandsschweißmaschinen in der Serienfertigung der blechverarbeitenden Industrie. S. 36–53 in Widerstandsschweißen III. Siemens-Sonderdruck aus der Fachbuchreihe „Schweißtechnik“ Bd. 45. Düsseldorf: DVS-Verlag, 1965

[10.10] Webs, A.: Berechnung der Spannungseinbrüche in Drehstromnetzen bei Belastung durch Widerstandsschweißmaschinen. ETZ-A 86 (1965) H. 7, S. 205–212

[10.11] Webs, A.: Berechnung der Spannungs- und Stromverhältnisse in Drehstromnetzen bei Belastung durch symmetrische Grundlast und einphasige Widerstandsschweißmaschinen. Schweißen & Schneiden 49 (1997) H. 1, S. 25–29

[10.12] Bochanky, L. u. Gerhardt, V.: Spannungsänderungen durch symmetrische und unsymmetrische Stoßlasten. Elektrie 32 (1978) H. 4, S. 185–189

[10.13] Kiank, H.: Probabilistische Bemessungskriterien für die Schweißnetzauslegung. etz 127 (2006) H. 4, S. 48–52

[10.14] Kiank, H.: Process-Depending Planning of Power Supply Systems for an Automobile Plant. Translation from "etz", Edition 22/2003, Pages 30 to 34. Siemens Special Edition. Order No.: E50001-U229-A209-X-7600

[10.15] DIN EN 12464-1 (DIN 12464-1): 2003-03 Licht und Beleuchtung – Beleuchtung von Arbeitsstätten – Teil 1: Arbeitsstätten in Innenräumen. Berlin · Wien · Zürich: Beuth Verlag

[10.16] CIE S 008/1: 2001-01 Lighting of indoor work places. International Commission of Illumination

[10.17] DIN VDE 0100-100 (VDE 0100-100): 2009-06 Errichten von Niederspannungsanlagen – Teil 100: Allgemeine Grundsätze, Bestimmungen allgemeiner Merkmale, Begriffe. Berlin · Offenbach: VDE Verlag

[10.18] IEC 60364-1: 2005-11 Low-voltage electrical installations – Part 1: Fundamental principles, assessment of general characteristics, definitions. Geneva/Switzerland: Bureau de la Commission Electrotechnique Internationale

[10.19] DIN EN 60598-1 (VDE 0711-1): 2009-09 Leuchten – Teil 1: Allgemeine Anforderungen und Prüfungen. Berlin · Offenbach: VDE Verlag

[10.20] IEC 60598-1: 2008-04 Luminaires – Part 1: General requirements and tests. Geneva/Switzerland: Bureau de la Commission Electrotechnique Internationale

[10.21] Putz, U. u. Schulz, W.: Unerwünschte Auswirkungen eines massiven Einsatzes von Energiesparlampen. ETG-Mitgliederinformation Nr. 2/2007, S. 8–10

[10.22] DIN VDE 0100-430 (VDE 0100-430): 1991-11 Errichten von Starkstromanlagen mit Nennspannungen bis 1.000 V. Schutzmaßnahmen – Schutz von Kabeln und Leitungen bei Überstrom. Berlin · Offenbach: VDE Verlag

[10.23] IEC 60364-4-43: 2008-08 Low-voltage electrical installations – Part 4–43: Protection for safety – Protection against overcurrent. Geneva/Switzerland: Bureau de la Commission Electrotechnique Internationale

[10.24] DIN VDE 0100-520 (VDE 0100-520): 2003-06 Errichten von Niederspannungsanlagen – Teil 5: Auswahl und Errichtung elektrischer Betriebsmittel – Kapitel 52: Kabel- und Leitungsdaten. Berlin · Offenbach: VDE Verlag

[10.25] IEC 60364-5-52: 2009-10 Low-voltage electrical installations – Part 5–52: Selection and erection of electrical equipment – Wiring systems. Geneva/Switzerland: Bureau de la Commission Electrotechnique Internationale

[10.26] Muhm, H.: Schadensverhütung in modernen Elektroinstallationen. np 42 (2003) H. 4, S. 36–41

[10.27] Simon, Th.: Oberschwingungen: Schienenverteiler in Gebäuden mit IuK-Anlagen. etz 126 (2005) H. 9, S. 44–49

[10.28] Schneider Electric GmbH, Ratingen/Germany: www.schneider-electric.com

[10.29] Heinze, R. u. Mayer, St.: Unterbrechungsfreie Stromversorgung. Techniken, Trends, Anwendungen. Berlin · Offenbach: VDE Verlag, 1999

[10.30] Masterguard GmbH, Erlangen/Germany: www.masterguard.com

[10.31] Wellßow, W. H.: USV-Anlagen-Technologien, Einsatzmöglichkeiten und Entwicklungstrends. etz 123 (2002) H. 19, S. 2–7

[10.32] Piller Power Systeme GmbH, Osterade/Germany: www.piller.com

[10.33] Dipl.-Ing. Hitzinger Gesellschaft m.b.H., Linz/Austria: www.hitzinger. at

[10.34] HITEC Power Protection BV, Almelo/Netherlands: www.hitecups.com

[10.35] Klein, St. H.: Herausforderung EMV: Aufbau richtlinienkonformer Netzwerke. de 79 (2004) H. 20, S. 40–45

[10.36] Budde, G.: Vagabundierende Ströme in Elektroanlagen und Gebäuden. de 79 (2004) H. 13–14, S. 48–50

[10.37] Fassbinder, St.: Wenn es hapert an der EMV (1). Verbesserte Stromversorgung einer Bank. de 80 (2005) H. 4, S. 43–46

[10.38] Steinkühler, B. u. Hergesell, H.-G.: Wavelet-Ströme lassen Rechner abstürzen. de 80 (2005) H. 21, S. 48–58

[10.39] DIN VDE 0100-410 (VDE 0100-410): 2007-06 Errichten von Niederspannungsanlagen – Teil 4-41: Schutzmaßnahmen-Schutz gegen elektrischen Schlag. Berlin · Offenbach: VDE Verlag

[10.40] IEC 60364-4-41: 2005-12 Low-voltage electrical installations – Part 4-41: Protection for safety – Protection against electric shock. Geneva/Switzerland: Bureau de la Commission Electrotechnique Internationale

[10.41] Hörmann, W.: Schutz gegen elektrischen Schlag beim Errichten von Niederspannungsanlagen. Teil 1: de 82 (2007) H. 13–14, S. 30–33. Teil 2: de 82 (2007) H. 15–16, S. 26–30

[10.42] DIN EN 60898-1 (VDE 0641-11): 2006-03 Elektrisches Installationsmaterial – Leitungsschutzschalter für Hausinstallationen und ähnliche Zwecke – Teil 1: Leitungsschutzschalter für Wechselstrom (AC). Berlin · Offenbach: VDE Verlag

[10.43] IEC 60898-1: 2003-07 Electrical accessories – Circuit-breakers for overcurrent protection for household and similar installations – Part 1: Circuit-breakers for a. c. operation. Geneva/Switzerland: Bureau de la Commission Electrotechnique Internationale

[10.44] DIN EN 60269-1 (VDE 0636-1): 2010-03 Niederspannungssicherungen – Teil 1: Allgemeine Anforderungen. Berlin · Offenbach: VDE Verlag

[10.45] IEC 60269-1: 2009-07 Low-voltage fuses – Part 1: General requirements. Geneva/Switzerland: Bureau de la Commission Electrotechnique Internationale

[10.46] DIN VDE 0100-600 (VDE 0100-600): 2008-06 Errichten von Niederspannungsanlagen – Teil 6: Prüfungen. Nationaler Anhang NA: Tabellen NA.1 und NA.2. Berlin · Offenbach: VDE Verlag

[10.47] DIN VDE 0100-540 (VDE 0100-540): 2007-06 Errichten von Niederspannungsanlagen – Teil 5-54: Auswahl und Errichtung elektrischer Betriebsmittel – Erdungsanlagen, Schutzleiter und Schutzpotentialausgleichsleiter. Berlin · Offenbach: VDE Verlag

[10.48] IEC 60364-5-54: 2002-06 Electrical installations of buildings – Part 5-54: Selection and erection of electrical equipment – Earthing arrangements, protective conductors and protective bonding conductors. Geneva/Switzerland: Bureau de la Commission Electrotechnique Internationale

[10.49] DIN EN 50310 (VDE 0800-2-310): 2006-10 Anwendung von Maßnahmen für Erdung und Potentialausgleich in Gebäuden mit Einrichtungen der Informationstechnik. Berlin · Offenbach: VDE Verlag

[10.50] Hörmann, W.; Nienhaus, H.; Schröder, B.: Schnelleinstieg in die neue DIN VDE 0100-410 (VDE 0100-410): 2007-06. Schutz gegen elektrischen Schlag. VDE-Schriftenreihe – Normen verständlich Bd. 140. Berlin · Offenbach: VDE Verlag, 2007

[10.51] Hofheinz, W.: Fehlerstrom-Überwachung in elektrischen Anlagen. Grundlagen, Anwendungen und Technik der Differenzstrommessung in Wechsel- und Gleichspannungssystemen. DIN EN 61140 (VDE 0140-1) und DIN VDE 0100-410 (VDE 0100-410) mit Überwachungsgeräten nach DIN EN 62020 (VDE 0663). VDE-Schriftenreihe – Normen verständlich Bd. 113. Berlin · Offenbach: VDE Verlag, 2008

[10.52] Hofheinz, W.: Schutztechnik mit Isolationsüberwachung. Grundlagen, Anwendungen und Wirkungsweisen ungeerdeter IT-Systeme in der Industrie, auf Schiffen, in Elektro- und Schienenfahrzeugen und im Bergbau. DIN EN 61140 (VDE 0140-1), DIN VDE 0100-410 (VDE 0100-410), DIN VDE 0100-600 (VDE 0100-600), DIN VDE 0118-1 (VDE 0118-1) und andere mit Isolationsüberwachungsgeräten nach DIN EN 61557-8 (VDE 0413-8). VDE-Schriftenreihe – Normen verständlich Bd. 114. Berlin · Offenbach: VDE Verlag, 2007

[10.53] Hofheinz, W.: Protective Measures with Insulation Monitoring. Application of Unearthed IT Power Systems in Industry, Mining, Railways, Marine/Oil and Electric/Rail Vehicles. Berlin · Offenbach: VDE Verlag, 2006

[10.54] Hofheinz, W.: Ungeerdete Niederspannungs-Stromerzeugungsanlagen als IT-Systeme. etz 130 (2009) H. 3, S. 68–75

[10.55] Hofheinz, W.: Aufbau und Wirkungsweise von Schutzmaßnahmen in IT-Systemen. etz 124 (2003) H. 23–24, S. 30–33

[10.56] DIN VDE 0100-710 (VDE 0100-710): 2002-11 Errichten von Niederspannungsanlagen – Anforderungen für Betriebsstätten, Räume und Anlagen besonderer Art – Teil 710: Medizinisch genutzte Bereiche. Berlin · Offenbach: VDE Verlag

[10.57] IEC 60364-7-710: 2002-11 Electrical installations of buildings – Part 7–710: Requirements for special installations or locations – Medical locations. Geneva/Switzerland: Bureau de la Commission Electrotechnique Internationale

[10.58] DIN EN 61557-2 (VDE 0413-2): 2008-02 Elektrische Sicherheit in Niederspannungsnetzen bis AC 1.000 V und DC 1.500 V – Geräte zum Prü-fen, Messen oder Überwachen von Schutzmaßnahmen – Teil 2: Isolationswiderstand. Berlin · Offenbach: VDE Verlag

[10.59] IEC 61557-2: 2007-01 Electrical safety in low voltage distribution systems up to 1,000 V a. c. and 1,500 V d. c. – Equipment for testing, measuring or monitoring of protective measures – Part 2: Insulation resistance. Geneva/Switzerland: Bureau de la Commission Electrotechnique Internationale

[10.60] DIN EN 62020 (VDE 0663): 2005-11 Elektrisches Installationsmaterial – Differenzstrom-Überwachungsgeräte für Hausinstallationen und ähnliche Verwendungen (RCMs). Berlin · Offenbach: VDE Verlag

[10.61] IEC 62020: 2003-11 Electrical accessories – Residual current monitors for household and similar uses (RCMs). Geneva/Switzerland: Bureau de la Commission Electrotechnique Internationale

[10.62] DIN EN 61557-9 (VDE 0413-9): 2009-11 Elektrische Sicherheit in Niederspannungsnetzen bis AC 1.000 V und DC 1.500 V – Geräte zum Prüfen, Messen oder Überwachen von Schutzmaßnahmen – Teil 9: Einrichtungen zur Isolationsfehlersuche in IT-Systemen. Berlin · Offenbach: VDE Verlag

[10.63] IEC 61557-9: 2009-01 Electrical safety in low voltage distribution systems up to 1,000 V a. c. and 1,500 V d. c. – Equipment for testing, measuring or monitoring of protective measures – Part 9: Equipment for insulation fault location in IT systems. Geneva/Switzerland: Bureau de la Commission Electrotechnique Internationale

[10.64] DIN EN 60947-2 (VDE 0660-101): 2010-04 Niederspannungsschaltgerä-te – Teil 2: Leistungsschalter. Berlin · Offenbach: VDE Verlag

[10.65] IEC 60947-2: 2009-05 Low-voltage switchgear and controlgear – Part 2: Circuit-breakers. Geneva/Switzerland: Bureau de la Commission Electrotechnique Internationale

[10.66] DIN EN 61543 (VDE 0664-30): 2006-06 Fehlerstromschutzeinrichtungen (RCDs) für Hausinstallationen und ähnliche Verwendung – Elektromagnetische Verträglichkeit. Berlin · Offenbach: VDE Verlag

[10.67] IEC 61543: 1995-04 Residual current-operated protective devices (RCDs) for household and similar use – Electromagnetic compatibility. Geneva/Switzerland: Bureau de la Commission Electrotechnique Internationale

[10.68] DIN EN 61008-1 (VDE 0664-10): 2010-01 Fehlerstrom-/Differenzstrom-Schutzschalter ohne eingebauten Überstromschutz (RCCBs) für Hausinstallationen und für ähnliche Anwendungen – Teil 1: Allgemeine Anforderungen. Berlin · Offenbach: VDE Verlag

[10.69] IEC 61008-1: 2010-02 Residual current-operated current circuit-breakers without integral overcurrent protection for household and similar uses (RCCBs) – Part 1: General rules. Geneva/Switzerland: Bureau de la Commission Electrotechnique Internationale

[10.70] DIN EN 61009-1 (VDE 0664-20): 2010-01 Fehlerstrom-/Differenzstrom-Schutzschalter mit eingebautem Überstromschutz (RCBOs) für Hausinstallationen und ähnliche Anwendungen – Teil 1: Allgemeine Anforderungen. Berlin · Offenbach: VDE Verlag

[10.71] IEC 61009-1: 2010-02 Residual current-operated current circuit-breakers with integral overcurrent protection for household and similar uses (RCBOs) – Part 1: General rules. Geneva/Switzerland: Bureau de la Commission Electrotechnique Internationale

[10.72] Hörmann, W.: Neue Norm zum Schutz gegen elektrischen Schlag. Erläuterungen zu DIN VDE 0100-410 (VDE 0100-410): 2007-06. Elektropraktiker 61 (2007) H. 9, S. 780–790

[10.73] Umlauft, D. u. Zankel, F.: TT- oder TN-System – ist ein Ende des Richtungsstreits in Sicht? VEÖ-Journal 3 (1996) H. 11, S. 44–49

[10.74] Biegelmeier, G. u. Krefter, K.-H.: TN- oder TT-System. Unterschiede in den Grenzrisiken für den Schutz gegen elektrischen Schlag. Elektrizitätswirtschaft 99 (2000) H. 4, S. 49–57

[10.75] Umlauft, D.: Genereller Übergang zum TN-System in Österreich bis 2008. etz 122 (2001) H. 23–24, S. 18–24

[10.76] Hörmann, W.: Blockheizkraftwerk im TT-System. de 84 (2009) H. 11, S. 31–34

[10.77] Hörmann, W.: TN-C-System oder TN-S-System? de 76 (2001) H. 13, S. 18–21

[10.78] Otto, K.-H..: Überprüfung von Anlagen in TN-?-Systemen. de 78 (2003) H. 12, S. 40–45

[10.79] Hörmann, W.: Umrüsten eines TN-C- in ein TN-S-System. PEN- und PE-Funktion und richtige Kennzeichnung. de 79 (2004) H. 18, S. 30–34

[10.80] Hörmann, W.: Umrüsten eines TN-C-Systems in ein TN-S-System. Normen der Reihe DIN VDE 0100 (VDE 0100), insbesondere DIN VDE 0100-540 (VDE 0100-540), DIN EN 60446 (VDE 0198), DIN EN 60439-1 (VDE 0660-500) und DIN VDE 0276-603 (VDE 0276-603). de 79 (2004) H. 20, S. 13–15

[10.81] Hörmann, W.: Umrüsten eines TN-C- in ein TN-S-System. DIN VDE 0100-510 (VDE 0100-510) und DIN VDE 0100-540 (VDE 0100-540). de 80 (2005) H. 8, S. 15–16

[10.82] Groß, H.: Vermeidung vagabundierender Ströme – TN-Netzsystem in Niederspannungsnetzen. S. 131–132 in ETG-Fachbericht 98. Technische Innovationen in Verteilungsnetzen. Vorträge der ETG-Fachtagung vom 1. 3.–2. 3. 2005 in Würzburg. Berlin · Offenbach: VDE Verlag

[10.83] IEC 60364-5-51: 2005-04 Electrical installations of buildings – Part 5-51: Selection and erection of electrical equipment – Common rules. Geneva/Switzerland: Bureau de la Commission Electrotechnique Internationale

[10.84] DIN VDE 0100-510 (VDE 0100-510): 2007-06 Errichten von Niederspannungsanlagen – Teil 5-51: Auswahl und Errichtung elektrischer Betriebsmittel – Allgemeine Bestimmungen. Berlin · Offenbach: VDE Verlag

[10.85] Kiank, H.: EMV und Personenschutz in mehrfachgespeisten NS-Industrienetzen. etz 126 (2005) H. 11, S. 44–49

[10.86] Dipl.-Ing. W. Bender GmbH & Co. KG, Gruenberg/Germany: www. bender-de.com

[10.87] Kloeppel, F. W.: Zur Gestaltung von Niederspannungs-Maschennetzen in Industriebetrieben. Elektrie 15 (1961) H. 4, S. 112–115

[10.88] Kloeppel, F. W.: Aktuelle Probleme der Gestaltung großer Industrienetze. Energietechnik 11 (1961) H. 4, S. 170–179

[10.89] Janisch, O.: Industrienetze – Gesichtspunkte für die Netzgestaltung, speziell für die chemische und Mineralöl-Industrie. ETZ-B 22 (1970) H. 25, S. 601–602

[10.90] Röper, P. u. Stenzel, J.: Versorgungszuverlässigkeit bei verschiedenen Niederspannungsnetzformen. Elektrizitätswirtschaft 78 (1979) H. 11, S. 421–424

Chapter 11

[11.1] DIN EN 60076-11 (VDE 0532-76-11): 2005-04 Leistungstransformatoren – Teil 11: Trockentransformatoren. Berlin · Offenbach: VDE Verlag

[11.2] IEC 60076-11: 2004-05 Power transformers – Part 11: Dry-type transformers. Geneva/Switzerland: Bureau de la Commission Electrotechnique Internationale

[11.3] Siemens AG, Kirchheim-Teck/Germany: www.siemens.com/energy

[11.4] Kiank, H.: Process-Depending Planning of Power Supply Systems for an Automobile Plant. Translation from "etz", Edition 22/2003, Pages 30 to 34. Siemens Special Edition. Order No.: E50001-U229-A209-X-7600

[11.5] Kiank, H.: Probabilistische Bemessungskriterien für die Schweißnetzauslegung. etz 127 (2006) H. 4, S. 48–52

[11.6] Biechl, H.: Drehstromtransformatoren. de 80 (2005) H. 21, S. 101–102

[11.7] Heuck, K. u. Dettmann, K.-D.: Elektrische Energieversorgung. Zweite, neubearbeitete Auflage. Braunschweig: Vieweg Verlag, 1991

[11.8] Fassbinder, S.: Oberschwingungen paralleler Transformatoren. de 81 (2006) H. 15–16, S. 15–16

[11.9] Leimgruber, H. u. Wolfrath, H.: Gesichtspunkte zum Aufbau elektrischer Gesamtanlagen in der metallverarbeitenden Industrie. Teil 2: Stromversorgung Niederspannung. Siemens-Bericht: E335-2-A-02, Erlangen, 1978

[11.10] Pantenburg, N. u. Wolfrath, H.: Planung von Niederspannungsnetzen in der metallverarbeitenden Industrie. Siemens-Energietechnik 4 (1982) H. 4, S. 206–210

[11.11] DIN EN 61439-1 (VDE 0660-600-1): 2010-06 Niederspannungs-Schaltgerätekombinationen – Teil 1: Allgemeine Festlegungen. Berlin · Offenbach: VDE Verlag

[11.12] IEC 61439-1: 2009-01 Low-voltage switchgear and controlgear assemblies – Part 1: General rules. Geneva/Switzerland: Bureau de la Commission Electrotechnique Internationale

[11.13] DIN EN 61439-2 (VDE 0660-600-2): 2010-06 Niederspannungs-Schaltgerätekombinationen – Teil 2: Energie-Schaltgerätekombinationen. Berlin · Offenbach: VDE Verlag

[11.14] IEC 61439-2: 2009-01 Low-voltage switchgear and controlgear assemblies – Part 2: Power switchgear and controlgear assemblies. Geneva/Switzerland: Bureau de la Commission Electrotechnique Internationale

[11.15] DIN EN 60439-1 Bbl. 2 (VDE 0660-500 Bbl. 2): 2009-05 Niederspannungs-Schaltgerätekombinationen – Teil 1: Typgeprüfte und partiell typ-geprüfte Kombinationen – Technischer Bericht: Verfahren für die Prüfung unter Störlichtbogenbedingungen. Berlin · Offenbach: VDE Verlag

[11.16] IEC/TR 61641: 2008-01 Enclosed low-voltage switchgear and controlgear assemblies – Guide for testing under conditions of arcing due to internal fault. Geneva/Switzerland: Bureau de la Commission Electrotechnique Internationale

[11.17] Siemens AG, Nuernberg/Germany: Totally Integrated Power^{TM}. Application Manual. Connecting the worlds of building construction and power distribution with integrated solutions for commercial and industrial buildings, 2nd Edition, 2005, Order No.: E20001-A70-M104-7600

[11.18] DIN EN 60529 (VDE 0470-1): 2000-09 Schutzarten durch Gehäuse (IP-Code). Berlin · Offenbach: VDE Verlag

[11.19] IEC 60529: 2001-02 Degrees of protection by enclosures (IP Code). Geneva/Switzerland: Bureau de la Commission Electrotechnique Internationale

[11.20] DIN VDE 0100-410 (VDE 0100-410): 2007-06 Errichten von Niederspannungsanlagen – Teil 4-41: Schutzmaßnahmen – Schutz gegen elektrischen Schlag. Berlin · Offenbach: VDE Verlag

[11.21] IEC 60364-4-41: 2005-12 Low-voltage electrical installations – Part 4-41: Protection for safety-protection against electric shock. Geneva/Switzerland: Bureau de la Commission Electrotechnique Internationale

[11.22] Siemens AG, Nürnberg: Schalten, Schützen, Verteilen in Niederspannungsnetzen. Handbuch mit Auswahlkriterien und Projektierungshinweisen für Schaltgeräte, Steuerungen und Schaltanlagen. 4. überarbeitete Auflage. München · Erlangen: Publicis MCD-Verlag, 1997

[11.23] Siemens AG, Nuernberg/Germany: www.siemens.com/sivacon

[11.24] Siemens AG, Nuernberg/Germany: Planning with SIVACON S8. Flexible Low-Voltage Switchboard. Planning Manual March 2008. Order No.: A5E0 1619128-01

[11.25] Siemens AG, Nuernberg/Germany: Monitoring and Control Devices SIMOCODE 3UF LOGO! Timing Relays Monitoring Relays Safety Relays Interface Converters. Reference Manual April 2009, Entry-ID: 35681283

[11.26] DIN EN 60947-4-1 (VDE 0660-102): 2006-04 Niederspannungsschaltgeräte – Teil 4-1: Schütze und Motorstarter. Berlin · Offenbach: VDE Verlag

[11.27] IEC 60947-4-1: 2009-09 Low-voltage switchgear and controlgear – Part 4-1: Contactors and motor-starters – Electromechanical contactors and motor-starters. Geneva/Switzerland: Bureau de la Commission Electrotechnique Internationale

[11.28] Siemens AG, Nuernberg/Germany: www.siemens.com/sirius

[11.29] Siemens AG, Nuernberg/Germany: ALPHA Distribution Boards and Terminal Blocks. Catalog ET A1 · 2010. Order No.: E86060-K8210-A101-B2-7600

[11.30] Siemens AG, Nürnberg: ALPHA 8HP Isolierstoff-Verteilersystem: Traggerüst und Kabelraumverkleidung. Variabler Aufbau als Standverteiler. Bestell-Nr.: E86060-K8210-E200-A1, 2008

[11.31] Siemens AG, Nuernberg/Germany: www.buildingtechnologies.siemens.com; Low-voltage power distribution and electrical installation technology; Busbar Trunking Systems

[11.32] Siemens AG, Nuernberg/Germany: SIVACON. Answers for industry. For Safe Power Flows SIVACON 8PS Busbar Trunking Systems. Order No.: E20001-A360-P309-V2-7600, 09/2008

[11.33] Siemens AG, Nuernberg/Germany: SIVACON. Answers for industry. Safe and Flexible Power Distribution in Multi-Story Functional Buildings. Order No.: E20001-A420-P309-X-7600, 04/2008

[11.34] Siemens AG, Nuernberg/Germany: SIVACON. Answers for industry. Safe and Flexible Power Distribution in the Automotive Industry. Order No.: E20001-A400-P309-X-7600, 01/2008

[11.35] DIN EN 62271-202 (VDE 0671-202): 2007-08 Hochspannungs-Schaltgeräte und -Schaltanlagen – Teil 202: Fabrikfertige Stationen für Hochspannung/Niederspannung. Berlin · Offenbach: VDE Verlag

[11.36] IEC 62271-202: 2006-06 High-voltage switchgear and controlgear – Part 202: High-voltage/Low-voltage prefabricated substation. Geneva/Switzerland: Bureau de la Commission Electrotechnique Internationale

[11.37] PEHLA – Gesellschaft für elektrische Hochleistungsprüfungen, Mannheim: www.pehla.com

[11.38] Siemens AG, Erlangen/Germany: SITRABLOC – The compact power station for industrial power supply systems. Order No.: E50001-U229-A140X-7600, 2001

[11.39] DIN VDE 0298-4 (VDE 0298-4): 2003-08 Verwendung von Kabeln und isolierten Leitungen für Starkstromanlagen – Teil 4: Empfohlene Werte für die Strombelastbarkeit von Kabeln und Leitungen für feste Verlegung in und an Gebäuden und von flexiblen Leitungen. Berlin · Offenbach: VDE Verlag

[11.40] IEC 60364-5-52: 2009-10 Low-voltage electrical installations – Part 5-52: Selection and erection of electrical equipment – Wiring systems. Geneva/Switzerland: Bureau de la Commission Electrotechnique Internationale

[11.41] DIN VDE 0276-1000 (VDE 0276-1000): 1995-06 Starkstromkabel – Teil 1000: Strombelastbarkeit, Allgemeines, Umrechnungsfaktoren. Berlin · Offenbach: VDE Verlag

[11.42] DIN VDE 0100-430 (VDE 0100-430): 1991-11 Errichten von Starkstromanlagen bis 1.000 V. Schutzmaßnahmen – Schutz von Kabeln und Leitungen bei Überstrom. Berlin · Offenbach: VDE Verlag

[11.43] IEC 60364-4-43: 2008-08 Low-voltage electrical installations – Part 4–43: Protection for safety – Protection against overcurrent. Geneva/Switzerland: Bureau de la Commission Electrotechnique Internationale

[11.44] DIN VDE 0100-520 (VDE 0100-520): 2003-06 Errichten von Niederspannungsanlagen – Teil 5: Auswahl und Errichtung elektrischer Betriebsmittel – Kapitel 52: Kabel- und Leitungsanlagen. Berlin · Offenbach: VDE Verlag

[11.45] E DIN IEC 60364-5-52 (VDE 0100-520): 2004-07 Errichten von Niederspannungsanlagen – Teil 5: Auswahl und Errichtung elektrischer Betriebsmittel – Kapitel 52: Kabel- und Leitungsanlagen. Berlin · Offenbach: VDE Verlag

[11.46] DIN IEC 60038 (VDE 0175): 2002-11 IEC Normenspannungen. Berlin · Offenbach: VDE Verlag

[11.47] IEC 60038: 2009-06 IEC standard voltages. Geneva/Switzerland: Bureau de la Commission Electrotechnique Internationale

[11.48] Heinhold, L. u. Stubbe, R.: Kabel und Leitungen für Starkstrom. Grundlagen und Produkt-Know-how für das Projektieren von Kabelanlagen. 5., wesentlich überarbeitete Auflage. Erlangen: Publicis MCD-Verlag, 1999

[11.49] DIN EN 60269-1 (VDE 0636-1): 2010-03 Niederspannungssicherungen – Teil 1: Allgemeine Anforderungen. Berlin · Offenbach: VDE Verlag

[11.50] IEC 60269-1: 2009-07 Low-voltage fuses – Part 1: General requirements. Geneva/Switzerland: Bureau de la Commission Electrotechnique Internationale

[11.51] DIN EN 60898-1 (VDE 0641-11): 2006-03 Elektrisches Installationsmaterial – Leitungsschutzschalter für Hausinstallationen und ähnliche Zwecke – Teil 1: Leitungsschutzschalter für Wechselstrom (AC). Berlin · Offenbach: VDE Verlag

[11.52] IEC 60898-1: 2003-07 Electrical accessories – Circuit-breakers for overcurrent protection for household and similar installations – Part 1: Circuit-breakers for a. c. operation. Geneva/Switzerland: Bureau de la Commission Electrotechnique Internationale

[11.53] DIN EN 60947-2 (VDE 0660-101): 2010-04 Niederspannungsschaltgeräte – Teil 2: Leistungsschalter. Berlin · Offenbach: VDE Verlag

[11.54] IEC 60947-2: 2009-05 Low-voltage switchgear and controlgear – Part 2: Circuit-breakers. Geneva/Switzerland: Bureau de la Commission Electrotechnique Internationale

[11.55] Siemens AG, Regensburg/Germany: BETA Low-Voltage Circuit Protection. Technical Information on Catalog ET B1. Catalog ET B1T-2007. Order No.: E86060-K8229-A101-A8-7600

[11.56] Albert, K. u. a.: Elektrischer Eigenbedarf. Energietechnik in Kraftwerken und Industrie. 2., korrigierte Auflage. Berlin · Offenbach: VDE Verlag

[11.57] DIN EN 60909-0 (VDE 0102): 2002-07 Kurzschlussströme in Drehstromnetzen. Berechnung der Ströme. Berlin · Offenbach: VDE Verlag

[11.58] IEC 60909-0: 2001-07 Short-circuit currents in three-phase a. c. systems – Part 0: Calculation of currents. Geneva/Switzerland: Bureau de la Commission Electrotechnique Internationale

[11.59] DIN EN 60909-0 Bbl. 4 (VDE 0102 Bbl. 4): 2009-08 Kurzschlussströme in Drehstromnetzen – Daten elektrischer Betriebsmittel für die Berechnung von Kurzschlussströmen. Berlin · Offenbach: VDE Verlag

[11.60] Heinhold, L. u. Stubbe, R.: Power Cables and their Application. Part 2: Tables Including Project Planning Data for Cables and Accessories, Details for the Determination of the Cross-Sectional Area. 3rd Edition. Munich · Erlangen: Publicis MCD, 1993

[11.61] DIN VDE 0100 Bbl. 5 (VDE 0100 Bbl. 5): 1995-11 Errichten von Starkstromanlagen mit Nennspannungen bis 1000 V – Maximal zulässige Längen von Kabeln und Leitungen unter Berücksichtigung des Schutzes bei indirektem Berühren, des Schutzes bei Kurzschluss und des Spannungsfalls. Berlin · Offenbach: VDE Verlag

[11.62] IEC 60228: 2004-11 Conductors of insulated cables. Geneva/Switzerland: Bureau de la Commission Electrotechnique Internationale

Chapter 12

[12.1] ZVEI-Zentralverband Elektrotechnik- und Elektronikindustrie e. V., Frankfurt a. M.: Energieeinsparung durch Blindleistungskompensation – Energiebezugskosten senken, Stromnetzverluste und CO_2-Emissionen senken. ZVEI-Publikation des Fachverbandes Energietechnik vom 18.04. 2008

[12.2] Just, W. u. Simon, Th.: Energieeffizienz durch Blindleistungskompensation. Gemeinsame Schritte für den Klimaschutz. Teil 1: ew 108 (2009) H. 20, S. 46–49. Teil 2: ew 108 (2009) H. 21, S. 54–58

[12.3] Bätz, H. u. a.: Elektroenergieanlagen. Berlin: Verlag Technik, 1989

[12.4] Floerke, H.: Leistungsbedarf elektrischer Anlagen. etz 104 (1983) H. 12, S. 586–589

[12.5] Schröder, K.-H.: Komplexe Energieversorgung von Territorien. Leipzig: Deutscher Verlag für Grundstoffindustrie, 1983

[12.6] Berechnung des elektrischen Leistungs- und Energiebedarfs für Stromversorgungsanlagen von Industrienetzen. VEM-Projektierungs-Vorschrift Ordnungs-Nr. 1.9/3.80. Berlin: Institut für Elektro-Anlagen, 1980

[12.7] Kloeppel, F. W.: Planung und Projektierung von Elektroenergieversorgungssystemen. 2., überarbeitete Auflage. Leipzig: Deutscher Verlag für Grundstoffindustrie, 1977

[12.8] Just, W. u. Hofman, W.: Blindstromkompensation in der Betriebspraxis. Ausführung, Energieeinsparung, Oberschwingungen, Spannungsqualität. 4., überarbeitete Auflage. Berlin · Offenbach: VDE Verlag, 2003

[12.9] Siemens AG, Berlin and Munich: Protection and Distribution in Low-Voltage Networks. Handbook with selection criteria and planning guidelines for switchgear, switchboards and distribution systems. 2nd revised edition. Publicis Corporate Publishing, Erlangen, 1994

[12.10] Siemens AG, Erlangen: Planung von Niederspannungsnetzen in der metallverarbeitenden Industrie. Druckschrift des Bereiches E375, Bestell-Nr.: A19100-E375-A333, 1979

[12.11] Modl GmbH, Pappenheim/Germany: www.modl.de

[12.12] Modl GmbH, Pappenheim/Germany: Product catalogue - 2008. Perfect Synergies by intelligent System Solutions.

[12.13] DIN EN 60831-1 (VDE 0560-46): 2003-08 Selbstheilende Leistungs-Parallelkondensatoren für Wechselstromanlagen mit einer Nennspannung bis 1 kV. Teil 1: Allgemeines – Leistungsanforderungen, Prüfung und Bemessung – Sicherheitsanforderungen – Anleitung für Errichtung und Betrieb. Berlin · Offenbach: VDE Verlag

[12.14] IEC 60831-1: 2002-11 Shunt power capacitors of the self-healing type for a. c. systems having a rated voltage up to and including 1000 V – Part 1: General – Performance, testing and rating – Safety requirements – Guide for installation and operation. Geneva/Switzerland: Bureau de la Commission Electrotechnique Internationale

[12.15] DIN VDE 0298-4 (VDE 0298-4): 2003-08 Verwendung von Kabeln und isolierten Leitungen für Starkstromanlagen – Teil 4: Empfohlene Werte für die Strombelastbarkeit von Kabeln und Leitungen für feste Verlegung in und an Gebäuden und von flexiblen Leitungen. Berlin · Offenbach: VDE Verlag

[12.16] IEC 60364-5-52: 2009-10 Low-voltage electrical installations – Part 5-52: Selection and erection of electrical equipment – Wiring systems. Geneva/Switzerland: Bureau de la Commission Electrotechnique Internationale

[12.17] BDEW · Bundesverband der Energie- und Wasserwirtschaft e. V., Berlin: www.bdew.de

[12.18] Tonfrequenz-Rundsteuerung – Empfehlung zur Vermeidung unzulässiger Rückwirkungen. 3., überarbeitete Ausgabe. Frankfurt a. M.: VWEW Energieverlag, 1997

[12.19] Fenn, B.: Intelligente Netze und Anlagen der Zukunft. Ein Beispiel aus der Praxis für Verteilungsnetze. ew 108 (2009) H. 6, S. 46–51

[12.20] Sofic, D.: Oberschwingungen im Generator-Inselbetrieb. de 84 (2009) H. 21, S. 32–35

[12.21] DIN VDE 0100-460 (VDE 0100-460): 2002-08 Errichten von Niederspannungsanlagen – Teil 4: Schutzmaßnahmen – Kapitel 46: Trennen und Schalten. Berlin · Offenbach: VDE Verlag

[12.22] IEC 60364-4-41: 2005-12 Low-voltage electrical installations – Part 4–41: Protection for safety – Protection against electric shock. Geneva/Switzerland: Bureau de la Commission Electrotechnique Internationale

[12.23] DIN EN 50160 (EN 50160): 2008-04 Merkmale der Spannung in öffentlichen Elektrizitätsversorgungsnetzen. Berlin · Wien · Zürich: Beuth Verlag

[12.24] D-A-CH-CZ-Richtlinie. Technische Regeln zur Beurteilung von Netzrückwirkungen. 2. Auflage 2007. Frankfurt a.M.: VWEW Energieverlag

[12.25] DIN EN 61000-2-2 (VDE 0839-2-2): 2003-02 Elektromagnetische Verträglichkeit (EMV) - Teil 2-2. Umgebungsbedingungen – Verträglichkeitspegel für niederfrequente leitungsgeführte Störgrößen und Signalübertragung in öffentlichen Niederspannungsnetzen. Berlin · Offenbach: VDE Verlag

[12.26] IEC 61000-2-2: 2002-03 Electromagnetic compatibility (EMC) – Part 2-2: Environment – Compatibility levels for low-frequency conducted disturbances and signalling in public low-voltage power supply systems. Geneva/Switzerland: Bureau de la Commission Electrotechnique Internationale

[12.27] IEC/TR 61000-3-6: 2008-02 Electromagnetic compatibility (EMC) – Part 3–6: Limits – Assessment of emission limits for the connection of distorting installations to MV, HV and EHV power system. Geneva/Switzerland: Bureau de la Commission Electrotechnique Internationale

[12.28] DIN EN 61000-2-4 (VDE 0839-2-4): 2003-05 Elektromagnetische Verträglichkeit (EMV) – Teil 2–4: Umgebungsbedingungen – Verträglichkeitspegel für niederfrequente leitungsgeführte Störgrößen in Industrieanlagen. Berlin · Offenbach: VDE Verlag

[12.29] IEC 61000-2-4: 2002-06 Electromagnetic compatibility (EMC) – Part 2–4: Environment – Compatibility levels in industrial plants for low-frequency conducted disturbances. Geneva/Switzerland: Bureau de la Commission Electrotechnique Internationale

[12.30] Pregizer, K.: Korrekte Leistungsangaben bei der Blindstromkompensation. etz 121 (2000) H. 10, S. 22–32

[12.31] Will, G.: Regeleinheiten und Filterkreise zur Blindleistungskompensation. Siemens-Energietechnik 2 (1980) H. 7, S. 251–255

[12.32] Hormann, W.; Just, W.; Schlabbach, J.: Netzrückwirkungen. Anlagentechnik für elektrische Verteilungsnetze, Band 14. 3., erweiterte Auflage. Berlin · Offenbach: VDE Verlag, 2008

Chapter 13

[13.1] DIN EN 60269-1 (VDE 0636-1): 2010-03 Niederspannungssicherungen – Teil 1: Allgemeine Anforderungen. Berlin · Offenbach: VDE Verlag

[13.2] IEC 60269-1: 2009-07 Low-voltage fuses – Part 1: General requirements. Geneva/Switzerland: Bureau de la Commission Electrotechnique Internationale

[13.3] DIN EN 60947-2 (VDE 0660-101): 2010-04 Niederspannungsschaltgeräte – Teil 2: Leistungsschalter. Berlin · Offenbach: VDE Verlag

[13.4] IEC 60947-2: 2009-05 Low-voltage switchgear and controlgear – Part 2: Circuit-breakers. Geneva/Switzerland: Bureau de la Commission Electrotechnique Internationale

[13.5] DIN EN 60898-1 (VDE 0641-11): 2006-03 Elektrisches Installationsmaterial – Leitungsschutzschalter für Hausinstallationen und ähnliche Zwecke – Teil 1: Leitungsschutzschalter für Wechselstrom (AC). Berlin · Offenbach: VDE Verlag

[13.6] IEC 60898-1: 2003-07 Electrical accessories – Circuit-breakers for over-current protection for household and similar installations – Part 1: Circuit-breakers for a. c. operation. Geneva/Switzerland: Bureau de la Commission Electrotechnique Internationale

[13.7] DIN EN 60947-4-1 (VDE 0660-102): 2006-04 Niederspannungsschaltgeräte – Teil 4-1: Schütze und Motorstarter – Elektromechanische Schütze und Motorstarter. Berlin · Offenbach: VDE Verlag

[13.8] IEC 60947-4-1: 2009-09 Low-voltage switchgear and controlgear – Part 4-1: Contactors and motor-starters – Electromechanical contactors and motor-starters. Geneva/Switzerland: Bureau de la Commission Electrotechnique Internationale

[13.9] DIN EN 60947-8 (VDE 0660-302): 2007-07 Niederspannungsschaltgerä-te – Teil 8: Auslösegeräte für den eingebauten thermischen Schutz (PTC) von rotierenden elektrischen Maschinen. Berlin · Offenbach: VDE Verlag

[13.10] IEC 60947-8: 2006-11 Low-voltage switchgear and controlgear – Part 8: Control units for built-in thermal protection (PTC) for rotating electrical machines. Geneva/Switzerland: Bureau de la Commission Electrotechnique Internationale

[13.11] Siemens AG, Regensburg/Germany: BETA Low-Voltage Circuit Protection. Technical Information on catalog ET B1. Catalog ET B1T-2007, Order No.: E86060-K8229-A101-A8-7600

[13.12] DIN EN 60617-7 (EN 60617-7): 1997-08 Graphische Symbole für Schaltpläne – Teil 7: Schaltzeichen für Schalt- und Schutzeinrichtungen. Berlin · Wien · Zürich: Beuth Verlag

[13.13] IEC 60617-DB-12M: 2001-11 Graphical symbols for diagrams – 12-month subscription to online database comprising parts 2 to 13 of IEC 60617. http://webstore.iec.ch/webstore/webstore.nsf/artnum/027836

[13.14] Siemens AG, Nuernberg/Germany: Circuit-Breakers with Communication Capability SENTRON WL and SENTRON VL. System Manual. Order No.: A5E0 1051353-01, 2009

[13.15] Mützel, T.; Berger, F.; Anheuser, M.: Methods of Electronic Short-Circuit Detection for Improving Current Limitation in Low-Voltage Systems. Paper 11.5 of Session 11: Relays & Switches, 23rd International Conference on Electrical Contacts (ICEC), 6.6.–9.6.2006 in Sendai/Japan

[13.16] Siemens AG, Nuernberg/Germany: Totally Integrated Power™. Application Manual. Connecting the worlds of building construction and power distribution with integrated solutions of commercial and industrial buildings, 2nd Edition, 2005, Order No.: E20001-A70-M104-X-7600

[13.17] Siemens AG, Nuernberg/Germany: Low-Voltage Controls and Distribution Brochure: SIRIUS Configuration – Selection Data for Fuseless Load Feeders. Order No.: E86060-T1815-A101-A3, 2009/01

[13.18] Seip, G. G.: Electrical Installations Handbook. Power Supply and Distribution, Protective Measures, Electromagnetic Compatibility, Electrical Installation Equipment and Systems, Application Examples for Electrical Installation Systems, Building Management. 3rd Edition. Erlangen: Publicis Corporate Publishing; Chichester: Wiley, 2000

[13.19] DIN VDE 0100-530 (VDE 0100-530): 2005-06 Errichten von Niederspannungsanlagen – Teil 530: Auswahl und Errichtung elektrischer Betriebsmittel – Schalt- und Steuergeräte. Berlin · Offenbach: VDE Verlag

[13.20] IEC 60364-5-53: 2002-06 Electrical installation of buildings – Part 5-53: Selection and erection of electrical equipment – Isolation, switching and control. Geneva/Switzerland: Bureau de la Commission Electrotechnique Internationale

[13.21] Siemens AG, Erlangen/Germany: www.siemens.com/sincal

[13.22] Siemens AG, Erlangen/Germany: www.siemens.com/sigrade

[13.23] Bessei, H.: Sicherungshandbuch. Starkstromsicherungen – Das Handbuch für Anwender von Niederspannungs- und Hochspannungssicherungen. ISBN 978-3-00-021360-1. Frankfurt a. M.: NH-HH-Recycling e. V., 2007

[13.24] Siemens AG, Nuernberg/Germany: www.siemens.com/simaris

Index

Q

R